CHARLES E. MERRILL PUBLISHING COMPANY
A Bell & Howell Company
Columbus   Toronto   London   Sydney

Thomas L. Floyd
Mayland Technical College

ELECTRIC CIRCUITS: ELECTRON-FLOW VERSION

Cover photograph by Mike Cummings.

Published by Charles E. Merrill Publishing Co.
A Bell & Howell Company
Columbus, Ohio 43216

This book was set in Times Roman and Memphis.
Cover Design Coordination: Tony Faiola
Text Designer: Ann Mirels
Production Coordination: Cherlyn B. Paul

Copyright © 1983, by Bell & Howell Company. All rights reserved.
No part of this book may be reproduced in any form, electronic, or mechanical, including photocopy, recording, or any information storage and retrieval system, without permission in writing from the publisher.

Library of Congress Catalog Card Number: 82–061580
International Standard Book Number: 0–675–20037–7
Printed in the United States of America
3 4 5 6 7 8 9 10 — 87 86 85 84

Merrill's International Series in
Electrical and Electronics Technology

BATESON	Introduction to Control System Technology, 2nd Edition
BOYLESTAD	Introductory Circuit Analysis, 4th Edition
	Student Guide to Accompany Introductory Circuit Analysis, 4th Edition
	Experiments in Circuit Analysis, 4th Edition
DAVIS	Industrial Electronics: Design and Application
DIXON	Network Analysis
FLOYD	Digital Fundamentals, 2nd Edition
	Principles of Electric Circuits
	Electric Circuits, Electron-flow Version
STANLEY, B.	Experiments in Electric Circuits
OPPENHEIMER	Semiconductor Logic and Switching Circuits, 2nd Edition
ROONEY	Analysis of Linear Circuits: Passive and Active Components
SCHWARTZ	Survey of Electronics, 2nd Edition
SEIDMAN	Electronics: Devices, Discrete and Integrated Circuits
TOCCI	Fundamentals of Electronic Devices, 3rd Edition
	Electronic Devices, 3rd Edition, Conventional Current Version
	Fundamentals of Pulse and Digital Circuits, 3rd Edition
	Introduction to Electric Circuit Analysis, 2nd Edition
TURNER	Fundamental Electronics
WARD	Applied Digital Electronics

In memory of my father, V. R. Floyd

This book is easy to read and understand. Sentences and paragraphs are short, and the material is presented in concise, easily digested segments. Frequent review questions and problems focus on the key ideas in each section. Self-tests appear at the end of each chapter, with worked out solutions at the end of the book. Concepts are presented in a down-to-earth manner, and there are abundant drawings, graphs, and photographs throughout.

Mathematics is used in a supportive role. Mathematical procedures are introduced as they are needed so that their application is clear. The level of the mathematics does not go beyond right-angle trigonometry and is such that a student with some high school algebra will be able to use the book.

Organization

The book consists of twenty-one chapters with a basically traditional dc/ac circuits coverage. The first nine chapters deal with basic circuit concepts, definitions, and analysis of resistive circuits with dc sources. Chapters 10 through 20 provide a coverage of time-varying wave forms, reactive components, and analysis of resistive and reactive circuits with time-varying sources. Chapter 21 covers basic test instruments, including the oscilloscope and other modern test equipment.

Suggestions for Use

The format of this book provides a great deal of flexibility in its use. It can be adapted to either the quarter system or the semester system very easily. The entire twenty-one chapters can be covered in a two quarter dc/ac circuits course. A two-semester coverage will, of course, allow for a somewhat slower pace or supplementary work.

Features

In addition to its readability, thorough coverage, and well-illustrated format, this book has several other notable features:

1. Electron flow orientation.
2. Emphasis on both frequency response and time response. Two chapters are completely devoted to pulse response of RC and RL circuits. Also, in addition to the sine wave, several nonsinusoidal wave forms are covered.
3. Trouble-shooting techniques and examples are found throughout the book.
4. Numerous examples and end-of-chapter problems are provided, with answers to selected odd-numbered problems at the end of the book.

PREFACE

5. Section review questions and problems are provided within the chapters, with answers at the end of the chapter.
6. Self-tests are included at the end of each chapter, with worked out solutions at the end of the book.
7. A list of formulas and a summary appear at the end of most chapters.
8. The use of the calculator is incorporated in the solution of several example problems to promote the skillful use of this instrument.
9. A circuit analysis program written in BASIC is given in Appendix C for optional use in conjunction with the circuit analysis methods presented in Chapter 9 and as an introduction to the potential of computer-aided analysis.
10. The basic oscilloscope controls and their use are presented in Chapter 21.
11. A comprehensive glossary is provided at the end of the book.
12. The appendices include a table of standard resistor values, batteries, a table of trigonometric functions, a table of exponential functions, and several important derivations.

Acknowledgments

This book would not have been possible without the contributions of many people.

First, I would like to express my appreciation to Jerry Cox, head of the Department of Electronics Technology at Mayland Technical College, for his many valuable suggestions and particularly for his help with the circuit analysis program, and to Mike Cummings for the very effective cover photograph.

Also, I would like to thank the reviewers who provided many comments and suggestions for improvement of the manuscript: Roger Everett, David M. Hata, Alexander Avtgis, Donald P. Leach, Lee Rosenthal, and Gerald E. Williams. Rich Brongel, Richard Falley, Mike Grub, Kevin Horne, Tim Hungate, Bob Lehman, John Morgan, Serge Silbey, Arthur Vuilleumier, and Jack L. Waintraub also provided valuable assistance.

To the people at Merrill, in particular Chris Conty and Cherlyn B. Paul, with whom it is always a pleasure to work, I wish to express my gratitude. My thanks also to Margaret Shaffer who did a great job of editing the manuscript, and to Bill Oltman, Charles Wheeler, Jerry Wilson, Dave Olejniczak, and James Bortner for checking the manuscript for the accuracy of electron flow notation.

Finally, I wish to thank all those industrial organizations that contributed the photographic material for use in this book.

All of these people helped to create this book, but the most important person is you, the user. This book was written for you to help meet your educational needs. If you are a student, I hope that this book will provide the assistance you need in preparing for your career. If you are an educator, I hope you will find that the book meets your requirements now and in the future. Best wishes for success.

CONTENTS

1
Introduction 1

- 1-1 History 2
- 1-2 Applications of Electronics 6
- 1-3 Electrical Units 10
- 1-4 Scientific Notation 11
- 1-5 Metric Prefixes 15
- Self-Test 16
- Problems 17
- Answers to Section Reviews 17

2
Current, Voltage, and Resistance 21

- 2-1 Atoms 22
- 2-2 Current 24
- 2-3 Voltage 25
- 2-4 Resistance 26
- 2-5 Types of Resistors 27
- 2-6 Conductors, Semiconductors, and Insulators 32
- 2-7 Wire Size 34
- 2-8 The Electric Circuit 36
- 2-9 Measurements 37
- 2-10 Protective Devices 39
- Formulas 43
- Summary 44
- Self-Test 45
- Problems 45
- Answers to Section Reviews 47

3
Ohm's Law 49

- 3-1 Statement of Ohm's Law 50
- 3-2 Calculating Current 52

3-3 Calculating Voltage 56
3-4 Calculating Resistance 60
3-5 Voltage and Current Are Linearly Proportional 62
Formulas 64
Summary 64
Self-Test 65
Problems 66
Answers to Section Reviews 68

4

Power and Energy 71

4-1 Power 72
4-2 Power Formulas 74
4-3 Resistor Power Ratings 76
4-4 Energy 77
4-5 Energy Loss and Voltage Drop 78
4-6 Power Supplies 80
Formulas 83
Summary 83
Self-Test 84
Problems 85
Answers to Section Reviews 87

5

Series Resistive Circuits 89

5-1 Resistors in Series 90
5-2 Current in a Series Circuit 93
5-3 Total Series Resistance 94
5-4 Applying Ohm's Law 98
5-5 Voltage Sources in Series 101
5-6 Kirchhoff's Voltage Law 104
5-7 Voltage Dividers 108
5-8 Power in a Series Circuit 114
5-9 Open and Closed Circuits 116
5-10 Trouble-Shooting Series Circuits 117
Formulas 118
Summary 119
Self-Test 120
Problems 122
Answers to Section Reviews 124

CONTENTS

6

Parallel Resistive Circuits   129

6-1   Resistors in Parallel   130
6-2   Voltage Drop in a Parallel Circuit   133
6-3   Total Parallel Resistance   135
6-4   Applying Ohm's Law   142
6-5   Current Sources in Parallel   144
6-6   Kirchhoff's Current Law   146
6-7   Current Dividers   150
6-8   Power in a Parallel Circuit   155
6-9   Effects of Open Paths   156
6-10 Trouble-Shooting Parallel Circuits   158
      Formulas   160
      Summary   161
      Self-Test   161
      Problems   163
      Answers to Section Reviews   166

7

Series-Parallel Combinations   171

7-1   Definition of a Series-Parallel Circuit   172
7-2   Circuit Identification   172
7-3   Analysis of Series-Parallel Circuits   176
7-4   Circuit Ground   186
7-5   Trouble-Shooting   190
7-6   Voltage Dividers with Resistive Loads   194
7-7   Ladder Networks   198
7-8   Wheatstone Bridge   201
      Formula   203
      Summary   203
      Self-Test   204
      Problems   206
      Answers to Section Reviews   210

8

Circuit Theorems and Conversion   213

8-1   The Voltage Source   214
8-2   The Current Source   216
8-3   Source Conversions   218

8-4	The Superposition Theorem	221
8-5	Thevenin's Theorem	229
8-6	Norton's Theorem	237
8-7	Millman's Theorem	241
8-8	Maximum Power Transfer Theorem	244
8-9	Delta-Wye (ΔY) and Wye-Delta (Y-Δ) Network Conversions	247

Formulas 250
Summary 251
Self-Test 251
Problems 253
Answers to Section Reviews 256

9

Circuit Analysis Methods 261

9-1	Branch Current Method	262
9-2	Determinants	265
9-3	Mesh Current Method	267
9-4	Node Voltage Method	273

Summary 276
Self-Test 276
Problems 277
Answers to Section Reviews 279

10

Signal Characteristics and Analysis 281

10-1	The Sine Wave	282
10-2	Period and Frequency	283
10-3	Amplitude Values of a Sine Wave	289
10-4	Phase Relationships	294
10-5	Equation for a Sine Wave	297
10-6	Phasors	300
10-7	Pulse Wave Forms	305
10-8	Triangular and Sawtooth Wave Forms	310
10-9	Harmonics	313

Formulas 314
Summary 316
Self-Test 317
Problems 319
Answers to Section Reviews 322

CONTENTS

11
Capacitance 325

- 11-1 The Basic Capacitor 326
- 11-2 Charging and Discharging a Capacitor 327
- 11-3 Unit of Capacitance 329
- 11-4 Characteristics of Capacitors 331
- 11-5 Types of Capacitors 336
- 11-6 Capacitor Labeling 341
- 11-7 Series Capacitors 343
- 11-8 Parallel Capacitors 348
- 11-9 Relationship of Capacitive Current and Voltage 350
- 11-10 Capacitive Reactance 351
- 11-11 Energy and Power in a Capacitor 354
- 11-12 Testing in a Capacitor 357
  - Formulas 358
  - Summary 359
  - Self-Test 360
  - Problems 362
  - Answers to Section Reviews 364

12
Frequency Response of RC Circuits 367

- 12-1 Sine Wave Response 368
- 12-2 Basic Phasor Algebra 368
- 12-3 Impedance of a Series RC Circuit 372
- 12-4 Phase Angle 375
- 12-5 Voltage and Current Magnitudes 379
- 12-6 The RC Lag Network 381
- 12-7 The RC Lead Network 385
- 12-8 The Series RC Circuit as a Filter 387
- 12-9 Power in an RC Circuit 391
- 12-10 The Parallel RC Circuit 394
  - Formulas 397
  - Summary 398
  - Self-Test 399
  - Problems 401
  - Answers to Section Reviews 404

13
Pulse Response of RC Circuits 407

- 13-1 Pulses 408
- 13-2 Time Constant 410
- 13-3 Exponential Curves 416
- 13-4 The RC Integrator 422
- 13-5 Integrator Response to a Rectangular Pulse 424
- 13-6 Integrator Response to Periodic Pulse Wave Forms 429
- 13-7 The RC Differentiator 435
- 13-8 Differentiator Response to a Rectangular Pulse 435
- 13-9 Differentiator Response to Periodic Pulse Wave Forms 440
- 13-10 Current in the Integrator and Differentiator 443
- 13-11 Relationship of Time Response and Frequency Response 445
  - Formulas 447
  - Summary 447
  - Self-Test 448
  - Problems 449
  - Answers to Section Reviews 452

14
Inductance 455

- 14-1 Magnetic Fields 456
- 14-2 Electromagnetism 458
- 14-3 Electromagnetic Properties 459
- 14-4 Electromagnetic Induction 461
- 14-5 The Basic Inductor 463
- 14-6 Faraday's Law 464
- 14-7 Self-Inductance 466
- 14-8 Characteristics of Inductors 468
- 14-9 Types of Inductors 470
- 14-10 Series Inductance 471
- 14-11 Parallel Inductance 472
- 14-12 Relationship of Inductive Current and Voltage 474
- 14-13 Inductive Reactance 476
- 14-14 Inductors in dc Circuits 479
- 14-15 Energy and Power in an Inductor 480
- 14-16 Testing an Inductor 481
  - Formulas 482
  - Summary 483
  - Self-Test 483
  - Problems 484
  - Answers to Section Reviews 486

15

Frequency Response of RL Circuits 489

- 15-1 Sine Wave Response 490
- 15-2 Impedance of a Series RL Circuit 490
- 15-3 Phase Angle 493
- 15-4 Voltage and Current Magnitudes 497
- 15-5 The RL Lag Network 498
- 15-6 The RL Lead Network 501
- 15-7 The Series RL Circuit as a Filter 504
- 15-8 Power in an RL Circuit 507
- 15-9 The Parallel RL Circuit 509
  - Formulas 511
  - Summary 512
  - Self-Test 513
  - Problems 514
  - Answers to Section Reviews 517

16

Pulse Response of RL Circuits 521

- 16-1 Time Constant 522
- 16-2 Induced Voltage in a Series RL Circuit 526
- 16-3 The RL Integrator 528
- 16-4 The RL Differentiator 532
- 16-5 Relationship of Time Response to Frequency Response 535
  - Formulas 537
  - Summary 537
  - Self-Test 538
  - Problems 538
  - Answers to Section Reviews 540

17

Transformers 543

- 17-1 Mutual Inductance 544
- 17-2 The Basic Transformer 545
- 17-3 Step-Up Transformer 548
- 17-4 Step-Down Transformer 549
- 17-5 Loading the Secondary 550
- 17-6 Reflected Impedance 551

17-7	Impedance Matching	553
17-8	The Transformer as an Isolation Device	555
17-9	Tapped Transformers	557
17-10	Multiple-Winding Transformers	558
17-11	Autotransformers	559
17-12	Transformer Construction	560

Formulas 562
Summary 562
Self-Test 563
Problems 564
Answers to Section Reviews 565

18
Complex Numbers in Reactive Circuits 569

18-1	Complex Numbers	570
18-2	Rectangular and Polar Forms	572
18-3	Arithmetic of Complex Numbers	577
18-4	Resistance and Reactance in Complex Form	579
18-5	Total Impedance in Complex Form	581
18-6	Impedance in RLC Circuits	584
18-7	Complex Analysis of Reactive Circuits	588
18-8	Series Equivalent Circuits	591

Formulas 597
Summary 598
Self-Test 598
Problems 600
Answers to Section Reviews 602

19
Resonance 605

19-1	Series Resonance	606
19-2	Parallel Resonance	611
19-3	Bandwidth of a Resonant Circuit	614
19-4	Quality Factor (Q) of a Resonant Circuit	618
19-5	Applications of Resonant Circuits	622

Formulas **622**
Summary **623**
Self-Test **624**
Problems **624**
Answers to Section Reviews **626**

20

Filters **629**

- 20-1 Low-Pass Filters **630**
- 20-2 Other Types of Low-Pass Filters **634**
- 20-3 High-Pass Filters **635**
- 20-4 Other Types of High-Pass Filters **638**
- 20-5 Band-Pass Filters **639**
- 20-6 Band-Stop Filters **642**
- 20-7 Filter Response Characteristics **646**
  Formulas **650**
  Summary **650**
  Self-Test **652**
  Problems **653**
  Answers to Section Reviews **655**

21

Test and Measurement Instruments **657**

- 21-1 Meter Movements **658**
- 21-2 The Ammeter **660**
- 21-3 The Voltmeter **663**
- 21-4 ac Meters **668**
- 21-5 The Ohmmeter **669**
- 21-6 Multimeters **671**
- 21-7 Signal Generators **673**
- 21-8 The Oscilloscope **678**
- 21-9 Oscilloscope Controls **682**
  Summary **685**
  Self-Test **685**
  Problems **686**
  Answers to Section Reviews **688**

Appendix A   Table of Standard 10% Resistor Values   **690**
Appendix B   Batteries   **691**
Appendix C   A Computer Program for Circuit Analysis   **694**
Appendix D   rms (Effective) Value of a Sine Wave   **698**
Appendix E   Average Value of a Half-Cycle Sine Wave   **699**
Appendix F   Reactance Derivations   **700**
Appendix G   Table of Trigonometric Functions   **701**
Appendix H   Table of Exponential Functions   **706**

Glossary   709
Answers to Selected Odd-Numbered Problems   719
Solutions to Self-Tests   725
Index   743

This chapter briefly examines the history of electricity and electronics and discusses some of the many areas of application. Also, to aid you throughout the book, the basics of scientific notation and metric prefixes are covered, along with the quantities and units commonly used in electronics.

1-1 History
1-2 Applications of Electronics
1-3 Electrical Units
1-4 Scientific Notation
1-5 Metric Prefixes

1 INTRODUCTION

1-1

HISTORY

One of the first important discoveries about static electricity is attributed to William Gilbert (1540–1603). Gilbert was an English physician who, in a book published in 1600, described how amber differs from magnetic loadstones in its attraction of certain materials. He found that when amber was rubbed with a cloth, it attracted only lightweight objects, whereas loadstones attracted only iron. Gilbert also discovered that other substances, such as sulfur, glass, and resin, behave like amber. He used the Latin word *elektron* for amber and originated the word *electrica* for the other substances that acted similarly to amber. The word *electricity* was used for the first time by Sir Thomas Browne (1605–1682), an English physician.

Another Englishman, Stephen Gray (1696–1736), discovered that some substances conduct electricity and some do not. Following Gray's lead, a Frenchman named Charles du Fay experimented with the conduction of electricity. These experiments led him to believe that there were two kinds of electricity. He called one type *vitreous electricity* and the other type *resinous electricity*. He found that objects charged with vitreous electricity repelled each other and those charged with resinous electricity attracted each other. It is known today that two types of electrical *charge* do exist. They are called *positive* and *negative*.

Benjamin Franklin (1706–90) conducted studies in electricity in the mid-1700s. He theorized that electricity consisted of a single *fluid*, and he was the first to use the terms *positive* and *negative*. In his famous kite experiment, Franklin showed that lightning is electricity.

Charles Augustin de Coulomb (1736–1806), a French physicist, in 1785 proposed the laws that govern the attraction and repulsion between electrically charged bodies. Today, the unit of electrical charge is called the *coulomb*.

Luigi Galvani (1737–98) experimented with current electricity in 1786. Galvani was a professor of anatomy at the University of Bologna in Italy. Electrical current was once known as *galvanism* in his honor.

In 1800, Alessandro Volta (1745–1827), an Italian professor of physics, discovered that the chemical action between moisture and two different metals produced electricity. Volta constructed the first battery, using copper and zinc plates separated by paper that had been moistened with a salt solution. This battery, called the *voltaic pile,* was the first source of steady electric current. Today, the unit of electrical potential energy is called the *volt* in honor of Volta.

A Danish scientist, Hans Christian Oersted (1777–1851), is credited with the discovery of electromagnetism, in 1820. He found that electrical current flowing through a wire caused the needle of a compass to move. This finding showed that a magnetic field exists around a current-carrying conductor and that the field is produced by the current.

The modern unit of electrical current is the *ampere* (also called *amp*) in honor of the French physicist André Ampère (1775–1836). In 1820, Ampère measured the magnetic effect of an electrical current. He found that two wires

HISTORY

carrying current can attract and repel each other, just as magnets can. By 1822, Ampère had developed the fundamental laws that are basic to current electricity.

One of the most well known and widely used laws in electrical circuits today is *Ohm's law*. It was formulated by Georg Simon Ohm (1789–1854), a German teacher, in 1826. Ohm's law gives us the relationship among the three important electrical quantities of resistance, voltage, and current.

Although it was Oersted who discovered electromagnetism, it was Michael Faraday (1791–1867) who carried the study further. Faraday was an English physicist who believed that if electricity could produce magnetic effects, then magnetism could produce electricity. In 1831 he found that a moving magnet caused an electric current in a coil of wire placed within the field of the magnet. This effect, known today as *electromagnetic induction*, is the basic principle of electric generators and transformers.

Joseph Henry (1797–1878), an American physicist, independently discovered the same principle in 1831, and it is in his honor that the unit of inductance is called the *henry*. The unit of capacitance, the *farad*, is named in honor of Michael Faraday.

In the 1860s, James Clerk Maxwell (1831–79), a Scottish physicist, produced a set of mathematical equations that expressed the laws governing electricity and magnetism. These are known as *Maxwell's equations*. Maxwell also predicted that electromagnetic waves (radio waves) that travel at the speed of light in space could be produced.

It was left to Heinrich Rudolph Hertz (1857–94), a German physicist, to actually produce these waves that Maxwell predicted. Hertz performed this work in the late 1880s. Today, the unit of frequency is called *hertz*.

The Beginning of Electronics

The early experiments in electronics involved electric currents flowing in glass tubes. One of the first to conduct such experiments was a German named Heinrich Geissler (1814–79). Geissler removed most of the air from a glass tube and found that the tube glowed when there was an electric current through it.

Around 1878, Sir William Crookes (1832–1919), a British scientist, experimented with tubes similar to those of Geissler. In his experiments, Crookes found that the current in the tubes seemed to consist of particles.

Thomas Edison (1847–1931), experimenting with the carbon-filament light bulb that he had invented, made another important finding. He inserted a small metal plate in the bulb. When the plate was positively charged, there was current from the filament to the plate. This device was the first *thermionic diode*. Edison patented it but never used it.

The electron was discovered in the 1890s. The French physicist Jean Baptiste Perrin (1870–1942) demonstrated that the current in a vacuum tube consists of negatively charged particles. Some of the properties of these particles were measured by Sir Joseph Thomson (1856–1940), a British physicist, in experiments he performed between 1895 and 1897. These negatively charged particles later became known as *electrons*. The charge on the electron was accurately measured by an

American physicist, Robert A. Millikan (1868–1953), in 1909. As a result of these discoveries, electrons could be controlled, and the electronic age was ushered in.

Putting the Electron to Work

A vacuum tube that allowed electrical current in only one direction was constructed in 1904 by John A. fleming, a British scientist. The tube was used to detect electromagnetic waves. Called the *Fleming valve,* it was the forerunner of the more recent vacuum diode tubes.

Major progress in electronics, however, awaited the development of a device that could boost, or *amplify*, a weak electromagnetic wave or radio signal. This device was the *audion,* patented in 1907 by Lee de Forest, an American. It was a triode vacuum tube capable of amplifying small electrical signals.

Two other Americans, Harold Arnold and Irving Langmuir, made great improvements in the triode tube between 1912 and 1914. About the same time, de Forest and Edwin Armstrong, an electrical engineer, used the triode tube in an *oscillator circuit*. In 1914, the triode was incorporated in the telephone system and made the transcontinental telephone network possible.

The tetrode tube was invented in 1916 by Walter Schottky, a German. The tetrode, along with the pentode (invented in 1926 by Tellegen, a Dutch engineer), provided great improvements over the triode. The first television picture tube, called the *kinescope*, was developed in the 1920s by Vladimir Zworykin, an American researcher.

During World War II, several types of microwave tubes were developed that made possible modern microwave radar and other communications systems. In 1939, the *magnetron* was invented in Britain by Henry Boot and John Randall. In the same year, the *klystron* microwave tube was developed by two Americans, Russell Varian and his brother Sigurd Varian. The *traveling-wave* tube was invented in 1943 by Rudolf Komphner, an Austrian-American.

The Computer

The computer probably has had more impact on modern technology than any other single type of electronic system. The first electronic digital computer was completed in 1946 at the University of Pennsylvania. It was called the Electronic Numerical Integrator and Computer (ENIAC). One of the most significant developments in computers was the *stored program* concept, developed in the 1940s by John von Neumann, an American mathematician.

Solid State Electronics

The crystal detectors used in the early radios were the forerunners of modern solid state devices. However, the era of solid state electronics began with the invention of the *transistor* in 1947 at Bell Labs. The inventors were Walter Brattain, John Bardeen, and William Shockley. Figure 1–1 shows these three men, along with the notebook entry describing the historic discovery.

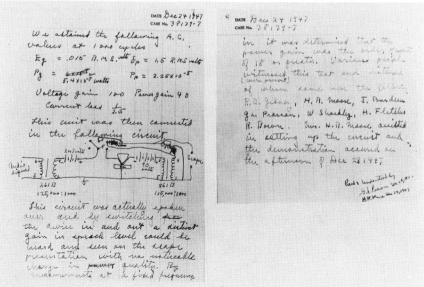

FIGURE 1–1 **A.** *Nobel Prize winners Drs. John Bardeen, William Shockley, and Walter Brattain, shown left to right, with apparatus used in their first investigations that led to the invention of the transistor. The trio received the 1956 Nobel Physics award for their invention of the transistor, which was announced by Bell Laboratories in 1948.* **B.** *The laboratory notebook entry of scientist Walter H. Brattain recorded the events of December 23, 1947, when the transistor effect was discovered at Bell Telephone Laboratories. The notebook entry describes the event and adds, "This circuit was actually spoken over and by switching the device in and out a distinct gain in speech level could be heard and seen on the scope presentation with no noticeable change in quality."* (**A** *and* **B**, *courtesy of Bell Laboratories*)

In the early 1960s, the integrated circuit was developed. It incorporated many transistors and other components on a single small *chip* of semiconductor material. Integrated circuit technology was developed and improved during the 1960s and 1970s, allowing more complex circuits to be built on smaller chips. The introduction of the microprocessor in the early 1970s created another electronics revolution: the entire processing portion of a computer placed on a single, small, silicon chip. Continued development brought about complete computers on a single chip by the late 1970s.

Review for 1-1

1. Who developed the first battery?
2. The unit of what electrical quantity is named after André Ampère?
3. What contribution did Georg Simon Ohm make to the study of electricity?
4. In what year was the transistor invented?
5. What major development followed invention of the transistor?

1-2

APPLICATIONS OF ELECTRONICS

Electronics is a diverse technological field with almost limitless applications. There is hardly an area of human endeavor that is not dependent to some extent on electronics. Some of the applications are discussed here in a general way to give you an idea of the scope of the field.

Computers

One of the most important electronic systems is the digital computer; its applications are broad and diverse. For example, computers have applications in business for record keeping, accounting, payrolls, inventory control, market analysis, and statistics, to name but a few.

Scientific fields utilize the computer to process huge amounts of data and to perform complex and lengthy calculations. In industry, the computer is used for controlling and monitoring intricate manufacturing processes. Communications, navigation, medical, military, and home uses are a few of the other areas in which the computer is used extensively.

The computer's success is based on its ability to perform mathematical operations extremely fast and to process and store large amounts of information.

Computers vary in complexity and capability, ranging from very large systems with vast capabilities down to a computer on a chip with much more limited performance. Figure 1-2 shows some typical computers of varying sizes.

APPLICATIONS OF ELECTRONICS

FIGURE 1–2 **A.** *TRS-80 Model II Microcomputer System. (Courtesy of Radio Shack)* **B.** *PDP-11V23 Computer. (Courtesy of Digital Equipment Corporation)* **C.** *V-8600 Computer. (Courtesy of NCR Corporation)*

Communications Systems

Communications electronics encompasses a wide range of specialized fields. Included are space and satellite communications, commercial radio and television, citizens' band and amateur radio, data communications, navigation systems, radar, telephone systems, military applications, and specialized radio applications such as police, aircraft, and so on. Computers are used to a great extent

FIGURE 1–3 *A digital long-distance telephone switching system, GTE No. 3 EAX (Electronic Automatic Exchange). (Courtesy of GTE Automatic Electric Incorporated)*

in many communications systems. Figure 1–3 shows a telephone switching system as an example of electronic communications.

Automation

Electronic systems are employed extensively in the control of manufacturing processes. Computers and specialized electronic systems are used in industry for various purposes, for example, control of ingredient mixes, operation of machine tools, product inspection, and control and distribution of power.

Medicine

Electronic devices and systems are finding ever-increasing applications in the medical field. The familiar *electrocardiograph* (ECG), used for the diagnosis of heart and other circulatory ailments, is a widely used medical electronic instrument. A closely related instrument is the *electromyograph,* which uses a cathode ray tube display rather than an ink trace.

The *diagnostic sounder* uses ultrasonic sound waves for various diagnostic procedures in neurology, for heart chamber measurement, and for detection of certain types of tumors. The *electroencephalograph* (EEG) is similar to the electrocardiograph. It records the electrical activity of the brain rather than heart activity. Another electronic instrument used in medical procedures is the *coagulograph*. This instrument is used in blood clot analysis.

Electronic instrumentation is also used extensively in intensive-care facilities. Heart rate, pulse, body temperature, respiration, and blood pressure can be monitored on a continuous basis. Monitoring equipment is also used a great deal

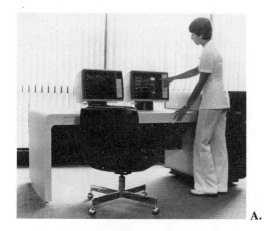
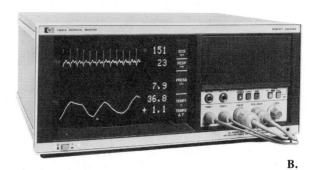
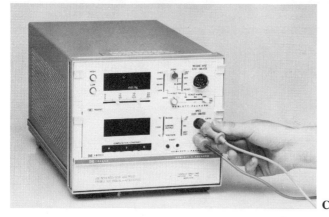
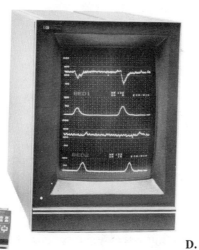

FIGURE 1–4 **A.** *Patient information center displays ECG wave forms and alphanumeric data.* **B.** *Neonatal monitor shows both wave form and computed numeric data of vital patient information.* **C.** *Cardiac output module computes cardiac output and measures continuous pulmonary artery blood pressure.* **D.** *Patient data memory system for fetal monitoring enables hospital staff to select and control information from a small, hand-held keyboard.* (**A–D**, *courtesy of Hewlett-Packard Company*)

in operating rooms. Some typical medical electronic equipment is pictured in Figure 1–4.

Consumer Products

Electronic products used directly by the consumers for information, entertainment, recreation, or work around the home are an important segment of the total electronics market. For example, the electronic calculator and digital watch are popular examples of consumer electronics. The small personal computer is used widely by hobbyists and is also becoming a common household appliance.

Electronic systems are used in automobiles to control and monitor engine functions, control braking, provide entertainment, and display useful information to the driver.

Most appliances such as microwave ovens, washers, and dryers are available with electronic controls. Home entertainment, of course, is largely electronic. Examples are television, radio, stereo, and recorders. Also, many new games for adults and children incorporate electronic devices.

Review for 1–2

1. Name some of the areas in which electronics is used.
2. Is television a consumer product?

1–3

ELECTRICAL UNITS

In electronics work, you must deal with measurable quantities. For example, you must be able to express how many volts are measured at a certain test point, how much current is flowing through a wire, or how much power a certain amplifier produces.

In this section you will learn the units and symbols for many of the electrical quantities that are used throughout the book. Definitions of the quantities are presented as they are needed in later chapters.

Symbols are used in electronics to represent both quantities and their units. One symbol is generally used to represent the quantity, and another to represent the unit of that quantity. For example, *P* stands for *power,* and W stands for *watts,* which is the unit of power. Table 1–1 lists the most important quantities, along with their SI units and symbols. The term *SI* is the French abbreviation for *International System* (*Système International* in French).

SCIENTIFIC NOTATION

TABLE 1-1 *Electrical quantities and units with SI symbols.*

Quantity	Symbol	Unit	Symbol
Capacitance	C	farad	F
Charge	Q	coulomb	C
Conductance	G	siemen	S
Current	I	ampere	A
Energy	$\mathcal{E}$	joule	J
Frequency	f	hertz	Hz
Impedance	Z	ohm	Ω
Inductance	L	henry	H
Power	P	watt	W
Reactance	X	ohm	Ω
Resistance	R	ohm	Ω
Time	t	second	s
Voltage	V	volt	V
Wavelength	λ	meter	m

Review for 1-3

1. What does *SI* stand for? *International System*
2. Without referring to Table 1-1, list as many electrical quantities as possible, including their symbols, units, and unit symbols.

1-4

SCIENTIFIC NOTATION

In electronics work, you will encounter both very small and very large numbers. For example, it is common to have electrical current values of only a few thousandths or even a few millionths of an ampere. On the other hand, you will find resistance values of several thousand or several million ohms. This range of values is typical of many other electrical quantities also.

Powers of Ten

Scientific notation uses *powers of ten,* a method that makes it much easier to express large and small numbers and to do calculations involving such numbers.

Table 1–2 lists some powers of ten, both positive and negative. The power of ten is expressed as an exponent of the base 10 in each case. The exponent indicates the number of decimal places to the right or left of the decimal point in the expanded number. If the power is *positive,* the decimal point is moved to the *right*. For example,

$$10^4 = 1 \times 10^4 = 1.0000. = 10,000.$$

If the power is *negative,* the decimal point is moved to the *left*. For example,

$$10^{-4} = 1 \times 10^{-4} = .0001. = 0.0001$$

TABLE 1–2 *Some positive and negative powers of ten.*

1,000,000	$= 10^6$	0.000001	$= 10^{-6}$
100,000	$= 10^5$	0.00001	$= 10^{-5}$
10,000	$= 10^4$	0.0001	$= 10^{-4}$
1,000	$= 10^3$	0.001	$= 10^{-3}$
100	$= 10^2$	0.01	$= 10^{-2}$
10	$= 10^1$	0.1	$= 10^{-1}$
1	$= 10^0$		

Example 1–1

Express each number as a positive power of ten.
(a) 200   (b) 5000   (c) 85,000   (d) 3,000,000

Solution:

In each case there are many possibilities for expressing the number in powers of ten. We do not show all possibilities in this example but include the most common powers of ten used in electrical work.
(a) $200 = 0.0002 \times 10^6 = 0.2 \times 10^3$
(b) $5000 = 0.005 \times 10^6 = 5 \times 10^3$
(c) $85,000 = 0.085 \times 10^6 = 8.5 \times 10^4 = 85 \times 10^3$
(d) $3,000,000 = 3 \times 10^6 = 3000 \times 10^3$

Example 1–2

Express each number as a negative power of ten.
(a) 0.2   (b) 0.005   (c) 0.00063   (d) 0.000015

Solution:

Again, all the possible ways to express each number as a power of ten are not given. The most commonly used powers are included, however.

SCIENTIFIC NOTATION

(a) $0.2 = 2 \times 10^{-1} = 200 \times 10^{-3} = 200{,}000 \times 10^{-6}$
(b) $0.005 = 5 \times 10^{-3} = 5000 \times 10^{-6}$
(c) $0.00063 = 0.63 \times 10^{-3} = 6.3 \times 10^{-4} = 630 \times 10^{-6}$
(d) $0.000015 = 0.015 \times 10^{-3} = 1.5 \times 10^{-5} = 15 \times 10^{-6}$

Example 1–3

Express each of the following powers of ten as a regular decimal number:
(a) 10^5 (b) 2×10^3 (c) 3.2×10^{-2} (d) 250×10^{-6}

Solution:

(a) $10^5 = 1 \times 10^5 = 100{,}000$ (b) $2 \times 10^3 = 2000$
(c) $3.2 \times 10^{-2} = 0.032$ (d) $250 \times 10^{-6} = 0.000250$

Calculating in Powers of Ten

The great convenience of scientific notation is in addition, subtraction, multiplication, and division of very small or very large numbers.

Rules for Addition: The rules for adding numbers of powers of ten are as follows:

1. Convert the numbers to be added to the *same* power of ten.
2. Add the numbers directly to get the sum.
3. Bring down the common power of ten, which is the power of ten of the sum.

Example 1–4

Add 2×10^6 and 5×10^7.

Solution:

1. Convert both numbers to the same power of ten:

$$(2 \times 10^6) + (50 \times 10^6)$$

2. Add $2 + 50 = 52$.
3. Bring down the common power of ten (10^6), and the sum is 52×10^6.

Rules for Subtraction: The rules for subtracting numbers in powers of ten are as follows:

1. Convert the numbers to be subtracted to the *same* power of ten.

2. Subtract the numbers directly to get the difference.
3. Bring down the common power of ten, which is the power of ten of the difference.

Example 1–5

Subtract 25×10^{-12} from 75×10^{-11}.

Solution:

1. Convert each number to the same power of ten:
$$(75 \times 10^{-11}) - (2.5 \times 10^{-11})$$
2. Subtract $75 - 2.5 = 72.5$.
3. Bring down the common power of ten (10^{-11}), and the difference is 72.5×10^{-11}.

Rules for Multiplication: The rules for multiplying numbers in powers of ten are as follows:

1. Multiply the numbers directly.
2. Add the powers of ten algebraically (the powers do not have to be the same).

Example 1–6

Multiply 5×10^{12} and 3×10^{-6}.

Solution:

Multiply the numbers, and algebraically add the powers:
$$(5 \times 10^{12})(3 \times 10^{-6}) = 15 \times 10^{[12+(-6)]} = 15 \times 10^{6}$$

Rules for Division: The rules for dividing numbers in powers of ten are as follows:

1. Divide the numbers directly.
2. Subtract the power of ten in the denominator from the power of ten in the numerator.

METRIC PREFIXES

> **Example 1–7**
>
> Divide 50×10^8 by 25×10^3.
>
> *Solution:*
>
> The division problem is written with a numerator and denominator as
> $$\frac{50 \times 10^8}{25 \times 10^3}$$
> Dividing the numbers and subtracting 3 from 8, we get
> $$\frac{50 \times 10^8}{25 \times 10^3} = 2 \times 10^{8-3} = 2 \times 10^5$$

Review for 1-4

1. Scientific notation uses powers of ten (T or F).
2. Express 100 as a power of ten.
3. Do the following operations:
   (a) $(1 \times 10^5) + (2 \times 10^5)$
   (b) $(3 \times 10^6)(2 \times 10^4)$
   (c) $(8 \times 10^3) \div (4 \times 10^2)$

1-5

METRIC PREFIXES

In electrical and electronics work, certain powers of ten are used more often than others. The most frequently used powers of ten are 10^9, 10^6, 10^3, 10^{-3}, 10^{-6}, 10^{-9}, and 10^{-12}.

It is common practice to use *metric prefixes* to represent these quantities. Table 1–3 lists the metric prefix for each of the commonly used powers of ten.

TABLE 1–3 *Metric prefixes and their symbols.*

Power of Ten	Value	Metric Prefix	Metric Symbol	Power of Ten	Value	Metric Prefix	Metric Symbol
10^9	one billion	giga	G	10^{-6}	one-millionth	micro	μ
10^6	one million	mega	M	10^{-9}	one-billionth	nano	n
10^3	one thousand	kilo	k	10^{-12}	one-trillionth	pico	p
10^{-3}	one-thousandth	milli	m				

Use of Metric Prefixes

Now we will use examples to illustrate use of metric prefixes. The number 2000 can be expressed in scientific notation as 2×10^3. Suppose we wish to represent 2000 watts (W) with a metric prefix. Since $2000 = 2 \times 10^3$, the metric prefix *kilo* (k) is used for 10^3. So we can express 2000 W as 2 kW (2 kilowatts).

As another example, 0.015 ampere (A) can be expressed as 15×10^{-3} A. The metric prefix *milli* (m) is used for 10^{-3}. So 0.015 A becomes 15 mA (15 milliamperes).

Example 1–8

Express each quantity using a metric prefix.
(a) 50,000 V   (b) 25,000,000 Ω   (c) 0.000036 A

Solution:

(a) 50,000 V = 50×10^3 V = 50 kV
(b) 25,000,000 Ω = 25×10^6 Ω = 25 MΩ
(c) 0.000036 A = 36×10^{-6} A = 36 μA

Review for 1–5

1. List the metric prefix for each of the following powers of ten: 10^6, 10^3, 10^{-3}, 10^{-6}, 10^{-9}, and 10^{-12}.
2. Use an appropriate metric prefix to express 0.000001 ampere.

Self-Test

1. List the units of the following electrical quantities: current, voltage, resistance, power, and energy.
2. List the symbol for each unit in Problem 1.
3. List the symbol for each quantity in Problem 1.
4. Express the following numbers in powers of ten:
   (a) 100   (b) 12,000   (c) 5,600,000   (d) 78,000,000
5. Express the following numbers in powers of ten:
   (a) 0.03   (b) 0.0005   (c) 0.00058   (d) 0.0000224
6. Express each of the following powers of ten as a regular decimal number:
   (a) 7×10^4 (b) 45×10^3 (c) 100×10^{-3} (d) 4×10^{-1}
7. Add 12×10^5 and 25×10^6.
8. Subtract 5×10^{-3} from 8×10^{-3}.

ANSWERS TO SECTION REVIEWS

9. Multiply 33×10^3 and 20×10^{-4}. 660×10^{-1}
10. Divide 4×10^2 by 2×10^{-3}. 2×10^5

Problems

1–1. Express each of the following numbers as a power of ten:
(a) 3000   (b) 75,000   (c) 2,000,000
  3×10^3 75×10^3 2×10^6

1–2. Express each number as a power of ten.
(a) 1/500   (b) 1/2000   (c) 1/5,000,000

1–3. Express each of the following numbers in three ways, using 10^3, 10^4, and 10^5:
(a) 8400   (b) 99,000   (c) 0.2×10^6

1–4. Express each of the following numbers in three ways, using 10^{-3}, 10^{-4}, and 10^{-5}:
(a) 0.0002   (b) 0.6   (c) 7.8×10^{-2}

1–5. Express each power of ten in regular decimal form.
(a) 2.5×10^{-6} (b) 50×10^2 (c) 3.9×10^{-1}

1–6. Express each power of ten in regular decimal form.
(a) 45×10^{-6} (b) 8×10^{-9} (c) 40×10^{-12}

1–7. Add the following numbers:
(a) $(92 \times 10^6) + (3.4 \times 10^7)$ (b) $(5 \times 10^3) + (85 \times 10^{-2})$
(c) $(560 \times 10^{-8}) + (460 \times 10^{-9})$

1–8. Perform the following subtractions:
(a) $(3.2 \times 10^{12}) - (1.1 \times 10^{12})$ (b) $(26 \times 10^8) - (1.3 \times 10^9)$
(c) $(150 \times 10^{-12}) - (8 \times 10^{-11})$

1–9. Perform the following multiplications:
(a) $(5 \times 10^3)(4 \times 10^5)$ (b) $(12 \times 10^{12})(3 \times 10^2)$
(c) $(2.2 \times 10^{-9})(7 \times 10^{-6})$

1–10. Divide the following:
(a) $(10 \times 10^3) \div (2.5 \times 10^2)$ (b) $(250 \times 10^{-6}) \div (50 \times 10^{-8})$
(c) $(4.2 \times 10^8) \div (2 \times 10^{-5})$

1–11. Express each of the following as a quantity having a metric prefix:
(a) 31×10^{-3} A   (b) 5.5×10^3 V   (c) 200×10^{-12} F
   31 mA 5.5 KV 200 pF

1–12. Express the following using metric prefixes:
(a) 3×10^{-6} F   (b) 3.3×10^6 Ω   (c) 350×10^{-9} A
   $3 \mu F$ 3.3 MΩ 350 NA

Answers to Section Reviews

Section 1–1:
1. Volta.   2. Current.   3. He established the relationship among current, voltage, and resistance as expressed in Ohm's law.   4. 1947.   5. Integrated circuits.

Section 1–2:
1. Computers, communications, automation, medicine, and consumer products.   **2.** Yes.

Section 1–3:
1. The abbreviation for Système International.   **2.** Refer to Table 1–1 after you have compiled your list.

Section 1–4:
1. T.   **2.** 10^2.   **3. (a)** 3×10^5 **(b)** 6×10^{10} **(c)** 2×10^1

Section 1–5:
1. Mega (M), kilo (k), milli (m), micro (μ), nano (n), and pico (p).   **2.** 1 μA (one microampere).

In this chapter you will learn about three of the most important quantities in the field of electricity and electronics: current, voltage, and resistance. No matter what type of electrical or electronic equipment you may deal with, these quantities will always be of primary importance.

To aid your understanding of electrical current, we will discuss the basic structure of the atom. The types of materials and the kinds of components that are basic to most electrical applications are also covered. Finally, the electric circuit is defined, and basic measurements of the electrical quantities are discussed.

2–1 Atoms
2–2 Current
2–3 Voltage
2–4 Resistance
2–5 Types of Resistors
2–6 Conductors, Semiconductors, and Insulators
2–7 Wire Size
2–8 The Electric Circuit
2–9 Measurements
2–10 Protective Devices

2
CURRENT, VOLTAGE, AND RESISTANCE

2-1
ATOMS

An atom is the smallest particle of an element that still retains the characteristics of that element. Different elements have different types of atoms. In fact, every element has a unique atomic structure.

Atoms have a *planetary* type of structure, consisting of a central *nucleus* surrounded by orbiting *electrons*. The nucleus consists of positively charged particles called *protons* and uncharged particles called *neutrons*. The electrons are the basic particles of *negative charge*.

Each type of atom has a certain number of electrons and protons that distinguishes the atom from all other atoms of other elements. For example, the simplest atom is that of hydrogen. It has one proton and one electron, as pictured in Figure 2–1A. The helium atom, shown in Figure 2–1B, has two protons and two neutrons in the nucleus, which is orbited by two electrons.

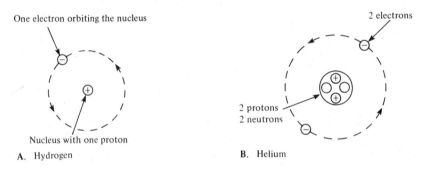

A. Hydrogen   B. Helium

FIGURE 2–1 *Hydrogen and helium atoms.*

Atomic Weight and Number

All elements are arranged in the *periodic table of the elements* in order according to their *atomic number,* which is the number of electrons in the orbits of the atom. The elements can also be arranged by their *atomic weight*, which is approximately the number of protons and neutrons in the nucleus. For example, hydrogen has an atomic number of *one* and an atomic weight of *one*. The atomic number of helium is *two*, and its atomic weight is *four*.

In their normal, or *neutral*, state, all atoms of a given element have the same number of electrons as protons. So the positive charges cancel the negative charges, and the atom has a net charge of zero.

The Copper Atom

Since copper is the most commonly used metal in electrical applications, let us examine its atomic structure. The copper atom has 29 electrons in orbit around the nucleus. They do not all occupy the same orbit, however. They move in orbits at varying distances from the nucleus. The orbits in which the

ATOMS

electrons revolve are called *shells*. The number of electrons in each shell follows a predictable pattern.

The first shell of any atom can have up to two electrons. The second shell can have up to eight electrons. The third shell can have up to 18 electrons. The fourth shell can have up to 32 electrons. A copper atom is shown in Figure 2-2. Notice that the fourth or outermost shell has only one electron, called the *valence* electron.

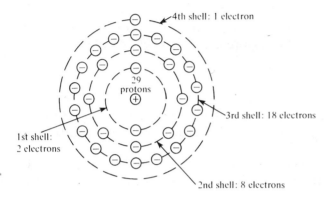

FIGURE 2-2 *The copper atom.*

Attraction between Positive and Negative Charges

The positively charged nucleus attracts the orbiting electrons. The electrons remain in a stable orbit because the centrifugal force counteracts the attractive force. The closer an electron is to the nucleus, the greater is the attractive force and the more tightly bound the electron is to the atom. The single electron in the outer shell of the copper atom is loosely bound to the atom because of its distance from the nucleus.

Free Electrons

When the 29th electron in the outer shell of the copper atom gains sufficient energy from the surrounding media, it can break away from the parent atom and become what is called a *free electron*. The free electrons in the copper material are capable of moving from one atom to another in the material. In other words, they drift randomly from atom to atom within the copper material. As you will see, the free electrons make electrical *current* possible.

Review for 2-1

1. What is the basic particle of negative charge? ELECTRON
2. Define *atom*. is smallest particle of an Element
3. What does a typical atom consist of? ELECTRO, PROTON, Nucleus

CHAPTER 2: CURRENT, VOLTAGE, AND RESISTANCE

4. Do all elements have the same types of atoms? *No*
5. What is a free electron? *Electron that have broken away an atom*

2–2 CURRENT

Electrical current is the net movement of free electrons from one point to another in a conductive material. In other words, current is the *rate of flow of electrons* in a conductor. Electrical charge is symbolized by the letter Q, current by the letter I.

Unit of Charge

The electron is the smallest unit of negative electrical charge. Electrical charge is measured in *coulombs*, abbreviated C. The charge possessed by one electron is 1.6×10^{-19} C.

One coulomb is defined to be the charge carried by 6.25×10^{18} electrons. (That's a lot of electrons!)

Formula for Current

Since current is the rate of electron flow, it can be stated as follows:

$$I = \frac{Q}{t} \tag{2-1}$$

where I is current, Q is charge, and t is time.

Ampere

Current is measured in a unit called the *ampere*, or *amp* for short, abbreviated by the letter A. *One ampere is the amount of current in a conductor when one coulomb of charge moves past a given point in one second.*

Example 2–1

Ten coulombs of charge flow past a given point in a wire in 2 seconds. How many amperes of current are there?

Solution:

$$I = \frac{Q}{t} = \frac{10 \text{ C}}{2 \text{ s}} = 5 \text{ A}$$

VOLTAGE

Review for 2–2

1. Define *current*. IS RATE OF FLOW ELECTRONS
2. How many electrons make up one coulomb of charge? 6.25 × 10¹⁸
3. If 20 coulombs flow past a point in a wire in 4 seconds, what is the current? 5A

2–3

VOLTAGE

Normally, in a conductive material such as copper wire, the free electrons are in random motion and have no net direction. In order to produce current, the free electrons *must* move in the same general direction. To produce motion in a given direction, energy must be imparted to the electrons. This energy comes from a source connected to a conductor.

The energy required to move a certain amount of charge from one point to another is called *voltage*, symbolized V. Voltage is also known as *electromotive force* (emf) or *potential difference*.

Formula for Voltage

Voltage is the amount of energy, $\mathcal{E}$, used to move an amount of charge, symbolized by Q:

$$V = \frac{\mathcal{E}}{Q} \qquad (2\text{–}2)$$

Energy is expressed in joules (J), and charge in coulombs.

Volt

Voltage is measured in a unit called the *volt*, abbreviated V. *One volt is the amount of potential difference between two points when one joule of energy is used to move one coulomb of charge from one point to the other.*

Example 2–2

If 50 joules of energy are used to move 10 coulombs of charge through a conductor, what is the voltage across the conductor?

Solution:

$$V = \frac{\mathcal{E}}{Q} = \frac{50 \text{ J}}{10 \text{ C}} = 5 \text{ V}$$

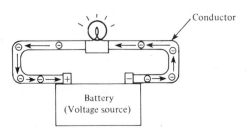

FIGURE 2–3 *Current in a conductor is produced by a voltage source.*

The Voltage Source

The battery is a typical energy source that provides voltage. Figure 2–3 shows a battery connected to a conductor to light a bulb. The voltage source has a positive terminal and a negative terminal as indicated. Energy is imparted to the electrons at the negative terminal; then the electrons move through the conductor and the bulb to the positive terminal. As a result, there is current through the conductor and the filament of the bulb. In other words, the negative terminal repels electrons away from it, and the positive terminal attracts electrons toward it. Keep in mind that *you must have voltage in order to have current.*

Review for 2–3

1. Define *voltage*. the energy required to move a charges
2. Voltage is required to produce current (T) or F).
3. One hundred joules of energy are used to move 20 coulombs of charge from the negative terminal of a battery through a wire to the positive terminal. What is the battery voltage? 5 V

2–4

RESISTANCE

Resistance is the opposition to current. It is used in electrical circuits to limit or control the amount of current.

When current flows through resistance, heat is produced. In certain applications, the main purpose of resistance is to produce heat (an electric heater is an example). In many other applications, the heat produced represents an unwanted loss of energy.

Ohm

Resistance, R, is measured in the unit of *ohms*, symbolized by the Greek letter omega, Ω. Figure 2–4 shows the schematic symbol for a 1-Ω resistance. *One ohm is the resistance when there is one ampere of current with one volt applied.*

TYPES OF RESISTORS

FIGURE 2–4 *Resistance symbol.*

Conductance

Later, in circuit analysis problems, you will find conductance to be useful. Conductance, symbolized by *G,* is simply the *reciprocal* of resistance, defined as follows:

$$G = \frac{1}{R} \tag{2-3}$$

The unit of conductance is the siemen, abbreviated S.

Review for 2–4

1. Define *resistance*. is the oppsition to current
2. What is the unit of resistance? ohm Ω
3. What two things does resistance do in an electrical circuit? limit or control current

2–5

TYPES OF RESISTORS

An electrical component having the property of resistance is called a *resistor*. There are many types of resistors in common use, but generally they can be placed in two main categories: *fixed* and *variable*.

Fixed Resistors

Fixed resistors have ohmic values set by the manufacturer and cannot be changed easily. Some common types of fixed resistors are shown in Figure 2–5. Various sizes and construction methods are used to control the heat-dissipating capabilities, the resistance value, and the precision.

Carbon-Composition Fixed Resistor

One of the most common types of fixed resistor is the carbon-composition, molded-case type. The basic construction is shown in Figure 2–6. This type of resistor is usually coded with colored bands to indicate its resistance value in ohms and its tolerance. The color-code band system is pictured in Figure 2–7, and the color code is listed in Table 2–1.

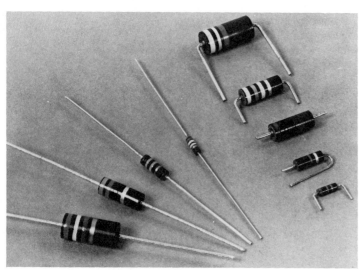
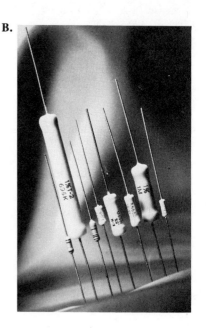
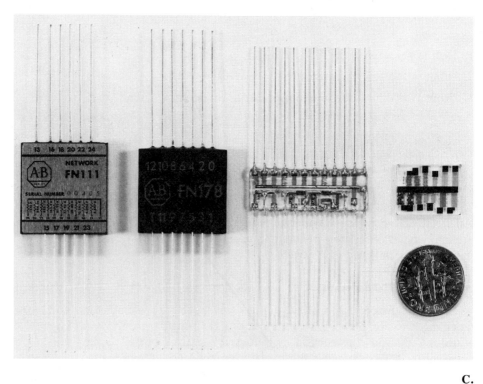

FIGURE 2–5 *Typical fixed resistors.* **A.** *Carbon-composition resistors. (Courtesy of Stackpole Carbon Company)* **B.** *Metal film resistors. (Courtesy of Stackpole Carbon Company)* **C.** *Resistor networks. (Courtesy of Allen-Bradley Company)*

TYPES OF RESISTORS

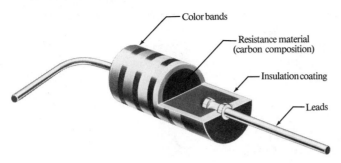

FIGURE 2–6 *Cutaway view of carbon-composition resistor. (Courtesy of Allen Bradley Company)*

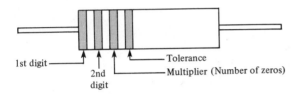

FIGURE 2–7 *Color-code bands on a resistor.*

TABLE 2–1 *Resistor color code.*

	Digit	Color
Resistance value, first three bands	0	Black
	1	Brown
	2	Red
	3	Orange
	4	Yellow
	5	Green
	6	Blue
	7	Violet
	8	Gray
	9	White
Tolerance, fourth band	5%	Gold
	10%	Silver
	20%	No band

    Beginning at the banded end, the first band is the first digit of the resistance value. The second band is the second digit. The third band is the number of zeros, or the *multiplier*. The fourth band indicates the tolerance. For example, a 5-% tolerance means that the *actual* resistance value is within ±5% of the color-coded value.

Example 2-3

Find the resistance value in ohms and the tolerance in percent for the resistor pictured in Figure 2–8.

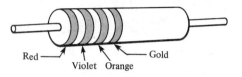

FIGURE 2–8

Solution:

> First band is red: 2
> Second band is violet: 7
> Third band is orange: 000
> Fourth band is gold: 5% tolerance

The resistance value is 27,000 Ω, ±5%.

Variable Resistors

Variable resistors are designed so that their resistance values can be changed easily with a manual or an automatic adjustment.

Two basic types of manually adjustable resistors are the *potentiometer* and the *rheostat*. Schematic symbols for these types are shown in Figure 2–9. The potentiometer is a *three-terminal device*, as indicated in Part A. Terminals 1 and 2 have a fixed resistance between them, which is the total resistance. Terminal 3 is connected to a *moving contact*. We can vary the resistance between 3 and 1 or between 3 and 2 by moving the contact up or down.

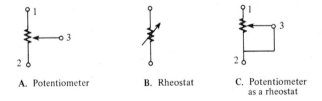

A. Potentiometer B. Rheostat C. Potentiometer as a rheostat

FIGURE 2–9 *Potentiometer and rheostat symbols.*

Figure 2–9B shows the rheostat as a *two-terminal* variable resistor. Part C shows how we can use a potentiometer as a rheostat by connecting terminal 3 to either terminal 1 or terminal 2. Some typical potentiometers are pictured in Figure 2–10.

TYPES OF RESISTORS

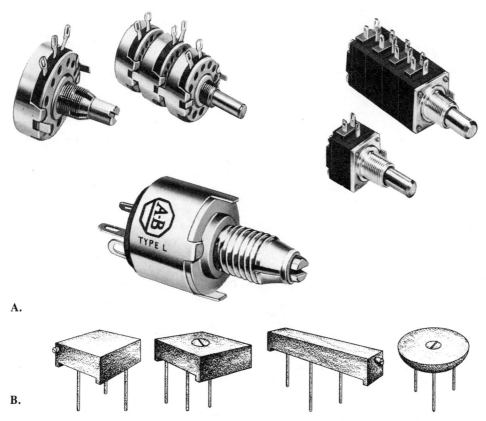

FIGURE 2–10 A. *Typical potentiometers. (Courtesy of Allen-Bradley Company)* **B.** *Trimmer potentiometers.*

Potentiometers and rheostats can be classified as *linear* or *tapered*. In a linear potentiometer, the resistance between either terminal and the moving contact varies linearly with the position of the moving contact. For example, one-half of a turn results in one-half the total resistance. Three-quarters of a turn results in three-quarters of the total resistance between the moving contact and one terminal, or one-quarter of the total resistance between the other terminal and the moving contact.

In the tapered potentiometer, the resistance varies nonlinearly with the position of the moving contact, so that one-half of a turn does not necessarily result in one-half the total resistance. This concept is illustrated in Figure 2–11, where a potentiometer with a total resistance of 100 Ω is used as an example. The nonlinear values are arbitrary.

Thermistors and Photoconductive Cells

A thermistor is a type of variable resistor that is temperature-sensitive. Its resistance changes inversely with temperature. That is, it has a *negative temperature coefficient*. When temperature increases, the resistance decreases, and vice versa.

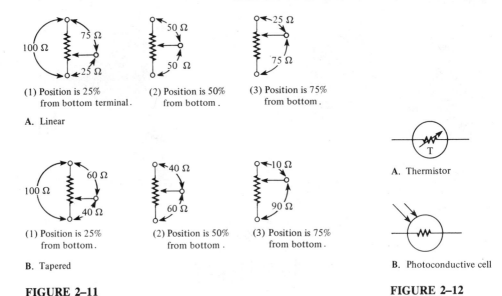

(1) Position is 25% from bottom terminal.
(2) Position is 50% from bottom.
(3) Position is 75% from bottom.

A. Linear

(1) Position is 25% from bottom.
(2) Position is 50% from bottom.
(3) Position is 75% from bottom.

B. Tapered

FIGURE 2–11

A. Thermistor

B. Photoconductive cell

FIGURE 2–12

The resistance of a photoconductive cell changes with a change in light intensity. This cell also has a negative temperature coefficient. Symbols for both of these devices are shown in Figure 2–12.

Review for 2–5

1. Name the two main categories of resistors, and briefly explain the difference between them. *FIX - IS A SET VALUE*
   VARIABLE - THE RESISTANCE CAN BE CHANGE
2. Some resistors have the resistance stamped on their bodies, and others are color coded (T) or F).
3. On a carbon-composition resistor, what does each of the four colored bands stand for? *ONE ¢ TWO DIGIT THREE multiper tolerance - FOURE*
   MAY OF the VALUE OF the RESISTOR
4. What is the main difference between a rheostat and a potentiometer?
5. What is a thermistor? *ONE THREE LEADS AND OTHER HAS TWO LEADS*
6. What does *negative temperature coefficient* mean?
   AS ONE THING INCREASE THE OTHER DECREASE

2–6

CONDUCTORS, SEMICONDUCTORS, AND INSULATORS

There are three categories of materials used in electronics: conductors, semiconductors, and insulators.

CONDUCTORS, SEMICONDUCTORS, AND INSULATORS

Conductors

Conductors are materials that readily permit current. They have a large number of free electrons in their structure. Most metals are good conductors. Silver is the best conductor, and copper is next. Copper is the most widely used conductive material because it is less expensive than silver.

Semiconductors

These materials are classed below the conductors in their ability to carry current. However, because of their unique characteristics, certain semiconductor materials are the basis for modern electronic devices such as the diode, transistor, and integrated circuit. Silicon and germanium are common semiconductor materials.

Insulators

Insulating materials are poor conductors of electric current. In fact, they are used to *prevent* current where it is not wanted. Compared to conductive materials, insulators have very few free electrons.

Although insulators do not normally carry current, they will break down and conduct if a sufficiently large voltage is applied. The *breakdown strength* of an insulator specifies how much voltage across a certain thickness of the material is required to break the insulating ability down. Table 2–2 lists some common insulating materials and their typical breakdown strengths in thousands of volts (kV) per centimeter of thickness.

TABLE 2–2 *Breakdown strengths of insulating materials.*

Material	Typical Breakdown Strength (kV/cm)
Mica	2000
Glass	900
Teflon®	600
Paper (paraffin)	500
Rubber	270
Bakelite®	150
Oil	140
Porcelain	70
Air	30

Review for 2–6

1. What is a conductor?
2. What are semiconductor materials used for?

3. What is an insulator? PREVENT CURRENT IN UNWANTED PLACES
4. What happens when an insulator breaks down? CURRENT FLOW THROUGH IT

2–7

WIRE SIZE

Wires are the most common form of conductive material used in electrical applications. They vary in size and are arranged according to standard *gage numbers*, called *American Wire Gage* (AWG) sizes. The larger the gage number is, the smaller the wire is. The AWG sizes are listed in Table 2–3.

TABLE 2–3 *American Wire Gage (AWG) sizes for solid round copper.*

AWG #	Area (CM)	Ω/1000 ft at 20C°	AWG #	Area (CM)	Ω/1000 ft at 20C°
0000	211,600	0.0490	19	1,288.1	8.051
000	167,810	0.0618	20	1,021.5	10.15
00	133,080	0.0780	21	810.10	12.80
0	105,530	0.0983	22	642.40	16.14
1	83,694	0.1240	23	509.45	20.36
2	66,373	0.1563	24	404.01	25.67
3	52,634	0.1970	25	320.40	32.37
4	41,742	0.2485	26	254.10	40.81
5	33,102	0.3133	27	201.50	51.47
6	26,250	0.3951	28	159.79	64.90
7	20,816	0.4982	29	126.72	81.83
8	16,509	0.6282	30	100.50	103.2
9	13,094	0.7921	31	79.70	130.1
10	10,381	0.9989	32	63.21	164.1
11	8,234.0	1.260	33	50.13	206.9
12	6,529.0	1.588	34	39.75	260.9
13	5,178.4	2.003	35	31.52	329.0
14	4,106.8	2.525	36	25.00	414.8
15	3,256.7	3.184	37	19.83	523.1
16	2,582.9	4.016	38	15.72	659.6
17	2,048.2	5.064	39	12.47	831.8
18	1,624.3	6.385	40	9.89	1049.0

As Table 2–3 shows, the size of a wire is also specified in terms of its *cross-sectional area*, as illustrated also in Figure 2–13. The unit of cross-sectional area is the *circular mil*, abbreviated CM. One circular mil is the area of a wire with a diameter of 0.001 inch (1 mil). The cross-sectional area is found by expressing the diameter in thousandths of an inch (mils) and squaring it, as follows:

$$A = d^2 \qquad (2\text{–}4)$$

where A is the cross-sectional area in circular mils and d is the diameter in mils.

WIRE SIZE

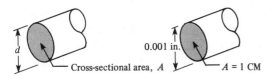

FIGURE 2–13 *Cross-sectional area of a wire.*

Example 2–4

What is the cross-sectional area of a wire with a diameter of 0.005 inch?

Solution:

$$d = 0.005 \text{ in.} = 5 \text{ mils}$$
$$A = d^2 = 5^2 = 25 \text{ CM}$$

Wire Resistance

Although copper wire conducts electricity extremely well, it still has some resistance, as do all conductors. The resistance of a wire depends on four factors: (1) type of material, (2) length of wire, (3) cross-sectional area, and (4) temperature.

Each type of conductive material has a characteristic called its *resistivity*, ρ. For each material, ρ is a constant value at a given temperature. The formula for the resistance of a wire of length l and cross-sectional area A is

$$R = \frac{\rho l}{A} \tag{2-5}$$

This formula tells us that resistance increases with resistivity and length, and decreases with cross-sectional area. For resistance to be calculated in ohms, the length must be in feet, the cross-sectional area in circular mils, and the resistivity in CM-Ω/ft.

Example 2–5

Find the resistance of a 100-ft length of copper wire with a cross-sectional area of 810.1 CM. The resistivity of copper is 10.4 CM-Ω/ft.

Solution:

$$R = \frac{\rho l}{A} = \frac{(10.4 \text{ CM-}\Omega\text{/ft})(100 \text{ ft})}{810.1 \text{ CM}} = 1.284 \text{ }\Omega$$

Table 2–3 lists the resistance of the various standard wire sizes in ohms per 1000 feet at 20°C. For example, a 1000-ft length of 14-gage copper wire has a resistance of 2.525 Ω. A 1000-ft length of 22-gage wire has a resistance of 16.14 Ω. For a given length, the smaller wire has more resistance. Thus, for a given voltage, larger wires can carry more current than smaller ones.

Review for 2–7

1. The larger the wire gage number is, the smaller the wire is ((T) or F).
2. What does "AWG" stand for? *American Wire Gauges*
3. Name the four factors that determine the resistance of a wire. *size – material, cross-sectional, temp*
4. What is the resistance of 500 feet of 18-gage copper wire? *3.1925*

2–8

THE ELECTRIC CIRCUIT

An electric circuit consists basically of a source, a load, and a current path. The *source* can be a battery or any other type of energy source that produces voltage. The *load* can be a simple resistor or any other type of electrical device or more complex circuit. The *current path* is the conductors connecting the source to the load. A simple circuit is shown in Figure 2–14.

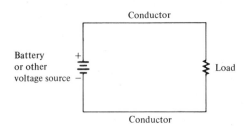

FIGURE 2–14 *A simple electrical circuit.*

Modern electronic circuits are fairly complex and may use many different types of components in various combinations, such as resistors, inductors, capacitors, diodes, transistors, and integrated circuits. This book will provide you with the underlying fundamentals you will need to understand the most complex electronic circuits or systems.

Closed Circuit and Open Circuit

A *closed* circuit is one in which the current has a complete path, as indicated in Figure 2–15A. An *open* circuit is one in which the current path is broken and, therefore, there is no current, as illustrated in Figure 2–15B. A *switch*,

MEASUREMENTS

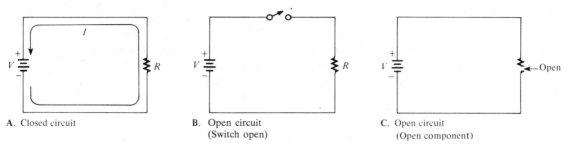

A. Closed circuit   B. Open circuit (Switch open)   C. Open circuit (Open component)

FIGURE 2–15 *Closed and open circuits.*

symbolized at the top of the diagram, is the device commonly used to open or close a circuit. An open circuit sometimes is a result of the failure of a component in a circuit, such as a burned-out resistor or lamp bulb, as illustrated in Figure 2–15C.

Short Circuit

A short circuit occurs when two points accidentally become connected, and there is current through the shorted contact. A short across a component such as a resistor will cause all of the current to go through the short, as illustrated in Figure 2–16.

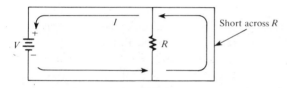

FIGURE 2–16 *Short circuit.*

Review for 2–8

1. What are the basic elements of an electric circuit?
2. What is an open circuit?
3. What is a closed circuit?
4. What is a short circuit?

2–9

MEASUREMENTS

Current, voltage, and resistance measurements are made commonly in electrical work. Certain types of instruments are used to measure these quantities. The instrument used to measure voltage is called a *voltmeter*. The instrument used to measure current is called an *ammeter*. The instrument used to measure resistance is called an *ohmmeter*. Test and measurement instruments are covered in Chapter 21. The purpose of this discussion is to familiarize you with the basic methods of measuring current, voltage, and resistance in a circuit.

Meter Symbols

Throughout the book, we will use certain symbols to represent different meters. You will encounter either of two types of both voltmeter and ammeter symbols, depending on which is more useful in a given diagram for conveying the information required. These symbols are shown in Figure 2–17.

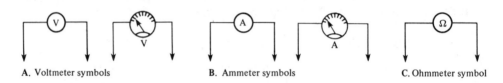

A. Voltmeter symbols   B. Ammeter symbols   C. Ohmmeter symbol

FIGURE 2–17 *Meter symbols.*

Measuring Current

Current is measured with an ammeter connected *in the current path,* as shown in Figure 2–18. As you will learn later, such a connection is called a *series* connection. The positive side of the meter (which is usually a red lead or post on the meter case) is connected toward the positive terminal of the voltage source. Either placement of the meter, as indicated in Figure 2–18, will measure the same current.

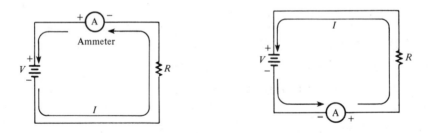

FIGURE 2–18 *Current measurement with an ammeter.*

Measuring Voltage

Voltage is measured with a voltmeter connected *across the current path,* as shown in Figure 2–19. As you will learn later, such a connection is called a *parallel* connection. The positive side of the meter must be connected toward the positive terminal of the voltage source. Either placement of the meter in Figure 2–19 will measure the same voltage.

PROTECTIVE DEVICES

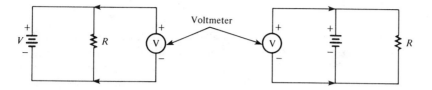

FIGURE 2–19 *Voltage measurement with a voltmeter.*

Measuring Resistance

Resistance is measured with an ohmmeter connected across the resistor, as shown in Figure 2–20. The resistor *must* be removed from the circuit or disconnected from the voltage source in some way, as indicated in the figure. Failure to disconnect the voltage source will result in damage to the ohmmeter.

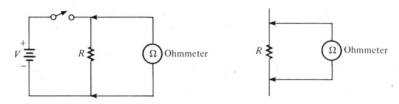

A. *R* disconnected from circuit with open switch   B. *R* removed from circuit

FIGURE 2–20 *Resistance measurement with an ohmmeter.*

Review for 2–9

1. Name the meters for measurement of current, voltage, and resistance. AMMETER, Voltmeter, ohmmeter
2. How is an ammeter connected in a circuit? IN SERIES
3. How is a voltmeter connected in a circuit? PARALLEL
4. What must you do in order to measure the value of a resistor? REMOVE it FROM the CIRCUIT

2–10
PROTECTIVE DEVICES

Protective devices are used in electrical and electronic circuits to protect the circuit from damage due to overcurrent, to prevent fire hazards due to excessive current, and to protect personnel from shock hazards.

Fuses

There are several types of fuses, each with various current ratings. The current rating is the maximum amount of current that the fuse can carry without opening. For example, a 20-A fuse will carry up to 20 A. If the current exceeds this amount, the fuse will blow and cause an open that stops the current. The schematic symbol for a fuse is shown in Figure 2–21A, and a fuse connected in a simple circuit is shown in Figure 2–21B.

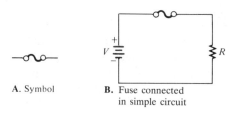

A. Symbol

B. Fuse connected in simple circuit

FIGURE 2–21 *Fuse.*

Two types of fuses are found in power applications such as residential wiring: the *plug* type and the *cartridge* type, shown in Figure 2–22.

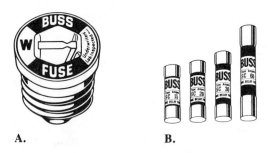

A.

B.

FIGURE 2–22 *Power fuses.* **A.** *Plug type.* **B.** *Cartridge type.* *(Courtesy of Bussmann Manufacturing, a division of McGraw-Edison Co.)*

Fuses are also commonly used to protect electronic instruments. The type shown in Figure 2–23A is normally used; it has lower ampere ratings than a power fuse. Common types of fuse holders are shown in Figure 2–23B and C.

Circuit Breakers

In power applications such as commercial, industrial, and residential wiring, circuit breakers are replacing fuses in new installations. A circuit breaker can be reset and reused repeatedly—an advantage over fuses, which must be replaced when they go out. Circuit breakers are also commonly used in electronic equipment. The schematic symbol for circuit breakers is shown in Figure 2–24.

There are two basic types of circuit breakers: magnetic and thermal. Some typical circuit breakers are pictured in Figure 2–25.

PROTECTIVE DEVICES

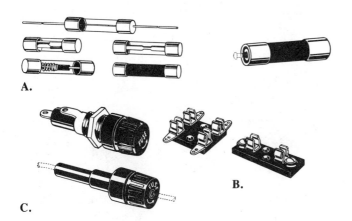

FIGURE 2–23 **A.** *Fuses* **B.,C.** *Fuse holders. (Courtesy of Bussmann Manufacturing, a division of McGraw-Edison Co.)*

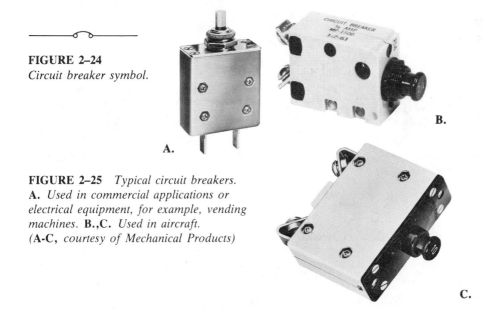

FIGURE 2–24
Circuit breaker symbol.

FIGURE 2–25 *Typical circuit breakers.*
A. *Used in commercial applications or electrical equipment, for example, vending machines.* **B.,C.** *Used in aircraft.*
(**A-C,** *courtesy of Mechanical Products*)

Switches

A switch can be considered a protective device. However, switches probably are classified more accurately as control devices used to turn current on or off. There are several types of switches that you should be familiar with: (1) the single-pole–single-throw (SPST); (2) the single-pole–double-throw (SPDT); (3) the double-pole–single-throw (DPST); (4) the double-pole–double-throw (DPDT); (5) the normally open push-button (NOPB); (6) the normally closed push-button (NCPB); and (7) the rotary switch.

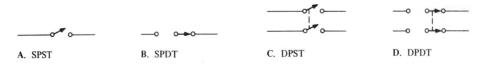

A. SPST   B. SPDT   C. DPST   D. DPDT

FIGURE 2–26 *Switch symbols.*

FIGURE 2–27 *Switches.* **A.** *Typical toggle-lever switches. (Courtesy of Eaton Corporation, Specialty Products Operations)* **B.** *Rocker switches. (Courtesy of Eaton Corporation, Specialty Products Operations)* **C.** *Rocker DIP switches. (Courtesy of Amp, Inc., and Grayhill, Inc., La Grange, Ill.)* **D.** *Push-button switches. (Courtesy of Eaton Corporation, Specialty Products Operations)* **E.** *Rotary-position switches. (Courtesy of Grayhill, Inc., La Grange, Ill.)*

FORMULAS

SPST Switch: This type of switch allows connection or disconnection between two contacts. In one position it is open, and in the other it is closed. Figure 2–26A shows a schematic symbol for this type of switch.

SPDT Switch: The symbol for this type of switch is shown in Figure 2–26B. With this switch, connection can be made between one contact and either of two others.

DPST Switch: This switch allows simultaneous connection or disconnection of two sets of contacts. The symbol is shown in Figure 2–26C. The dashed line indicates that the contact arms are grouped together so that both move with a single switch action.

DPDT Switch: This switch provides connection from one set of contacts to either of two other sets. The symbol is shown in Figure 2–26D. These types of switches are normally found in toggle (Figure 2–27A), slide, or rocker configurations (Figure 2–27B and C).

Push-Button Switches: In the NOPB switch, connection is made between two contacts when the button is depressed, and the connection is broken when the button is released. In the NCPB switch, connection between the two contacts is broken when the button is depressed and is made again when the button is released. Typical push-button switches are pictured in Figure 2-27D. The symbols for NOPB and NCPB switches are shown in Figure 2–28A and B.

Rotary Switch: In a rotary switch, a knob is turned to make connection between one contact and any of several others, as symbolized in Figure 2–28C. The one shown is a five-position switch; however, switches with many more contacts are available. Typical rotary switches are shown in Figure 2–27E.

A. NOPB   B. NCPB   C. Rotary (6-position)

FIGURE 2–28 *Push-button and rotary switch symbols.*

Review for 2–10

1. What is the difference between a fuse and a circuit breaker? *C.B. is reuseable, fuse is not*
2. List the main types of switches.

Formulas

$$I = \frac{Q}{t} \qquad (2\text{–}1)$$

$$V = \frac{\mathscr{E}}{Q} \qquad (2\text{--}2)$$

$$G = \frac{1}{R} \qquad (2\text{--}3)$$

$$A = d^2 \qquad (2\text{--}4)$$

$$R = \frac{\rho l}{A} \qquad (2\text{--}5)$$

Summary

1. An atom is the smallest particle of an element that retains the characteristics of that element.
2. An atom consists of a positively charged nucleus surrounded by orbiting electrons.
3. The electron is the basic particle of negative charge.
4. The proton is the particle of positive charge.
5. The neutron is an uncharged particle.
6. The nucleus of an atom contains protons and neutrons.
7. The atomic number is the number of electrons in an atom.
8. The atomic weight is the number of protons and neutrons in an atom.
9. Free electrons are loosely bound to the atom and provide for current in a material.
10. Current is the rate of flow of electrons.
11. The unit of charge is the coulomb (C).
12. One ampere (A) is the current when one coulomb passes a point in one second.
13. Voltage is the amount of energy used to move an amount of charge between two points in a circuit.
14. One volt (V) is the amount of potential difference when one joule (J) is used to move one coulomb.
15. Resistance is the opposition to current.
16. One ohm (Ω) is the resistance when there is one ampere with one volt applied.
17. Conductance is the reciprocal of resistance.
18. The unit of conductance is the siemen (S).
19. Larger wires have smaller gage numbers.
20. The resistance of a wire depends on its material, size, length, and temperature.
21. An open circuit prevents current. It is a break in the current path.
22. A closed circuit allows current. It completes a current path.

PROBLEMS

23. A short is an unintentional (usually), low-resistance path that bypasses the normal current path between two points.
24. An ammeter measures current, a voltmeter measures voltage, and an ohmmeter measures resistance.

Self-Test

1. How many electrons are in an atom with an atomic number of three?   3
2. How many protons and neutrons are in an atom with an atomic weight of six?   6
3. Fifty coulombs of charge flow past a point in a circuit in 5 seconds. What is the current?   10 A
4. If 2 A of current exists in a circuit, how many coulombs pass a given point in 10 seconds?   20 C
5. Five hundred joules of energy are used to move 100 C of charge through a resistor. What is the voltage across the resistor?   5 V
6. What is the conductance of a 10-Ω resistor?

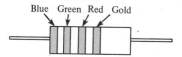

Blue Green Red Gold

FIGURE 2–29

7. Figure 2–29 shows a color-coded resistor. What are the resistance and the tolerance values?   6,500 Ω   ±5%
8. The adjustable contact of a linear potentiometer is set at the center of its adjustment. If the total resistance is 1000 Ω, what is the resistance between the end terminal and the contact?   500 Ω
9. What is the maximum voltage that a 10-cm thickness of mica can withstand?   20,000
10. What is the resistance of a 10-ft length of copper wire with a diameter of 0.003 in. (ρ = 10.4)?   11.56   9 mils

Problems

2–1. Six-tenths coulomb flows past a point in 3 seconds (3 s). What is the current?   .2 A
2–2. How long does it take 10 C to move past a point if the current is 5 A?   2 s
2–3. How many coulombs pass a point in 0.1 s if the current is 1.5 A?   .15 C
2–4. What is the voltage of a battery that uses 800 J of energy to move 40 C of charge through a resistor?   20 V
2–5. How much energy does a 12-V battery use to move 2.5 C through a circuit?   30 J

2–6. Determine the current in each of the following cases:
(a) 75 C in 1 s 75A (b) 10 C in 0.5 s 20A
(c) 5 C in 2 s 2.5A (d) 12 C in 12 s 1A

2–7. Determine the voltage in each of the following cases:
(a) 10 J/C 10V (b) 5 J/2 C 2.5V
(c) 100 J/25 C 4V (d) 1 J/5 C .2V

2–8. Find the conductance for each of the following resistance values:
(a) 5 Ω (b) 25 Ω (c) 100 Ω (d) 1000 Ω

2–9. Find the resistance corresponding to the following conductances:
(a) 0.1 S (b) 0.5 S (c) 0.02 S (d) 3 S

2–10. Find the resistance and the tolerance for each color code:

(a) First band is orange;
second band is orange;
third band is black;
fourth band is gold.

(b) First band is yellow;
second band is violet;
third band is orange;
fourth band is silver.

2–11. Find the minimum and the maximum resistance for the two resistors in Problem 2–10.

2–12. A piece of rubber is 50 cm thick. How many volts will it withstand before it breaks down?

2–13. An air capacitor has a plate separation of 75 cm. What is its breakdown voltage?

2–14. What is the cross-sectional area of a wire having a diameter of 0.008 in.?

2–15. Determine the resistance of a 150-ft length of copper wire with a diameter of 10 mils.

2–16. Find the resistance of 50 ft of the following gage copper wire at 20°C:
(a) 00 (b) 6 (c) 10 (d) 12 (e) 14
(f) 18 (g) 22 (h) 28 (i) 40

FIGURE 2–30

2–17. Show the placement of an ammeter and a voltmeter to measure the total current and the source voltage in Figure 2–30.

Answers to Section Reviews

Section 2–1:
1. Electron **2.** The smallest particle of an element that retains the unique characteristics of the element. **3.** A positively charged nucleus surrounded by orbiting electrons. **4.** No. **5.** An outer-shell electron that has drifted away from the parent atom.

Section 2–2:
1. Rate of flow of charge. **2.** 6.25×10^{18}. **3.** 5 A.

Section 2–3:
1. The amount of energy per unit charge. **2.** T. **3.** 5 V.

Section 2–4:
1. Opposition to current. **2.** Ohm (Ω). **3.** It limits current and dissipates heat.

Section 2–5:
1. Fixed, value cannot be changed. Variable, value easily changed. **2.** T. **3.** First band, first digit. Second band, second digit. Third band, multiplier. Fourth band, tolerance. **4.** A rheostat is a two-terminal device that controls current. A potentiometer is a three-terminal device that controls voltage. **5.** A temperature-sensitive resistor. **6.** The resistance changes inversely with the temperature.

Section 2–6:
1. A good carrier of electrons. **2.** Transistors and integrated circuits. **3.** A material that does not conduct electric current. **4.** It conducts.

Section 2–7:
1. T. **2.** American Wire Gage. **3.** Material, cross-sectional area, length, and temperature. **4.** 3.1925 Ω.

Section 2–8:
1. Energy source, load, and interconnections. **2.** A break in the current path. **3.** A complete current path. **4.** A low-resistance connection that bypasses the normal current path.

Section 2–9:
1. Ammeter, voltmeter, and ohmmeter, respectively. **2.** In series. **3.** In parallel. **4.** Remove the resistor from the circuit.

Section 2–10:
1. A circuit breaker is resettable, and a fuse is not. **2.** SPST, DPST, SPDT, DPDT, NOPB, NCPB, and rotary.

Georg Simon Ohm found that voltage, current, and resistance are related in a specific way. This basic relationship, known as Ohm's law, is one of the most important laws in electrical and electronics work. In this chapter you will learn Ohm's law and how to use it in solving circuit problems.

3–1 Statement of Ohm's Law
3–2 Calculating Current
3–3 Calculating Voltage
3–4 Calculating Resistance
3–5 Voltage and Current Are Linearly Proportional

3
OHM'S LAW

3-1

STATEMENT OF OHM'S LAW

Ohm's law tells us how current, voltage, and resistance are related. Georg Ohm determined experimentally that if the voltage across a resistor is increased, the current will also increase, and, likewise, if the voltage is decreased, the current will decrease. For example, if the voltage is doubled, the current will double. If the voltage is halved, the current will also be halved. This relationship is illustrated in Figure 3–1, with meter indications of voltage and current.

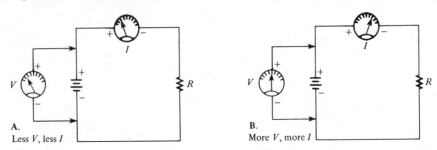

FIGURE 3–1 *Effect of changing the voltage with the same resistance in both circuits.*

Ohm's law also states that if the voltage is kept constant, *less* resistance results in *more* current, and, also, *more* resistance results in *less* current. For example, if the resistance is halved, the current doubles. If the resistance is doubled, the current is halved. This concept is illustrated by the meter indications in Figure 3–2, where the resistance is varied and the voltage is constant.

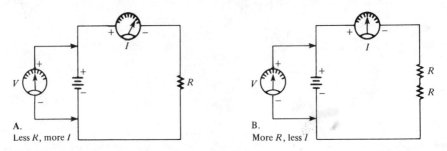

FIGURE 3–2 *Effect of changing the resistance with the same voltage in both circuits.*

Formula for Current

Ohm's law can be stated as follows:

$$I = \frac{V}{R} \tag{3-1}$$

This formula describes what was indicated by the circuits of Figures 3–1 and 3–2. For a constant value of R, if the value of V is increased, the value of I increases; if V is decreased, I decreases. Also notice in Equation (3–1) that if V is constant and R is increased, I decreases. Similarly, if V is constant and R is decreased, I increases.

STATEMENT OF OHM'S LAW

51

Using Equation (3–1), we can calculate the *current* if the values of voltage and resistance are known.

Formula for Voltage

Ohm's law can also be stated another way. By multiplying both sides of Equation (3–1) by R, we obtain an *equivalent* form of Ohm's law, as follows:

$$V = IR \qquad (3\text{–}2)$$

With this equation, we can calculate *voltage* if the current and resistance are known.

Formula for Resistance

There is a third *equivalent* way to state Ohm's law. By multiplying both sides of Equation (3–1) by R/I, we obtain

$$R = \frac{V}{I} \qquad (3\text{–}3)$$

This form of Ohm's law is used to determine *resistance* if voltage and current values are known.

Remember, *the three formulas you have learned in this section are all equivalent*. They are simply three different ways of expressing Ohm's law.

A Memory Aid

The simple circle diagram shown in Figure 3–3 is often used as an aid to remembering the three formulas for Ohm's law. It may help you.

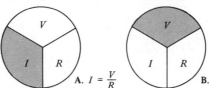

FIGURE 3–3 *Ohm's law memory aid.*

FIGURE 3–4

Here is the way to use the diagram. You can remember the formula $I = V/R$ by visualizing the circle as shown in Figure 3–4A, where V appears *over* R. You can remember the formula $V = IR$ by visualizing the circle as shown in Figure 3–4B, where I appears *next to* R. Finally, you can recall $R = V/I$ by visualizing the circle as shown in Figure 3–4C, where V appears *over* I.

Review for 3–1

1. Ohm's law defines how three basic quantities are related. What are these quantities? CURRENT RESISTANCE VOLTAGE
2. What is the Ohm's law formula for current? $I = \frac{E}{R}$

CHAPTER 3: OHM'S LAW

3. What is the Ohm's law formula for voltage? E = IR
4. What is the Ohm's law formula for resistance? R = E/I
5. If the voltage across a fixed value resistor is tripled, does the current increase or decrease, and by how much? INCREASE TRIPLE
6. If the voltage across a fixed resistor is cut in half, how much will the current change? HALF
7. There is a fixed voltage across a resistor, and you measure a current of 1 A. If you replace the resistor with one that has twice the resistance value, how much current will you measure? .5
8. In a circuit the voltage is doubled and the resistance is cut in half. Would you observe any change in the current value? Yes

3–2

CALCULATING CURRENT

In this section you will learn to determine current values when you know the values of voltage and resistance. In these problems, the formula $I = V/R$ is used. In order to get current in *amperes*, you must express the value of V in *volts* and the value of R in *ohms*.

Example 3–1

How many amperes of current are there in the circuit of Figure 3–5?

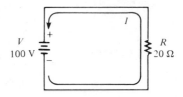

FIGURE 3–5

Solution:

Substitute into the formula $I = V/R$, 100 V for V and 20 Ω for R. Divide 20 Ω into 100 V as follows:

$$I = \frac{V}{R} = \frac{100 \text{ V}}{20 \text{ Ω}} = 5 \text{ A}$$

There are 5 A of current in this circuit.

CALCULATING CURRENT

Example 3-2

If the resistance in Figure 3-5 is changed to 50 Ω, what is the new value of current?

Solution:

We still have 100 V. Substituting 50 Ω into the formula for I gives 2 A as follows:

$$I = \frac{V}{R} = \frac{100 \text{ V}}{50 \text{ Ω}} = 2 \text{ A}$$

Larger Units of Resistance

In electronics work, resistance values of thousands of ohms or even millions of ohms are common. As you learned in Chapter 1, large values of resistance are indicated by the metric system prefixes *kilo* (k) and *mega* (M). Thus, thousands of ohms are expressed in kilohms (kΩ), and millions of ohms in megohms (MΩ). The following examples will illustrate use of kilohms and megohms when using Ohm's law to calculate current.

Example 3-3

Calculate the current in Figure 3-6.

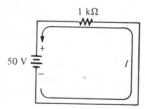

FIGURE 3-6

Solution:

Remember that 1 kΩ is the same as 1×10^3 Ω. Substituting 50 V for V and 1×10^3 Ω for R gives the current in amperes as follows:

$$I = \frac{V}{R} = \frac{50 \text{ V}}{1 \times 10^3 \text{ Ω}} = 50 \times 10^{-3} \text{ A}$$
$$= 0.05 \text{ A}$$

Notice in Example 3–3 that 50×10^{-3} A equals 50 milliamperes (50 mA). This fact can be used to advantage when we divide *volts by kilohms*. The current will always be in *milliamperes*, as Example 3–4 will illustrate.

Example 3–4

How many milliamperes are there in the circuit of Figure 3–7?

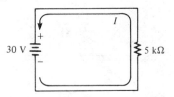

FIGURE 3–7

Solution:

When we divide *volts* by *kilohms,* we get *milliamperes.* In this case, 30 V divided by 5 kΩ gives 6 mA as follows:

$$I = \frac{V}{R} = \frac{30 \text{ V}}{5 \text{ k}\Omega} = 6 \text{ mA}$$

If *volts* are applied when resistance values are in *megohms,* the current is in *microamperes* (μA), as Example 3–5 shows.

Example 3–5

Determine the amount of current in the circuit of Figure 3–8.

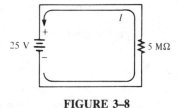

FIGURE 3–8

Solution:

Recall that 5 MΩ equals 5×10^6 Ω. Substituting 25 V for V and 5×10^6 Ω for R gives the following result:

CALCULATING CURRENT

$$I = \frac{V}{R} = \frac{25\text{ V}}{5 \times 10^6 \text{ }\Omega} = 5 \times 10^{-6}\text{ A}$$

Notice that 5×10^{-6} A equals 5 microamperes (5 µA).

Example 3–6

Change the value of R in Figure 3–8 to 2 MΩ. What is the new value of current?

Solution:

When we divide *volts* by *megohms*, we get *microamperes*. In this case, 25 V divided by 2 MΩ gives 12.5 µA as follows:

$$I = \frac{V}{R} = \frac{25\text{ V}}{2\text{ M}\Omega} = 12.5\text{ µA}$$

Larger Units of Voltage

Small voltages, usually less than 50 volts, are common in transistor circuits. Occasionally, however, large voltages are encountered. For example, the high-voltage supply in a television receiver is around 20,000 volts (20 kilovolts, or 20 kV), and transmission voltages generated by the power companies may be as high as 345,000 V (345 kV). We will work two examples using voltage values in the kilovolt range.

Example 3–7

How much current is produced by a voltage of 24 kV across a 12-kΩ resistance?

Solution:

Since we are dividing kV by kΩ, the units cancel, and we get amperes:

$$I = \frac{V}{R} = \frac{24\text{ kV}}{12\text{ k}\Omega}$$

$$= \frac{24 \times 10^3 \text{ V}}{12 \times 10^3 \text{ V}} = 2\text{ A}$$

Example 3–8

How much current is there through 100 MΩ when 50 kV are applied?

Solution:

In this case, we divide 50 kV by 100 MΩ to get the current. Using 50×10^3 V for 50 kV and 100×10^6 Ω for 100 MΩ, we obtain the current as follows:

$$I = \frac{V}{R} = \frac{50 \text{ kV}}{100 \text{ M}\Omega} = \frac{50 \times 10^3 \text{ V}}{100 \times 10^6 \text{ }\Omega}$$

$$= 0.5 \times 10^{-3} \text{ A} = 0.5 \text{ mA}$$

Remember that the power of ten in the denominator is subtracted from the power of ten in the numerator. So 50 was divided by 100, giving 0.5, and 6 was subtracted from 3, giving 10^{-3}.

Review for 3–2

In Problems 1–4, calculate I when
1. $V = 10$ V and $R = 5$ Ω.   = 2 A
2. $V = 100$ V and $R = 500$ Ω. = .02 A
3. $V = 5$ V and $R = 2.5$ kΩ.
4. $V = 15$ V and $R = 5$ MΩ.
5. If a 5-MΩ resistor has 20 kV across it, how much current is there?
6. How much current will 10 kV across 2 kΩ produce?

3–3

CALCULATING VOLTAGE

In this section you will learn to determine voltage values when the current and resistance are known. In these problems, the formula $V = IR$ is used. To obtain voltage in *volts,* you must express the value of I in *amperes* and the value of R in *ohms.*

CALCULATING VOLTAGE

Example 3–9

In the circuit of Figure 3–9, how much voltage is needed to cause 5 A of current?

FIGURE 3–9

Solution:

Substitute 5 A for I and 100 Ω for R into the formula $V = IR$ as follows:

$$V = IR = (5\ A)(100\ \Omega) = 500\ V$$

Thus, 500 V are required to produce 5 A of current through a 100-Ω resistor.

Smaller Units of Current

In the following two examples, we will work with milliampere (mA) and microampere (μA) current values.

Example 3–10

How much voltage will be measured across the resistor in Figure 3–10?

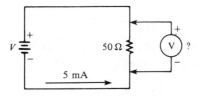

FIGURE 3–10

Solution:

Note that 5 mA equals 5×10^{-3} A. Substituting the values for I and R into

> **Example 3–10** (continued)
>
> the formula $V = IR$, we get the following result:
> $$V = IR = (5 \text{ mA})(50 \text{ }\Omega)$$
> $$= (5 \times 10^{-3} \text{ A})(50 \text{ }\Omega) = 250 \times 10^{-3} \text{ V}$$
>
> Since 250×10^{-3} V equals 250 mV, when *milliamperes* are multiplied by *ohms*, we get *millivolts*.

> **Example 3–11**
>
> Suppose that there are 8 µA through a 10-Ω resistor. How much voltage is across the resistor?
>
> *Solution:*
>
> Note that 8 µA equals 8×10^{-6} A. Substituting the values for I and R into the formula $V = IR$, we get the voltage as follows:
> $$V = IR = (8 \text{ µA})(10 \text{ }\Omega)$$
> $$= (8 \times 10^{-6} \text{ A})(10 \text{ }\Omega) = 80 \times 10^{-6} \text{ V}$$
>
> Since 80×10^{-6} V equals 80 µV, when *microamperes* are multiplied by *ohms*, we get *microvolts*.

These examples have demonstrated that when we multiply *milliamperes* and *ohms*, we get *millivolts*. When we multiply *microamperes* and *ohms*, we get *microvolts*.

Larger Units of Resistance

In the following examples we will work with resistance values in the kilohm (kΩ) and megohm (MΩ) range.

> **Example 3–12**
>
> The circuit in Figure 3–11 has a current of 10 mA. What is the voltage?

CALCULATING VOLTAGE

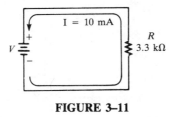

FIGURE 3–11

Solution:

Note that 10 mA equals 10×10^{-3} A and that 3.3 kΩ equals 3.3×10^{3} Ω. Substituting these values into the formula $V = IR$, we get

$$V = IR = (10 \text{ mA})(3.3 \text{ k}\Omega)$$
$$= (10 \times 10^{-3} \text{ A})(3.3 \times 10^{3} \text{ }\Omega) = 33 \text{ V}$$

Since 10^{-3} and 10^{3} cancel, *milliamperes* cancel *kilohms* when multiplied, and the result is *volts*.

Example 3–13

If there are 50 μA through a 5-MΩ resistor, what is the voltage?

Solution:

Note that 50 μA equals 50×10^{-6} A and that 5 MΩ is 5×10^{6} Ω. Substituting these values into $V = IR$, we get

$$V = IR = (50 \text{ }\mu\text{A})(5 \text{ M}\Omega)$$
$$= (50 \times 10^{-6} \text{ A})(5 \times 10^{6} \text{ }\Omega) = 250 \text{ V}$$

Since 10^{-6} and 10^{6} cancel, *microamperes* cancel *megohms* when multiplied, and the result is *volts*.

Review for 3–3

In Problems 1–7, calculate V when
1. $I = 1$ A and $R = 10$ Ω.
2. $I = 8$ A and $R = 470$ Ω.
3. $I = 3$ mA and $R = 100$ Ω.

4. $I = 25 \ \mu A$ and $R = 50 \ \Omega$.
5. $I = 2 \ mA$ and $R = 1.8 \ k\Omega$.
6. $I = 5 \ mA$ and $R = 100 \ M\Omega$.
7. $I = 10 \ \mu A$ and $R = 2 \ M\Omega$.
8. How much voltage is required to produce 100 mA through 4.7 kΩ?
9. What voltage do you need to cause 3 mA of current in a 3-kΩ resistance?
10. A battery produces 2 A of current into a 6-Ω resistive load. What is the battery voltage?

3–4

CALCULATING RESISTANCE

In this section you will learn to determine resistance values when the current and voltage are known. In these problems, the formula $R = V/I$ is used. To get resistance in *ohms*, you must express the value of I in *amperes* and the value of V in *volts*.

Example 3–14

In the circuit of Figure 3–12, how much resistance is needed to draw 3 A of current from the battery?

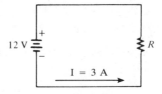

FIGURE 3–12

Solution:

Substitute 12 V for V and 3 A for I into the formula $R = V/I$:

$$R = \frac{V}{I} = \frac{12 \ V}{3 \ A} = 4 \ \Omega$$

Smaller Units of Current

In the following examples we will use current values in the milliampere (mA) and microampere (μA) range when calculating the resistance.

CALCULATING RESISTANCE

Example 3-15

Suppose that the ammeter in Figure 3-13 indicates 5 mA of current and the voltmeter reads 150 V. What is the value of R?

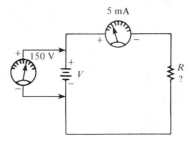

FIGURE 3-13

Solution:

Note that 5 mA equals 5×10^{-3} A. Substituting the voltage and current values into the formula $R = V/I$, we get

$$R = \frac{V}{I} = \frac{150 \text{ V}}{5 \text{ mA}} = \frac{150 \text{ V}}{5 \times 10^{-3} \text{ A}}$$

$$= 30 \times 10^3 \; \Omega = 30 \text{ k}\Omega$$

Thus, if *volts* are divided by *milliamperes*, the resistance will be in *kilohms*.

Example 3-16

Suppose that the value of the resistor in Figure 3-13 is changed. If the battery voltage is still 150 V and the ammeter reads 75 µA, what is the new resistor value?

Solution:

Note that 75 µA equals 75×10^{-6} A. Substituting V and I values into the equation for R, we get

$$R = \frac{V}{I} = \frac{150 \text{ V}}{75 \text{ µA}} = \frac{150 \text{ V}}{75 \times 10^{-6} \text{ A}}$$

$$= 2 \times 10^6 \; \Omega = 2 \text{ M}\Omega$$

Thus, if *volts* are divided by *microamperes,* the resistance has units of *megohms*.

Review for 3–4

In Problems 1–5, calculate R when

1. $V = 10$ V and $I = 2$ A.
2. $V = 250$ V and $I = 10$ A.
3. $V = 20$ kV and $I = 5$ A.
4. $V = 15$ V and $I = 3$ mA.
5. $V = 5$ V and $I = 2$ μA.
6. You have a resistor across which you measure 25 V, and your ammeter indicates 50 mA of current. What is the resistor's value in kilohms? In ohms?

3–5

CURRENT AND VOLTAGE ARE LINEARLY PROPORTIONAL

Ohm's law brings out a very important relationship between current and voltage: They are *linearly proportional.* You may have already recognized this relationship from our discussions in the previous sections.

When we say that the current and voltage are linearly proportional, we mean that if one is increased or decreased by a certain percentage, the other will increase or decrease by the same percentage, assuming that the resistance is constant in value. For example, if the voltage across a resistor is tripled, the current will triple.

Example 3–17

Show that if the voltage in the circuit of Figure 3–14 is increased to three times its present value, the current will triple in value.

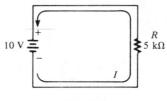

FIGURE 3–14

Solution:

With 10 V, the current is

$$I = \frac{V}{R} = \frac{10 \text{ V}}{5 \text{ k}\Omega} = 2 \text{ mA}$$

If the voltage is increased to 30 V, the current will be

CURRENT AND VOLTAGE ARE LINEARLY PROPORTIONAL

$$I = \frac{V}{R} = \frac{30 \text{ V}}{5 \text{ k}\Omega} = 6 \text{ mA}$$

The current went from 2 mA to 6 mA (tripled) when the voltage was tripled to 30 V.

The Graph of Current-Voltage Relationship

Let us take a constant value of resistance, for example, 10 Ω, and calculate the current for several values of voltage ranging from 10 V to 100 V. The values obtained are shown in Figure 3-15A. The graph of the V values versus the I values is shown in Figure 3-15B. Note that it is a straight line graph. This graph tells us that a change in voltage results in a linearly proportional change in current.

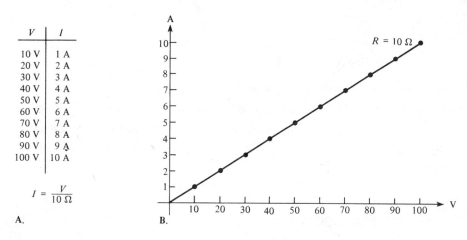

FIGURE 3-15 *Graph of voltage versus current for $R = 10 \text{ }\Omega$.*

No matter what value R is, assuming that R is constant, the graph of V versus I will always be a straight line. Example 3-18 will illustrate a use for the linear relationship between voltage and current in a resistive circuit.

Example 3-18

Assume that you are measuring the current in a circuit that is operating with 25 V. The ammeter reads 50 mA. Later, you notice that the current has dropped to 40 mA. Assuming that the resistance did not change, you must conclude that the voltage has changed. How much has the voltage changed, and what is its new value?

Solution:

The current has dropped from 50 mA to 40 mA, which is a decrease of

Example 3–18 (continued)

20%. Since the voltage is linearly proportional to the current, the voltage has decreased by the same percentage that the current did. Taking 20% of 25 V, we get

$$\text{Change in voltage} = (0.2)(25 \text{ V}) = 5 \text{ V}$$

Subtracting this change from the original voltage, we get the new voltage as follows:

$$\text{New voltage} = 25 \text{ V} - 5 \text{ V} = 20 \text{ V}$$

Notice that we did not need the resistance value in order to find the new voltage.

Review for 3–5

1. What does *linearly proportional* mean?
2. In a circuit, $V = 2$ V and $I = 10$ mA. If V is changed to 1 V, what will I equal?
3. If $I = 3$ A at a certain voltage, what will it be if the voltage is doubled?
4. By how many volts must you increase a 12-V source in order to increase the current in a circuit by 50%?

Formulas

$$I = \frac{V}{R} \tag{3-1}$$

$$V = IR \tag{3-2}$$

$$R = \frac{V}{I} \tag{3-3}$$

Summary

1. There are three forms of Ohm's law, all of which are equivalent.
2. Use $I = V/R$ when calculating the current.
3. Use $V = IR$ when calculating the voltage.
4. Use $R = V/I$ when calculating the resistance.
5. Voltage and current are linearly proportional.

Self-Test

1. Write the three forms of Ohm's law from memory.
2. Which of the three formulas is used for finding current? Voltage? Resistance?
3. Calculate I for $V = 10$ V and $R = 5$ Ω.
4. Calculate I for $V = 75$ V and $R = 10$ kΩ.
5. Calculate V for $I = 2.5$ A and $R = 20$ Ω.
6. Calculate V for $I = 30$ mA and $R = 2.2$ kΩ.
7. Calculate R for $V = 12$ V and $I = 6$ A.
8. Calculate R for $V = 9$ V and $I = 100$ μA.
9. Determine the current if $V = 50$ kV and $R = 2.5$ kΩ.
10. Determine the current if $V = 15$ mV and $R = 10$ kΩ.
11. Determine the voltage if $I = 4$ μA and $R = 100$ kΩ.
12. Determine the voltage if $I = 100$ mA and $R = 1$ MΩ.
13. Determine the resistance if $V = 5$ V and $I = 2$ mA.
14. If a 1.5-V battery is connected across a 5-kΩ resistor, what is the current?
15. Determine the current in each circuit of Figure 3–16.

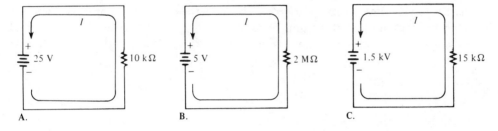

FIGURE 3–16

16. Assign a voltage value to each source in the circuits of Figure 3–17 to obtain the indicated amounts of current.

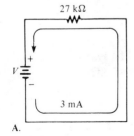 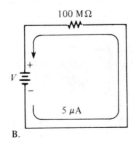 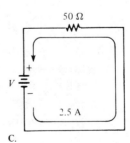

A.   B.   C.

FIGURE 3–17

17. Choose the correct value of resistance to get the current values indicated in each circuit of Figure 3–18.

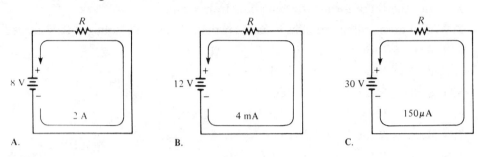

FIGURE 3–18

18. You are measuring the current in a circuit that is operated on a 10-V battery. The ammeter reads 50 mA. Later, you notice that the current has dropped to 30 mA. Eliminating the possibility of a resistance change, you must conclude that the voltage has changed. How much has the voltage of the battery changed, and what is its new value?

19. If you wish to increase the amount of current in a resistor from 100 mA to 150 mA by changing the 20-V source, by how many volts should you change the source? To what new value should you set it?

20. By varying the rheostat (variable resistor) in the circuit of Figure 3–19, you can change the amount of current. The setting of the rheostat is such that the current is 750 mA. What is the ohmic value of this setting? To adjust the current to 1 A, to what ohmic value must you set the rheostat?

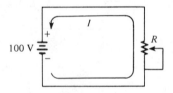

FIGURE 3–19

Problems

3–1. Determine the current in each case.
 (a) $V = 5$ V, $R = 1\ \Omega$ (b) $V = 15$ V, $R = 10\ \Omega$
 (c) $V = 50$ V, $R = 100\ \Omega$ (d) $V = 30$ V, $R = 15$ kΩ
 (e) $V = 250$ V, $R = 5$ MΩ

3–2. Determine the current in each case.
 (a) $V = 9$ V, $R = 2.7$ kΩ (b) $V = 5.5$ V, $R = 10$ kΩ
 (c) $V = 40$ V, $R = 68$ kΩ (d) $V = 1$ kV, $R = 2$ kΩ
 (e) $V = 66$ kV, $R = 10$ MΩ

PROBLEMS

3–3. Calculate the voltage for each value of I and R.
 (a) $I = 2$ A, $R = 18$ Ω
 (b) $I = 5$ A, $R = 50$ Ω
 (c) $I = 2.5$ A, $R = 600$ Ω
 (d) $I = 0.6$ A, $R = 47$ Ω
 (e) $I = 0.1$ A, $R = 500$ Ω

3–4. Calculate the voltage for each value of I and R.
 (a) $I = 1$ mA, $R = 10$ Ω
 (b) $I = 50$ mA, $R = 33$ Ω
 (c) $I = 3$ A, $R = 5$ kΩ
 (d) $I = 1.6$ mA, $R = 2.2$ kΩ
 (e) $I = 250$ μA, $R = 1$ kΩ
 (f) $I = 500$ mA, $R = 1.5$ MΩ
 (g) $I = 850$ μA, $R = 10$ MΩ
 (h) $I = 75$ μA, $R = 50$ Ω

3–5. Calculate the resistance for each value of V and I.
 (a) $V = 10$ V, $I = 2$ A
 (b) $V = 90$ V, $I = 45$ A
 (c) $V = 50$ V, $I = 5$ A
 (d) $V = 5.5$ V, $I = 10$ A
 (e) $V = 150$ V, $I = 0.5$ A

3–6. Calculate R for each set of V and I values.
 (a) $V = 10$ kV, $I = 5$ A
 (b) $V = 7$ V, $I = 2$ mA
 (c) $V = 500$ V, $I = 250$ mA
 (d) $V = 50$ V, $I = 500$ μA
 (e) $V = 1$ kV, $I = 1$ mA

3–7. A 10-Ω resistor is connected across a 12-V battery. How much current is there through the resistor?

3–8. Three amperes of current are measured through a 27-Ω resistor connected across a voltage source. How much voltage does the source produce?

3–9. Six volts are applied across a resistor. A current of 2 mA is measured. What is the value of the resistor?

3–10. The filament of a light bulb in the circuit of Figure 3–20A has a certain amount of resistance, represented by an equivalent resistance in Figure 3–20B. If the bulb operates with 120 V and 0.8 A of current, what is the resistance of its filament?

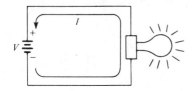
A.

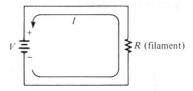
B.

FIGURE 3–20

3–11. A certain electrical device has an unknown resistance. You have available a 12-V battery and an ammeter. How would you determine the value of the unknown resistance? Draw the necessary circuit connections.

3–12. A variable voltage source is connected to the circuit of Figure 3–21. Start at 0 V and increase the voltage in 10-V steps up to 100 V. Determine the current at each voltage point, and plot a graph of V versus I. Is the graph a straight line? What does the graph indicate?

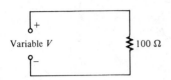

FIGURE 3–21

3–13. In a certain circuit, $V = 1$ V and $I = 5$ mA. Determine the current for each of the following voltages in the same circuit:
  (a) $V = 1.5$ V (b) $V = 2$ V
  (c) $V = 3$ V   (d) $V = 4$ V
  (e) $V = 10$ V

3–14. Figure 3–22 is a graph of voltage versus current for three resistance values. Determine R_1, R_2, and R_3.

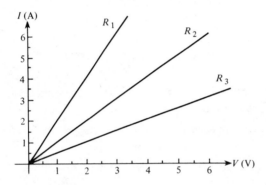

FIGURE 3–22

3–15. Which circuit in Figure 3–23 has the most current? The least current?

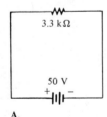 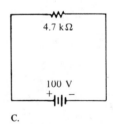

A. B. C.

FIGURE 3–23

Answers to Section Reviews

Section 3–1:
1. Current, voltage, and resistance. 2. $I = V/R$. 3. $V = IR$. 4. $R = V/I$.

ANSWERS TO SECTION REVIEWS

5. Increases by three times. **6.** Reduces to one-half of original value. **7.** 0.5 A.
8. Yes, it would increase by four times.

Section 3–2:
1. 2 A. **2.** 0.2 A. **3.** 2 mA. **4.** 3 μA. **5.** 4 mA. **6.** 5 A.

Section 3–3:
1. 10 V. **2.** 3760 V. **3.** 300 mV or 0.3 V. **4.** 1250 μV or 1.25 mV. **5.** 3.6 V. **6.** 500 kV. **7.** 20 V. **8.** 470 V. **9.** 9 V. **10.** 12 V.

Section 3–4:
1. 5 Ω. **2.** 25 Ω. **3.** 4 kΩ. **4.** 5 kΩ. **5.** 2.5 MΩ. **6.** 0.5 kΩ and 500 Ω.

Section 3–5:
1. The same percentage change occurs in two quantities. **2.** 5 mA. **3.** 6 A. **4.** 6 V.

Energy is the ability to do work, and power is the rate at which energy is used. Current carries electrical energy through a circuit. As the electrons pass through the resistance of the circuit, they give up their energy when they collide with atoms in the resistive material. The electrical energy given up by the electrons is converted into heat energy. The rate at which the electrical energy is lost is the power in the circuit.

In this chapter you will learn how to determine the amount of power in a resistive circuit and how to choose resistors that can handle the power. Also, we will discuss some important characteristics of devices that supply electrical energy.

4–1 Power
4–2 Power Formulas
4–3 Resistor Power Ratings
4–4 Energy
4–5 Energy Loss and Voltage Drop
4–6 Power Supplies

4
ENERGY
AND POWER

4–1
POWER

As mentioned, *power is the rate at which energy is used.* In other words, power is a certain amount of energy used in a certain length of time, expressed as follows:

$$\text{Power} = \frac{\text{energy}}{\text{time}} \tag{4-1}$$

The symbol for energy (work) is $\mathcal{E}$, the symbol for time is t, and the symbol for power is P. Using these symbols, we can rewrite Equation (4–1) more concisely:

$$P = \frac{\mathcal{E}}{t} \tag{4-2}$$

Units

Energy is measured in *joules* (J), time is measured in *seconds* (s), and power is measured in *watts* (W).

Energy in *joules* divided by time in *seconds* gives power in *watts*. For example, if 50 J of energy are used in 2 s, the power is 50 J/2 s = 25 W. Equation (4–3) expresses power in terms of units:

$$\text{Watts} = \frac{\text{joules}}{\text{seconds}} \tag{4-3}$$

One Watt

By definition, *one watt* is the amount of power when *one joule* of energy is consumed in *one second*. Thus, the number of joules consumed in one second is always equal to the number of watts. For example, if 75 J are used in 1 s, the power is 75 W.

Example 4–1

An amount of energy equal to 100 J is used in 10 s. What is the power in watts?

Solution:

$$P = \frac{\text{energy}}{\text{time}} = \frac{100 \text{ J}}{10 \text{ s}} = 10 \text{ W}$$

POWER

Example 4–2

If 1000 J are used in 1 s, what is the power?

Solution:

$$P = \frac{1000 \text{ J}}{1 \text{ s}} = 1000 \text{ W}$$

Smaller and Larger Units

Amounts of power much less than one watt are common in certain areas of electronics. As with small current and voltage values, metric prefixes are used to designate small amounts of power. Thus, *milliwatts* (mW) and *microwatts* (μW) are commonly found in some applications.

In the electrical utilities field, *kilowatts* (kW) and *megawatts* (MW) are common units. Radio and television stations also use large amounts of power to transmit signals.

Example 4–3

Convert the following powers from watts to the appropriate metric units:
(a) 0.045 W   (b) 0.000012 W   (c) 3500 W   (d) 10,000,000 W

Solution:

(a) 0.045 W = 45 mW   (b) 0.000012 W = 12 μW
(c) 3500 W = 3.5 kW   (d) 10,000,000 W = 10 MW

Review for 4–1

1. Define *power*. *is the rate at which ENERGY*
2. Write the formula for power in terms of energy and time. $W = \frac{E}{s}$
3. Define *watt*. *unit it express in*
4. If 25 J of energy are used in 5 s, what is the rate of energy consumption in watts? *5 W*
5. If 100 J are used in 10 ms, what is the power? *10 mA*
6. Express each of the following values of power in the most appropriate units:
   (a) 68,000 W   (b) 0.005 W   (c) 0.000025 W

4-2

POWER FORMULAS

When there is current through a resistance, energy is dissipated. In this section you will learn three formulas for calculating power in a resistance. First, recall from Chapter 2 that voltage can be expressed in terms of energy and charge as $V = \mathcal{E}/Q$. Also recall that current can be expressed in terms of charge and time as $I = Q/t$. When V and I are multiplied, the result is

$$VI = \left(\frac{\mathcal{E}}{Q}\right)\left(\frac{Q}{t}\right)$$

$$= \frac{\mathcal{E}}{t} = \text{power}$$

The power in a resistor is the product of the voltage across the resistor and the current through it:

$$P = VI \qquad (4-4)$$

Ohm's law states that $V = IR$. If IR is substituted for V in the formula $P = VI$, we get a second form of the power equation:

$$P = VI$$
$$= (IR)I$$
$$P = I^2R \qquad (4-5)$$

Ohm's law also states that $I = V/R$. If V/R is substituted for I in the formula $P = VI$, we get a third form of the power equation:

$$P = VI$$
$$= V\left(\frac{V}{R}\right)$$
$$P = \frac{V^2}{R} \qquad (4-6)$$

Equations (4-4), (4-5), and (4-6) are all equivalent.

Calculation of Power

To calculate the power in a resistance, you can use any one of the three power formulas, depending on what information you have. For example, in Figure 4-1, assume that you know the values of current and voltage. In this case you calculate the power with the formula $P = VI$. If you know I and R, use the formula $P = I^2R$. If you know V and R, use the formula $P = V^2/R$.

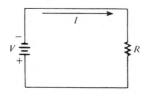

FIGURE 4-1

POWER FORMULAS

Example 4-4

Calculate the power in each of the three circuits of Figure 4-2.

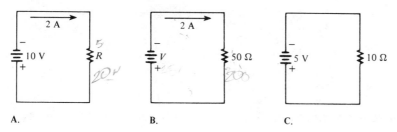

A.   B.   C.

FIGURE 4-2

Solution:

In circuit A, V and I are known. The power is determined as follows:

$$P = VI = (10 \text{ V})(2 \text{ A}) = 20 \text{ W}$$

In circuit B, I and R are known. The power is determined as follows:

$$P = I^2 R = (2 \text{ A})^2 (50 \text{ }\Omega) = 200 \text{ W}$$

In circuit C, V and R are known. The power is determined as follows:

$$P = \frac{V^2}{R} = \frac{(5 \text{ V})^2}{10 \text{ }\Omega} = 2.5 \text{ W}$$

Example 4-5

A 100-W light bulb operates on 120 V. How much current does it require?

Solution:

Use the formula $P = VI$ and solve for I as follows:

$$I = \frac{P}{V} = \frac{100 \text{ W}}{120 \text{ V}} = 0.833 \text{ A}$$

Review for 4-2

1. Write the three power formulas from memory.
2. If there are 10 V across a resistor and a current of 3 A through it, what is the power?

3. How much power does the source in Figure 4–3 generate? What is the power in the resistor? Are the two values the same? Why?

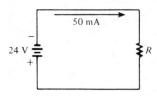

FIGURE 4–3

4. If there is a current of 5 A through a 50-Ω resistor, what is the power?
5. How much power is produced by 20 mA through a 5-kΩ resistor?
6. Five volts are applied to a 10-Ω resistor. What is the power?
7. How much power does a 2-kΩ resistor with 8 V across it produce?
8. What is the resistance of a 75-W bulb that takes 0.5 A?

4–3

RESISTOR POWER RATINGS

As you have learned, a resistor dissipates energy, and, as a result, it heats up. A resistor must be able to dissipate a sufficient amount of heat, depending on how much power it is required to handle. The amount of power that a resistor can handle is determined by the physical size and shape of the resistor. *The larger the surface area of a resistor, the more power it can handle.* Figure 4–4 shows carbon-composition resistors with standard available power ratings.

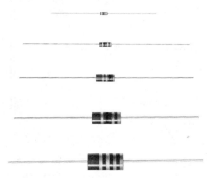

FIGURE 4–4 *Carbon-composition resistors with standard power ratings of 1/8 W, 1/4 W, 1/2 W, 1 W, and 2 W. (Courtesy of Allen-Bradley Company)*

When a resistor is used in a circuit, its power rating must be greater than the power that it actually handles. For example, if a carbon-composition resistor such as those shown in Figure 4–4 is to handle 0.75 W in a circuit, its rating should be

ENERGY

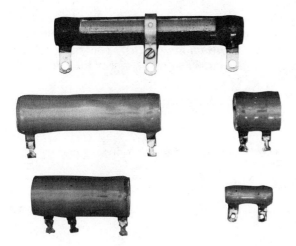

FIGURE 4–5 *Typical power resistors.*

1 watt or greater. Figure 4–5 shows some resistors with power ratings higher than those available in the standard carbon composition resistors.

Review for 4–3

1. Name two important values associated with a resistor.
2. How does the physical size of a resistor determine the amount of power that it can handle?
3. List the standard power ratings of carbon-composition resistors.
4. A resistor must handle 0.3 W. What size carbon resistor should be used to dissipate the energy properly?

4–4
ENERGY

Since power is the rate of energy usage, power utilized over a period of time represents energy consumption. If we multiply *power* and *time,* we have *energy:*

$$\text{Energy} = (\text{power})(\text{time})$$
$$\mathcal{E} = Pt \tag{4–7}$$

Unit of Energy

Earlier, the joule was defined as a unit of energy. However, there is another way of expressing energy. Since power is expressed in *watts* and time in *seconds,* we can use a unit of energy called the *wattsecond* (Ws).

When you pay your electric bill, you are charged on the basis of the amount of *energy* you use. Because power companies deal in huge amounts of energy, the most practical unit is the *kilowatthour* (kWh). You have used a kilowatthour of energy when you have used 1000 watts of power for one hour.

Example 4-6

Determine the number of kilowatthours for each of the following energy consumptions:
(a) 1400 W for 1 h   (b) 2500 W for 2 h   (c) 100,000 W for 5 h

Solution:

(a) 1400 W = 1.4 kW
    Energy = (1.4 kW)(1 h) = 1.4 kWh
(b) 2500 W = 2.5 kW
    Energy = (2.5 kW)(2 h) = 5 kWh
(c) 100,000 W = 100 kW
    Energy = (100 kW)(5 h) = 500 kWh

Review for 4-4

1. Distinguish between energy and power.
2. Write the formula for energy in terms of power and time.
3. What is the most practical unit of energy?
4. If you use 100 W of power for 10 h, how much energy (in kilowatthours) have you consumed?
5. Convert 2000 Wh to kilowatthours.
6. Convert 360,000 Ws to kilowatthours.

4-5

ENERGY LOSS AND VOLTAGE DROP

When there is current through a resistance, energy is dissipated in the form of heat. This heat loss is caused by collisions of the free electrons within the atomic structure of the resistive material. When a collision occurs, heat is given off, and the electron loses some of its acquired energy.

ENERGY LOSS AND VOLTAGE DROP

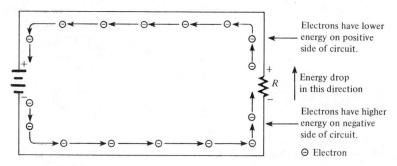

FIGURE 4-6

In Figure 4-6, electrons are flowing out of the negative terminal of the battery. They have acquired energy from the battery and are at their highest energy level at the negative side of the circuit. As the electrons move through the resistor, they *lose* energy. The electrons emerging from the upper end of the resistor are at a lower energy level than those entering the lower end. The drop in energy level through the resistor creates a potential difference, or *voltage drop,* across the resistor having the polarity shown in Figure 4-6.

The upper end of the resistor in Figure 4-6 is *less negative* (more positive) than the lower end. When we follow the flow of electrons, this change corresponds to a voltage drop from a more negative potential to a less negative potential. When we use *electron current,* as is the practice in this book, we think of voltage drop as being from a more negative potential to a less negative (positive) potential. This concept is illustrated in Figure 4-7.

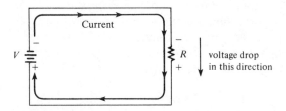

FIGURE 4-7 *Voltage drop with electron flow.*

Review for 4-5

1. What is the basic reason for energy loss in a resistor?
2. What is a voltage drop?
3. What is the polarity of a voltage drop when electron flow is used?

4-6

POWER SUPPLIES

A *power supply* is a device that provides power to a load. A *load* is any electrical device or circuit that is connected to the output of the power supply and draws current from the supply.

Figure 4-8 shows a block diagram of a power supply with a loading device connected to it. The load can be anything from a light bulb to a computer. The power supply produces a voltage across its two output terminals and provides current through the load, as indicated in the figure. The product $V_{out}I$ is the amount of power produced by the supply and consumed by the load. For a given output voltage (V_{out}), more current drawn by the load means more power from the supply.

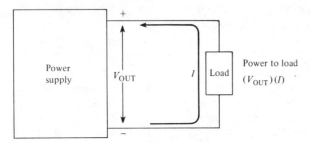

FIGURE 4-8 *Block diagram of power supply and load.*

Power supplies range from simple batteries to accurately regulated electronic circuits where an accurate output voltage is automatically maintained. A battery is a dc power supply that converts chemical energy into electrical energy. (See Appendix B.) Electronic power supplies normally convert 115 V ac from a wall outlet into a regulated dc or ac voltage. Figure 4-9A shows some typical batteries, and Figure 4-9B and C illustrates typical regulated power supplies used in the laboratory or shop.

Ampere-Hour Ratings of Batteries

Batteries convert chemical energy into electrical energy. Because of their limited source of chemical energy, batteries have a certain *capacity* which limits the amount of time over which they can produce a given power level. This capacity is measured in *ampere-hours (Ah)*. *The ampere-hour rating determines the length of time that a battery can deliver a certain amount of current to a load at the rated voltage.*

A rating of one ampere-hour means that a battery can deliver *one* ampere of current to a load for *one* hour at the rated voltage output. This same battery can deliver two amperes for half an hour. The more current the battery is required to deliver, the shorter is the life of the battery. In practice, a battery usually is rated for a specified current level and output voltage. For example, a 12-V automobile battery may be rated for 70 Ah at 3.5 A. This means that it can produce 3.5 A for 20 hours.

POWER SUPPLIES

A.

B.

C.

FIGURE 4–9 **A.** *Sealed Nicad® nickel-cadmium cells and batteries for consumer applications. (Courtesy of Gould, Inc. Portable Battery Division, St. Paul, Minn.)* **B., C.** *Typical regulated power supplies. (B., courtesy of B&K-Precision Test Instruments, Dynascan Corp. C., courtesy of Heath/Schlumberger Instruments)*

Example 4–7

For how many hours can a battery deliver 2 A if it is rated at 70 Ah?

Solution:

The ampere-hour rating is the current times the hours:

$$70 \text{ Ah} = (2 \text{ A})(x \text{ h})$$

Solving for x hours, we get

$$x = \frac{70 \text{ Ah}}{2 \text{ A}} = 35 \text{ h}$$

Power Supply Efficiency

An important characteristic of electronic power supplies is efficiency. *Efficiency is the ratio of the output power to the input power:*

$$\text{Efficiency} = \frac{\text{output power}}{\text{input power}}$$

$$\text{Efficiency} = \frac{P_{\text{out}}}{P_{\text{in}}} \tag{4–8}$$

Efficiency is usually expressed as a percentage. For example, if the input power is 100 W and the output power is 50 W, the efficiency is (50 W/100 W)(100) = 50%.

All power supplies require that power be put into them. For example, an electronic power supply might use the ac power from a wall outlet as its input. Its output may be regulated dc or ac. The output power is *always* less than the input power because some of the total power must be used internally to operate the power supply circuitry. This amount is normally called the *power loss*. The output power is the input power minus the amount of internal power loss:

$$P_{out} = P_{in} - P_{loss} \tag{4-9}$$

High efficiency means that little power is lost and there is a higher proportion of output power for a given input power.

Example 4–8

A certain power supply unit requires 25 W of input power. It can produce an output power of 20 W. What is its efficiency, and what is the power loss?

Solution:

$$\text{Efficiency} = \left(\frac{P_{out}}{P_{in}}\right)100 = \left(\frac{20 \text{ W}}{25 \text{ W}}\right)100$$

$$= 80\%$$

$$P_{loss} = P_{in} - P_{out} = 25 \text{ W} - 20 \text{ W}$$

$$= 5 \text{ W}$$

Review for 4–6

1. When a loading device draws an increased amount of current from a power supply, does this change represent a greater or a smaller load on the supply?

2. A power supply produces an output voltage of 10 V. If the supply provides 0.5 A to a load, what is the power output?

3. If a battery has an ampere-hour rating of 100 Ah, how long can it provide 5 A to a load?

4. If the battery in Problem 3 is a 12-V device, what is its power output for the specified value of current?

5. An electronic power supply used in the lab operates with an input power of 1 W. It can provide an output power of 750 mW. What is its efficiency?

SUMMARY

Formulas

$$P = \frac{\mathcal{E}}{t} \tag{4-2}$$

$$P = VI \tag{4-4}$$

$$P = I^2 R \tag{4-5}$$

$$P = \frac{V^2}{R} \tag{4-6}$$

$$\mathcal{E} = Pt \tag{4-7}$$

$$\text{Efficiency} = \frac{P_{out}}{P_{in}} \tag{4-8}$$

$$P_{out} = P_{in} - P_{loss} \tag{4-9}$$

Summary

1. Power is the rate at which energy is used.
2. One watt equals one joule per second.
3. Watt is the unit of power, joule is a unit of energy, and second is a unit of time.
4. Use $P = VI$ to calculate power when voltage and current are known.
5. Use $P = I^2R$ to calculate power when current and resistance are known.
6. Use $P = V^2/R$ to calculate power when voltage and resistance are known.
7. The power rating of a resistor determines the maximum power that it can handle safely.
8. Resistors with a larger physical size can dissipate more power than smaller ones.
9. A resistor should have a power rating higher than the maximum power that it is expected to handle in the circuit.
10. Energy is equal to power multiplied by time.
11. The kilowatthour is a unit of energy.
12. One kilowatthour = 1000 watts used for one hour.
13. When electron flow direction is used, the polarity of a voltage drop is from minus $(-)$ to plus $(+)$.
14. A power supply is an energy source used to operate electrical and electronic devices.
15. A battery is one type of power supply that converts chemical energy into electrical energy.
16. An electronic power supply converts commercial energy (ac from the power company) to regulated dc or ac at various voltage levels.

17. The output power of a supply is the output voltage times the load current.
18. A load is a device that draws current from the power supply.
19. The capacity of a battery is measured in ampere-hours (Ah).
20. One ampere-hour equals one ampere used for one hour, or any other combination of amperes and hours that has a product of one.
21. Efficiency of a power supply is equal to the output power divided by the input power. Multiply by 100 to get percentage of efficiency.
22. A power supply with a high efficiency wastes less power than one with a lower efficiency.

Self-Test

1. Two hundred joules of energy are consumed in 10 s. What is the power?
2. If it takes 300 ms to use 10,000 J of energy, what is the power?
3. How many watts are there in 50 kW?
4. How many milliwatts are there in 0.045 W?
5. If a resistive load draws 350 mA from a 10-V battery, how much power is handled by the resistor? How much power is delivered by the battery?
6. If a 1000-Ω resistor is connected across a 50-V source, what is the power?
7. A 5-kΩ resistor draws 500 mA from a voltage source. How much power is handled by the resistor?
8. What is the power of a light bulb that operates on 115 V and 2 A of current?
9. If 15 W of power are handled by a resistor, how many joules of energy are used in one minute?
10. If you have used 500 W of power for 24 h, how many kilowatthours have you used?
11. How many watthours represent 75 W used for 10 h?
12. If your average daily power usage is 750 W, how many kilowatthours will be read on your meter at the end of April? The reading at the end of March was 1000 kWh.
13. Convert 1,500,000 Ws to kWh.
14. For each circuit in Figure 4–10, assign the proper polarity for the voltage drop across the resistor.

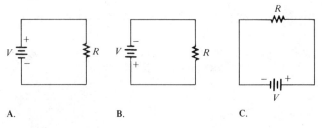

A. B. C.

FIGURE 4–10

PROBLEMS

15. Two 25-V power supplies are sitting on the lab bench. Power supply 1 is providing 100 mA to a load. Power supply 2 is supplying 0.5 A to a load. Which supply has the greater load? Which resistive load has the smaller ohmic value?

16. What is the power output of each power supply in Problem 15?

17. A 12-V battery is connected to a 600-Ω load. Under these conditions, it is rated at 50 Ah. How long can it supply current to the load?

18. A given power supply is capable of providing 8 A for 2.5 h. What is its ampere-hour rating?

19. A laboratory power supply requires 250 mW internally. If it takes 2 W of input power, what is the output power?

20. A power supply produces a 0.5-W output with an input of 0.6 W. What is its percentage of efficiency?

Problems

4-1. What is the power when energy is consumed at the rate of 350 J/s?

4-2. How many watts are used when 7500 J of energy are consumed in 5 h?

4-3. How many watts does 1000 J in 50 ms equal?

4-4. Convert the following to kilowatts:
   (a) 1000 W   (b) 3750 W
   (c) 160 W   (d) 50,000 W

4-5. Convert the following to megawatts:
   (a) 1,000,000 W   (b) 3×10^6 W
   (c) 15×10^7 W (d) 8700 kW

4-6. Convert the following to milliwatts:
   (a) 1 W   (b) 0.4 W
   (c) 0.002 W   (d) 0.0125 W

4-7. Convert the following to microwatts:
   (a) 2 W   (b) 0.0005 W
   (c) 0.25 mW   (d) 0.00667 mW

4-8. Convert the following to watts:
   (a) 1.5 kW   (b) 0.5 MW
   (c) 350 mW   (d) 9000 μW

4-9. If a 75-V source is supplying 2 A to a load, what is the ohmic value of the load?

4-10. If a resistor has 5.5 V across it and 3 mA through it, what is the power?

4-11. An electric heater works on 115 V and draws 3 A of current. How much power does it use?

4-12. How much power is produced by 500 mA of current through a 4.7-kΩ resistor?

4-13. Calculate the power handled by a 10-kΩ resistor carrying 100 μA.

4-14. If there are 60 V across a 600-Ω resistor, what is the power?

4-15. A 50-Ω resistor is connected across the terminals of a 1.5-V battery. What is the power in the resistor?

4-16. If a resistor is to carry 2 A of current and handle 100 W of power, how many ohms must it be? Assume that the voltage can be adjusted to any required value.

4-17. A 6.8-kΩ resistor has burned out in a circuit. You must replace it with another resistor with the same ohmic value. If the resistor carries 10 mA, what should its power rating be? Assume that you have available carbon-composition resistors in all the standard power ratings.

4-18. A certain type of power resistor comes in the following ratings: 3 W, 5 W, 8 W, 12 W, 20 W. Your particular application requires a resistor that can handle approximately 8 W. Which rating would you use? Why?

4-19. A particular electronic device uses 100 mW of power. If it runs for 24 h, how many joules of energy does it consume?

4-20. How many watthours does 50 W used for 12 h equal? How many kilowatthours?

4-21. A certain appliance uses 300 W. If it is allowed to run continuously for 30 days, how many kilowatthours of energy does it consume?

4-22. At the end of a 31-day period, your utility bill shows that you have used 1500 kWh. What is your average daily power consumption?

4-23. Convert 5×10^6 wattminutes to kWh.

4-24. Convert 6700 wattseconds to kWh.

4-25. If a power supply has an output voltage of 50 V and provides 2 A to a load, what is its output power?

4-26. A 50-Ω load consumes 1 W of power. What is the output voltage of the power supply?

4-27. A battery can provide 1.5 A of current for 24 h. What is its ampere-hour rating?

4-28. How much continuous current can be drawn from an 80-Ah battery for 10 h?

4-29. If a battery is rated at 650 mAh, how much current will it provide for 48 h?

4-30. If the input power is 500 mW and the output power is 400 mW, how much power is lost? What is the efficiency of this power supply?

4-31. To operate at 85% efficiency, how much output power must a source produce if the input power is 5 W?

4-32. A certain power supply provides a continuous 2 W to a loading device. It is operating at 60% efficiency. In a 24-h period, how many watthours does it consume?

Answers to Section Reviews

Section 4–1:
1. Power is the rate at which energy is used. 2. Power = energy/time. 3. One watt is one joule of energy consumed in one second. 4. 5 W 5. 10 kW 6. (a) 68 kW (b) 5 mW (c) 25 μW.

Section 4–2:
1. $P = VI$, $P = I^2R$, $P = V^2/R$. 2. 30 W. 3. 1.2 W, 1.2 W. Yes, all energy produced by source is dissipated by resistance. 4. 1250 W. 5. 2 W. 6. 2.5 W. 7. 32 mW. 8. 300 Ω.

Section 4–3:
1. Ohmic value, power rating. 2. A larger surface area dissipates more energy. 3. 0.125 W, 0.25 W, 0.5 W, 1 W, 2 W. 4. 0.5 W or greater.

Section 4–4:
1. Energy is power times time. 2. Energy = (power)(time). 3. Kilowatt-hour. 4. 1 kWh. 5. 2 kWh. 6. 0.1 kWh.

Section 4–5:
1. Collisions of the electrons within the atomic structure. 2. Potential difference between two points due to energy loss. 3. Minus to plus.

Section 4–6:
1. Greater. 2. 5 W. 3. 20 h. 4. 60 W. 5. 75%.

Resistors can be connected in a circuit in two basic ways: in series or parallel. This chapter discusses circuits with series resistance. You will learn how total resistance is determined and how voltages divide in series circuits. A new circuit law will be introduced, and you will see what happens when voltage sources are connected in series. Also, you will apply Ohm's law to series circuit problems and will learn how to calculate total power in a series circuit.

5–1 Resistors in Series
5–2 Current in a Series Circuit
5–3 Total Series Resistance
5–4 Applying Ohm's Law
5–5 Voltage Sources in Series
5–6 Kirchhoff's Voltage Law
5–7 Voltage Dividers
5–8 Power in a Series Circuit
5–9 Open and Closed Circuits
5–10 Trouble-Shooting Series Circuits

SERIES RESISTIVE CIRCUITS

5–1
RESISTORS IN SERIES

Resistors in series are connected end-to-end or in a "string," as shown in Figure 5–1. Figure 5–1A shows two resistors connected in series between point A and point B. Part B of the figure shows three in series, and Part C shows four in series. Of course, there can be any number of resistors in a series connection.

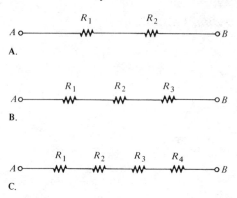

FIGURE 5–1 *Resistors in series.*

The only way for electrons to get from point A to point B in any of the connections of Figure 5–1 is to go through *each* of the resistors. The following is an important way to identify a series connection: *A series connection provides only one path for current between two points in a circuit so that the current is the same through each series resistor.*

Identifying Series Connections

In an actual circuit diagram, a series connection may not always be as easy to identify as those in Figure 5–1. For example, Figure 5–2 shows series resistors drawn in other ways. Remember, *if there is only one current path between two points, the resistors between those two points are in series,* no matter how they appear in a diagram.

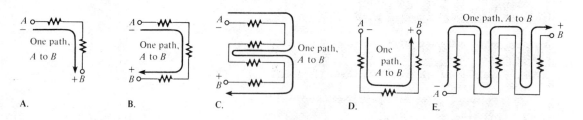

FIGURE 5–2 *Some examples of series connections. Notice that the current must be the same at all points.*

RESISTORS IN SERIES

Example 5–1

Suppose that there are five resistors positioned on a circuit board as shown in Figure 5–3. Wire them together in series so that, starting from the negative (−) terminal, R_1 is first, R_2 is second, R_3 is third, and so on. Draw a schematic diagram showing this connection.

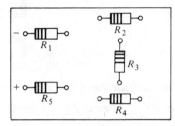

FIGURE 5–3

Solution:

The wires are connected as shown in Figure 5–4A, which is the *assembly diagram*. The *schematic diagram* is shown in Figure 5–4B. Note that the schematic diagram does not necessarily show the actual physical arrangement of the resistors as does the assembly diagram. The purpose of the *schematic* is to show how components are connected *electrically*. The purpose of the *assembly* diagram is to show how components are arranged *physically*.

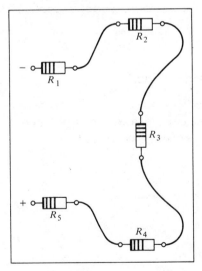

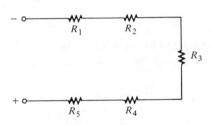

B. Schematic diagram

A. Assembly diagram

FIGURE 5–4

Example 5–2

Describe how the resistors on the printed circuit (PC) board in Figure 5–5 are related electrically.

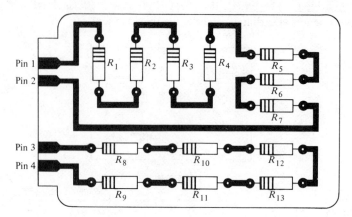

FIGURE 5–5

Solution:

Resistors R_1 through R_7 are in series with each other. This series combination is connected between pins 1 and 2 on the PC board.
    Resistors R_8 through R_{13} are in series with each other. This series combination is connected between pins 3 and 4 on the PC board.

Review for 5–1

1. How are the resistors connected in a series circuit?
2. How can you identify a series connection?
3. Complete the schematic diagrams for the circuits in each part of Figure 5–6 by connecting the resistors in series in numerical order from *A* to *B*.

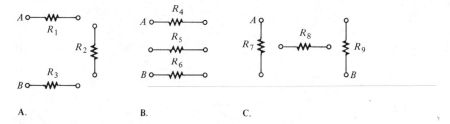

A.   B.   C.

FIGURE 5–6

4. Now connect each *group* of series resistors in Figure 5–6 in series.

5-2
CURRENT IN A SERIES CIRCUIT

The same current exists through all points in a series circuit. Figure 5-7 shows three series resistors connected to a voltage source. *At any point in this circuit, the current entering that point must equal the current leaving that point,* as illustrated by the current directional arrows at points, *A*, *B*, *C*, and *D*.

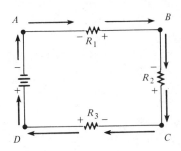

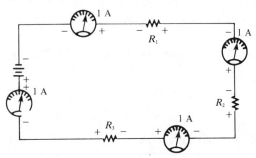

FIGURE 5-7

FIGURE 5-8 *Current is the same at all points in a series circuit.*

Notice also that the current out of each of the resistors must equal the current in, because there is no place where part of the current can branch off and go somewhere else. Therefore, the current in each section of the circuit is the same as the current in all other sections. It has only one path going from the negative (−) side of the source to the positive (+) side.

Let us assume that the battery in Figure 5-7 supplies one ampere of current to the series resistance. One ampere is out of the negative terminal. If we connect ammeters at several points in the circuit as shown in Figure 5-8, *each* meter will read one ampere.

Review for 5-2

1. In a series circuit with a 10-Ω and a 5-Ω resistor in series, 1 A is through the 10-Ω resistor. How much current is through the 5-Ω resistor?
2. A milliammeter is connected between points *A* and *B* in Figure 5-9. It measures 50 mA. If you move the meter and connect it between points *C* and *D*, how much current will it indicate? Between *E* and *F*?

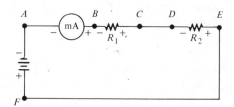

FIGURE 5-9

3. In Figure 5–10, how much current does ammeter 1 indicate? How much current does ammeter 2 indicate?

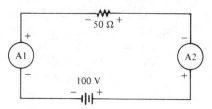

FIGURE 5–10

4. What statement can you make about the amount of current in a series circuit?

5–3

TOTAL SERIES RESISTANCE

The total resistance of a series connection is equal to the sum of the resistances of each individual resistor. This fact is understandable because each of the resistors in series offers opposition to the current in direct proportion to its ohmic value. A greater number of resistors connected in series creates *more opposition* to current. More opposition to current implies a higher ohmic value of resistance. Thus, every time a resistor is added in series, the total resistance increases.

Series Resistor Values Add

Figure 5–11 illustrates how series resistances add to *increase* the total resistance. Figure 5–11A has a single 10-Ω resistor. Figure 5–11B shows another 10-Ω resistor connected in series with the first one, making a total resistance of 20 Ω. If a third 10-Ω resistor is connected in series with the first two, as shown in Figure 5–11C, the total resistance becomes 30 Ω.

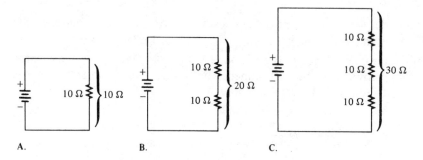

FIGURE 5–11 *Total resistance increases with each additional series resistor.*

Series Resistance Formula

For *any number* of individual resistors connected in series, the total resistance is the sum of each of the individual values:

TOTAL SERIES RESISTANCE

$$R_T = R_1 + R_2 + R_3 + \cdots + R_n \qquad (5\text{-}1)$$

where R_T is the total resistance and R_n is the last resistor in the series string (n can be any positive integer equal to the number of resistors in series). For example, if we have four resistors in series ($n = 4$), the total resistance formula is

$$R_T = R_1 + R_2 + R_3 + R_4$$

If we have six resistors in series ($n = 6$), the total resistance formula is

$$R_T = R_1 + R_2 + R_3 + R_4 + R_5 + R_6$$

To illustrate the calculation of total series resistance, let us take the circuit of Figure 5–12 and determine its R_T. (V_S is the source voltage.)

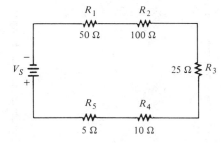

FIGURE 5–12

The circuit of Figure 5–12 has five resistors in series. To get the total resistance, we simply add the values as follows:

$$R_T = 50\ \Omega + 100\ \Omega + 25\ \Omega + 10\ \Omega + 5\ \Omega = 190\ \Omega$$

The equation illustrates an important point: In Figure 5–12, the order in which the resistances are added does not matter; we still get the same total. Also, we can physically change the positions of the resistors in the circuit without affecting the total resistance.

Example 5–3

What is the total resistance (R_T) in the circuit of Figure 5–13?

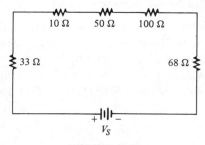

FIGURE 5–13

Solution:

Sum all the values as follows:

Example 5-3 (continued)

$$R_T = 68 \text{ } \Omega + 100 \text{ } \Omega + 50 \text{ } \Omega + 10 \text{ } \Omega + 33 \text{ } \Omega$$
$$= 261 \text{ } \Omega$$

Example 5-4

Calculate R_T for the circuit of Figure 5-14.

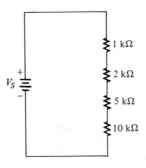

FIGURE 5-14

Solution:

Sum the resistor values:

$$R_T = 10 \text{ k}\Omega + 5 \text{ k}\Omega + 2 \text{ k}\Omega + 1 \text{ k}\Omega$$
$$= 18 \text{ k}\Omega$$

Notice that we can change the positions of the resistors without changing the total resistance.

Equal-Value Series Resistors

When a circuit has more than one resistor of the *same* value in series, there is a shortcut method to obtain the total resistance: Simply multiply the *ohmic value* of the resistors having the same value by the *number* of resistors that are in series. This method is essentially the same as adding the values. For example, five 100-Ω resistors in series have an R_T of $5(100 \text{ } \Omega) = 500 \text{ } \Omega$. In general, the formula is expressed as

$$R_T = nR \tag{5-2}$$

where n is the number of equal-value resistors and R is the value.

TOTAL SERIES RESISTANCE

Example 5–5

Find the R_T of eight 20-Ω resistors in series.

Solution:

We find R_T by adding the values as follows:

$R_T = 20\ \Omega + 20\ \Omega + 20\ \Omega + 20\ \Omega + 20\ \Omega + 20\ \Omega + 20\ \Omega + 20\ \Omega$
$= 160\ \Omega$

However, it is much easier to multiply:

$$R_T = 8(20\ \Omega) = 160\ \Omega$$

Review for 5–3

1. The following resistors (one each) are in series: 1-Ω, 2-Ω, 3-Ω, and 4-Ω. What is the total resistance?
2. Calculate R_T for each circuit in Figure 5–15.

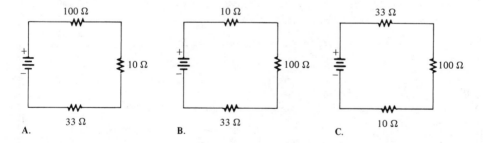

FIGURE 5–15

3. The following resistors are in series: one 100-Ω, two 50-Ω, four 12-Ω, and one 330-Ω. What is the total resistance?
4. Suppose that you have one resistor each of the following values: 1-kΩ, 2.5-kΩ, 5-kΩ, and 500-Ω. To get a total resistance of 10-kΩ, you need one more resistor. What should its value be?
5. What is the R_T for twelve 50-Ω resistors in series?
6. What is the R_T for twenty 5-Ω resistors and thirty 8-Ω resistors in series?

5-4

APPLYING OHM'S LAW

In this section we will solve several circuit problems to see how Ohm's law can be applied to series circuit analysis.

Example 5-6

Find the current in the circuit of Figure 5-16.

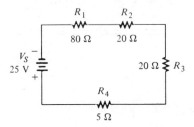

FIGURE 5-16

Solution:

The current is determined by the voltage and the *total resistance*. First, we calculate the total resistance as follows:

$$R_T = R_1 + R_2 + R_3 + R_4$$
$$= 80\ \Omega + 20\ \Omega + 20\ \Omega + 5\ \Omega$$
$$= 125\ \Omega$$

Next, using Ohm's law, we calculate the current as follows:

$$I = \frac{V_S}{R_T} = \frac{25\ \text{V}}{125\ \Omega}$$
$$= 0.2\ \text{A}$$

Remember, the *same* current exists at all points in the circuit. Thus, *each* resistor has 0.2 A through it.

Example 5-7

In the circuit of Figure 5-17, there is 1 mA of current. For this amount of current, what must the source voltage V_S be?

APPLYING OHM'S LAW

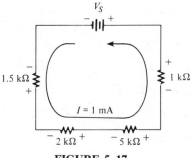

FIGURE 5–17

Solution:

In order to calculate V_S, we must determine R_T as follows:

$$R_T = 1.5 \text{ k}\Omega + 2 \text{ k}\Omega + 5 \text{ k}\Omega + 1 \text{ k}\Omega$$
$$= 9.5 \text{ k}\Omega$$

Now use Ohm's law to get V_S:

$$V_S = IR_T = (1 \text{ mA})(9.5 \text{ k}\Omega)$$
$$= 9.5 \text{ V}$$

Example 5–8

Calculate the voltage across each resistor in Figure 5–18, and find the value of V_S.

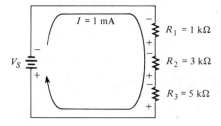

FIGURE 5–18

Solution:

By Ohm's law, the voltage across each resistor is equal to its resistance multiplied by the current through it. Using the Ohm's law formula $V = IR$, we determine the voltage across each of the resistors. Keep in mind that the current is the same through each series resistor.

Example 5–8 (continued)

Voltage across R_1:
$$V_1 = IR_1 = (1 \text{ mA})(1 \text{ k}\Omega) = 1 \text{ V}$$

Voltage across R_2:
$$V_2 = IR_2 = (1 \text{ mA})(3 \text{ k}\Omega) = 3 \text{ V}$$

Voltage across R_3:
$$V_3 = IR_3 = (1 \text{ mA})(5 \text{ k}\Omega) = 5 \text{ V}$$

The source voltage V_S is equal to the current times the *total resistance:*

$$R_T = 1 \text{ k}\Omega + 3 \text{ k}\Omega + 5 \text{ k}\Omega$$
$$= 9 \text{ k}\Omega$$
$$V_S = (1 \text{ mA})(9 \text{ k}\Omega)$$
$$= 9 \text{ V}$$

Notice that if you add the voltage drops of the resistors, they total 9 V, which is the same as the source voltage.

Review for 5–4

1. A 10-V battery is connected across three 100-Ω resistors in series. What is the current through each resistor?
2. How much voltage is required to produce 5 A through the circuit of Figure 5–19?

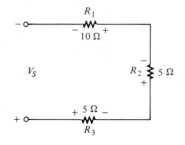

FIGURE 5–19

3. How much voltage is dropped across each resistor in Figure 5–19?
4. There are four equal-value resistors connected in series with a 5-V source. Five milliamperes of current are measured. What is the value of each resistor?

5-5
VOLTAGE SOURCES IN SERIES

A voltage source is an energy source that provides a constant voltage to a load. Batteries and dc power supplies are practical examples.

When two or more voltage sources are in series, the total voltage is equal to the algebraic sum of the individual source voltages. The *algebraic sum* means that the polarities of the sources must be included when the sources are combined in series. Sources with opposite polarities have voltages with opposite signs.

When the sources are all in the same direction in terms of their polarities, as in Figure 5–20A, all of the voltages have the same sign when added, and we get a total of 4.5 V with terminal A more positive than terminal B:

$$V_{AB} = +4.5 \text{ V}$$

In Figure 5–20B, the middle source is opposite to the other two; so its voltage has an opposite sign when added to the others. For this case the total voltage is

$$V_{AB} = +1.5 \text{ V} - 1.5 \text{ V} + 1.5 \text{ V} = +1.5 \text{ V}$$

Terminal A is 1.5 V more positive than terminal B.

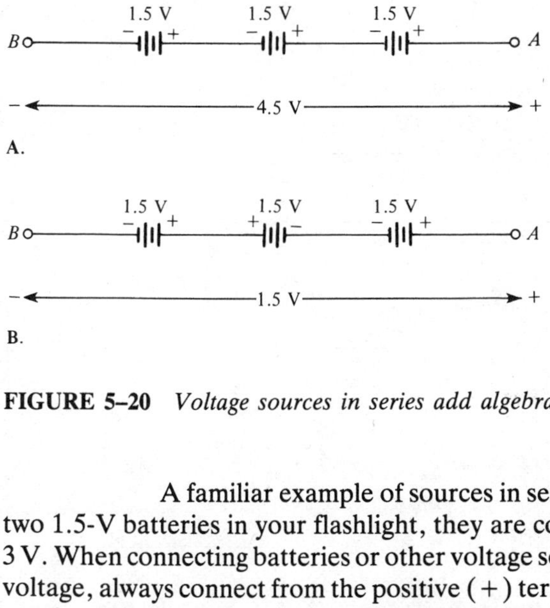

FIGURE 5–20 *Voltage sources in series add algebraically.*

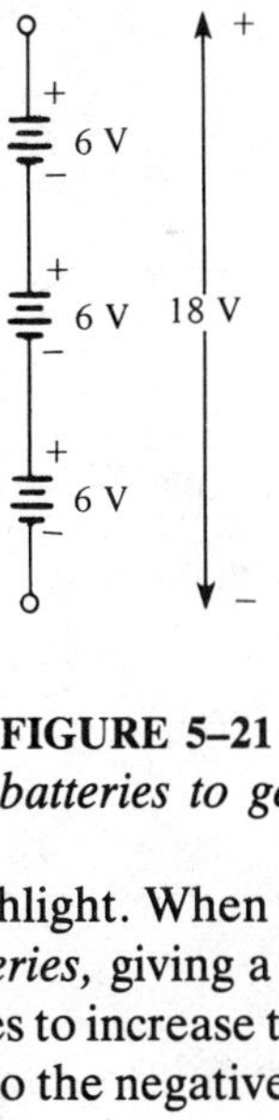

FIGURE 5–21 *Connection of three 6-V batteries to get 18 V.*

A familiar example of sources in series is the flashlight. When you put two 1.5-V batteries in your flashlight, they are connected in *series,* giving a total of 3 V. When connecting batteries or other voltage sources in series to increase the total voltage, always connect from the positive (+) terminal of one to the negative (−) of another. Such a connection is illustrated in Figure 5–21.

Example 5–9

What is the total source voltage (V_{ST}) in Figure 5–22?

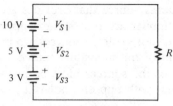

FIGURE 5–22

Solution:

The polarity of each source is the same (the sources are connected in the same direction in the circuit). So we sum the three voltages to get the total:

$$V_{ST} = V_{S1} + V_{S2} + V_{S3} = 10\text{ V} + 5\text{ V} + 3\text{ V}$$
$$= 18\text{ V}$$

The three individual sources can be replaced by a single equivalent source of 18 V with its polarity as shown in Figure 5–23.

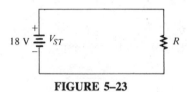

FIGURE 5–23

Example 5–10

Determine V_{ST} in Figure 5–24.

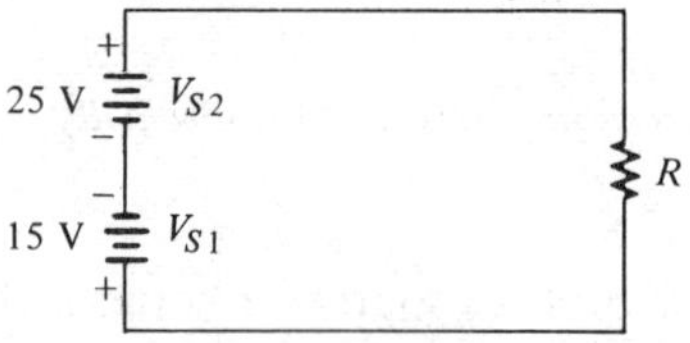

FIGURE 5–24

Solution:

These sources are connected in *opposing* directions. If you go counterclockwise around the circuit, you go from plus to minus through

VOLTAGE SOURCES IN SERIES

V_{S2}, and minus to plus through V_{S1}. The total voltage is the *difference* of the two source voltages (algebraic sum of oppositely signed values). The total voltage has the same polarity as the larger-value source. Here we will choose V_{S2} to be positive:

$$V_{ST} = V_{S2} - V_{S1} = 25\text{ V} - 15\text{ V}$$
$$= 10\text{ V}$$

The two sources in Figure 5–24 can be replaced by a 10-V equivalent one with polarity as shown in Figure 5-25.

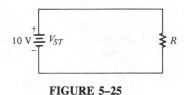

FIGURE 5–25

Review for 5–5

1. Four 1.5-V flashlight batteries are connected in series plus to minus. What is the total voltage of all four cells?
2. How many 12-V batteries must be connected in series to produce 60 V? Sketch a schematic showing the battery connections.
3. The resistive circuit in Figure 5–26 is used to bias a transistor amplifier. Show how to connect two 15-V power supplies in order to get 30 V across the two resistors.

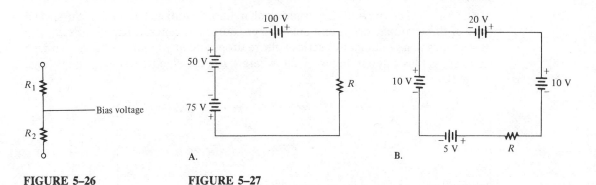

FIGURE 5–26   **FIGURE 5–27**

4. Determine the total voltage in each circuit of Figure 5–27.
5. Sketch the *equivalent single source* circuit for each circuit of Figure 5–27.

5-6

KIRCHHOFF'S VOLTAGE LAW

Kirchhoff's voltage law states that *the sum of all the voltages around a closed path is zero.* In other words, *the sum of the voltage drops equals the total source voltage.*

For example, in the circuit of Figure 5–28, there are three voltage drops and one voltage source. If we sum all of the voltages around the circuit, we get

$$V_S - V_1 - V_2 - V_3 = 0 \qquad (5\text{--}3)$$

Notice that the source voltage has a sign opposite to that of the voltage drops. Thus, the algebraic sum equals zero. Equation (5–3) can be written another way by transposing of the voltage drop terms to the right side of the equation:

$$V_S = V_1 + V_2 + V_3 \qquad (5\text{--}4)$$

In words, Equation (5–4) says that the source voltage equals the sum of the voltage drops. Both Equations (5–3) and (5–4) are equivalent ways of expressing Kirchhoff's voltage law for the circuit in Figure 5–28.

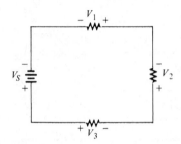

FIGURE 5–28

The previous illustration of Kirchhoff's voltage law was for the special case of three voltage drops and one voltage source. In general, Kirchhoff's voltage law applies to any number of series voltage drops and any number of series voltage sources, as illustrated in Figure 5–29, where V_{ST} is the algebraic sum of the voltage sources.

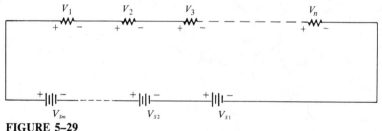

FIGURE 5–29

The *general form* of Kirchhoff's voltage law is

$$V_{ST} = V_{S1} + V_{S2} + \ldots + V_{Sm} = V_1 + V_2 + \ldots + V_n \qquad (5\text{--}5)$$

KIRCHHOFF'S VOLTAGE LAW

Polarities of Voltage Drops

Always remember that the total source voltage has a polarity in the circuit *opposite* to that of the voltage drops. As we go around the series circuit of Figure 5–30 in the clockwise direction, the voltage drops have polarities as shown. The negative (−) side of each resistor is the one *nearest* the negative terminal of the source as we follow the current path. Notice that the positive (+) side of each resistor is the one *nearest* the positive terminal of the source as we follow the current path.

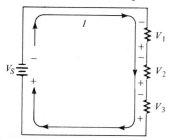

FIGURE 5–30

Let us examine polarities in a series loop from the electron current standpoint. Refer to Figure 5–30. *The current is out of the negative side of the source* and through the resistors as shown. *The current is into the negative side of each resistor and out of the positive side.* When current enters a resistor, it makes that side negative; and when it leaves the resistor, it makes that side positive, thus causing a voltage drop across the resistor.

A voltage drop can be thought of as a change in voltage from negative to positive in the direction of electron flow through a resistor.

Some texts use *conventional current,* which is in a direction opposite to electron current. The use of electron or conventional current is normally based on personal preference or as a convenience in some circuit concepts. The direction of conventional current is from positive to negative.

For purposes of analysis, we will use electron current in this text. You should, however, learn to think in terms of conventional current also.

We will now work some examples using Kirchhoff's voltage law to solve circuit problems.

Example 5–11

Determine the applied voltage V_S in Figure 5–31 where the two voltage drops are given.

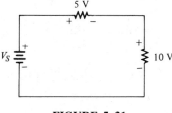

FIGURE 5–31

Example 5–11 (continued)

Solution:

By Kirchhoff's voltage law, the source voltage (applied voltage) must equal the sum of the voltage drops. Adding the voltage drops gives us the value of the source voltage:

$$V_s = 5 \text{ V} + 10 \text{ V} = 15 \text{ V}$$

Example 5–12

Determine the unknown voltage drop, V_3, in Figure 5–32.

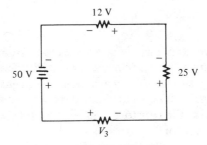

FIGURE 5–32

Solution:

By Kirchhoff's voltage law, the algebraic sum of all the voltages around the circuit is zero:

$$V_s - V_1 - V_2 - V_3 = 0$$

The value of each voltage drop except V_3 is known. Substitute these values into the equation as follows:

$$50 \text{ V} - 12 \text{ V} - 25 \text{ V} - V_3 = 0$$

Next combine the known values:

$$13 \text{ V} - V_3 = 0$$

Transpose 13 V to the right side of the equation, and cancel the minus signs:

$$-V_3 = -13 \text{ V}$$
$$V_3 = 13 \text{ V}$$

The voltage drop across R_3 is 13 V, and its polarity is as shown in Figure 5–32.

KIRCHHOFF'S VOLTAGE LAW

Example 5–13

Find the value of R_4 in Figure 5–33.

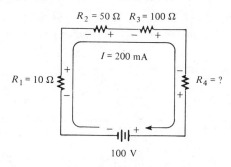

FIGURE 5–33

Solution:

In this problem we must use both Ohm's law *and* Kirchhoff's voltage law. Follow this procedure carefully.

First find the voltage drop across each of the *known* resistors. Use Ohm's law:

$$V_1 = IR_1 = (200 \text{ mA})(10 \text{ }\Omega) = 2 \text{ V}$$
$$V_2 = IR_2 = (200 \text{ mA})(50 \text{ }\Omega) = 10 \text{ V}$$
$$V_3 = IR_3 = (200 \text{ mA})(100 \text{ }\Omega) = 20 \text{ V}$$

Next, use Kirchhoff's voltage law to find V_4, the voltage drop across the *unknown* resistor:

$$V_S - V_1 - V_2 - V_3 - V_4 = 0$$
$$100 \text{ V} - 2 \text{ V} - 10 \text{ V} - 20 \text{ V} - V_4 = 0$$
$$68 \text{ V} - V_4 = 0$$
$$V_4 = 68 \text{ V}$$

Now that we know V_4, we can use Ohm's law to calculate R_4 as follows:

$$R_4 = \frac{V_4}{I} = \frac{68 \text{ V}}{200 \text{ mA}} = 340 \text{ }\Omega$$

Review for 5–6

1. State Kirchhoff's voltage law in two ways.
2. A 50-V source is connected to a series resistive circuit. What is the total of the voltage drops in this circuit?

3. Two equal-value resistors are connected in series across a 10-V battery. What is the voltage drop across each resistor?

4. In a series circuit with a 25-V source, there are three resistors. One voltage drop is 5 V, and the other is 10 V. What is the value of the third voltage drop?

5. The individual voltage drops in a series string are as follows: 1 V, 3 V, 5 V, 8 V, and 7 V. What is the total voltage applied across the series string?

5–7

VOLTAGE DIVIDERS

A series circuit is a *voltage divider*. In this section you will see what this term means and why voltage dividers are an important application of series circuits.

To illustrate how a series string of resistors acts as a voltage divider, we will examine Figure 5–34 where there are two resistors in series. As you already know, there are two voltage drops: one across R_1 and one across R_2. We call these voltage drops V_1 and V_2, respectively, as indicated in the diagram.

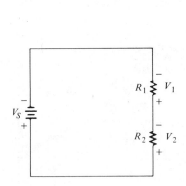
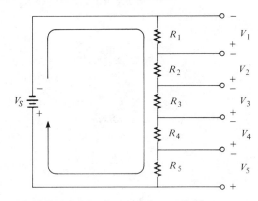

FIGURE 5–34   **FIGURE 5–35** *Series voltage divider.*

Since the current is the same through each resistor, the voltage drops are proportional to the ohmic values of the resistors. For example, if the value of R_2 is twice that of R_1, then the value of V_2 is twice that of V_1. In other words, the total voltage drop *divides* among the series resistors in amounts directly proportional to the resistance values.

For example, in Figure 5–34, if V_S is 10 V, R_1 is 50 Ω, and R_2 is 100 Ω, then V_1 is one-third the total voltage, or 3.33 V, because R_1 is one-third the *total* resistance. Likewise, V_2 is two-thirds V_S, or 6.67 V.

Voltage Divider Formula

With a few calculations, a formula for determining how voltages divide among series resistors can be developed. Let us assume that we have several resistors in series as shown in Figure 5–35. This figure shows five resistors, but there can be any number.

VOLTAGE DIVIDERS

Let us call the voltage drop across any one of the resistors V_x, where x represents the number of a particular resistor (1, 2, 3, and so on). By Ohm's law, the voltage drop across any of the resistors in Figure 5–35 can be written as follows:

$$V_x = IR_x$$

where $x = 1, 2, 3, 4,$ or 5 and R_x is any one of the series resistors.

The current is equal to the source voltage divided by the total resistance. For our example circuit of Figure 5–35, the total resistance is $R_1 + R_2 + R_3 + R_4 + R_5$, and

$$I = \frac{V_S}{R_T}$$

Substituting V_S/R_T for I in the expression for V_x, we get

$$V_x = \left(\frac{V_S}{R_T}\right)R_x$$

By rearranging, we get

$$V_x = \left(\frac{R_x}{R_T}\right)V_S \qquad (5\text{–}6)$$

Equation (5–6) is the *general voltage divider formula*. It tells us the following: *The voltage drop across any resistor or combination of resistors in a series circuit is equal to the ratio of that resistance value to the total resistance, multiplied by the source voltage.*

The following examples will illustrate use of the voltage divider formula.

Example 5–14

Determine the voltage across R_1 and the voltage across R_2 in the voltage divider in Figure 5–36.

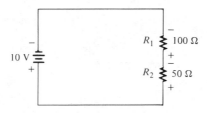

FIGURE 5–36

Solution:

Use the voltage divider formula, $V_x = (R_x/R_T)V_S$. In this problem we are looking for V_1; so $V_x = V_1$ and $R_x = R_1$. The total resistance is

$$R_T = R_1 + R_2 = 100 \text{ Ω} + 50 \text{ Ω} = 150 \text{ Ω}$$

R_1 is 100 Ω and V_S is 10 V. Substituting these values into the voltage divider formula, we have

Example 5–14 (continued)

$$V_1 = \left(\frac{R_1}{R_T}\right)V_S = \left(\frac{100 \text{ }\Omega}{150 \text{ }\Omega}\right)10 \text{ V}$$

$$= 6.67 \text{ V}$$

There are two ways to find the value of V_2 in this problem: Kirchhoff's voltage law or the voltage divider formula.

First, using Kirchhoff's voltage law, we know that $V_S = V_1 + V_2$. By substituting the values for V_S and V_1, we can solve for V_2 as follows:

$$V_2 = 10 \text{ V} - 6.67 \text{ V} = 3.33 \text{ V}$$

A second way is to use the voltage divider formula to find V_2 as follows:

$$V_2 = \left(\frac{R_2}{R_T}\right)V_S = \left(\frac{50 \text{ }\Omega}{150 \text{ }\Omega}\right)10 \text{ V}$$

$$= 3.33 \text{ V}$$

We get the same result either way.

Example 5–15

Calculate the voltage drop across each resistor in the voltage divider of Figure 5–37.

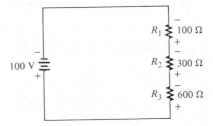

FIGURE 5–37

Solution:

Look at the circuit for a moment and consider the following: The total resistance is 1000 Ω. We can examine the circuit and determine that 10% of the total voltage is across R_1 because it is 10% of the total resistance (100 Ω is 10% of 1000 Ω). Likewise, we see that 30% of the total voltage is dropped across R_2 because it is 30% of the total resistance (300 Ω is 30% of 1000 Ω). Finally, R_3 drops 60% of the total voltage because 600 Ω is 60% of 1000 Ω.

Because of the convenient values in this problem, it is easy to figure the voltages mentally. Such is not always the case, but sometimes a little thinking will produce a result more efficiently and eliminate some calculating.

VOLTAGE DIVIDERS

Although we have already reasoned through this problem, the calculations will verify our results:

$$V_1 = \left(\frac{R_1}{R_T}\right)V_S = \left(\frac{100\ \Omega}{1000\ \Omega}\right)100\ V$$
$$= 10\ V$$

$$V_2 = \left(\frac{R_2}{R_T}\right)V_S = \left(\frac{300\ \Omega}{1000\ \Omega}\right)100\ V$$
$$= 30\ V$$

$$V_3 = \left(\frac{R_3}{R_T}\right)V_S = \left(\frac{600\ \Omega}{1000\ \Omega}\right)100\ V$$
$$= 60\ V$$

Notice that the sum of the voltage drops is equal to the source voltage, in accordance with Kirchhoff's voltage law. This check is a good way to verify your results.

Example 5–16

Determine the voltages between the following points in the voltage divider of Figure 5–38:
(a) A to B (b) A to C (c) B to C (d) B to D (e) C to D

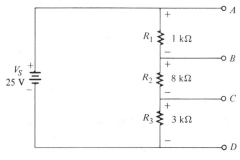

FIGURE 5–38

Solution:

First determine R_T:

$$R_T = 1\ k\Omega + 8\ k\Omega + 3\ k\Omega = 12\ k\Omega$$

Now apply the voltage divider formula to obtain each required voltage.

(a) The voltage A to B is the voltage drop across R_1. The calculation is as follows:

$$V_{AB} = \left(\frac{R_1}{R_T}\right)V_S = \left(\frac{1\ k\Omega}{12\ k\Omega}\right)25\ V$$
$$= 2.08\ V$$

Example 5–16 (continued)

(b) The voltage from A to C is the combined voltage drop across both R_1 and R_2. In this case, R_x in the general formula [Equation (5–6)] is $R_1 + R_2$. The calculation is as follows:

$$V_{AC} = \left(\frac{R_1 + R_2}{R_T}\right)V_S = \left(\frac{9 \text{ k}\Omega}{12 \text{ k}\Omega}\right)25 \text{ V}$$
$$= 18.75 \text{ V}$$

(c) The voltage from B to C is the voltage drop across R_2. The calculation is as follows:

$$V_{BC} = \left(\frac{R_2}{R_T}\right)V_S = \left(\frac{8 \text{ k}\Omega}{12 \text{ k}\Omega}\right)25 \text{ V}$$
$$= 16.67 \text{ V}$$

(d) The voltage from B to D is the combined voltage drop across both R_2 and R_3. In this case, R_x in the general formula is $R_2 + R_3$. The calculation is as follows:

$$V_{BD} = \left(\frac{R_2 + R_3}{R_T}\right)V_S = \left(\frac{11 \text{ k}\Omega}{12 \text{ k}\Omega}\right)25 \text{ V}$$
$$= 22.92 \text{ V}$$

(e) Finally, the voltage from C to D is the voltage drop across R_3. The calculation is as follows:

$$V_{CD} = \left(\frac{R_3}{R_T}\right)V_S = \left(\frac{3 \text{ k}\Omega}{12 \text{ k}\Omega}\right)25 \text{ V}$$
$$= 6.25 \text{ V}$$

If you connect this voltage divider in the lab, you can verify each of the calculated voltages by connecting a voltmeter between the appropriate points in each case.

The Potentiometer as an Adjustable Voltage Divider

Recall from Chapter 2 that a potentiometer is a variable resistor with three terminals. A potentiometer connected to a voltage source is shown in Figure 5–39A. Notice that the two end terminals are labeled 1 and 2. The adjustable terminal or wiper is labeled 3. The potentiometer acts as a voltage divider. We can illustrate this concept better by separating the total resistance into two parts, as shown in Figure 5–39B. The resistance between terminal 1 and terminal 3 (R_{13}) is one part, and the resistance between terminal 3 and terminal 2 (R_{32}) is the other part. So this potentiometer actually is a two-resistor voltage divider that can be manually adjusted.

Figure 5–40 shows what happens when the wiper terminal (3) is moved. In Part A of Figure 5–40, the wiper is exactly centered, making the two

VOLTAGE DIVIDERS

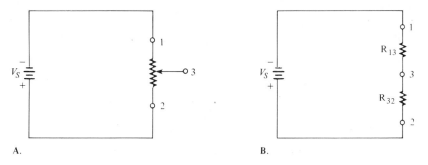

FIGURE 5-39 *The potentiometer as a voltage divider.*

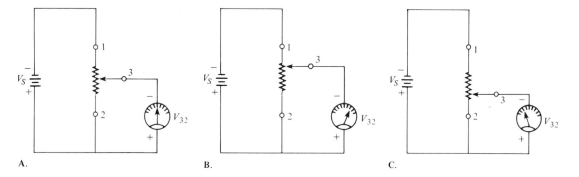

FIGURE 5-40 *Adjusting the voltage divider.*

resistances equal. If we measure the voltage across terminals 3 to 2 as indicated by the voltmeter symbol, we have one-half of the total source voltage. When the wiper is moved up, as in Figure 5-40B, the resistance between terminals 3 and 2 increases, and the voltage across it increases proportionally. When the wiper is moved down, as in Figure 5-40C, the resistance between terminals 3 and 2 decreases, and the voltage decreases proportionally.

Voltage Divider Applications

The volume control of radio or TV receivers is a common application of a potentiometer used as a voltage divider. Since the loudness of the sound is dependent on the amount of voltage associated with the audio signal, you can increase or decrease the volume by adjusting the potentiometer, that is, by turning the knob of the volume control on the set.

Another application for a voltage divider is in setting the dc operating voltage (*bias*) in transistor amplifiers. Figure 5-41 shows a voltage divider used for this purpose. You will study transistor amplifiers and biasing in a later course; so it is important that you understand the basics of voltage dividers at this point.

These examples are only two out of an almost endless number of applications of voltage dividers.

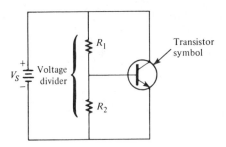

FIGURE 5-41 *The voltage divider as a bias circuit for a transistor amplifier.*

Review for 5-7

1. What is a voltage divider?
2. How many resistors can there be in a series voltage divider circuit?
3. Write the general formula for voltage dividers.
4. If two series resistors of equal value are connected across a 10-V source, how much voltage is there across each resistor?
5. A 50-Ω resistor and a 75-Ω resistor are connected as a voltage divider. The source voltage is 100 V. Sketch the circuit, and determine the voltage across each of the resistors.
6. The circuit of Figure 5-42 is an adjustable voltage divider. If the potentiometer is linear, where would you set the wiper in order to get 5 V from B to A and 5 V from C to B?

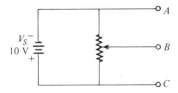

FIGURE 5-42

5-8

POWER IN A SERIES CIRCUIT

The total amount of power in a series resistive circuit is equal to the sum of the powers in each resistor in series:

$$P_T = P_1 + P_2 + P_3 + \cdots + P_n \qquad (5\text{-}7)$$

where n is the number of resistors in series, P_T is the total power, and P_n is the power in the last resistor in series. In other words, the powers are additive.

POWER IN A SERIES CIRCUIT

The power formulas that you learned in Chapter 4 are, of course, directly applicable to series circuits. Since the current is the same through each resistor in series, the following formulas are used to calculate the total power:

$$P_T = V_S I$$
$$P_T = I^2 R_T$$
$$P_T = \frac{V_S^2}{R_T}$$

where V_S is the total source voltage across the series connection and R_T is the total resistance. Example 5–17 illustrates how to calculate total power in a series circuit.

Example 5–17

Determine the total amount of power in the series circuit in Figure 5–43.

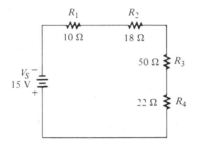

FIGURE 5–43

Solution:

We know that the source voltage is 15 V. The total resistance is

$$R_T = 10\ \Omega + 18\ \Omega + 50\ \Omega + 22\ \Omega = 100\ \Omega$$

The easiest formula to use is $P_T = V_S^2/R_T$ since we know both V_S and R_T:

$$P_T = \frac{V_S^2}{R_T} = \frac{(15\ \text{V})^2}{100\ \Omega} = \frac{225\ \text{V}^2}{100\ \Omega}$$
$$= 2.25\ \text{W}$$

If the power of each resistor is determined separately and all of these powers are added, the same result is obtained. We will work through another calculation to illustrate this formula.

First, find the current as follows:

$$I = \frac{V_S}{R_T} = \frac{15\ \text{V}}{100\ \Omega}$$
$$= 0.15\ \text{A}$$

Next, calculate the power for each resistor using $P = I^2 R$:

Example 5–17 (continued)

$$P_1 = (0.15 \text{ A})^2(10 \text{ }\Omega) = 0.225 \text{ W}$$
$$P_2 = (0.15 \text{ A})^2(18 \text{ }\Omega) = 0.405 \text{ W}$$
$$P_3 = (0.15 \text{ A})^2(50 \text{ }\Omega) = 1.125 \text{ W}$$
$$P_4 = (0.15 \text{ A})^2(22 \text{ }\Omega) = 0.495 \text{ W}$$

Now, add these powers to get the total power:

$$P_T = 0.225 \text{ W} + 0.405 \text{ W} + 1.125 \text{ W} + 0.495 \text{ W}$$
$$= 2.25 \text{ W}$$

This result shows that the sum of the individual powers is equal to the total power as determined by one of the power formulas.

Review for 5–8

1. If you know the power in each resistor in a series circuit, how can you find the total power?
2. The resistors in a series circuit have the following powers: 2 W, 5 W, 1 W, and 8 W. What is the total power in the circuit?
3. A circuit has a 100-Ω, a 300-Ω, and a 600-Ω resistor in series. There is a current of 1 A through the circuit. What is the total power?

5–9

OPEN AND CLOSED CIRCUITS

An *open* circuit is one in which the current path is interrupted and no current flows. In an actual circuit, an open can be caused by a switch connected in series, as shown in Figure 5–44. Here, the switch symbol represents a simple *single-pole–single-throw* switch (see Chapter 2). In one position the switch disconnects and causes an *open* circuit. In the *open* position, there is no current and the light is off, as illustrated in Figure 5–44A. When the switch is moved to the *closed* position, contact is made, allowing current and the light to turn on, as shown in Figure 5–44B.

Sometimes a component failure will cause a circuit to open. There are many ways such a failure can occur, but we will use a familiar example to illustrate. In Figure 5–45, there are two light bulbs in series. As long as the bulbs are good, there is a *closed* circuit and current, keeping the lights on, as in Part A of the figure. A common failure in a light bulb is filament burnout. When burnout occurs, an *open* results and current ceases, as in Part B. Since there is no current, both bulbs are off.

TROUBLE-SHOOTING SERIES CIRCUITS

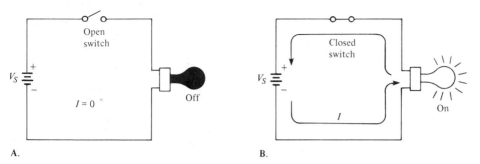

FIGURE 5–44 *Opening and closing a circuit with a switch.*

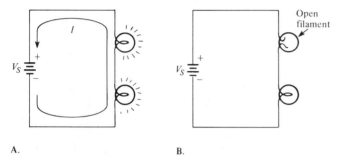

FIGURE 5–45 *Example of open circuit caused by a component failure.*

Review for 5–9

1. Define *closed circuit*.
2. Define *open circuit*.
3. What happens when a series circuit opens?
4. Name two general ways in which an open circuit can occur in practice.
5. What is the main purpose of a switch?

5–10
TROUBLE-SHOOTING SERIES CIRCUITS

The most common failure of a resistor is an *open*. A resistor will overheat when the power is greater than the power rating of the resistor. When the excess power is sufficient, the resistor will be damaged by the excessive heat, and an open resistor can result.

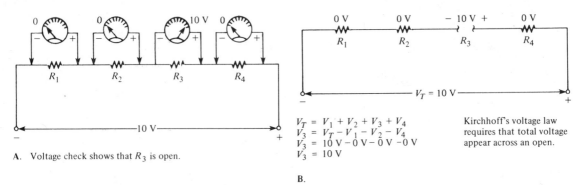

FIGURE 5–46 *Trouble-shooting a series circuit.*

How to Check for an Open Resistor

Sometimes a visual check will reveal a charred resistor. However, it is possible for a resistor to open without showing visible signs of damage. In this situation, a *voltage check* of the series circuit is required. The general procedure is as follows: Measure the voltage across each resistor in series. *The voltage across all of the good resistors will be zero. The voltage across the open resistor will equal the total voltage across the series combination.*

The above condition occurs because an open resistor will prevent current through the series circuit. With no current, there can be no voltage drop across any of the good resistors. Since $IR = 0$, in accordance with *Ohm's law*, the voltage on each side of a good resistor is the same. The total voltage must then appear across the open resistor in accordance with *Kirchhoff's voltage law*, as illustrated in Figure 5–46.

Review for 5–10

1. When a resistor fails, it will normally open (T or F).
2. The total voltage across a string of series resistors is 24 V. If one of the resistors is open, how much voltage is there across it? How much is there across each of the good resistors?

Formulas

$$R_T = R_1 + R_2 + R_3 + \cdots + R_n \qquad (5\text{–}1)$$

$$R_T = nR \qquad (5\text{–}2)$$

$$V_S - V_1 - V_2 - V_3 = 0 \qquad (5\text{–}3)$$

SUMMARY

$$V_S = V_1 + V_2 + V_3 \quad (5\text{-}4)$$

$$V_{ST} = V_{S1} + V_{S2} + \ldots + V_{Sm} = V_1 + V_2 + V_3 + \ldots + V_n \quad (5\text{-}5)$$

$$V_x = \left(\frac{R_x}{R_T}\right) V_S \quad (5\text{-}6)$$

$$P_T = P_1 + P_2 + P_3 + \cdots + P_n \quad (5\text{-}7)$$

Summary

1. Resistors in series are connected end-to-end.
2. A series connection has only *one* path for current.
3. There is the same amount of current at all points in a series circuit.
4. The total series resistance is the sum of all resistors in the series circuit.
5. The total resistance between any two points in a series circuit is equal to the sum of all resistors connected in series between those two points.
6. If all of the resistors in a series connection are of equal ohmic value, the total resistance is the number of resistors multiplied by the ohmic value.
7. Voltage sources in series add algebraically.
8. First statement of Kirchhoff's voltage law: The sum of all the voltages around a closed path is zero.
9. Second statement of Kirchhoff's voltage law: The sum of the voltage drops equals the total source voltage. (Both 8 and 9 say the same thing.)
10. The voltage drops in a circuit are always opposite in polarity to the total source voltage.
11. Current is out of the negative side of a source and into the positive side.
12. Current into a resistor makes that end of the resistor negative. Current out of a resistor makes that end positive.
13. A voltage drop is considered to be from a more negative voltage to a more positive voltage.
14. A voltage divider is a series arrangement of resistors.
15. A voltage divider is so named because the voltage drop across any resistor in the series circuit is divided down from the total voltage by an amount proportional to that resistance value in relation to the total resistance.
16. A potentiometer can be used as an adjustable voltage divider.
17. The total power in a resistive circuit is the sum of all the individual powers of the resistors making up the series circuit.
18. An open circuit is one in which the current is interrupted.
19. A closed circuit is one having a complete current path.

Self-Test

1. Sketch a series circuit having four resistors in series with a voltage source.
2. There are 5 amperes of current into a series string of 10 resistors. How much is the current out of the sixth resistor? The tenth resistor?
3. Connect each set of resistors in Figure 5–47 in series between points A and B.

A.

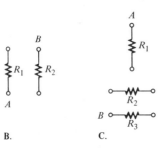

B.   C.

FIGURE 5–47

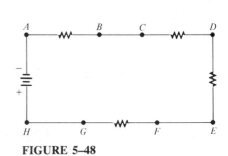

FIGURE 5–48

4. If you measured the current between each of the points marked in Figure 5–48, how much current would you read in each case? The total current out of the source is 3 A.
5. If a 75-Ω resistor and a 470-Ω resistor are connected in series, what is the total resistance?
6. Eight 56-Ω resistors are in series. Determine the total resistance.
7. Suppose that you need a total resistance of 20 kΩ. The resistors that you have available are two 1-kΩ, one 5-kΩ, and one 3-kΩ. What other single resistor do you need to make the required total?
8. Three 1-kΩ resistors are connected in series, and 5 V are applied across the series circuit. How much current is there through each resistor?
9. Which circuit in Figure 5–49 has more current?

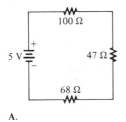

A.

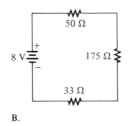

B.

FIGURE 5–49

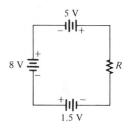

FIGURE 5–50

10. Six resistors of equal value are in series with a 12-V source. The current through the circuit is 2 mA. What is the total resistance? Determine the value of each resistor.
11. Determine the total voltage in the circuit in Figure 5–50.

SELF-TEST

12. Two 9-V batteries are connected in opposite directions in a series circuit. What is the total voltage?

13. Determine the value and the polarity of the total voltage in each circuit of Figure 5–51. Sketch the equivalent single-source circuit.

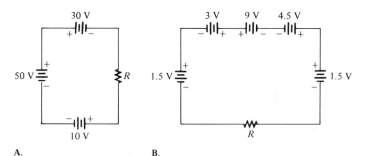

FIGURE 5–51

14. Five equal-value resistors are in series with a voltage source. The voltage drop across the first resistor is 2 V. What is the value of the source voltage?

15. A circuit with a 50-V source has three resistors in series. The voltage drop of the first resistor is 10 V. The voltage drop of the second resistor is 15 V. What is the voltage drop of the third resistor?

16. Determine the unknown resistance (R_3) in the circuit in Figure 5–52.

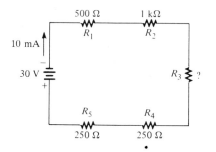   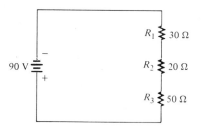

FIGURE 5–52   FIGURE 5–53

17. The following resistors (one each) are connected in series: 10-Ω, 50-Ω, 100-Ω, and 40-Ω. The total voltage is 20 V. Using the voltage divider formula, calculate the voltage drop across the 50-Ω resistor.

18. Calculate the voltage across each resistor in the voltage divider of Figure 5–53.

19. You have a 9-V battery available and need 3 V to bias a transistor amplifier for proper operation. You must use two resistors to form a voltage divider that will divide the 9 V down to 3 V. Determine the ideal values needed to achieve this if their total must be 100 kΩ.

20. In a series circuit with three resistors, one handles 2.5 W, one handles 5 W, and one handles 1.2 W. What is the total power in the circuit?

Problems

5–1. Connect the resistors in Figure 5–54 in series, with R_1 closest to the negative terminal, R_2 next, and so forth.

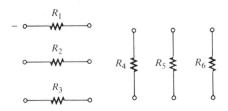

FIGURE 5–54

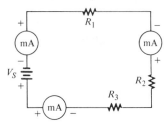

FIGURE 5–55

5–2. What is the current through each resistor of a series circuit if the total voltage is 12 V and the total resistance is 120 Ω?

5–3. The current out of the source in Figure 5–55 is 5 mA. How much current does each milliammeter in the circuit indicate?

5–4. The following resistors (one each) are connected in a series circuit: 1-Ω, 2-Ω, 5-Ω, 12-Ω, and 22-Ω. Determine the total resistance.

5–5. Find the total resistance of each of the following groups of series resistors:
(a) 560-Ω and 1000-Ω
(b) 47-Ω and 56-Ω
(c) 1.5-kΩ, 2.2-kΩ, and 10-kΩ
(d) 1-MΩ, 470-kΩ, 1-kΩ, 2.5-MΩ

5–6. Calculate R_T for each circuit of Figure 5–56.

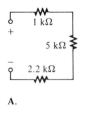

FIGURE 5–56

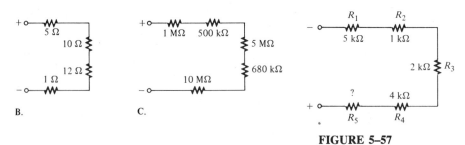

FIGURE 5–57

5–7. If the total resistance in Figure 5–57 is 18 kΩ, what is the value of R_5?

5–8. What is the total resistance of twelve 5.6-kΩ resistors in series?

5–9. Six 50-Ω resistors, eight 100-Ω resistors, and two 22-Ω resistors are all connected in series. What is the total resistance?

5–10. You have the following resistor values available to you in the lab in unlimited quantities: 10 Ω, 100 Ω, 470 Ω, 560 Ω, 680 Ω, 1 kΩ, 2.2 kΩ, and 5.6 kΩ. All of the other standard values are out of stock. A project that you are working on requires an 18-kΩ resistance. What combinations of the available values would you use in series to achieve this total resistance?

5–11. What is the current in each circuit of Figure 5–58?

PROBLEMS

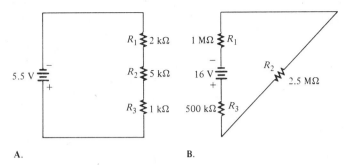

A.   B.

FIGURE 5–58

5–12. Three 470-Ω resistors are connected in series with a 500-V source. How much current is in the circuit?

5–13. Four equal-value resistors are in series with a 5-V battery, and 2.5 mA are measured. What is the value of each resistor?

5–14. *Series aiding* is a term sometimes used to describe voltage sources of the same polarity in series. If a 5-V and a 9-V source are connected in this manner, what is the total voltage?

5–15. The term *series opposing* means that sources are in series with opposite polarities. If a 12-V and a 3-V battery are series opposing, what is the total voltage?

5–16. Determine the total voltage in each circuit of Figure 5–59.

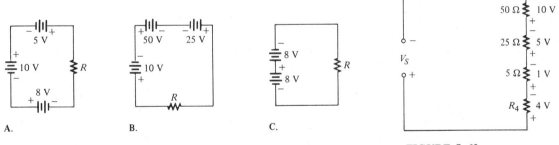

A.   B.   C.

FIGURE 5–59   **FIGURE 5–60**

5–17. The following voltage drops are measured across three resistors in series: 5.5 V, 8.2 V, and 12.3 V. What is the value of the source voltage to which these resistors are connected?

5–18. Five resistors are in series with a 20-V source. The voltage drops across four of the resistors are 1.5 V, 5.5 V, 3 V, and 6 V. How much voltage is dropped across the fifth resistor?

5–19. In the circuit of Figure 5–60, determine the resistance of R_4.

5–20. The total resistance of a circuit is 500 Ω. What percentage of the total voltage appears across a 25-Ω resistor that makes up part of the total series resistance?

5–21. Determine the voltage between points A and B in each voltage divider of Figure 5–61.

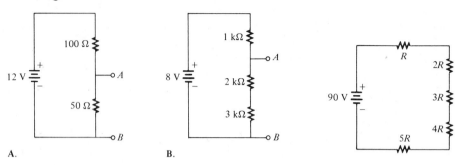

A. **B.**

FIGURE 5–61 **FIGURE 5–62**

5–22. What is the voltage across each resistor in Figure 5–62? R is the lowest-value resistor, and all others are multiples of that value as indicated.

5–23. Determine the voltage at each point in Figure 5–63 with respect to the negative side of the battery.

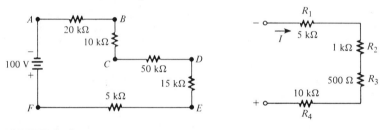

FIGURE 5–63 **FIGURE 5–64**

5–24. If there are 10 V across R_1 in Figure 5–64, what is the voltage across each of the other resistors?

5–25. Five series resistors each handle 50 mW. What is the total power?

5–26. What is the total power in the circuit in Figure 5–64?

Answers to Section Reviews

Section 5–1:
1. End-to-end. **2.** There is a single current path. **3.** See Figure 5–65. **4.** See Figure 5–66.

ANSWERS TO SECTION REVIEWS

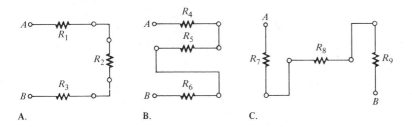

FIGURE 5–65

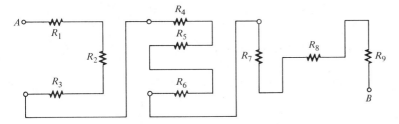

FIGURE 5–66

Section 5–2:
1. 1 A. **2.** 50 mA between C and D; 50 mA between E and F. **3.** 2 A, 2 A. **4.** Current is the same at all points.

Section 5–3:
1. 10 Ω. **2.** A, 143 Ω; B, 143 Ω; C, 143 Ω. **3.** 578 Ω. **4.** 1 kΩ. **5.** 600 Ω. **6.** 340 Ω.

Section 5–4:
1. 0.033 A or 33 mA. **2.** 100 V. **3.** $V_1 = 50$ V, $V_2 = 25$ V, $V_3 = 25$ V. **4.** 250 Ω.

Section 5–5:
1. 6 V. **2.** Five. See Figure 5–67. **3.** See Figure 5–68. **4.** A, 75 V; B, 15 V. **5.** See Figure 5–69.

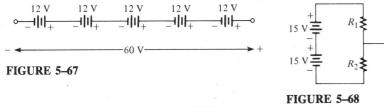

FIGURE 5–67

FIGURE 5–68

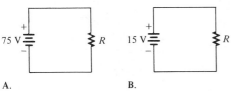

FIGURE 5–69

Section 5–6:

1. **(a)** The sum of the voltages around a closed path is zero. **(b)** The sum of the voltage drops equals the total source voltage. **2.** 50 V. **3.** 5 V. **4.** 10 V. **5.** 24 V.

Section 5–7:

1. Two or more resistors in a series connection in which the voltage taken across any resistor or combination of resistors is proportional to the value of that resistance. **2.** Any number. **3.** $V_x = (R_x/R_T)V_s$. **4.** 5 V. **5.** 40 V across the 50-Ω; 60 V across the 75-Ω. See Figure 5–70. **6.** At the midpoint.

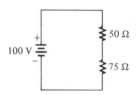

FIGURE 5–70

Section 5–8:

1. Add the power in each resistor. **2.** 16 W. **3.** 1000 W or 1 kW.

Section 5–9:

1. A circuit with a complete current path. **2.** A circuit with an interrupted current path. **3.** Current ceases. **4.** Switch or other component failure. **5.** To turn current on or off in a circuit.

Section 5–10:

1. T. **2.** 24 V, 0 V.

In Chapter 5, you learned about series circuits. In this chapter you will study circuits with parallel resistors. You will learn how to recognize parallel combinations and how to determine their total resistance.

Kirchhoff's current law will be introduced. You will see how current divides in parallel circuits and how Ohm's law and Kirchhoff's current law are used to solve circuit problems. Also, total power and the effects of open and closed current paths on a parallel circuit will be discussed.

6–1 Resistors in Parallel
6–2 Voltage Drop in a Parallel Circuit
6–3 Total Parallel Resistance
6–4 Applying Ohm's Law
6–5 Current Sources in Parallel
6–6 Kirchhoff's Current Law
6–7 Current Dividers
6–8 Power in a Parallel Circuit
6–9 Effects of Open Paths
6–10 Trouble-Shooting Parallel Circuits

6 PARALLEL RESISTIVE CIRCUITS

6-1

RESISTORS IN PARALLEL

When two or more components are connected across the same voltage source, they are in parallel. A parallel circuit provides more than one path for current. Each parallel path is called a *branch*. Two resistors connected in parallel are shown in Figure 6–1A.

In Figure 6–1B, the current out of the source divides when it gets to point A. Part of it goes through R_1 and part through R_2. If additional resistors are connected in parallel with the first two, more current paths are provided, as shown in Figure 6–1C.

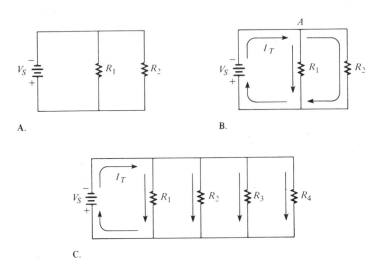

FIGURE 6–1 *Resistors in parallel.*

Identifying Parallel Connections

In Figure 6–1, the resistors obviously are connected in parallel. Often, in actual circuit diagrams, the parallel relationship is not so clear. It is important that you learn to recognize parallel connections regardless of how they may be drawn.

A rule for identifying parallel circuits is as follows: *If there is more than one current path (branch) between two points, and if the voltage between those two points appears across each of the branches, then there is a parallel circuit between those two points.* Figure 6–2 shows parallel resistors drawn in different ways between two points labeled A and B. Notice that in each case, the current "travels" two paths going from A to B, and the voltage across each branch is the same. Although these figures show only two parallel paths, there can be any number of resistors in parallel.

RESISTORS IN PARALLEL

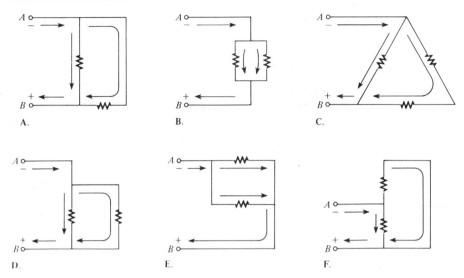

FIGURE 6–2 *Circuits with two parallel paths.*

Example 6–1

Suppose that there are five resistors positioned on a circuit board as shown in Figure 6–3. Wire them together in parallel between the negative (−) and the positive (+) terminals. Draw a schematic diagram showing this connection.

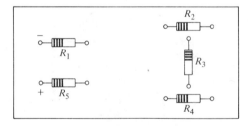

FIGURE 6–3

Solution:
Wires are connected as shown in the assembly diagram of Figure 6–4A. The schematic diagram is shown in Figure 6–4B. Again, note that the schematic does not necessarily have to show the actual physical arrangement of the resistors. The purpose of the schematic is to show how components are connected electrically.

Example 6–1 (continued)

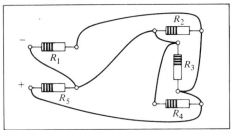

A. Assembly diagram. Compare this to the same resistor arrangement connected in series in Figure 5-4A.

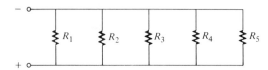

B. Schematic diagram.

FIGURE 6–4

Example 6–2

Describe how the resistors on the PC board in Figure 6–5 are related electrically.

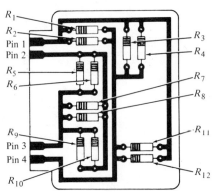

FIGURE 6–5

Solution:

Resistors R_1 through R_4 and R_{11} and R_{12} are all in parallel. This parallel combination is connected to pins 1 and 4.
 Resistors R_5 through R_{10} are all in parallel. This combination is connected to pins 2 and 3.

VOLTAGE DROP IN A PARALLEL CIRCUIT

Review for 6–1

1. How are the resistors connected in a parallel circuit?
2. How do you identify a parallel connection?
3. Complete the schematic diagrams for the circuits in each part of Figure 6–6 by connecting the resistors in parallel between points A and B.

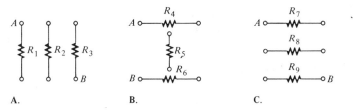

A. B. C.

FIGURE 6–6

4. Now connect each *group* of parallel resistors in Figure 6–6 in parallel with each other.

6–2

VOLTAGE DROP IN A PARALLEL CIRCUIT

As mentioned, each path in a parallel circuit is sometimes called a *branch*. The voltage across any branch of a parallel combination is equal to the voltage across each of the other branches in parallel.

To illustrate voltage drop in a parallel circuit, let us examine Figure 6–7A. Points A, B, C, and D along the top of the parallel circuit are *electrically the same point* because the voltage is the same along this line. You can think of all of these points as being connected by a single wire to the negative terminal of the battery. The points E, F, G, and H along the bottom of the circuit are all at a potential equal to the positive terminal of the source. Thus, each voltage across each parallel resistor is the same, and each is equal to the source voltage.

Figure 6–7B is the same circuit as in Part A, drawn in a slightly different way. Here the tops of the resistors are connected to a single point, which is the negative battery terminal. The bottoms of the resistors are all connected to the same point, which is the positive battery terminal. The resistors are still all in parallel across the source.

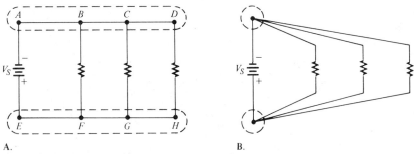

A. B.

FIGURE 6–7 *Voltage across parallel branches is the same.*

Example 6–3

Determine the voltage across each resistor in Figure 6–8.

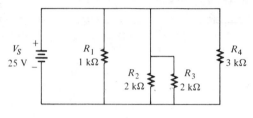

FIGURE 6–8

Solution:

The four resistors are in parallel; so the voltage drop across each one is equal to the applied source voltage:

$$V_S = V_1 = V_2 = V_3 = V_4 = 25 \text{ V}$$

Review for 6–2

1. A 10-Ω and a 20-Ω resistor are connected in parallel with a 5-V source. What is the voltage across each of the resistors?   5 V

2. A voltmeter is connected across R_1 in Figure 6–9. It measures 118 V. If you move the meter and connect it across R_2, how much voltage will it indicate? What is the source voltage?   118

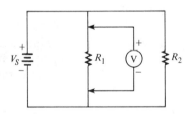

FIGURE 6–9

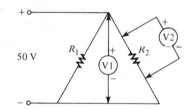

FIGURE 6–10

3. In Figure 6–10, how much voltage does voltmeter 1 indicate? Voltmeter 2?   50   50

4. What can you say about the amount of voltage across a parallel circuit?

6-3 TOTAL PARALLEL RESISTANCE

When resistors are connected in parallel, the total resistance of the circuit decreases. The total resistance of a parallel combination is always less than the value of the smallest resistor. For example, if a 10-Ω resistor and a 100-Ω resistor are connected in parallel, the total resistance is less than 10 Ω. The exact value must be calculated, and you will learn how to do so later in this section.

More Current Paths Mean Less Opposition

As you know, when resistors are connected in parallel, the current has more than one path. The number of current paths is equal to the number of parallel branches.

For example, in Figure 6–11A, there is only one current path since it is a series circuit. A certain amount of current, I_1, is through R_1. If resistor R_2 is connected in parallel with R_1, as shown in Figure 6–11B, an additional amount of current, I_2, flows through R_2. The total current coming from the source has increased with the addition of the parallel branch. An increase in the *total current* from the source means that the total resistance has decreased, in accordance with Ohm's law. Additional resistors connected in parallel will further reduce the resistance and increase the total current.

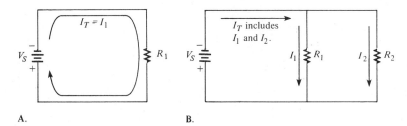

FIGURE 6–11 *Resistors in parallel reduce total resistance and increase total current.*

Conductances in Parallel

Recall from Chapter 2 that conductance (G) is a measure of a resistor's ability to *conduct* current and is measured in Siemens. Thus, it is the inverse (reciprocal) of the opposition (resistance) to current:

$$G = \frac{1}{R} \qquad (6\text{–}1)$$

Every resistor has a certain value of conductance, just as it has a certain value of resistance. Starting with the concept of total parallel conductance, you will see how the formula for calculating total parallel resistance is developed.

The key idea to keep in mind here is that an *increase* in conductance allows more current, and therefore an increase in G corresponds to a *decrease* in R. When resistors are connected in parallel, current is made easier because a parallel circuit provides multiple current paths. Thus, conductance *increases* as parallel paths are added, as illustrated in Figure 6–12.

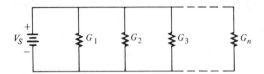

FIGURE 6–12 *Conductances add in parallel because there is more total current as more paths are added.*

Each time a new resistor is connected in parallel in a parallel circuit, the conductance of the circuit increases by the value of the conductance of that new resistor. Therefore, *the total conductance of a parallel circuit is equal to the sum of the conductances of the individual resistors:*

$$G_T = G_1 + G_2 + G_3 + \cdots + G_n \qquad (6\text{–}2)$$

where G_T is the total conductance, G_n is the conductance of the last resistor in parallel, and n is the number of parallel branches.

Example 6–4 illustrates how to add parallel conductances to obtain the total conductance.

Example 6–4

Determine the total conductance of the circuit of Figure 6–13.

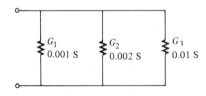

FIGURE 6–13

Solution:

The total conductance G_T is the sum of each individual conductance. In this circuit there are three conductances in parallel, and G_T is calculated as follows:

$$G_T = 0.001 \text{ S} + 0.002 \text{ S} + 0.01 \text{ S} = 0.013 \text{ S}$$

where "S" stands for *siemens,* which is the SI unit of conductance.

TOTAL PARALLEL RESISTANCE

Formula for Total Parallel Resistance

Now, using the relationship between resistance and conductance, $G = 1/R$, we can rewrite Equation (6–2) as follows:

$$\frac{1}{R_T} = \frac{1}{R_1} + \frac{1}{R_2} + \frac{1}{R_3} + \cdots + \frac{1}{R_n} \qquad (6\text{–}3)$$

If we take the reciprocal of both sides of Equation (6–3), we get the general formula for total parallel resistance:

$$R_T = \frac{1}{(1/R_1) + (1/R_2) + (1/R_3) + \cdots + (1/R_n)} \qquad (6\text{–}4)$$

Equation (6–4) says that the total parallel resistance is found by adding the conductances $(1/R)$ of all of the resistors in parallel and then taking the reciprocal of this sum. Example 6–5 shows how to use this formula in a specific case.

Example 6–5

Calculate the total parallel resistance between points A and B of the circuit in Figure 6–14.

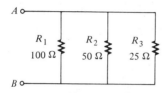

FIGURE 6–14

Solution:

First, find the conductance of each of the three resistors as follows:

$$G_1 = \frac{1}{R_1} = \frac{1}{100\ \Omega} = 0.01\ \text{S}$$

$$G_2 = \frac{1}{R_2} = \frac{1}{50\ \Omega} = 0.02\ \text{S}$$

$$G_3 = \frac{1}{R_3} = \frac{1}{25\ \Omega} = 0.04\ \text{S}$$

Next, calculate R_T by adding G_1, G_2, and G_3 and taking the reciprocal of the sum as follows:

$$R_T = \frac{1}{0.01\ \text{S} + 0.02\ \text{S} + 0.04\ \text{S}} = \frac{1}{0.07\ \text{S}}$$

$$= 14.29\ \Omega$$

Notice that the value of R_T (14.29 Ω) is smaller than the smallest value in parallel, which is R_3 (25 Ω).

Calculator Solution

The parallel resistance formula is easily solved on an electronic calculator. The general procedure is to enter the value of R_1 and then take its reciprocal by pressing the $1/x$ key. Next, press the $+$ key; then enter the value of R_2 and take its reciprocal. Repeat this procedure until all of the resistor values have been entered and the reciprocal of each has been added. The final step is to press the $1/x$ key to convert $1/R_T$ to R_T. The total parallel resistance is now on the display. This calculator procedure is illustrated in Example 6–6.

Example 6–6

Show the steps required for a calculator solution of Example 6–5.

Solution:

Step 1: Enter 100. Display shows 100.
Step 2: Press $1/x$ key. Display shows 0.01.
Step 3: Press $+$ key. Display shows 0.01.
Step 4: Enter 50. Display shows 50.
Step 5: Press $1/x$ key. Display shows 0.02.
Step 6: Press $+$ key. Display shows 0.03.
Step 7: Enter 25. Display shows 25.
Step 8: Press $1/x$ key. Display shows 0.04.
Step 9: Press $=$ key. Display shows 0.07.
Step 10: Press $1/x$ key. Display shows 14.2857 . . . .

The number displayed in Step 9 is the total *conductance* in siemens. The number displayed in Step 10 is the total *resistance* in ohms.

Two Resistors in Parallel

Equation (6–4) is a general formula for finding the total resistance for any number of resistors in parallel. It is often useful to consider only two resistors in parallel because this setup occurs commonly in practice. Also, any number of resistors in parallel can be broken down into *pairs* as an alternate way to find the R_T. Based on Equation (6–4), the formula for two resistors in parallel is

$$R_T = \frac{1}{(1/R_1) + (1/R_2)}$$

Combining the terms in the denominator, we get

$$R_T = \frac{1}{(R_1 + R_2)/(R_1 R_2)}$$

This equation can be rewritten as follows:

TOTAL PARALLEL RESISTANCE

$$R_T = \frac{R_1 R_2}{R_1 + R_2} \qquad (6\text{–}5)$$

Equation (6–5) says that *the total resistance for two resistors in parallel is equal to the product of the two resistors divided by the sum of the two resistors.* This equation is sometimes referred to as the "product over the sum" formula. Example 6–7 will illustrate how to use it.

Example 6–7

Calculate the total resistance between the positive and negative terminals of the source of the circuit in Figure 6–15.

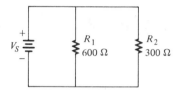

FIGURE 6–15

Solution:

Use Equation (6–5) as follows:

$$R_T = \frac{R_1 R_2}{R_1 + R_2} = \frac{(600 \ \Omega)(300 \ \Omega)}{600 \ \Omega + 300 \ \Omega}$$

$$= \frac{180{,}000 \ \Omega^2}{900 \ \Omega} = 200 \ \Omega$$

Resistors of Equal Value in Parallel

Another special case of parallel circuits is the parallel connection of several resistors having the same ohmic value. There is a shortcut method of calculating R_T when this case occurs.

If several resistors in parallel have the same resistance, they can be assigned the same symbol R. For example, $R_1 = R_2 = R_3 = \cdots = R_n = R$. Starting with Equation (6–4), we can develop a special formula for finding R_T:

$$R_T = \frac{1}{(1/R) + (1/R) + (1/R) + \cdots + (1/R)}$$

Notice that in the denominator, the same term, $1/R$, is added n times (n is the number of equal resistors in parallel). Therefore, the formula can be written as

$$R_T = \frac{1}{(n/R)}$$

Rewriting, we obtain

$$R_T = \frac{R}{n} \quad (6\text{--}6)$$

Equation (6–6) says that when any number of resistors (n), all having the same resistance (R), are connected in parallel, R_T is equal to the resistance divided by the number of resistors in parallel. Example 6–8 shows how to use this formula.

Example 6–8

Find the total resistance between points A and B in Figure 6–16.

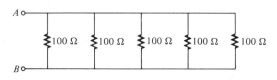

FIGURE 6–16

Solution:

There are five 100-Ω resistors in parallel. Use Equation (6–6) as follows:

$$R_T = \frac{R}{n} = \frac{100\ \Omega}{5} = 20\ \Omega$$

Determining an Unknown Parallel Resistor

Sometimes it is necessary to determine the values of resistors that are to be combined to produce a desired total resistance. For example, consider the case where *two* parallel resistors are used to obtain a desired total resistance. One resistor value is arbitrarily chosen, and then the second resistor value is calculated using Equation (6–7). This equation is derived from the formula for two parallel resistors as follows:

$$R_T = \frac{R_x R_A}{R_x + R_A}$$

$$R_T(R_x + R_A) = R_x R_A$$

$$R_T R_x + R_T R_A = R_x R_A$$

$$R_A R_x - R_T R_x = R_T R_A$$

$$R_x(R_A - R_T) = R_T R_A$$

$$R_x = \frac{R_A R_T}{R_A - R_T} \quad (6\text{--}7)$$

where R_x is the unknown resistor and R_A is the selected value. Example 6–9 illustrates use of this formula.

TOTAL PARALLEL RESISTANCE

Example 6–9

Suppose that you wished to obtain a resistance of 200 Ω by combining two resistors in parallel. There is a 300-Ω resistor available. What other value is needed?

Solution:

$$R_T = 200 \; \Omega \quad \text{and} \quad R_A = 300 \; \Omega$$

$$R_x = \frac{R_A R_T}{R_A - R_T} = \frac{(300 \; \Omega)(200 \; \Omega)}{300 \; \Omega - 200 \; \Omega}$$

$$= 600 \; \Omega$$

Notation for Parallel Resistors

Sometimes, for convenience, parallel resistors are designated by two parallel vertical marks. For example, R_1 in parallel with R_2 can be written as $R_1 \| R_2$. Also, when several resistors are in parallel with each other, this notation can be used. For example, $R_1 \| R_2 \| R_3 \| R_4 \| R_5$ indicates that R_1 through R_5 are all in parallel.

This notation is also used with resistance values. For example,

$$10 \; \text{k}\Omega \| 5 \; \text{k}\Omega$$

means that 10 kΩ are in parallel with 5 kΩ.

Review for 6–3

1. Does the total resistance increase or decrease as more resistors are connected in parallel?
2. The total parallel resistance is always less than <u>smallest resistor</u>
3. From memory, write the general formula for R_T with any number of resistors in parallel. $R_T = \frac{R_1 R_2}{R_1 + R_2} = R_T = \frac{1}{\frac{1}{R_1} + \frac{1}{R_2} + \frac{1}{R_3}}$
4. Write the special formula for two resistors in parallel.
5. Write the special formula for any number of equal-value resistors in parallel. $R_T = R/N$
6. Calculate R_T for Figure 6–17.

FIGURE 6–17

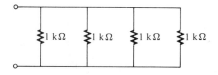

FIGURE 6–18

7. Determine R_T for Figure 6–18.

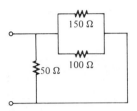

FIGURE 6–19

8. Find R_T for Figure 6–19.

6–4

APPLYING OHM'S LAW

In this section you will see how Ohm's law can be applied to parallel circuit problems. First you will learn how to find the total current in a parallel circuit, and then you will learn how to determine branch currents using Ohm's law. Let us start with Example 6–10. Then, in Example 6–11, we will use Ohm's law to find branch currents.

Example 6–10

Find the total current produced by the battery in Figure 6–20.

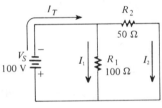

FIGURE 6–20

Solution:

The battery "sees" a total parallel resistance which determines the amount of current that it generates. First, we calculate R_T:

$$R_T = \frac{R_1 R_2}{R_1 + R_2} = \frac{(100\ \Omega)(50\ \Omega)}{100\ \Omega + 50\ \Omega}$$

$$= \frac{5000\ \Omega^2}{150\ \Omega} = 33.33\ \Omega$$

The battery voltage is 100 V. Use Ohm's law to find I_T:

$$I_T = \frac{100\ \text{V}}{33.33\ \Omega} = 3\ \text{A}$$

APPLYING OHM'S LAW

Example 6–11

Determine the current through each resistor in the parallel circuit of Figure 6–21.

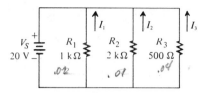

FIGURE 6–21

Solution:

The voltage across each resistor (branch) is equal to the source voltage. That is, the voltage across R_1 is 20 V, the voltage across R_2 is 20 V, and the voltage across R_3 is 20 V. The current through each resistor is determined as follows:

$$I_1 = \frac{V_S}{R_1} = \frac{20 \text{ V}}{1 \text{ k}\Omega} = 20 \text{ mA}$$

$$I_2 = \frac{V_S}{R_2} = \frac{20 \text{ V}}{2 \text{ k}\Omega} = 10 \text{ mA}$$

$$I_3 = \frac{V_S}{R_3} = \frac{20 \text{ V}}{500 \text{ }\Omega} = 40 \text{ mA}$$

In Example 6–12, we use Ohm's law to determine the unknown voltage across a parallel circuit.

Example 6–12

Find the voltage across the parallel circuit in Figure 6–22.

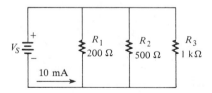

FIGURE 6–22

Solution:

We know the total current into the parallel circuit. We need to know the

Example 6–12 (continued)

total resistance, and then we can apply Ohm's law to get the voltage. The total resistance is

$$R_T = \frac{1}{(1/R_1) + (1/R_2) + (1/R_3)} = \frac{1}{0.005 \text{ S} + 0.002 \text{ S} + 0.001 \text{ S}}$$

$$= \frac{1}{0.008 \text{ S}} = 125 \text{ }\Omega$$

$$V_S = I_T R_T = (10 \text{ mA})(125 \text{ }\Omega)$$

$$= 1.25 \text{ V}$$

Review for 6–4

1. A 10-V battery is connected across three 60-Ω resistors that are in parallel. What is the total current from the battery?
2. How much voltage is required to produce 2 A of current through the circuit of Figure 6–23?

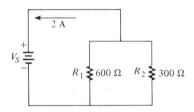

FIGURE 6–23

3. How much current is there through each resistor of Figure 6–23?
4. There are four equal-value resistors in parallel with a 12-V source, and 6 mA of current flow from the source. What is the value of each resistor?
5. A 1-kΩ and a 2-kΩ resistor are connected in parallel. A total of 100 mA is through the parallel combination. How much voltage is dropped across the resistors?

6–5

CURRENT SOURCES IN PARALLEL

A current source is an energy source that provides a constant value of current to a load. A transistor can be used as a current source, and thus current sources are important in circuit analysis. At this point, you are not prepared to study transistor circuits in detail, but you do need to understand how current sources act in circuits.

CURRENT SOURCES IN PARALLEL

The only practical case that we need to consider is current sources connected in parallel.

The general rule to remember is that the total current produced by current sources in parallel is equal to the algebraic sum of the individual current sources. The *algebraic sum* means that you must consider the direction of current flow when combining the sources in parallel. For example, in Figure 6–24A, the three current sources in parallel provide current in the same direction (into point A). So the total current into point A is $I_T = 1\text{ A} + 2\text{ A} + 2\text{ A} = 5\text{ A}$.

In Figure 6–24B, the 1-A source provides current in a direction opposite to the other two. The total current into point A in this case is $I_T = 2\text{ A} + 2\text{ A} - 1\text{ A} = 3\text{ A}$.

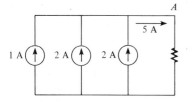

A.

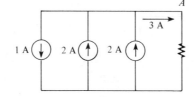

B.

FIGURE 6–24

Example 6–13

Determine the current through R_L in Figure 6–25.

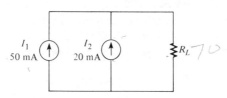

FIGURE 6–25

Solution:

The two current sources are in the same direction; so the current through R_L is

$$I_L = I_1 + I_2 = 50\text{ mA} + 20\text{ mA} = 70\text{ mA}$$

Review for 6–5

1. Four 0.5-A current sources are connected in parallel in the same direction. What current will be produced through a load resistor?

2. How many 1-A current sources must be connected in parallel to produce a total current output of 3 A? Sketch a schematic showing the sources connected.

3. In a transistor amplifier circuit, the transistor can be represented by a 10-mA current source, as shown in Figure 6–26. The transistors act in parallel, as in a differential amplifier. How much current is there through the resistor R_E?

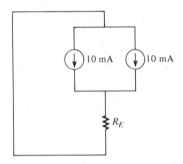

FIGURE 6–26

6–6

KIRCHHOFF'S CURRENT LAW

Kirchhoff's current law states that *the sum of the currents into a junction is equal to the sum of the currents out of that junction.* A *junction* is any point in a circuit where two or more circuit paths come together. In a parallel circuit, a junction is where the parallel branches connect together. Another way to state Kirchhoff's current law is to say that *the total current into a junction is equal to the total current out of that junction.*

For example, in the circuit of Figure 6–27, point A is one junction and point B is another. Let us start at the negative terminal of the source and follow the current. The total current I_T is out of the source and *into* the junction at point A. At

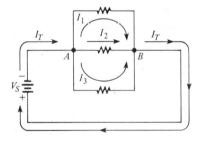

FIGURE 6–27

KIRCHHOFF'S CURRENT LAW

this point, the current splits up among the three branches as indicated. Each of the three branch currents (I_1, I_2, and I_3) is *out of* junction A. Kirchhoff's current law says that the total current into junction A is equal to the total current out of junction A; that is

$$I_T = I_1 + I_2 + I_3$$

Now, following the currents in Figure 6–27 through the three branches, you see that they come back together at point B. Currents I_1, I_2, and I_3 are into junction B, and I_t is out. Kirchhoff's current law formula at this junction is therefore the same as at junction A:

$$I_T = I_1 + I_2 + I_3$$

General Formula for Kirchhoff's Current Law

The previous discussion was a specific case to illustrate Kirchhoff's current law. Now let us look at the general case. Figure 6–28 shows a *generalized* circuit junction where a number of branches are connected to a point in the circuit. Currents $I_{in(1)}$ through $I_{in(n)}$ are into the junction (*n* can be any number). Currents $I_{out(1)}$ through $I_{out(m)}$ are out of the junction (*m* can be any number, but not necessarily equal to *n*). By Kirchhoff's current law, the sum of the currents into a junction must equal the sum of the currents out of the junction. With reference to Figure 6–28, the general formula for Kirchhoff's current law is

$$I_{in(1)} + I_{in(2)} + \cdots + I_{in(n)} = I_{out(1)} + I_{out(2)} + \cdots + I_{out(m)} \qquad (6\text{–}8)$$

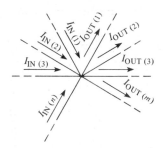

FIGURE 6–28 *Generalized circuit junction.*

If all of the terms on the right side of Equation (6–8) are brought over to the left side, their signs change to negative, and a zero is left on the right side. Kirchhoff's current law is sometimes stated in this way: *The algebraic sum of all the currents entering and leaving a junction is equal to zero.* This statement is just another, equivalent way of stating what we have just discussed. Some examples will illustrate the use of Kirchhoff's current law.

Example 6–14

The branch currents in the circuit of Figure 6–29 are known. Determine the total current entering junction A and the total current leaving junction B.

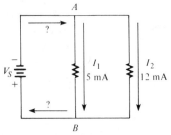

FIGURE 6–29

Solution:

The total current out of junction A is the sum of the two branch currents. So the total current into A is

$$I_T = I_1 + I_2 = 5 \text{ mA} + 12 \text{ mA} = 17 \text{ mA}$$

The total current entering point B is the sum of the two branch currents. So the total current out of B is

$$I_T = I_1 + I_2 = 5 \text{ mA} + 12 \text{ mA} = 17 \text{ mA}$$

Example 6–15

Determine the current through R_2 in Figure 6–30.

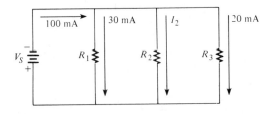

FIGURE 6–30

Solution:

The total current into the junction of the three branches is known. Two of the branch currents are known. The current equation at this junction is

$$I_T = I_1 + I_2 + I_3$$

KIRCHHOFF'S CURRENT LAW

Solving for I_2, we get

$$I_2 = I_T - I_1 - I_3 = 100 \text{ mA} - 30 \text{ mA} - 20 \text{ mA}$$
$$= 50 \text{ mA}$$

Example 6–16

Find I_1 in Figure 6–31.

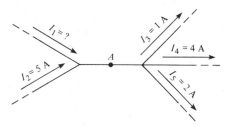

FIGURE 6–31

Solution:

Currents I_1 and I_2 are into junction A, and I_3, I_4, are out. Kirchhoff's current equation at this junction is

$$I_1 + I_2 = I_3 + I_4 + I_5$$

Solving for I_1, we have

$$I_1 = I_3 + I_4 + I_5 - I_2$$

Now we substitute the known current values into the equation and get I_1 as follows:

$$I_1 = 1 \text{ A} + 4 \text{ A} + 2 \text{ A} - 5 \text{ A} = 2 \text{ A}$$

Review for 6–6

1. State Kirchhoff's current law in two ways.
2. There is a total current of 2.5 A into the junction of three parallel branches. What is the sum of all three branch currents?
3. In Figure 6–32, 100 mA and 300 mA are into the junction. What is the amount of current out of the junction?

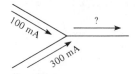

FIGURE 6–32

4. Determine I_1 in the circuit of Figure 6–33.

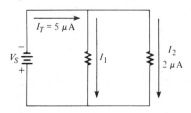

FIGURE 6–33

5. Two branch currents enter a junction, and two branch currents leave the same junction. One of the currents into the junction is 1 A, and one of the currents out of the junction is 3 A. The total current into and out of the junction is 8 A. Determine the value of the unknown current into the junction and the value of the unknown current out of the junction.

6–7

CURRENT DIVIDERS

A parallel circuit is a *current divider*. In this section you will see what this term means and why current dividers are an important application of parallel circuits.

To understand how a parallel connection of resistors acts as a current divider, look at Figure 6–34, where there are two resistors in parallel. As you already know, there is a current through R_1 and a current through R_2. We call these branch currents I_1 and I_2, respectively, as indicated in the diagram.

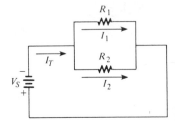

FIGURE 6–34 *Total current divides between two branches.*

Since the same voltage is across each of the resistors in parallel, the branch currents are inversely proportional to the ohmic values of the resistors. For example, if the value of R_2 is twice that of R_1, then the value of I_2 is one-half that of I_1. In other words, *the total current divides among parallel resistors in a manner inversely proportional to the resistance values.* The branches with higher resistance have less

CURRENT DIVIDERS

current, and the branches with lower resistance have more current, in accordance with Ohm's law.

General Current Divider Formula

With a few steps, a formula for determining how currents divide among parallel resistors can be developed. Let us assume that we have several resistors in parallel, as shown in Figure 6–35. This figure shows five resistors, but there can be any number.

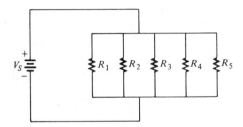

FIGURE 6–35 *Current divider.*

Let us call the current through any one of the parallel resistors I_x, where x represents the number of a particular resistor (1, 2, 3, and so on). By Ohm's law, the current through any one of the resistors in Figure 6–35 can be written as follows:

$$I_x = \frac{V_s}{R_x}$$

where $x = 1, 2, 3, 4,$ or 5. The source voltage V_s appears across each of the parallel resistors, and R_x represents any one of the parallel resistors. The total source voltage V_s is equal to the total current times the total parallel resistance:

$$V_s = I_T R_T$$

Substituting $I_T R_T$ for V_s in the expression for I_x, we get

$$I_x = \frac{I_T R_T}{R_x}$$

By rearranging, we get

$$I_x = \left(\frac{R_T}{R_x}\right) I_T \tag{6-9}$$

Equation (6–9) is the general current divider formula. It tells us that the current (I_x) through any branch equals the total parallel resistance (R_T), divided by the resistance (R_x) of that branch and multiplied by the total current (I_T) into the parallel junction. This formula applies to a parallel circuit with any number of branches.

Example 6–17 will illustrate the use of Equation (6–9).

Example 6-17

Determine the current through each resistor in the circuit of Figure 6-36.

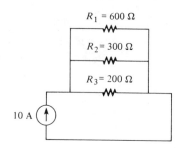

FIGURE 6-36

Solution:

First calculate the total parallel resistance:

$$R_T = \frac{1}{(1/R_1) + (1/R_2) + (1/R_3)}$$

$$= \frac{1}{(1/600\ \Omega) + (1/300\ \Omega) + (1/200\ \Omega)} = 100\ \Omega$$

The total current is 10 A. Using Equation (6-9), we calculate each branch current as follows:

$$I_1 = \left(\frac{R_T}{R_1}\right)I_T = \left(\frac{100\ \Omega}{600\ \Omega}\right)10\ \text{A}$$
$$= 1.67\ \text{A}$$

$$I_2 = \left(\frac{R_T}{R_2}\right)I_T = \left(\frac{100\ \Omega}{300\ \Omega}\right)10\ \text{A}$$
$$= 3.33\ \text{A}$$

$$I_3 = \left(\frac{R_T}{R_3}\right)I_T = \left(\frac{100\ \Omega}{200\ \Omega}\right)10\ \text{A}$$
$$= 5.00\ \text{A}$$

Current Divider Formula for Two Branches

For the special case of two parallel branches, Equation (6-9) can be modified. The reason for doing so is that two parallel resistors are often found in practical circuits. We will start by restating Equation (6-5), the formula for the total resistance of two parallel branches:

CURRENT DIVIDERS

$$R_T = \frac{R_1 R_2}{R_1 + R_2} \qquad (6\text{--}5)$$

When we have two parallel resistors, as in Figure 6–37, we may want to find the current through either or both of the branches. To do so, we need two special formulas.

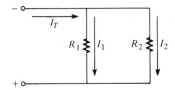

FIGURE 6–37

Using Equation (6–9), we write the formulas for I_1 and I_2 as follows:

$$I_1 = \left(\frac{R_T}{R_1}\right) I_T$$

$$I_2 = \left(\frac{R_T}{R_2}\right) I_T$$

Substituting $R_1 R_2/(R_1 + R_2)$ for R_T, we get

$$I_1 = \left(\frac{(\cancel{R_1} R_2)/(R_1 + R_2)}{\cancel{R_1}}\right) I_T$$

$$I_2 = \left(\frac{(R_1 \cancel{R_2})/(R_1 + R_2)}{\cancel{R_2}}\right) I_T$$

Canceling as shown, we get the final formulas:

$$I_1 = \left(\frac{R_2}{R_1 + R_2}\right) I_T \qquad (6\text{--}10)$$

$$I_2 = \left(\frac{R_1}{R_1 + R_2}\right) I_T \qquad (6\text{--}11)$$

When there are only *two* resistors in parallel, these equations are a little easier to use than Equation (6–9), because it is not necessary to know R_T.

Note that in Equations (6–10) and (6–11), the current in one of the branches is equal to the *opposite* branch resistance over the *sum* of the two resistors, all times the total current. In all applications of the current divider equations, you must know the total current going into the parallel branches. Example 6–18 will illustrate the use of these special current divider formulas.

Example 6–18

Find I_1 and I_2 in Figure 6–38.

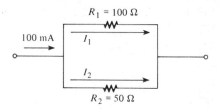

FIGURE 6–38

Solution:

Using Equation (6–10), we determine I_1 as follows:

$$I_1 = \left(\frac{R_2}{R_1 + R_2}\right)I_T = \left(\frac{50\ \Omega}{150\ \Omega}\right)100\ \text{mA}$$

$$= 33.33\ \text{mA}$$

Using Equation (6–11), we determine I_2 as follows:

$$I_2 = \left(\frac{R_1}{R_1 + R_2}\right)I_T = \left(\frac{100\ \Omega}{150\ \Omega}\right)100\ \text{mA}$$

$$= 66.67\ \text{mA}$$

Review for 6–7

1. Write the general current divider formula.
2. Write the two special formulas for calculating each branch current for a two-branch circuit.
3. A parallel circuit has the following resistors in parallel: 200-Ω, 100-Ω, 75-Ω, 50-Ω, and 22-Ω. Which resistor has the most current through it? The least current?
4. Determine the current through R_3 in Figure 6–39.

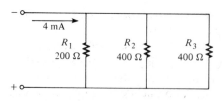

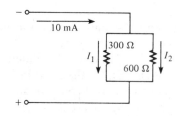

FIGURE 6–39

FIGURE 6–40

5. Find I_1 and I_2 in the circuit of Figure 6–40.

6-8 POWER IN A PARALLEL CIRCUIT

The total amount of power in a parallel resistive circuit is equal to the sum of the powers in each resistor in parallel. Equation (6–12) states this in a concise way for any number of resistors in parallel:

$$P_T = P_1 + P_2 + P_3 + \cdots + P_n \qquad (6\text{–}12)$$

where P_T is the total power and P_n is the power in the last resistor in parallel. As you can see, the power losses are additive, just as in the series circuit.

The power formulas that you learned in Chapter 4 are directly applicable to parallel circuits. The following formulas are used to calculate the total power P_T:

$$P_T = VI_T$$
$$P_T = I_T^2 R_T$$
$$P_T = \frac{V^2}{R_T}$$

where V is the voltage across the parallel circuit, I_T is the total current into the parallel circuit, and R_T is the total resistance of the parallel circuit. Example 6–19 shows how total power can be calculated in a parallel circuit.

Example 6–19

Determine the total amount of power in the parallel circuit in Figure 6–41.

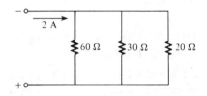

FIGURE 6–41

Solution:

We know that the total current is 2 A. The total resistance is

$$R_T = \frac{1}{(1/60 \ \Omega) + (1/30 \ \Omega) + (1/20 \ \Omega)} = 10 \ \Omega$$

The easiest formula to use is $P_T = I_T^2 R_T$ since we know both I_T and R_T. Thus,

$$P_T = I_T^2 R_T = (2 \text{ A})^2 (10 \ \Omega) = 40 \text{ W}$$

Example 6–19 (continued)

To demonstrate that if the power in each resistor is determined and if all of these values are added together, you get the same result, we will work through another calculation. First, find the voltage across each branch of the circuit:

$$V = I_T R_T = (2 \text{ A})(10 \text{ }\Omega) = 20 \text{ V}$$

Remember that the voltage across all branches is the same.

Next, calculate the power for each resistor using $P = V^2/R$:

$$P_1 = \frac{(20 \text{ V})^2}{60 \text{ }\Omega} = 6.67 \text{ W}$$

$$P_2 = \frac{(20 \text{ V})^2}{30 \text{ }\Omega} = 13.33 \text{ W}$$

$$P_3 = \frac{(20 \text{ V})^2}{20 \text{ }\Omega} = 20 \text{ W}$$

Now, add these powers to get the total power:

$$P_T = 6.67 \text{ W} + 13.33 \text{ W} + 20 \text{ W}$$
$$= 40 \text{ W}$$

This calculation shows that the sum of the individual powers is equal to the total power as determined by one of the power formulas.

Review for 6–8

1. If you know the power in each resistor in a parallel circuit, how can you find the total power?
2. The resistors in a parallel circuit have the following powers: 2 W, 5 W, 1 W, and 8 W. What is the total power in the circuit?
3. A circuit has a 1-kΩ, a 2-kΩ, and a 4-kΩ resistor in parallel. There is a total current of 1 A into the parallel circuit. What is the total power?

6–9

EFFECTS OF OPEN PATHS

Recall that an open circuit is one in which the current path is interrupted and there is no current. In this section we will examine what happens when a branch of a parallel circuit opens.

EFFECTS OF OPEN PATHS

If a switch is connected in a branch of a parallel circuit, as shown in Figure 6–42, an open or a closed path can be made by the switch. When the switch is closed, as in Figure 6–42A, R_1 and R_2 are in parallel. The total resistance is 50 Ω (two 100-Ω resistors in parallel). There is current through both resistors. If the switch is opened, as in Figure 6–42B, R_1 is effectively removed from the circuit, and the total resistance is 100 Ω. There is current now only through R_2.

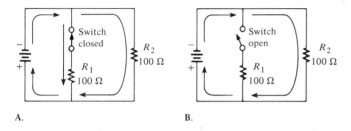

FIGURE 6–42

In general, *when an open circuit occurs in a parallel branch, the total resistance increases, the total current decreases, and curent continues through the remaining parallel paths.* The decrease in total current equals the amount of current that was previously in the open branch. The other branch currents remain the same. remain the same.

Consider the lamp circuit in Figure 6–43. There are four bulbs in parallel with a 120-V source. In Part A, there is current through each bulb. Now suppose that one of the bulbs burns out, creating an open path as shown in Figure 6–43B. This light will go out because there is no current through the open path. Notice, however, that current continues through all the other parallel bulbs, and they continue to glow. The open branch does not change the voltage across the parallel branches; it remains at 120 V.

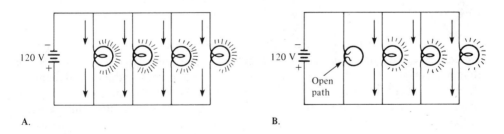

FIGURE 6–43

You can see that a parallel circuit has an advantage over a series connection in lighting systems because if one or more of the parallel bulbs burn out, the others will stay on. In a *series* circuit, when one bulb goes out, all of the others go out also because the current path is *completely* interrupted.

Review for 6-9

1. In a parallel circuit, what happens when one of the branches opens?
2. If several light bulbs are connected in parallel and one of the bulbs opens (burns out), will the others continue to glow?
3. In a parallel circuit, if one of the branch resistances opens, does the total resistance increase or decrease?
4. If a parallel branch opens, does the total current increase or decrease?
5. There is one ampere of current in each branch of a parallel circuit. If one branch opens, what is the current in each of the remaining branches?

6-10

TROUBLE-SHOOTING PARALLEL CIRCUITS

When a resistor in a parallel circuit opens, the open resistor cannot be located by measurement of the voltage across the branches, because the same voltage exists across all the branches. Thus, there is no way to tell which resistor is open by simply measuring voltage. The good resistors will always have the same voltage as the open one, as illustrated in Figure 6-44.

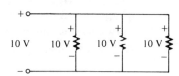

FIGURE 6-44 *Parallel branches will all have the same voltage.*

If a visual inspection does not reveal the open resistor, it must be located by current measurements. In practice, measuring current is more difficult than measuring voltage because you must insert the ammeter *in* the line to measure the current. Thus, a wire or a PC connection must be cut or disconnected, or one end of a component must be lifted off the circuit board, in order to connect the ammeter in the line. This procedure, of course, is not required when voltage measurements are made, because the meter leads are simply connected *across* a component.

Where to Measure the Current

In a parallel circuit, the total current should be measured. When a parallel resistor opens, I_T is less than its normal value. Once I_T and the voltage across the branches are known, a few calculations will determine the open resistor when all the resistors are of different ohmic values.

Consider the two-branch circuit in Figure 6-45A. If one of the resistors opens, the total current will equal the current in the good resistor. Ohm's law quickly tells us what the current in each resistor should be:

TROUBLE-SHOOTING PARALLEL CIRCUITS

$$I_1 = \frac{50 \text{ V}}{500 \text{ }\Omega} = 0.1 \text{ A}$$

$$I_2 = \frac{50 \text{ V}}{100 \text{ }\Omega} = 0.5 \text{ A}$$

If R_2 is open, the total current is 0.1 A, as indicated in Figure 6–45B. If R_1 is open, the total current is 0.5 A, as indicated in Figure 6–45C.

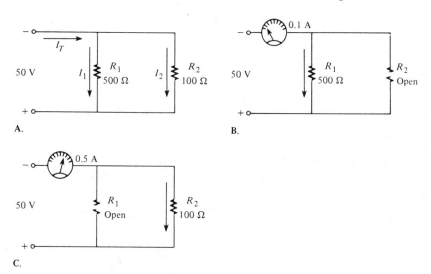

FIGURE 6–45

This procedure can be extended to any number of branches having unequal resistances. If the parallel resistances are all equal, the current in each branch must be checked until a branch is found with no current. This is the open resistor.

Example 6–20

In Figure 6–46, there is a total current of 32 mA, and the voltage across the parallel branches is 20 V. Is there an open resistor, and, if so, which one is it?

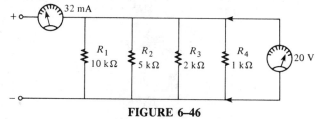

FIGURE 6–46

Solution:

Calculate the current in each branch:

Example 6–20 (continued)

$$I_1 = \frac{V}{R_1} = \frac{20 \text{ V}}{10 \text{ k}\Omega} = 2 \text{ mA}$$

$$I_2 = \frac{V}{R_2} = \frac{20 \text{ V}}{5 \text{ k}\Omega} = 4 \text{ mA}$$

$$I_3 = \frac{V}{R_3} = \frac{20 \text{ V}}{2 \text{ k}\Omega} = 10 \text{ mA}$$

$$I_4 = \frac{V}{R_4} = \frac{20 \text{ V}}{1 \text{ k}\Omega} = 20 \text{ mA}$$

The total current *should* be

$$I_T = I_1 + I_2 + I_3 + I_4 = 2 \text{ mA} + 4 \text{ mA} + 10 \text{ mA} + 20 \text{ mA}$$
$$= 36 \text{ mA}$$

The actual *measured* current is 32 mA, as stated, which is 4 mA less than normal, indicating that the branch carrying 4 mA is open. Thus, R_2 must be open.

Review for 6–10

1. If a parallel branch opens, what changes can be detected in the circuit voltage and the currents, assuming that the parallel circuit is across an ideal voltage source?
2. What happens to the total resistance if one branch opens?

Formulas

$$G = \frac{1}{R} \tag{6-1}$$

$$G_T = G_1 + G_2 + G_3 + \cdots + G_n \tag{6-2}$$

$$R_T = \frac{1}{(1/R_1) + (1/R_2) + (1/R_3) + \cdots + (1/R_n)} \tag{6-4}$$

$$R_T = \frac{R_1 R_2}{R_1 + R_2} \tag{6-5}$$

$$R_T = \frac{R}{n} \tag{6-6}$$

$$R_x = \frac{R_A R_T}{R_A - R_T} \tag{6-7}$$

$$I_{\text{in}(1)} + I_{\text{in}(2)} + \cdots + I_{\text{in}(n)} = I_{\text{out}(1)} + I_{\text{out}(2)} + \cdots + I_{\text{out}(m)} \tag{6-8}$$

$$I_x = \left(\frac{R_T}{R_x}\right)I_T \quad (6\text{–}9)$$

$$I_1 = \left(\frac{R_2}{R_1 + R_2}\right)I_T \quad (6\text{–}10)$$

$$I_2 = \left(\frac{R_1}{R_1 + R_2}\right)I_T \quad (6\text{–}11)$$

$$P_T = P_1 + P_2 + P_3 + \cdots + P_n \quad (6\text{–}12)$$

Summary

1. Resistors in parallel are connected across the same points.
2. A parallel combination has more than one path for current.
3. The number of current paths equals the number of resistors in parallel.
4. The total parallel resistance is less than the lowest-value resistor.
5. The voltages across all branches of a parallel circuit are the same.
6. Current sources in parallel add algebraically.
7. One way to state Kirchhoff's current law: The algebraic sum of all the currents at a junction is zero.
8. Another way to state Kirchhoff's current law: The sum of the currents into a junction (total current in) equals the sum of the currents out of the junction (total current out).
9. A parallel circuit is a current divider, so called because the total current entering the parallel junction divides up into each of the branches.
10. If all of the branches of a parallel circuit have equal resistance, the currents through all of the branches are equal.
11. The total power in a parallel resistive circuit is the sum of all of the individual powers of the resistors making up the parallel circuit.
12. The total power for a parallel circuit can be calculated with the power formulas using values of total current, total resistance, or total voltage.
13. If one of the branches of a parallel circuit opens, the total resistance increases, and therefore the total current decreases.
14. If a branch of a parallel circuit opens, there is still current through the remaining branches.

Self-Test

1. Sketch a parallel circuit having four resistors and a voltage source.
2. There are 5 amperes of current into a parallel circuit having two branches of equal resistance. How much is the current through each of the branches?

3. Connect each set of resistors in Figure 6–47 in parallel between points A and B.

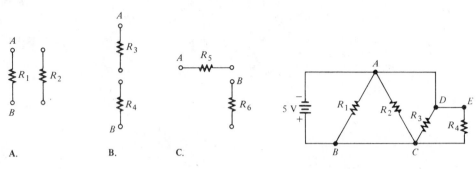

A. B. C.

FIGURE 6–47

FIGURE 6–48

4. If you measured the voltage across each of the points marked in Figure 6–48, how much voltage would you read in each case?

5. If an 80-Ω resistor and a 150-Ω resistor are connected in parallel, what is the total resistance?

6. The following resistors are in parallel: 1000-Ω, 800-Ω, 500-Ω, 200-Ω, and 100-Ω. What is the total resistance?

7. Eight 56-Ω resistors are connected in parallel. Determine the total resistance.

8. Suppose that you need a total resistance of 100 Ω. The only resistors that are immediately available are one 200-Ω and several 400-Ω resistors. How would you connect these resistors to get a total of 100 Ω?

9. Three 600-Ω resistors are connected in parallel, and 5 V are applied across the parallel circuit. How much current is there out of the source?

10. Which circuit of Figure 6–49 has more total current?

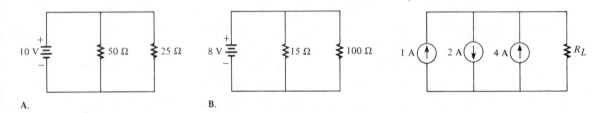

A.

B.

FIGURE 6–49

FIGURE 6–50

11. Six resistors of equal value are in parallel, with 12 V across them. The total current is 3 mA. Determine the value of each resistor. What is the total resistance?

12. Determine the current through R_L in the multiple-current-source circuit of Figure 6–50.

13. Two 1-A current sources are connected in opposite directions in parallel. What is the total current produced?

PROBLEMS

14. Each branch in a five-branch parallel circuit has 25 mA of current through it. What is the total current into the parallel circuit?

15. The total current into the junction of three parallel resistors is 0.5 A. One branch has a current of 0.1 A, and another branch has a current of 0.2 A. What is the current through the third branch?

16. Determine the unknown resistances in the circuit of Figure 6–51.

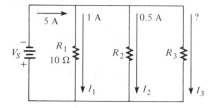

FIGURE 6–51

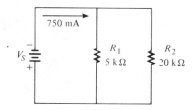

FIGURE 6–52

17. The following resistors are connected in parallel: 2-kΩ, 6-kΩ, 3-kΩ, and 1-kΩ. The total current is 1 A. Using the general current divider formula, calculate the current through each of the resistors.

18. Determine the current through each resistor in Figure 6–52.

19. Determine the total power in the circuit of Figure 6–52.

20. There are two resistors in parallel across a 10-V source. One is 200 Ω, and the other is 500 Ω. You measure a total current of 20 mA. What is wrong with the circuit?

21. What resistance value in parallel with 100 Ω produces a total resistance of 40 Ω?

Problems

6–1. Connect the resistors in Figure 6–53 in parallel across the battery.

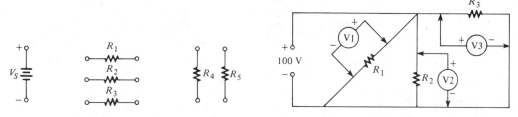

FIGURE 6–53

FIGURE 6–54

6–2. What is the current through each parallel resistor if the total voltage is 12 V and the total resistance is 600 Ω? There are four resistors, all of equal value.

6–3. The source voltage in Figure 6–54 is 100 V. How much voltage does each of the meters read?

6–4. The following resistors are connected in parallel: 1-MΩ, 2-MΩ, 5-MΩ, 12-MΩ, and 20-MΩ. Determine the total resistance.

6–5. Find the total resistance for each following group of parallel resistors:
 (a) 560-Ω and 1000-Ω.
 (b) 47-Ω and 56-Ω.
 (c) 1.5-kΩ, 2.2-kΩ, 10-kΩ.
 (d) 1-MΩ, 470-kΩ, 1-kΩ, 2.5-MΩ.

6–6. Calculate R_T for each circuit in Figure 6–55.

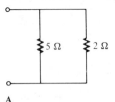
A.
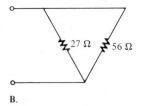
B.
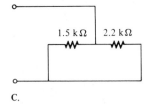
C.

FIGURE 6–55

6–7. If the total resistance in Figure 6–56 is 200 Ω, what is the value of R_2?

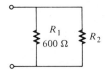

FIGURE 6–56

6–8. What is the total resistance of twelve 6-kΩ resistors in parallel?

6–9. Five 50-Ω, ten 100-Ω, and two 10-Ω resistors are all connected in parallel. What is the total resistance?

6–10. What is the total current in each circuit of Figure 6–57?

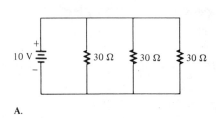
A.
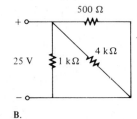
B.

FIGURE 6–57

6–11. Three 33-Ω resistors are connected in parallel with a 100-V source. How much current is there from the source?

6–12. Four equal-value resistors are connected in parallel. Five volts are applied across the parallel circuit, and 2.5 mA are measured from the source. What is the value of each resistor?

6–13. A 10-mA and a 20-mA current source are connected in parallel in the same direction. What is the total current that can be provided to a load?

PROBLEMS

6–14. Determine the current through R_L in each circuit in Figure 6–58.

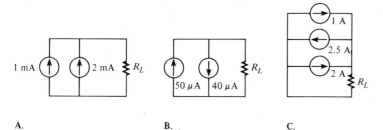

FIGURE 6–58

6–15. The following currents are measured in the same direction in a three-branch parallel circuit: 250-mA, 300-mA, and 800-mA. What is the value of the current into the junction of these three branches?

6–16. There are five hundred milliamperes (500 mA) into five parallel resistors. The currents through four of the resistors are 50 mA, 150 mA, 25 mA, and 100 mA. How much current is there through the fifth resistor?

6–17. In the circuit of Figure 6–59, determine the resistances R_2, R_3, and R_4.

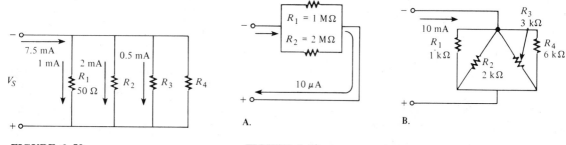

FIGURE 6–59 **FIGURE 6–60**

6–18. The total resistance of a parallel circuit is 25 Ω. How much current is there through a 200-Ω resistor that makes up part of the parallel circuit if the total current is 100 mA?

6–19. Determine the current in each branch of the current dividers of Figure 6–60.

6–20. What is the current through each resistor in Figure 6–61? R is the lowest-value resistor, and all others are multiples of that value as indicated.

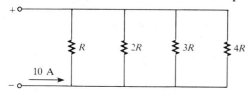

FIGURE 6–61

6–21. Five parallel resistors each handle 40 mW. What is the total power?

6–22. Determine the total power in each circuit of Figure 6–60.

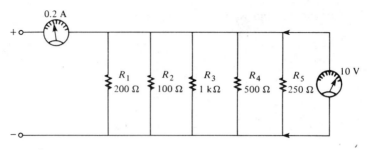

FIGURE 6-62

6-23. Six light bulbs are connected in parallel across 110 V. Each bulb is rated at 75 W. How much current is there through each bulb, and what is the total current?

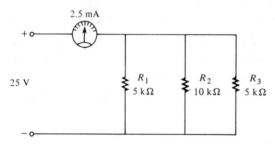

FIGURE 6-63

6-24. If one of the bulbs burns out in Problem 6-23, how much current is there through each of the remaining bulbs? What will the total current be?

6-25. In Figure 6-62, the current and voltage measurements are indicated. Has a resistor opened, and, if so, which one?

6-26. What is wrong with the circuit in Figure 6-63?

Answers to Section Reviews

Section 6-1:
1. Between the same two points. 2. More than one current path between two given points. 3. See Figure 6-64. 4. See Figure 6-65.

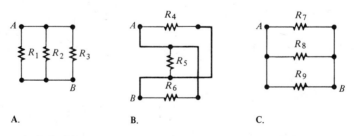

FIGURE 6-64

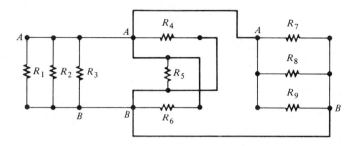

FIGURE 6–65

Section 6–2:
1. 5 V. 2. 118 V, 118 V. 3. 50 V and 50 V, respectively. 4. Voltage is the same across all branches.

Section 6–3:
1. Decrease. 2. the smallest resistance value. 3. The equation is as follows:

$$R_T = \frac{1}{(1/R_1) + (1/R_2) + \cdots + (1/R_n)}$$

4. $R_T = \dfrac{(R_1 R_2)}{(R_1 + R_2)}$

5. $R_T = R/n$. 6. 667 Ω. 7. 250 Ω. 8. 27.27 Ω.

Section 6–4:
1. 0.5 A. 2. 400 V. 3. 0.667 A through the 600-Ω; 1.33 A through the 300-Ω.
4. 8 kΩ. 5. 66.67 V.

Section 6–5:
1. 2 A. 2. Three. See Figure 6–66. 3. 20 mA.

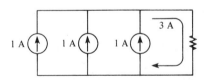

FIGURE 6–66

Section 6–6:
1. (a) The algebraic sum of all the currents at a junction is zero. (b) The sum of the currents entering a junction equals the sum of the currents leaving that junction.
2. 2.5 A. 3. 400 mA. 4. 3 μA. 5. 7 A entering, 5 A leaving.

Section 6–7:
1. The equation is as follows:

$$I_x = \left(\frac{R_T}{R_x}\right) I_T$$

2. The equations are as follows:

$$I_1 = \left(\frac{R_2}{R_1 + R_2}\right)I_T$$

$$I_2 = \left(\frac{R_1}{R_1 + R_2}\right)I_T$$

3. The 22-Ω has most current; the 200-Ω has least current. **4.** 1 mA. **5.** $I_1 = 6.67$ mA, $I_2 = 3.33$ mA.

Section 6–8:
1. Add the power of each resistor. **2.** 16 W. **3.** 571.43 W.

Section 6–9:
1. Current continues in the other branches. **2.** Yes. **3.** Increase. **4.** Decrease. **5.** 1 A.

Section 6–10:
1. There is no change in voltage. The total current decreases. **2.** It increases.

Circuits having some combination of *both* series and parallel resistors are often found in practice. In this chapter we will examine different combinations of series-parallel circuits. You will learn to apply the circuit laws and principles learned in previous chapters to these more complex circuits.

The concept of circuit ground will be introduced, and some problems in trouble-shooting for open and shorted paths will be presented to give you experience in practical situations. Special types of series-parallel circuits called *ladder networks* and *bridges* are also covered.

7–1 Definition of Series-Parallel Circuit
7–2 Circuit Identification
7–3 Analysis of Series-Parallel Circuits
7–4 Circuit Ground
7–5 Trouble-Shooting
7–6 Voltage Dividers with Resistive Loads
7–7 Ladder Networks
7–8 Wheatstone Bridge

7
SERIES-PARALLEL COMBINATIONS

7–1

DEFINITION OF SERIES-PARALLEL CIRCUIT

A *series-parallel circuit* consists of a combination of both series and parallel current paths. Figure 7–1A illustrates a simple series-parallel combination of resistors. Observe that R_2 and R_3 are in parallel and that this parallel combination is in series with R_1.

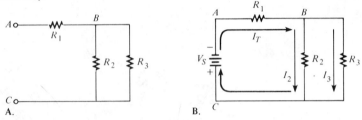

FIGURE 7–1 A simple series-parallel circuit.

If the circuit of Figure 7–1A is connected to a voltage source, the *total* current is through the resistor R_1 and it divides at point B. Part of it is through each of the parallel resistors R_2 and R_3. The current is shown in Figure 7–1B. The voltage across R_2 is the same as the voltage across R_3.

Review for 7–1

1. Define *series-parallel resistive circuit*.
2. A certain series-parallel circuit is described as follows: R_1 and R_2 are in parallel. This parallel combination is in series with another parallel combination of R_3 and R_4. Sketch the circuit.
3. In the circuit of Figure 7–2, describe the series-parallel relationships of the resistors.

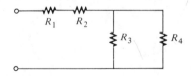

FIGURE 7–2

7–2

CIRCUIT IDENTIFICATION

In this section you will learn to distinguish the series and the parallel relationships of various circuit arrangements. Keep the following rules in mind as you analyze various circuits:

CIRCUIT IDENTIFICATION

1. When the total current between two points is through a resistor or a combination of resistors, then that resistor or combination of resistors is in series with any other resistive combination that also appears between those two points.

2. When the total current between two points divides and goes through more than one branch, then those branches are in parallel with each other, if the same voltage appears across each one.

Now let us examine several examples of series-parallel circuits in order to learn to recognize the component relationships.

Example 7–1

Identify the series-parallel relationships in Figure 7–3

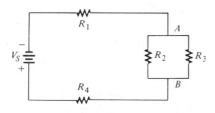

FIGURE 7–3

Solution:

Starting at the negative terminal of the source, follow the current paths. All of the current produced by the source must go through R_1. Therefore, R_1 is in series with the rest of the circuit.

The total current takes two paths when it gets to point A. Part of it goes through R_2, and part of it through R_3; the same voltage is across both. Therefore, R_2 and R_3 are in parallel with each other. This parallel combination is in series with R_1.

At point B, the currents through R_2 and R_3 come together again. Thus, the total current is through R_2, and so R_4 is in series with R_1 and the parallel combination of R_2 and R_3. The current is shown in Figure 7–4, where I_T is the total current.

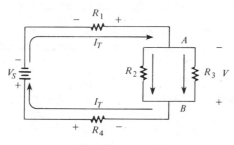

FIGURE 7–4

Example 7–1 (continued)

In summary, R_1 and R_4 are in series with the parallel combination of R_2 and R_3.

Example 7–2

Identify the series-parallel relationships in Figure 7–5.

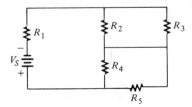

FIGURE 7–5

Solution:

Sometimes it is easier to see a particular circuit arrangement if it is drawn in a different way. In this case, the circuit schematic is redrawn in Figure 7–6, which better illustrates the series-parallel relationships. Now you can see that R_2 and R_3 are in parallel with each other and also that R_4 and R_5 are in parallel with each other. Both parallel combinations are in series with each other and with R_1.

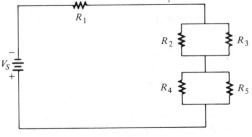

FIGURE 7–6

Example 7–3

Describe the series-parallel combination between points A and D in Figure 7–7.

CIRCUIT IDENTIFICATION

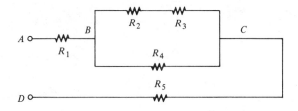

FIGURE 7–7

Solution:

Between points B and C, there are two parallel paths. The lower path consists of R_4, and the upper path consists of a *series* combination of R_2 and R_3. This parallel combination is in series with both R_1 and R_5.

In summary, R_1 and R_5 are in series with the parallel combination of R_4 and $R_2 + R_3$.

Example 7–4

Describe the total resistance between each pair of points in Figure 7–8.

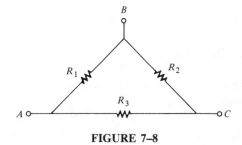

FIGURE 7–8

Solution:

From point A to B: R_1 is in *parallel* with the *series* combination of R_2 and R_3.

From point A to C: R_3 is in *parallel* with the *series* combination of R_1 and R_2.

From point B to C: R_2 is in *parallel* with the *series* combination of R_1 and R_3.

Review for 7–2

1. Describe the resistor combination in Figure 7–9, shown on page 176.

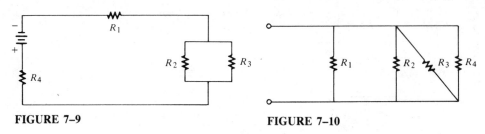

FIGURE 7–9 FIGURE 7–10

2. Which resistors are in parallel in Figure 7–10?
3. Describe the parallel arrangements in Figure 7–11.

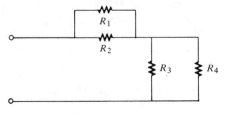

FIGURE 7–11

4. Are the parallel combinations in Figure 7–11 in series?

7–3

ANALYSIS OF SERIES-PARALLEL CIRCUITS

Several quantities are important when you have a circuit that is a series-parallel configuration of resistors. In this section you will learn how to determine total resistance, total current, branch currents, and the voltage across any portion of a circuit.

Determining Total Resistance

In Chapter 5, you learned how to determine total series resistance. In Chapter 6, you learned how to determine total parallel resistance.

To find the total resistance R_T of a series-parallel combination, simply define the series and parallel relationships, and then perform the calculations that you have previously learned. We will solve example problems to illustrate the general approach.

Example 7–5

Determine R_T of the circuit in Figure 7–12 between points A and B.

ANALYSIS OF SERIES-PARALLEL CIRCUITS

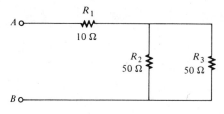

FIGURE 7-12

Solution:

First calculate the equivalent parallel resistance of R_2 and R_3. Since R_2 and R_3 are equal in value, we can use Equation (6-6):

$$R_{eq} = \frac{R}{n} = \frac{50 \, \Omega}{2} = 25 \, \Omega$$

Notice that we used the term R_{eq} here to designate the total resistance of a *portion* of a circuit in order to distinguish it from the total resistance R_T of the *complete* circuit.

Now, since R_1 is in series with R_{eq}, their values are added as follows:

$$R_T = R_1 + R_{eq} = 10 \, \Omega + 25 \, \Omega = 35 \, \Omega$$

Example 7-6

Find the total resistance between the negative and positive terminals of the battery in Figure 7–13.

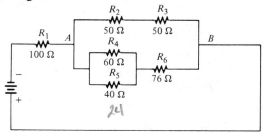

FIGURE 7-13

Solution:

In the *upper branch*, R_2 is in series with R_3. We will call the series combination R_{eq1}. It is equal to $R_2 + R_3$:

$$R_{eq1} = R_2 + R_3 = 50 \, \Omega + 50 \, \Omega = 100 \, \Omega$$

Example 7–6 (continued)

In the *lower branch*, R_4 and R_5 are in parallel with each other. We will call this parallel combination R_{eq2}. It is calculated as follows:

$$R_{eq2} = \frac{R_4 R_5}{R_4 + R_5} = \frac{(60\ \Omega)(40\ \Omega)}{60\ \Omega + 40\ \Omega} = 24\ \Omega$$

Also in the *lower branch*, the parallel combination of R_4 and R_5 is in series with R_6. This series-parallel combination is designated R_{eq3} and is calculated as follows:

$$R_{eq3} = R_6 + R_{eq2} = 76\ \Omega + 24\ \Omega = 100\ \Omega$$

Figure 7–14 shows the original circuit in a simplified *equivalent* form.

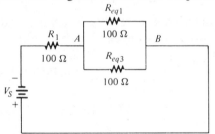

FIGURE 7–14

Now we can find the equivalent resistance between points A and B. It is R_{eq1} in parallel with R_{eq3}. Since these resistances are equal, the equivalent resistance is calculated as follows:

$$R_{AB} = \frac{100\ \Omega}{2} = 50\ \Omega$$

Finally, the total resistance is R_1 in series with R_{AB}:

$$R_T = R_1 + R_{AB} = 100\ \Omega + 50\ \Omega = 150\ \Omega$$

Determining Total Current

Once the total resistance and the source voltage are known, we can find total current in a circuit by applying Ohm's law. Total current is the total source voltage divided by the total resistance:

$$I_T = \frac{V_S}{R_T}$$

For example, let us find the total current in the circuit of Example 7–6 (Figure 7–13). Assume that the source voltage is 30 V. The calculation is as follows:

ANALYSIS OF SERIES-PARALLEL CIRCUITS

$$I_T = \frac{V_S}{R_T} = \frac{30 \text{ V}}{150 \text{ }\Omega} = 0.2 \text{ A}$$

In this case the source "sees" 150 Ω and therefore produces a current of 0.2 A.

Determining Branch Currents

Using the *current divider formula*, or *Kirchhoff's current law*, or *Ohm's law*, or combinations of these, we can find the current in any branch of a series-parallel circuit. In some cases it may take repeated application of the formula to find a given current. Working through some examples will help give you an understanding of the procedure.

Example 7–7

Find the current through R_2 and the current through R_3 in Figure 7–15.

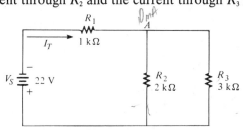

FIGURE 7–15

Solution:

First we need to know how much current is into the junction (point A) of the parallel branches. This is the total circuit current. To find I_T, we need to know R_T:

$$R_T = R_1 + \frac{R_2 R_3}{R_2 + R_3} = 1 \text{ k}\Omega + \frac{(2 \text{ k}\Omega)(3 \text{ k}\Omega)}{2 \text{ k}\Omega + 3 \text{ k}\Omega}$$

$$= 1 \text{ k}\Omega + 1.2 \text{ k}\Omega = 2.2 \text{ k}\Omega$$

$$I_T = \frac{V_S}{R_T} = \frac{22 \text{ V}}{2.2 \text{ k}\Omega}$$

$$= 10 \text{ mA}$$

Using the current divider rule for two branches as given in Chapter 6, we find the current through R_2 as follows:

$$I_2 = \left(\frac{R_3}{R_2 + R_3}\right) I_T = \left(\frac{3 \text{ k}\Omega}{5 \text{ k}\Omega}\right) 10 \text{ mA}$$

$$= 6 \text{ mA}$$

Example 7-7 (continued)

Now we can use Kirchhoff's current law to find the current through R_3 as follows:

$$I_T = I_2 + I_3$$
$$I_3 = I_T - I_2 = 10 \text{ mA} - 6 \text{ mA}$$
$$= 4 \text{ mA}$$

Example 7-8

Determine the current through R_4 in Figure 7-16 if $V_S = 50$ V.

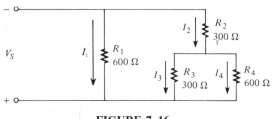

FIGURE 7-16

Solution:

First, the current (I_2) into the junction of R_3 and R_4 must be found. Once we know this current, we can use the current divider formula to find I_4.

Notice that there are two main branches in the circuit. The left-most branch consists of only R_1. The right-most branch has R_2 in series with R_3 and R_4 in parallel. The voltage across both of these main branches is the same and equal to 50 V. We can find the current (I_2) into the junction of R_3 and R_4 by calculating the equivalent resistance (R_{eq}) of the right-most main branch and then applying Ohm's law, because this current is the total current through this main branch. Thus,

$$R_{eq} = R_2 + \frac{R_3 R_4}{R_3 + R_4}$$

$$= 300 \text{ }\Omega + \frac{(600 \text{ }\Omega)(300 \text{ }\Omega)}{900 \text{ }\Omega} = 500 \text{ }\Omega$$

$$I_2 = \frac{V_S}{R_{eq}} = \frac{50 \text{ V}}{500 \text{ }\Omega}$$

$$= 0.1 \text{ A}$$

Using the current divider formula, we calculate I_4 as follows:

ANALYSIS OF SERIES-PARALLEL CIRCUITS

$$I_4 = \left(\frac{R_3}{R_3 + R_4}\right)I_2 = \left(\frac{300\ \Omega}{900\ \Omega}\right)0.1\ \text{A}$$

$$= 0.033\ \text{A}$$

Determining Voltage Drops

It is often necessary to find the voltages across certain parts of a series-parallel circuit. We can find these voltages by using the voltage divider formula given in Chapter 5, or Kirchhoff's voltage law, or Ohm's law, or combinations of each. Some examples will illustrate.

Example 7–9

Determine the voltage drop from A to B in Figure 7–17, and then find the voltage across R_1.

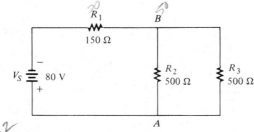

FIGURE 7–17

Solution:

Note that R_2 and R_3 are in parallel in this circuit. Since they are equal in value, their equivalent resistance is

$$R_{eq} = \frac{500\ \Omega}{2} = 250\ \Omega$$

As shown in the equivalent circuit (Figure 7–18), R_1 is in series with R_{eq}. The total circuit resistance as seen from the source is

$$R_T = R_1 + R_{eq} = 150\ \Omega + 250\ \Omega = 400\ \Omega$$

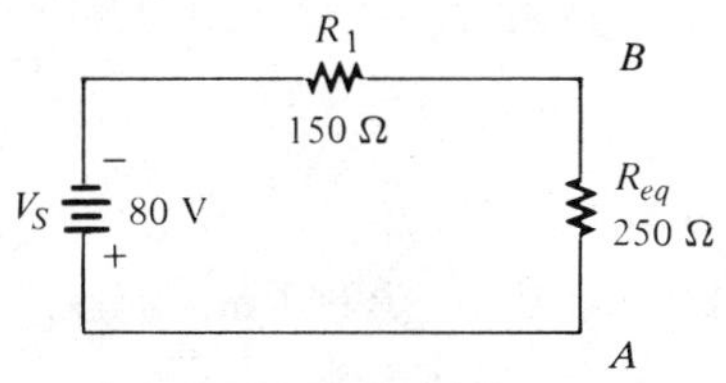

FIGURE 7–18

Example 7-9 (continued)

Now we can use the voltage divider formula to find the voltage across the parallel combination of Figure 7–17 (between points A and B). Let us call it V_{AB}:

$$V_{AB} = \left(\frac{R_{eq}}{R_T}\right)V_S = \left(\frac{250\ \Omega}{400\ \Omega}\right)80\ V$$
$$= 50\ V$$

We can now use Kirchhoff's voltage law to find V_1 as follows:

$$V_S = V_1 + V_{AB}$$
$$V_1 = V_S - V_{AB} = 80\ V - 50\ V$$
$$= 30\ V$$

Example 7-10

Determine the voltages across each resistor in the circuit of Figure 7–19.

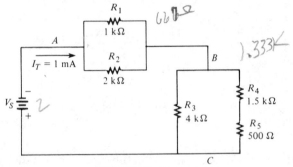

FIGURE 7–19

Solution:

The source voltage is not given, but the total current is known. Since R_1 and R_2 are in parallel, they each have the same voltage. The current through R_1 is

$$I_1 = \left(\frac{R_2}{R_1 + R_2}\right)I_T = \left(\frac{2\ k\Omega}{3\ k\Omega}\right)1\ mA$$
$$= 0.667\ mA$$

The voltages are

$$V_1 = I_1 R_1 = (0.667\ mA)(1\ k\Omega)$$
$$= 0.667\ V$$
$$V_2 = V_1 = 0.667\ V$$

ANALYSIS OF SERIES-PARALLEL CIRCUITS

The current through R_3 is

$$I_3 = \left(\frac{R_4 + R_5}{R_3 + R_4 + R_5}\right)I_T = \left(\frac{2\ k\Omega}{6\ k\Omega}\right)1\ mA$$
$$= 0.333\ mA$$

The voltage across R_3 is

$$V_3 = I_3 R_3 = (0.333\ mA)(4\ k\Omega)$$
$$= 1.332\ V$$

The currents through R_4 and R_5 are the same because these resistors are in series:

$$I_4 = I_5 = I_T - I_3$$
$$= 1\ mA - 0.333\ mA = 0.667\ mA$$

The voltages across R_4 and R_5 are

$$V_4 = I_4 R_4 = (0.667\ mA)(1.5\ k\Omega)$$
$$= 1.001\ V$$
$$V_5 = I_5 R_5 = (0.667\ mA)(500\ \Omega)$$
$$= 0.334\ V$$

The small difference in V_3 and the sum of V_4 and V_5 is a result of rounding to three places.

Example 7–11

Determine the voltage drop across each resistor in Figure 7–20.

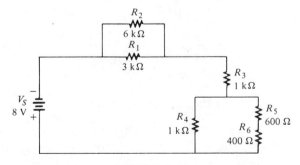

FIGURE 7–20

Solution:

Because we know the total voltage, we will solve this problem using the voltage divider formula. First reduce each parallel combination to an equivalent resistance. Since R_1 and R_2 are in parallel, we combine their values:

Example 7–11 (continued)

$$R_{eq1} = \frac{R_1 R_2}{R_1 + R_2} = \frac{(3 \text{ k}\Omega)(6 \text{ k}\Omega)}{9 \text{ k}\Omega}$$
$$= 2 \text{ k}\Omega$$

Since R_4 is in parallel with the series combination of R_5 and R_6, we combine these values to obtain

$$R_{eq2} = \frac{(R_4)(R_5 + R_6)}{R_4 + R_5 + R_6} = \frac{(1 \text{ k}\Omega)(1 \text{ k}\Omega)}{2 \text{ k}\Omega}$$
$$= 500 \text{ }\Omega$$

The equivalent circuit is drawn in Figure 7–21.

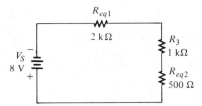

FIGURE 7–21

Applying the voltage divider formula to solve for the voltages, we get the following results:

$$V_{eq1} = \left(\frac{R_{eq1}}{R_T}\right)V_S = \left(\frac{2 \text{ k}\Omega}{3.5 \text{ k}\Omega}\right)8 \text{ V}$$
$$= 4.57 \text{ V}$$

$$V_{eq2} = \left(\frac{R_{eq2}}{R_T}\right)V_S = \left(\frac{500 \text{ }\Omega}{3.5 \text{ k}\Omega}\right)8 \text{ V}$$
$$= 1.14 \text{ V}$$

$$V_3 = \left(\frac{R_3}{R_T}\right)V_S = \left(\frac{1 \text{ k}\Omega}{3.5 \text{ k}\Omega}\right)8 \text{ V}$$
$$= 2.29 \text{ V}$$

V_{eq1} equals the voltage across both R_1 and R_2:

$$V_1 = V_2 = V_{eq1} = 4.57 \text{ V}$$

V_{eq2} is the voltage across R_4 and across the series combination of R_5 and R_6:

$$V_4 = V_{eq2} = 1.14 \text{ V}$$

Now apply the voltage divider formula to the series combination of R_5 and R_6 to get V_5 and V_6:

ANALYSIS OF SERIES-PARALLEL CIRCUITS

$$V_5 = \left(\frac{R_5}{R_5 + R_6}\right)V_{eq2} = \left(\frac{600 \, \Omega}{1000 \, \Omega}\right)1.14 \text{ V}$$
$$= 0.684 \text{ V}$$

$$V_6 = \left(\frac{R_6}{R_5 + R_6}\right)V_{eq2} = \left(\frac{400 \, \Omega}{1000 \, \Omega}\right)1.14 \text{ V}$$
$$= 0.456 \text{ V}$$

As you have seen in this section, the analysis of series-parallel circuits can be approached in many ways, depending on what information you need and what circuit values you know. The examples in this section do not represent an exhaustive coverage. They are meant only to give you an idea of how to approach series-parallel circuit analysis.

If you know Ohm's law, Kirchhoff's laws, the voltage divider formula, and the current divider formula, and if you know how to apply these laws, you can solve most resistive circuit analysis problems. The ability to recognize series and parallel combinations is, of course, essential.

Review for 7-3

1. List the circuit laws and formulas that may be necessary in the analysis of a series-parallel circuit.
2. Find the total resistance between A and B in the circuit of Figure 7-22.

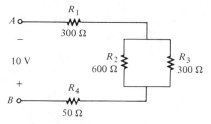

FIGURE 7-22

3. Find I_3 in Figure 7-22.
4. Find V_2 in Figure 7-22.
5. Determine R_T and I_T in Figure 7-23 as "seen" by the source.

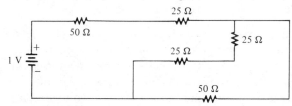

FIGURE 7-23

7-4

CIRCUIT GROUND

Voltage is relative. That is, the voltage at one point in a circuit is always measured relative to another point. For example, if we say that there are +100 V at a certain point in a circuit, we mean that the point is 100 V more positive than some *reference point* in the circuit. This *reference point* in a circuit is usually called *ground*.

The term *ground* derives from the method used in ac power lines, in which one side of the line is neutralized by connecting it to a water pipe or a metal rod driven into the ground. This method of grounding is called *earth ground*.

In most electronic equipment, the metal chassis that houses the assembly or a large conductive area on a printed circuit board is used as the *common* or *reference point,* called the *chassis ground* or *circuit ground*. This ground provides a convenient way of connecting all common points within the circuit back to one side of the battery or other energy source. The chassis or circuit ground does not necessarily have to be connected to the earth ground. However, in many cases it is earth-grounded in order to prevent a shock hazard due to a potential difference between chassis and earth grounds.

In summary, ground is the reference point in electronic circuits. It has a potential of zero volts (0 V) *with respect to all other points in the circuit that are referenced to it,* as illustrated in Figure 7-24. In Part A, the negative side of the source is grounded, and all voltages indicated are positive with respect to ground. In Part B, the positive side of the source is grounded. The voltages at all other points are therefore negative with respect to ground. Note the symbol for ground in the circuit diagram.

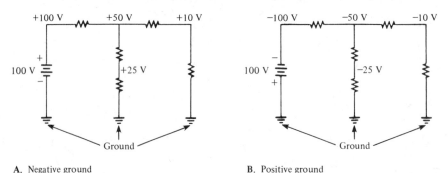

A. Negative ground

B. Positive ground

FIGURE 7-24 *Example of negative and positive ground.*

Measurement of Voltages with Respect to Ground

When voltages are measured in a circuit, one meter lead is connected to the circuit ground, and the other to the point at which the voltage is to be measured. In negative ground, the negative meter terminal is connected to the circuit ground. Black leads are normally used for negative voltages or negative ground. The positive terminal of the voltmeter is then connected to the positive voltage point. Red leads are normally used for the positive voltage connection. Measurement of positive voltage is illustrated in Figure 7-25, where the meter reads the voltage at point *A* with respect to ground.

CIRCUIT GROUND

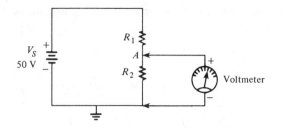

FIGURE 7-25 *Measuring a positive voltage with respect to ground.*

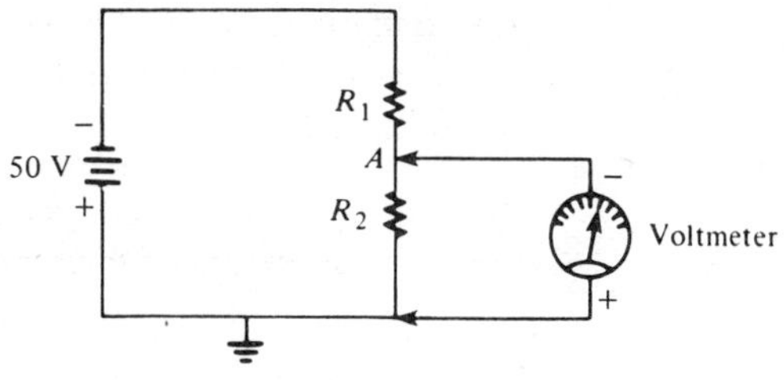

FIGURE 7-26 *Measuring a negative voltage with respect to ground.*

For a circuit with a positive ground, the positive voltmeter lead is connected to ground, and the negative lead is connected to the negative voltage point, as indicated in Figure 7-26. Here the meter reads the voltage at point A with respect to ground.

When voltages must be measured at several points in a circuit, the ground lead can be clipped to ground at one point in the circuit and left there. The other lead is then moved from point to point as the voltages are measured. This method is illustrated in Figure 7-27.

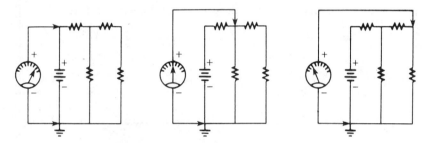

FIGURE 7-27 *Measuring voltages at several points in a circuit.*

Measurement of Voltage across an Ungrounded Resistor

Voltage can normally be measured across a resistor, as shown in Figure 7-28, even though neither side of the resistor is connected to circuit ground.

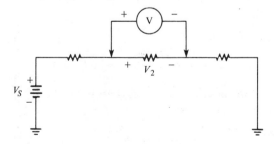

FIGURE 7-28

In some cases, when the meter is not isolated from the power line ground, the negative lead of the meter will ground one side of the resistor and alter the operation of the circuit. In this situation, another method must be used, as illustrated in Figure 7–29. The voltages on each side of the resistor are measured *with respect to ground*. The *difference* of these two measurements is the voltage drop across the resistor.

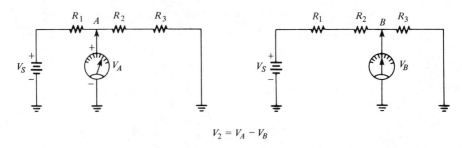

$$V_2 = V_A - V_B$$

FIGURE 7–29

Example 7–12

Determine the voltages of each of the indicated points in each circuit of Figure 7–30. Assume that 25 V are dropped across each resistor.

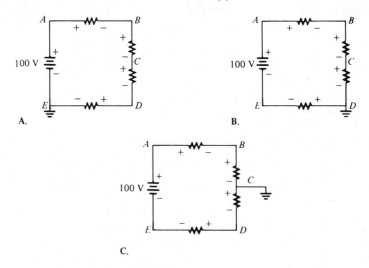

FIGURE 7–30

Solution:

In Circuit A, the voltage polarities are as shown. Point E is ground. The voltages with respect to ground are as follows:

$$V_E = 0 \text{ V}$$
$$V_D = +25 \text{ V}$$
$$V_C = +50 \text{ V}$$

CIRCUIT GROUND

$$V_B = +75 \text{ V}$$
$$V_A = +100 \text{ V}$$

In Circuit B, the voltage polarities are as shown. Point D is ground. The voltages with respect to ground are as follows:

$$V_E = -25 \text{ V}$$
$$V_D = 0 \text{ V}$$
$$V_C = +25 \text{ V}$$
$$V_B = +50 \text{ V}$$
$$V_A = +75 \text{ V}$$

In Circuit C, the voltage polarities are as shown. Point C is ground. The voltages with respect to ground are as follows:

$$V_E = -50 \text{ V}$$
$$V_D = -25 \text{ V}$$
$$V_C = 0 \text{ V}$$
$$V_B = +25 \text{ V}$$
$$V_A = +50 \text{ V}$$

Review for 7–4

1. The common point in a circuit is called _____.
2. All voltages in a circuit are referenced to ground (T or F).
3. The housing or chassis is often used as circuit ground (T or F).
4. What is the symbol for ground?
5. What does *earth ground* mean?
6. In Figure 7–31, how would you connect a voltmeter to measure the voltage at point A with respect to ground?

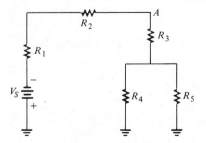

FIGURE 7–31

7–5

TROUBLE-SHOOTING

Trouble-shooting is the process of identifying and locating a failure or problem in a circuit. Some trouble-shooting techniques have already been discussed in relation to both series circuits and parallel circuits. Now we will extend these methods to the series-parallel networks.

Opens and *shorts* are typical problems that occur in electric circuits. As mentioned before, if a resistor burns out, it will normally produce an *open circuit*. Broken wires and contacts can also be causes of open paths. Pieces of foreign material, such as solder splashes, broken insulation on wires, and so on, can often lead to shorts in a circuit. A *short* is a zero resistance path between two points. We will now work some examples of trouble-shooting in series-parallel resistive circuits.

Example 7–13

From the indicated voltmeter reading, determine if there is a fault in Figure 7–32. If there is a fault, identify it as either a short or an open.

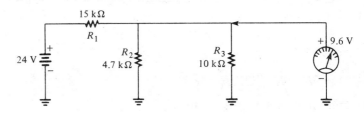

FIGURE 7–32

Solution:

First determine what the voltmeter *should* be indicating. Since R_2 and R_3 are in parallel, their equivalent resistance is

$$R_{eq} = \frac{R_2 R_3}{R_2 + R_3} = \frac{(4.7 \text{ k}\Omega)(10 \text{ k}\Omega)}{14.7 \text{ k}\Omega}$$
$$= 3.2 \text{ k}\Omega$$

1.3 mA

The voltage across the equivalent parallel combination is determined by the voltage divider formula as follows:

$$V_{R_{eq}} = \left(\frac{R_{eq}}{R_1 + R_{eq}}\right) V_s = \left(\frac{3.2 \text{ k}\Omega}{18.2 \text{ k}\Omega}\right) 24 \text{ V}$$
$$= 4.22 \text{ V}$$

4.22

Thus, 4.22 V is the voltage reading that you *should* get on the meter. But the meter reads 9.6 V instead. This value is incorrect, and, because it is higher than it should be, R_2 or R_3 is probably open. Why? Because if either of these two resistors is open, the resistance across which the meter is

TROUBLE-SHOOTING

191

connected is larger than expected. A higher resistance will drop a higher voltage in this circuit, which is, in effect, a voltage divider.

Let us start by assuming that R_2 is open. If it is, the voltage across R_3 is as follows:

$$V_{R_3} = \left(\frac{R_3}{R_1 + R_3}\right)V_s = \left(\frac{10 \text{ k}\Omega}{25 \text{ k}\Omega}\right)24 \text{ V}$$
$$= 9.6 \text{ V}$$

This calculation shows that R_2 is open. Replace R_2 with a new resistor.

Example 7–14

Suppose that you measure 24 V with the voltmeter in Figure 7–33. Determine if there is a fault, and, if there is, isolate it.

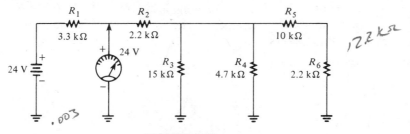

FIGURE 7–33

Solution:

There is no voltage drop across R_1 because both sides of the resistor are at +24 V. Either no current is flowing through R_1 from the source, which tells us that R_2 is open in the circuit, or R_1 is shorted.

If R_1 were open, the meter would not read 24 V. The most logical failure is in R_2. If it is open, then there is no current from the source. To verify this, measure across R_2 with the voltmeter as shown in Figure 7–34. If R_2 is open, the meter will indicate 24 V. The right side of R_2 will be at zero volts because there is no current through any of the other resistors to cause a voltage drop.

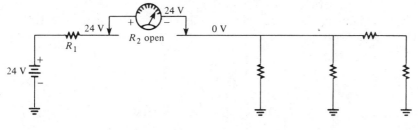

FIGURE 7–34

Example 7–15

The two voltmeters in Figure 7–35 indicate the voltages shown. Determine if there are any opens or shorts in the circuit and, if so, where they are located.

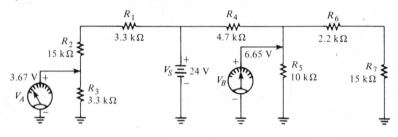

FIGURE 7–35

Solution:

First let us see if the voltmeter readings are correct. R_1, R_2, and R_3 act as a voltage divider on the left side of the source. The voltage across R_3 is calculated as follows:

$$V_{R_3} = \left(\frac{R_3}{R_1 + R_2 + R_3}\right)V_S$$

$$= \left(\frac{3.3 \text{ k}\Omega}{21.6 \text{ k}\Omega}\right)24 \text{ V} = 3.67 \text{ V}$$

The voltmeter A reading (V_A) is correct.

Now let us see if the voltmeter B reading (V_B) is correct. The part of the circuit to the right of the source also acts as a voltage divider. The series-parallel combination of R_5, R_6, and R_7 is in series with R_4. The equivalent resistance of the R_5, R_6, and R_7 combination is figured as follows:

$$R_{eq} = \frac{(R_6 + R_7)R_5}{R_5 + R_6 + R_7}$$

$$= \frac{(17.2 \text{ k}\Omega)(10 \text{ k}\Omega)}{27.2 \text{ k}\Omega} = 6.3 \text{ k}\Omega$$

where R_5 is in parallel with R_6 and R_7 in series. R_{eq} and R_4 form a voltage divider. Voltmeter B is measuring the voltage across R_{eq}. Is it correct? We check as follows:

$$V_{R_{eq}} = \left(\frac{R_{eq}}{R_4 + R_{eq}}\right)V_S$$

$$= \left(\frac{6.3 \text{ k}\Omega}{11 \text{ k}\Omega}\right)24 \text{ V} = 13.75 \text{ V}$$

TROUBLE-SHOOTING

Thus, the actual measured voltage at this point is incorrect. Some further thought will help to isolate the problem.

We know that R_4 is not open, because if it were, the meter would read 0 V. If there were a short across it, the meter would read 24 V. Since the actual voltage is much less than it should be, R_{eq} must be less than the calculated value. The most likely problem is a short across R_7. If there is a short from the top of R_7 to ground, R_6 is effectively in parallel with R_5. In this case, R_{eq} is

$$R_{eq} = \frac{R_5 R_6}{R_5 + R_6}$$

$$= \frac{(2.2 \text{ k}\Omega)(10 \text{ k}\Omega)}{12.2 \text{ k}\Omega} = 1.8 \text{ k}\Omega$$

Then V_{eq} is

$$V_{eq} = \left(\frac{1.8 \text{ k}\Omega}{6.5 \text{ k}\Omega}\right) 24 \text{ V} = 6.65 \text{ V}$$

This value for V_{eq} agrees with the voltmeter B reading. So there is a short across R_7. If this were an actual circuit, you would try to find the physical cause of the short.

Review for 7-5

1. Name two types of common circuit faults.
2. In Figure 7-36, one of the resistors in the circuit is open. Based on the meter reading, determine which is the bad resistor.

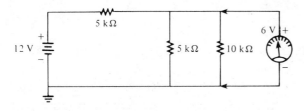

FIGURE 7-36

3. For the following faults in Figure 7-37 (p. 194), what voltage would be measured at point A?
   (a) No faults
   (b) R_1 open
   (c) Short across R_5
   (d) R_3 and R_4 open

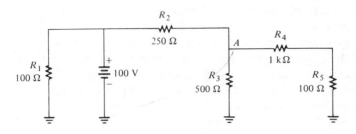

FIGURE 7–37

7–6

VOLTAGE DIVIDERS WITH RESISTIVE LOADS

Voltage dividers were discussed in Chapter 5. In this section we will discuss the effect of resistive loads on voltage dividers.

The simple voltage divider in Figure 7–38A produces an output voltage of 5 V because the two resistors are of equal value. This voltage is the *unloaded output voltage*. If a load resistor R_L is connected across the output, as shown in Figure 7–38B, the output voltage will be reduced by an amount that depends on the value R_L. The larger R_L is compared to R_2, the less the output voltage is reduced. Actually, as you can see in Figure 7–38B, a loaded voltage divider is a series-parallel circuit. The load resistor always acts in parallel with the resistance across which it is connected.

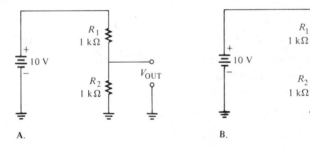

FIGURE 7–38

Example 7–16

Determine both the unloaded and the loaded output voltages of the voltage divider in Figure 7–39A for the following two values of load resistance: 10 kΩ and 100 kΩ.

VOLTAGE DIVIDERS WITH RESISTIVE LOADS

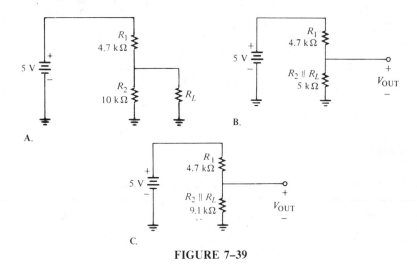

FIGURE 7–39

Solution:

The *unloaded* output voltage is

$$V_{OUT} = \left(\frac{10 \text{ k}\Omega}{14.7 \text{ k}\Omega}\right) 5 \text{ V} = 3.4 \text{ V}$$

With the 10-kΩ load resistor connected, R_L is in parallel with R_2, which gives 5 kΩ, as shown by the equivalent circuit in Figure 7–39B.
The *loaded* output voltage is

$$V_{OUT} = \left(\frac{5 \text{ k}\Omega}{9.7 \text{ k}\Omega}\right) 5 \text{ V} = 2.6 \text{ V}$$

With the 100-kΩ load, the resistance from output to ground is

$$\frac{R_2 R_L}{R_2 + R_L} = \frac{(10 \text{ k}\Omega)(100 \text{ k}\Omega)}{110 \text{ k}\Omega} = 9.1 \text{ k}\Omega$$

as shown in Figure 7–39C.
The *loaded* output voltage is

$$V_{OUT} = \left(\frac{9.1 \text{ k}\Omega}{13.8 \text{ k}\Omega}\right) 5 \text{ V} = 3.3 \text{ V}$$

Notice that with the larger value of R_L, the output is reduced from its unloaded value by much less than it is with the smaller R_L. This problem illustrates the loading effect of R_L on the voltage divider.

Voltage dividers are sometimes useful in obtaining various voltages from a power supply. For example, suppose that we wished to derive 12 V and 6 V from a 24-V supply. To do so requires a voltage divider with two *taps*, as shown in

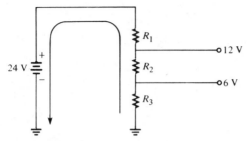

FIGURE 7–40

Figure 7–40. In this example, R_1 must equal $R_2 + R_3$, and R_2 must equal R_3. The actual values of the resistors are set by the amount of current that is to be drawn from the source under unloaded conditions. This current, called the *bleeder current*, represents a continuous drain on the source. With these ideas in mind, in Example 7–17 we will design a voltage divider to meet certain specified requirements.

Example 7–17

A power supply requires 12 V and 6 V and a 24-V battery. The unloaded current drain on this battery is not to exceed 1 mA. Determine the values of the resistors. Also determine the output voltage at the 12-V tap when both outputs are loaded with 100 kΩ each.

Solution:

A circuit as shown in Figure 7–40 is required. In order to have an unloaded current of 1 mA, the total resistance must be as follows:

$$R_T = \frac{V_S}{I} = \frac{24 \text{ V}}{1 \text{ mA}} = 24 \text{ k}\Omega$$

To get 12 V, R_1 must equal $R_2 + R_3 = 12$ kΩ. To get 6 V, R_2 must equal $R_3 = 6$ kΩ.

Now if the 100-kΩ loads are connected across the outputs as shown in Figure 7–41, the loaded output voltages are as determined below.

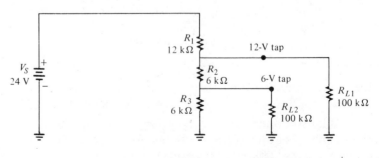

FIGURE 7–41

VOLTAGE DIVIDERS WITH RESISTIVE LOADS

The equivalent resistance from the 12-V tap to ground is the 100-kΩ load resistor R_{L1} in parallel with the combination of R_2 in series with the parallel combination of R_3 and R_{L2}. This solution is as follows.

R_3 in parallel with R_{L2}:

$$R_{eq1} = \frac{(6 \text{ k}\Omega)(100 \text{ k}\Omega)}{106 \text{ k}\Omega} = 5.66 \text{ k}\Omega$$

R_2 in series with R_{eq1}:

$$R_{eq2} = 6 \text{ k}\Omega + 5.66 \text{ k}\Omega = 11.66 \text{ k}\Omega$$

R_{L1} in parallel with R_{eq2}:

$$R_{eq3} = \frac{(100 \text{ k}\Omega)(11.66 \text{ k}\Omega)}{111.66 \text{ k}\Omega} = 10.44 \text{ k}\Omega$$

R_{eq3} is the equivalent resistance from the 12-V tap to ground. The equivalent circuit from the 12-V tap to ground is shown in Figure 7–42. Using this equivalent circuit, we calculate the *loaded* voltage at the 12-V tap by using the voltage divider formula as follows:

$$V_{12} = \left(\frac{R_{eq3}}{R_T}\right)V_S = \left(\frac{10.44 \text{ k}\Omega}{22.44 \text{ k}\Omega}\right)24 \text{ V} = 11.17 \text{ V}$$

As you can see, the output voltage at the 12-V tap decreases slightly from its unloaded value when the 100-kΩ loads are connected. Smaller values of load resistance would result in a greater decrease in the output voltage.

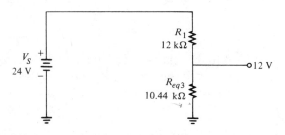

FIGURE 7–42

Review for 7–6

1. A load resistor is connected to an output tap on a voltage divider. What effect does the load resistor have on the output voltage at this tap?

2. A larger-value load resistor will cause the output voltage to change less than a smaller-value one will. (T or F)

3. For the voltage divider in Figure 7–43, determine the unloaded output voltage. Also determine the output voltage with a 10-kΩ load resistor connected across the output. (See Figure 7–43 on page 198.)

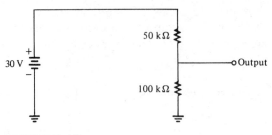

FIGURE 7-43

7-7

LADDER NETWORKS

A ladder network is a special type of series-parallel circuit. One form is commonly used to scale down voltages to certain weighted values for digital-to-analog conversion. You will study this process in later courses. In this section we will examine a basic resistive ladder of limited complexity, as shown in Figure 7-44.

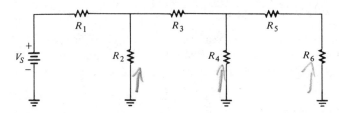

FIGURE 7-44 *Basic 3-step ladder circuit.*

One approach to the analysis of a ladder network is to simplify it one step at a time, *starting at the side farthest from the source.* In this way the current in any branch or the voltage at any point can be determined. Example 7-18 will illustrate.

Example 7-18

Determine each branch current and the voltage at each point in the ladder circuit of Figure 7-45.

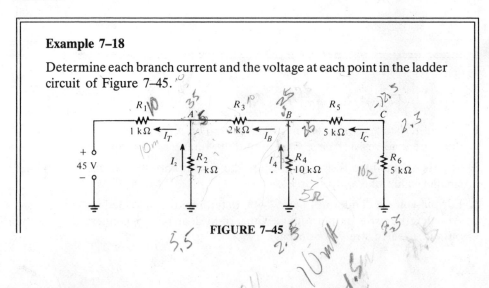

FIGURE 7-45

LADDER NETWORKS

Solution:

To find the branch currents, we must know the total current from the source (I_T). To obtain I_T, we must find the total resistance "seen" by the source.

We determine R_T in a step-by-step process, starting at the right of the circuit diagram. First notice that R_5 and R_6 are in series across R_4. So the resistance from point B to ground is as follows:

$$R_B = \frac{R_4(R_5 + R_6)}{R_4 + (R_5 + R_6)} = \frac{(10 \text{ k}\Omega)(10 \text{ k}\Omega)}{20 \text{ k}\Omega}$$

$$= 5 \text{ k}\Omega$$

Using R_B (the resistance from point B to ground), the equivalent circuit is shown in Figure 7–46.

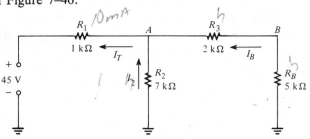

FIGURE 7–46

Next, the resistance from point A to ground (R_A) is R_2 in parallel with the series combination of R_3 and R_B. It is calculated as follows:

$$R_A = \frac{R_2(R_3 + R_B)}{R_2 + (R_3 + R_B)} = \frac{(7 \text{ k}\Omega)(7 \text{ k}\Omega)}{14 \text{ k}\Omega}$$

$$= 3.5 \text{ k}\Omega$$

Using R_A, the equivalent circuit of Figure 7–46 is further simplified to the circuit in Figure 7–47.

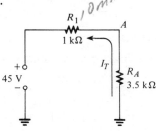

FIGURE 7–47

Finally, the total resistance "seen" by the source is R_1 in series with R_A:

Example 7–18 (continued)

$$R_T = R_1 + R_A = 1 \text{ k}\Omega + 3.5 \text{ k}\Omega = 4.5 \text{ k}\Omega$$

The total circuit current is

$$I_T = \frac{V_S}{R_T} = \frac{45 \text{ V}}{4.5 \text{ k}\Omega} = 10 \text{ mA}$$

As indicated in Figure 7–46, I_T is out of point A and divides between R_2 (I_2) and the branch containing $R_3 + R_B$ (I_B). Since the branch resistances are equal in this particular example, half the total current is through R_2 and half out of point B. So I_2 is 5 mA and I_B is 5 mA.

If the branch resistances are not equal, the current divider formula is used. As indicated in Figure 7–45, I_B is out of point B and is divided equally between R_4 and the branch containing $R_5 + R_6$ because the branch resistances are equal. So I_4, I_5, and I_6 are all equal to 2.5 mA.

To determine V_A, V_B, and V_C, we apply Ohm's law as follows:

$$V_A = I_2 R_2 = (5 \text{ mA}) (7 \text{ k}\Omega)$$
$$= 35 \text{ V}$$
$$V_B = I_4 R_4 = (2.5 \text{ mA}) (10 \text{ k}\Omega)$$
$$= 25 \text{ V}$$
$$V_C = I_6 R_6 = (2.5 \text{ mA}) (5 \text{ k}\Omega)$$
$$= 12.5 \text{ V}$$

As you can see, the circuit values in this example have been chosen so that the computations are easily done. However, the same approach applies for more cumbersome values and more complex ladder circuits.

Review for 7–7

1. Sketch a basic four-step ladder network.
2. Determine the total circuit resistance presented to the source by the ladder network of Figure 7–48.

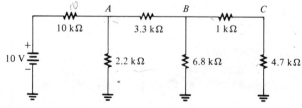

FIGURE 7–48

WHEATSTONE BRIDGE

3. What is the total current in Figure 7–48?
4. What is the current through the 2.2-kΩ resistor in Figure 7–48?
5. What is the voltage at point A with respect to ground in Figure 7–48?

7–8

WHEATSTONE BRIDGE

The *bridge circuit* is widely used in measurement devices and other applications which you will learn later. For now, we will consider the *balanced bridge,* which can be used to measure unknown resistance values. This circuit, shown in Figure 7-49A, is known as a *Wheatstone bridge.* Figure 7–49B is the same circuit drawn in a slightly different way.

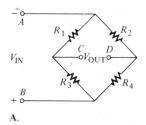
A.

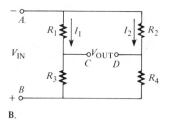
B.

FIGURE 7–49 *Wheatstone bridge.*

A bridge is said to be *balanced* when the voltage across the output terminals C and D is *zero;* that is, $V_{AC} = V_{AD}$. If V_{AC} equals V_{AD}, then $I_1 R_1 = I_2 R_2$, since one side of both R_1 and R_2 is connected to point A. Also, $I_1 R_3 = I_2 R_4$, since one side of both R_3 and R_4 connects to point B. Because of these equalities, we can write the ratios of the voltage as follows:

$$\frac{I_1 R_1}{I_1 R_3} = \frac{I_2 R_2}{I_2 R_4}$$

The currents cancel to give

$$\frac{R_1}{R_3} = \frac{R_2}{R_4}$$

Solving for R_1, we get

$$R_1 = R_3 \left(\frac{R_2}{R_4}\right) \qquad (7\text{–}1)$$

How can this formula be used to determine an unknown resistance? First, let us make R_3 a *variable* resistor and call it R_V. Also, we set the ratio R_2/R_4 to a known value. If R_V is adjusted until the bridge is balanced, the product of R_V and the ratio R_2/R_4 is equal to R_1, which is our unknown resistor (R_{unk}). Equation (7–1) is restated in Equation (7–2), using the new subscripts:

$$R_{unk} = R_V\left(\frac{R_2}{R_4}\right) \qquad (7\text{-}2)$$

The bridge is balanced when the voltage across the output terminals equals zero ($V_{AC} = V_{AD}$). A *galvanometer* (a meter that measures small currents in either direction) is connected between the output terminals. Then R_V is adjusted until the galvanometer shows zero current ($V_{AC} = V_{AD}$), indicating a balanced condition. The setting of R_V multiplied by the ratio R_2/R_4 gives the value of R_{unk}. Figure 7–50 shows this arrangement. For example, if $R_2/R_4 = \frac{1}{10}$ and $R_V = 680\ \Omega$, then $R_{unk} = (680\ \Omega)(\frac{1}{10}) = 68\ \Omega$.

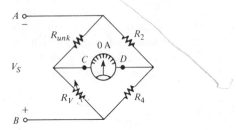

FIGURE 7–50 Balanced Wheatstone bridge.

Example 7–19

What is R_{unk} under the balanced bridge conditions shown in Figure 7–51?

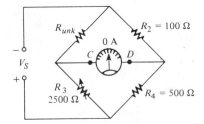

FIGURE 7–51

Solution:

$$R_{unk} = R_V\left(\frac{R_2}{R_4}\right) = 2500\ \Omega\left(\frac{100\ \Omega}{500\ \Omega}\right)$$
$$= 500\ \Omega$$

Review for 7–8

1. Sketch a basic Wheatstone bridge circuit.
2. Under what condition is the bridge balanced?

SUMMARY

3. What formula is used to determine the value of the unknown resistance when the bridge is balanced?
4. What is the unknown resistance for the values shown in Figure 7–52?

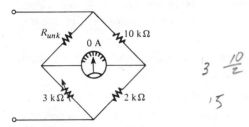

FIGURE 7–52

Formula

$$R_{unk} = R_V\left(\frac{R_2}{R_4}\right) \tag{7-2}$$

Summary

1. A series-parallel circuit is a combination of both series paths and parallel paths.
2. To determine total resistance in a series-parallel circuit, identify the series and parallel relationships, and then apply the formulas for series resistance and parallel resistance from Chapters 5 and 6.
3. To find the total current, divide the total voltage by the total resistance.
4. To determine branch currents, apply the current divider formula, or Kirchhoff's current law, or Ohm's law. Consider each circuit problem individually to determine the most appropriate method.
5. To determine voltage drops across any portion of a series-parallel circuit, use the voltage divider formula, or Kirchhoff's voltage law, or Ohm's law. Consider each circuit problem individually to determine the most appropriate method.
6. *Ground* is the common or reference point in a circuit.
7. All voltages in a circuit are referenced to ground unless otherwise specified.
8. Ground is zero volts with respect to all points referenced to it in the circuit.
9. *Negative ground* is the term used when the negative side of the source is grounded.
10. *Positive ground* is the term used when the positive side of the source is grounded.
11. *Trouble-shooting* is the process of identifying and locating a fault in a circuit.
12. *Open circuits* and *short circuits* are typical circuit faults.
13. Resistors normally open when they burn out.

14. When a load resistor is connected across a voltage divider output, the output voltage decreases.
15. The load resistor should be large compared to the resistance across which it is connected, in order that the loading effect may be minimized. A *10-times* value is sometimes used as a rule of thumb, but the value depends on the accuracy required for the output voltage.
16. To find total resistance of a ladder network, start at the point farthest from the source and reduce the resistance in steps.
17. A Wheatstone bridge can be used to measure an unknown resistance.
18. A bridge is *balanced* when the output voltage is *zero*. The balanced condition produces zero current through a load connected across the output terminals of the bridge.

Self-Test

1. Identify the series-parallel relationships in Figure 7–53.

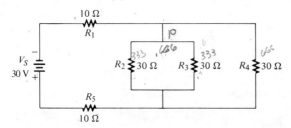

FIGURE 7–53

2. For the circuit of Figure 7–53, determine the following:
   (a) Total resistance as "seen" by the source
   (b) Total current drawn from the source
   (c) Current through R_3
   (d) Voltage across R_4

3. For the circuit in Figure 7–54, determine the following:
   (a) Total resistance   (b) Total current
   (c) Current through R_1 (d) Voltage across R_6

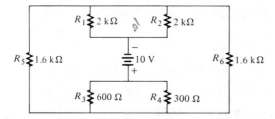

FIGURE 7–54

SELF-TEST

4. In Figure 7–55, find the following:
   (a) Total resistance between terminals A and B
   (b) Total current drawn from a 6-V source connected from A to B
   (c) Current through R_5
   (d) Voltage across R_2

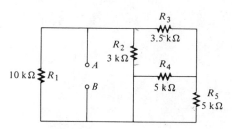

FIGURE 7–55

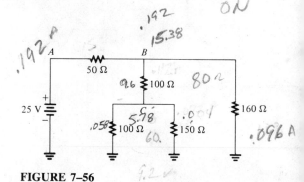

FIGURE 7–56

5. Determine the voltages with respect to ground in Figure 7–56.
6. If R_2 in Figure 7–57 opens, what voltages will be read at points A, B, and C?

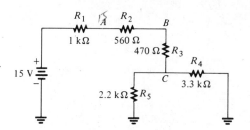

FIGURE 7–57

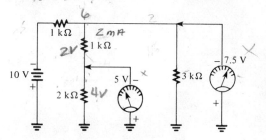

FIGURE 7–58

7. Check the meter readings in Figure 7–58 and locate any fault that may exist.
8. Determine the unloaded output voltage in Figure 7–59. If a 200-kΩ load is connected, what is the loaded output voltage?

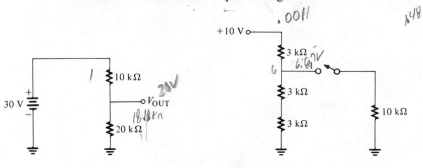

FIGURE 7–59   FIGURE 7–60

9. In Figure 7–60, determine the voltage at point A when the switch is open. Also determine the voltage at point A when the switch is closed.

10. For the ladder network in Figure 7–61, determine the following:
   (a) Total resistance
   (b) Total current
   (c) Current through R_3
   (d) Current through R_4
   (e) Voltage at point A
   (f) Voltage at Point B

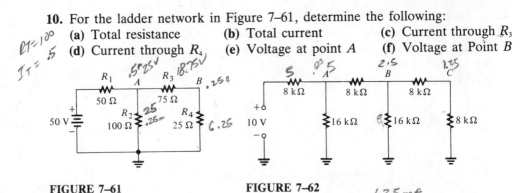

FIGURE 7–61

FIGURE 7–62

11. Calculate V_A, V_B, and V_C for the ladder in Figure 7–62.

12. In the bridge circuit of Figure 7–63, what is the value of the unknown resistance when the other values are as shown?

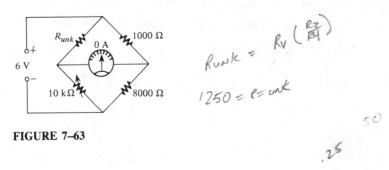

FIGURE 7–63

Problems

7–1. Visualize and sketch the following series-parallel combinations:
   (a) R_1 in series with the parallel combination of R_2 and R_3
   (b) R_1 in parallel with the series combination of R_2 and R_3
   (c) R_1 in parallel with a branch containing R_2 in series with a parallel combination of four other resistors

7–2. Visualize and sketch the following series-parallel circuits:
   (a) A parallel combination of three branches, each containing two series resistors
   (b) A series combination of three parallel circuits, each containing two resistors

7–3. A certain circuit is composed of two parallel resistors. The total resistance is 667 Ω. One of the resistors is 1 kΩ. What is the other resistor?

7–4. In each circuit of Figure 7–64, identify the series and parallel relationships of the resistors viewed from the source.

7–5. For each circuit in Figure 7–65, identify the series and parallel relationships of the resistors viewed from the source.

PROBLEMS

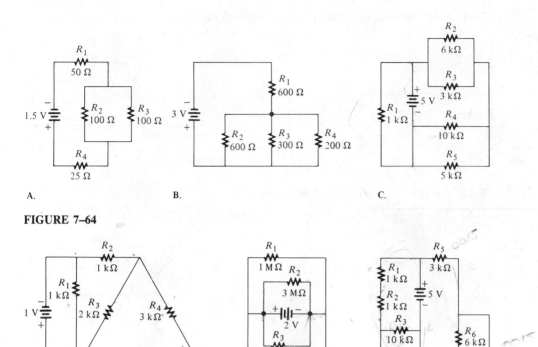

FIGURE 7-64

FIGURE 7-65

7-6. For each circuit in Figure 7-64, determine the total resistance presented to the source.

7-7. Repeat Problem 7-6 for each circuit in Figure 7-65.

7-8. Determine the voltage at each point with respect to ground in Figure 7-66.

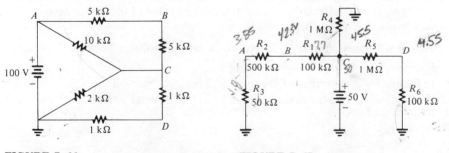

FIGURE 7-66

FIGURE 7-67

7-9. Determine the voltage at each point with respect to ground in Figure 7-67.

7-10. In Figure 7-67, how would you determine the voltage across R_2 by measuring without connecting a meter directly across the resistor?

7–11. Is the voltmeter reading in Figure 7–68 correct?

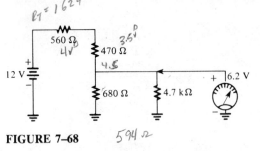

FIGURE 7–68

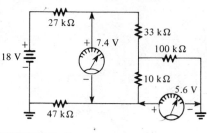

FIGURE 7–69

7–12. Are the meter readings in Figure 7–69 correct?

7–13. There is one fault in Figure 7–70. Based on the meter indications, determine what the fault is.

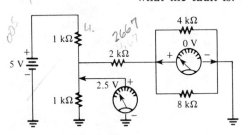

FIGURE 7–70

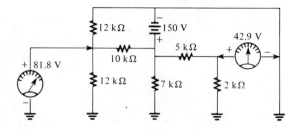

FIGURE 7–71

7–14. Look at the meters in Figure 7–71 and determine if there is a fault in the circuit. If there is a fault, identify it.

7–15. A voltage divider consists of two 50-kΩ resistors and a 15-V source. Calculate the unloaded output voltage. What will the output voltage be if a load resistor of 1 MΩ is connected to the output?

7–16. A 12-V battery output is divided down to obtain two output voltages. Three 3.3-kΩ resistors are used to provide the two taps. Determine the output voltages. If a 10-kΩ load is connected to the higher of the two outputs, what will its loaded value be?

7–17. Which will cause a smaller decrease in output voltage for a given voltage divider, a 10-kΩ load or a 50-kΩ load?

7–18. In Figure 7–72, determine the continuous current drain on the battery with no load across the two terminals. With a 10-kΩ load, what is the battery current?

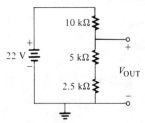

FIGURE 7–72

PROBLEMS

7–19. Determine the resistance values for a voltage divider that must meet the following specifications: The current drain when the voltage divider is unloaded is not to exceed 5 mA. The source voltage is to be 10 V. A 5-V output and a 2.5-V output are required. Sketch the circuit. Determine the effect on the output voltages if a 1-kΩ load is connected to each tap.

7–20. For the circuit shown in Figure 7–73, calculate the following:
  (a) Total resistance across the source
  (b) Total current from the source
  (c) Current through the 900-Ω resistor
  (d) Voltage from point A to point B

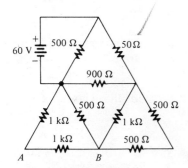
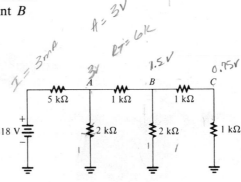

FIGURE 7–73 **FIGURE 7–74**

7–21. Determine the total resistance and the voltage at points A, B, and C in the ladder network of Figure 7–74.

7–22. Determine the total resistance between terminals A and B of the ladder network in Figure 7–75. Also calculate the current in each branch with 10 V between A and B.

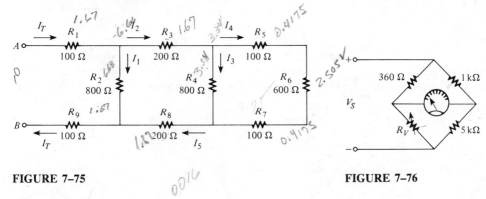

FIGURE 7–75 **FIGURE 7–76**

7–23. What is the voltage across each resistor in Figure 7–75?

7–24. A resistor of unknown value is connected to a Wheatstone bridge circuit. The bridge parameters are set as follows: $R_V = 18$ kΩ and $R_2/R_4 = 0.02$. What is R_{unk}?

7–25. A bridge network is shown in Figure 7–76. To what value must R_V be set in order to balance the bridge?

Answers to Section Reviews

Section 7–1:
1. A circuit consisting of both series and parallel connections. 2. See Figure 7–77.
3. R_1 and R_2 are in series with the parallel combination of R_3 and R_4.

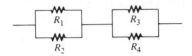

FIGURE 7–77

Section 7–2:
1. R_1 and R_4 are in series with each other and with the parallel combination of R_2 and R_3. 2. All resistors. 3. R_1 and R_2 are in parallel. R_3 and R_4 are in parallel. 4. Yes.

Section 7–3:
1. Voltage and current divider formulas, Kirchhoff's laws, and Ohm's law. 2. 550 Ω.
3. 0.012 A. 4. 3.6 V. 5. 100 Ω, 0.01 A.

Section 7–4:
1. Ground. 2. T. 3. T. 4. See Figure 7–78. 5. A connection to earth through a metal rod or a water pipe. 6. Negative terminal to A, positive terminal to ground.

FIGURE 7–78

Section 7–5:
1. Opens and shorts. 2. The 10-kΩ resistor. 3. (a) 57.9 V (b) 57.9 V (c) 57.1 V (d) 100 V.

Section 7–6:
1. It decreases the output voltage. 2. T. 3. 20 V, 4.62 V.

Section 7–7:
1. See Figure 7–79. 2. 11.64 kΩ. 3. 0.859 mA. 4. 0.639 mA. 5. 1.41 V.

Section 7–8:
1. See Figure 7–80. 2. $V_A = V_B$. 3. $R_1(R_1/R_2)$. 4. 15 kΩ.

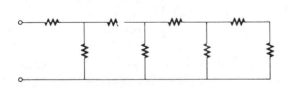
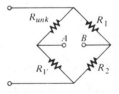

FIGURE 7–79   FIGURE 7–80

There are many ways to simplify circuits to make analysis easier. In this chapter you will learn several important circuit theorems that can be applied to circuit problems.

Voltage source and current source are discussed in this chapter; they are important because they are the two types of energy sources found in electronic circuits. Also, conversion from one type of source to the other is discussed.

8–1 The Voltage Source
8–2 The Current Source
8–3 Source Conversions
8–4 The Superposition Theorem
8–5 Thevenin's Theorem
8–6 Norton's Theorem
8–7 Millman's Theorem
8–8 Maximum Power Transfer Theorem
8–9 Delta-Wye (Δ-Y) and Wye-Delta (Y-Δ) Network Conversions

8
CIRCUIT THEOREMS
AND CONVERSIONS

8-1

THE VOLTAGE SOURCE

Figure 8–1A is the familiar symbol for an ideal dc voltage source. The voltage across its terminals A and B remains fixed regardless of the value of load resistance that may be connected across its output. Figure 8–1B shows a load resistor R_L connected. All of the source voltage, V_S, is dropped across R_L. R_L can be changed to any value and the voltage will remain fixed. The ideal voltage source has an internal resistance of zero.

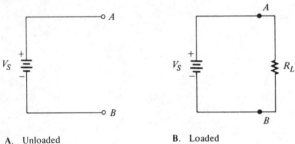

A. Unloaded   B. Loaded

FIGURE 8–1 *Ideal dc voltage source.*

In reality, no voltage source is ideal. That is, all have some internal resistance. This concept can be represented by a resistor in series with an ideal source, as shown in Figure 8–2A, where R_S is the internal source resistance and V_S is the source voltage. With no load, the output voltage (voltage from A to B) is V_S. This voltage is sometimes called the *open circuit voltage*.

Loading of the Voltage Source

When a load resistor is connected across the output terminals, as shown in Figure 8–2B, all of the source voltage does not appear across R_L. Some of the voltage is dropped across R_S, because the practical (nonideal) voltage source with a load acts as a *voltage divider*.

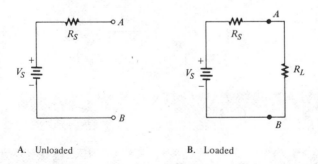

A. Unloaded   B. Loaded

FIGURE 8–2 *Practical voltage source.*

THE VOLTAGE SOURCE

If R_S is very small compared to R_L, the source approaches ideal, because almost all of the source voltage, V_S, appears across the larger resistance R_L. Very little voltage is dropped across the internal resistance, R_S. If R_L changes, most of the source voltage remains across the output as long as R_L is much larger than R_S. As a result, very little change occurs in the output voltage. The larger R_L is compared to R_S, the less change (loss) there is in the output voltage. As a rule, R_L should be at least ten times R_S ($R_L \geq 10R_S$).

Example 8–1 illustrates the effect of changes in R_L on the output voltage when R_L is much greater than R_S. Example 8–2 shows the effect of smaller load resistances.

Example 8–1

Calculate the voltage output of the source in Figure 8–3 for the following values of R_L: 100 Ω, 500 Ω, and 1 kΩ.

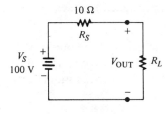

FIGURE 8–3

Solution:

For $R_L = 100$ Ω:

$$V_{OUT} = \left(\frac{R_L}{R_S + R_L}\right)V_S = \left(\frac{100 \text{ }\Omega}{110 \text{ }\Omega}\right)100 \text{ V}$$
$$= 91 \text{ V}$$

For $R_L = 500$ Ω:

$$V_{OUT} = \left(\frac{500 \text{ }\Omega}{510 \text{ }\Omega}\right)100 \text{ V}$$
$$= 98 \text{ V}$$

For $R_L = 1$ kΩ:

$$V_{OUT} = \left(\frac{1000 \text{ }\Omega}{1010 \text{ }\Omega}\right)100 \text{ V}$$
$$= 99 \text{ V}$$

Notice that the output voltage is within 10% of the source voltage, V_S, for all three values of R_L, because R_L is at least ten times R_S.

Example 8–2

Determine V_{OUT} for $R_L = 10\ \Omega$ and $R_L = 1\ \Omega$ in Figure 8–3.

Solution:

For $R_L = 10\ \Omega$:

$$V_{OUT} = \left(\frac{R_L}{R_s + R_L}\right)V_s = \left(\frac{10\ \Omega}{20\ \Omega}\right)100\ V$$
$$= 50\ V$$

For $R_L = 1\ \Omega$:

$$V_{OUT} = \left(\frac{R_L}{R_s + R_L}\right)V_s = \left(\frac{1\ \Omega}{11\ \Omega}\right)100\ V$$
$$= 9.1\ V$$

Notice in Example 8–2 that the output voltage decreases significantly as R_L is made smaller compared to R_s. This example illustrates the requirement that R_L must be much larger than R_s in order to maintain the output voltage near its open circuit value.

Review for 8–1

1. What is the symbol for the ideal voltage source?
2. Sketch a practical voltage source.
3. What is the internal resistance of the ideal voltage source?
4. What effect does the load have on the output voltage of the practical voltage source?

8–2

THE CURRENT SOURCE

Figure 8–4A shows a symbol for the ideal current source. The arrow indicates the direction of current, and I_s is the value of the source current. An ideal current source produces a *fixed* or constant value of current through a load, regardless of the value of the load. This concept is illustrated in Figure 8–4B, where a load resistor is connected to the current source between terminals A and B. The ideal current source has an infinitely large internal resistance.

Transistors act basically as current sources, and for this reason knowledge of the current source concept is important. You will find that the equivalent model of a transistor does contain a current source.

THE CURRENT SOURCE

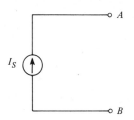

A. Unloaded

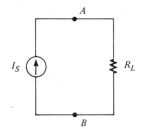

B. Loaded

FIGURE 8–4 *Ideal current source.*

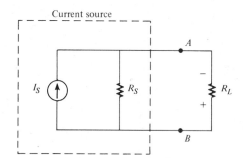

FIGURE 8–5 *Practical current source with load.*

Although the ideal current source can be used in most analysis work, no actual device is ideal. A practical current source representation is shown in Figure 8–5. Here the internal resistance appears in parallel with the ideal current source.

If the internal source resistance, R_S, is much *larger* than a load resistor, the practical source approaches ideal. The reason is illustrated in the practical current source shown in Figure 8–5. Part of the current I_S is through R_S, and part through R_L. R_S and R_L act as a current divider. If R_S is much larger than R_L, most of the current will be through R_L, and very little through R_S. As long as R_L remains much smaller than R_S, the current through it will stay almost constant, no matter how much R_L changes.

If we have a constant current source, we normally assume that R_S is so much larger than the load that it can be neglected. This simplifies the source to ideal, making the analysis easier.

Example 8–3 illustrates the effect of changes in R_L on the load current when R_L is much smaller than R_S. Generally, R_L should be at least ten times smaller $(10R_L \leq R_S)$.

Example 8–3

Calculate the load current in Figure 8–6 for the following values of R_L: 100 Ω, 500 Ω, and 1 kΩ.

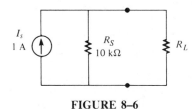

FIGURE 8–6

Solution:

For $R_L = 100$ Ω:

$$I_L = \left(\frac{R_S}{R_S + R_L}\right)I_S = \left(\frac{10 \text{ k}\Omega}{10.1 \text{ k}\Omega}\right)1 \text{ A}$$
$$= 0.99 \text{ A}$$

Example 8–3 (continued)

For $R_L = 500 \, \Omega$:

$$I_L = \left(\frac{10 \text{ k}\Omega}{10.5 \text{ k}\Omega}\right) 1 \text{ A}$$

$$= 0.95 \text{ A}$$

For $R_L = 1 \text{ k}\Omega$:

$$I_L = \left(\frac{10 \text{ k}\Omega}{11 \text{ k}\Omega}\right) 1 \text{ A}$$

$$= 0.91 \text{ A}$$

Notice that the load current I_L is within 10% of the source current for each value of R_L.

Review for 8–2

1. What is the symbol for an ideal current source?
2. Sketch the practical current source.
3. What is the internal resistance of the ideal current source?
4. What effect does the load have on the load current of the practical current source?

8–3

SOURCE CONVERSIONS

In circuit analysis, it is sometimes useful to convert a voltage source to an *equivalent* current source, or vice versa.

Converting a Voltage Source into a Current Source

The source voltage, V_S, divided by the source resistance, R_S, gives the value of the equivalent source current:

$$I_S = \frac{V_S}{R_S} \tag{8–1}$$

As illustrated in Figure 8–7, the directional arrow for the current points from plus to minus. The equivalent current source is the source in parallel with R_S.

Equivalency of two sources means that for the same load resistance, the same load voltage and current are produced by both sources. This concept is called *terminal equivalency*.

SOURCE CONVERSIONS

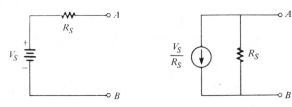

A. Voltage source B. Current source

FIGURE 8–7 Conversion of voltage source to equivalent current source.

We can show that the voltage source and the current source in Figure 8–7 are equivalent by connecting a load resistor to each, as shown in Figure 8–8, and then calculating the load current as follows: For the voltage source,

$$I_L = \frac{V_S}{R_S + R_L}$$

For the current source,

$$I_L = \left(\frac{R_S}{R_S + R_L}\right)\frac{V_S}{R_S} = \frac{V_S}{R_S + R_L}$$

As you see, both expressions for I_L are the same. These equations prove that the sources are equivalent as far as the load or terminals AB are concerned.

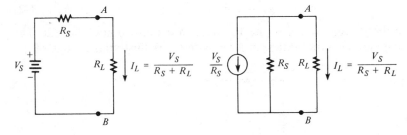

A. Loaded voltage source B. Loaded current source

FIGURE 8–8 Equivalent sources with loads.

Example 8–4

Convert the voltage source in Figure 8–9 to an equivalent current source.

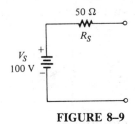

FIGURE 8–9

Example 8-4 (continued)

Solution:

$$I_s = \frac{V_s}{R_s} = \frac{100 \text{ V}}{50 \text{ }\Omega} = 2 \text{ A}$$

$$R_s = 50 \text{ }\Omega$$

The equivalent current source is shown in Figure 8-10.

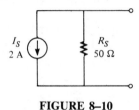

FIGURE 8-10

Converting a Current Source into a Voltage Source

The source current, I_s, multiplied by the source resistance, R_s, gives the value of the equivalent source voltage:

$$V_s = I_s R_s \qquad (8\text{-}2)$$

The polarity of the voltage source is plus to minus in the direction of current. The equivalent voltage source is the voltage in series with R_s, as illustrated in Figure 8-11.

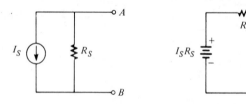

A. Current source B. Voltage source

FIGURE 8-11 *Conversion of current source to equivalent voltage source.*

Example 8-5

Convert the current source in Figure 8-12 to an equivalent voltage source.

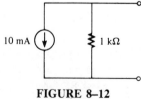

FIGURE 8-12

THE SUPERPOSITION THEOREM

Solution:

$$V_s = I_s R_s = (10 \text{ mA})(1 \text{ k}\Omega)$$
$$= 10 \text{ V}$$
$$R_s = 1 \text{ k}\Omega$$

The equivalent voltage source is shown in Figure 8–13.

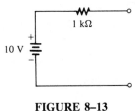

FIGURE 8–13

Review for 8-3

1. Write the formula for converting a voltage source to a current source.
2. Write the formula for converting a current source to a voltage source.
3. Convert the voltage source in Figure 8–14 to an equivalent current source.

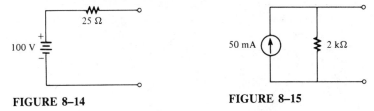

FIGURE 8–14   **FIGURE 8–15**

4. Convert the current source in Figure 8–15 to an equivalent voltage source.

8-4

THE SUPERPOSITION THEOREM

The *superposition theorem* is useful in the analysis of circuits with more than one source. It provides a method of determining the current in any branch of a multiple-source circuit. A statement of the superposition theorem is as follows:

The current in any given branch of a multiple-source circuit can be found by determining the currents in that particular branch produced by each source acting alone, with all other sources replaced by their internal resistances. The total current in the branch is the algebraic sum of the individual source currents in that branch.

Thus, for a circuit with more than one source, proceed as follows to find the total current in any given branch:

1. Leave *one* of the sources in the circuit, and reduce *all* others to zero. Reduce the voltage source to zero by putting a short between its terminals; any internal series resistance remains. Reduce the current source to zero by opening its terminals; any internal parallel resistance remains.
2. Find the current in the branch of interest due to the one remaining source.
3. Repeat Steps 1 and 2 for each source in turn. When you finish, you will have a number of current values equal to the number of sources in the circuit.
4. Add all of the individual current values algebraically. All currents in one direction will have a plus (+) sign, and all currents in the other direction will have a minus (−) sign.

Some examples will clarify this procedure.

Example 8–6

Find the current in R_2 of Figure 8–16 by using the superposition theorem.

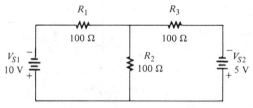

FIGURE 8–16

Solution:

First, by shorting V_{S2}, find the current in R_2 due to voltage source V_{S1}, as shown in Figure 8–17.

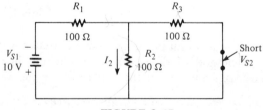

FIGURE 8–17

To find I_2, we can use the current divider formula from Chapter 6. Looking from V_{S1},

THE SUPERPOSITION THEOREM

$$R_T = R_1 + \frac{R_3}{2} = 100 \ \Omega + 50 \ \Omega$$
$$= 150 \ \Omega$$

$$I_T = \frac{V_{S1}}{R_T} = \frac{10 \ \text{V}}{150 \ \Omega} = 0.0667 \ \text{A}$$
$$= 66.7 \ \text{mA}$$

The current in R_2 due to V_{S1} is

$$I_2 = \left(\frac{R_3}{R_2 + R_3}\right) I_T = \left(\frac{100 \ \Omega}{200 \ \Omega}\right) 66.7 \ \text{mA}$$
$$= 33.3 \ \text{mA}$$

Note that this current flows *downward* through R_2.

Next find the current in R_2 due to voltage source V_{S2} by shorting V_{S1}, as shown in Figure 8–18.

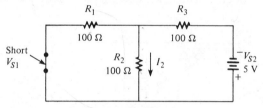

FIGURE 8–18

Looking from V_{S2},

$$R_T = R_3 + \frac{R_1}{2} = 100 \ \Omega + 50 \ \Omega$$
$$= 150 \ \Omega$$

$$I_T = \frac{V_{S2}}{R_T} = \frac{5 \ \text{V}}{150 \ \Omega} = 0.0333 \ \text{A}$$
$$= 33.3 \ \text{mA}$$

The current in R_2 due to V_{S2} is

$$I_2 = \left(\frac{R_1}{R_1 + R_2}\right) I_T = \left(\frac{100 \ \Omega}{200 \ \Omega}\right) 33.3 \ \text{mA}$$
$$= 16.7 \ \text{mA}$$

Note that this current is *downward* through R_2.

Both component currents are *downward* through R_2; so they have the same algebraic sign. Therefore, we add the values to get the total current through R_2.

$$I_2 \text{ (total)} = I_2 \text{ (due to } V_{S1}) + I_2 \text{ (due to } V_{S2})$$
$$= 33.3 \ \text{mA} + 16.7 \ \text{mA} = 50 \ \text{mA}$$

Example 8–7

Find the current through R_2 in the circuit of Figure 8–19.

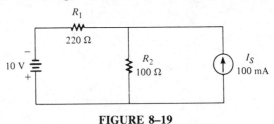

FIGURE 8–19

Solution:

First find the current in R_2 *due to* V_S by replacing I_S with an *open*, as shown in Figure 8–20. Notice that all of the current produced by V_S is through R_2.

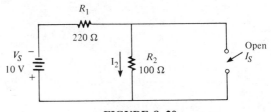

FIGURE 8–20

Looking from V_S,

$$R_T = R_1 + R_2 = 320 \ \Omega$$

The current through R_2 due to V_S is

$$I_2 = \frac{V_S}{R_T} = \frac{10 \text{ V}}{320 \ \Omega} = 0.031 \text{ A}$$
$$= 31 \text{ mA}$$

Note that this current is *downward* through R_2.

Next find the current through R_2 due to I_S by replacing V_S with a *short*, as shown in Figure 8–21.

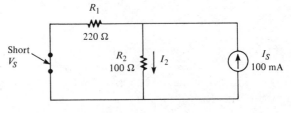

FIGURE 8–21

THE SUPERPOSITION THEOREM

Using the current divider formula, we get the current through R_2 due to I_s as follows:

$$I_2 = \left(\frac{R_1}{R_1 + R_2}\right)I_s = \left(\frac{220 \ \Omega}{320 \ \Omega}\right)100 \ \text{mA}$$
$$= 69 \ \text{mA}$$

Note that this current also is *downward* through R_2.

Both currents are in the same direction through R_2; so we add them to get the total:

$$I_2 \ (\text{total}) = I_2 \ (\text{due to } V_s) + I_2 \ (\text{due to } I_s)$$
$$= 31 \ \text{mA} + 69 \ \text{mA} = 100 \ \text{mA}$$

Example 8–8

Find the current through the 100-Ω resistor in Figure 8–22.

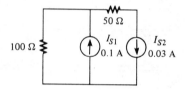

FIGURE 8–22

Solution:

First find the current through the 100-Ω resistor due to current source I_{s1} by replacing source I_{s2} with an *open,* as shown in Figure 8–23. As you can see, the entire 0.1 A from the current source I_{s1} is *downward* through the 100-Ω resistor.

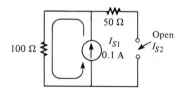

FIGURE 8–23

Next find the current through the 100-Ω resistor due to source I_{s2} by replacing source I_{s1} with an *open*, as indicated in Figure 8–24. Notice that all of the 0.03 A from source I_{s2} is *upward* through the 100-Ω resistor.

Example 8–8 (continued)

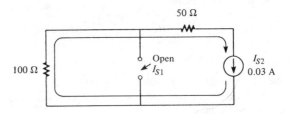

FIGURE 8–24

To get the total current through the 100-Ω resistor, we subtract the smaller current from the larger because they are in opposite directions. The resulting total current is in the direction of the larger current from source I_{s1}:

$$I_{100\Omega} \text{ (total)} = I_{100\Omega} \text{ (due to } I_{s1}\text{)} - I_{100\Omega} \text{ (due to } I_{s2}\text{)}$$
$$= 0.1 \text{ A} - 0.03 \text{ A} = 0.07 \text{ A}$$

The resulting current is downward through the resistor.

Example 8–9

Find the total current through R_3 in Figure 8–25.

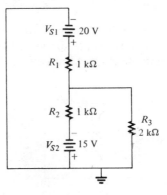

FIGURE 8–25

Solution:

First find the current through R_3 due to source V_{s1} by replacing source V_{s2} with a short, as shown in Figure 8–26.

THE SUPERPOSITION THEOREM

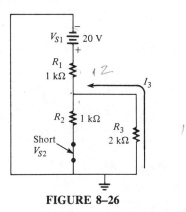

FIGURE 8–26

Looking from V_{S1},

$$R_T = R_1 + \frac{R_2 R_3}{R_2 + R_3}$$

$$= 1 \text{ k}\Omega + \frac{(1 \text{ k}\Omega)(2 \text{ k}\Omega)}{3 \text{ k}\Omega} = 1.67 \text{ k}\Omega$$

$$I_T = \frac{V_{S1}}{R_T} = \frac{20 \text{ V}}{1.67 \text{ k}\Omega}$$

$$= 12 \text{ mA}$$

Now apply the current divider formula to get the current through R_3 due to source V_{S1} as follows:

$$I_3 = \left(\frac{R_2}{R_2 + R_3}\right)I_T = \left(\frac{1 \text{ k}\Omega}{3 \text{ k}\Omega}\right)12 \text{ mA}$$

$$= 4 \text{ mA}$$

Notice that this current is *upward* through R_3.

Next find I_3 due to source V_{S2} by replacing source V_{S1} with a short, as shown in Figure 8–27.

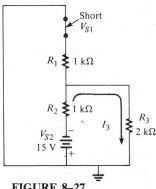

FIGURE 8–27

Example 8–9 (continued)

Looking from V_{S2},

$$R_T = R_2 + \frac{R_1 R_3}{R_1 + R_3}$$

$$= 1 \text{ k}\Omega + \frac{(2 \text{ k}\Omega)(1 \text{ k}\Omega)}{3 \text{ k}\Omega} = 1.67 \text{ k}\Omega$$

$$I_T = \frac{V_{S2}}{R_T} = \frac{15 \text{ V}}{1.67 \text{ k}\Omega}$$

$$= 9 \text{ mA}$$

Now apply the current divider formula to find the current through R_3 due to source V_{S2} as follows:

$$I_3 = \left(\frac{R_1}{R_1 + R_3}\right) I_T = \left(\frac{1 \text{ k}\Omega}{3 \text{ k}\Omega}\right) 9 \text{ mA}$$

$$= 3 \text{ mA}$$

Notice that this current is *downward* through R_3.

Calculation of the total current through R_3 is as follows:

$$I_3 \text{ (total)} = I_3 \text{ (due to } V_{S1}) - I_3 \text{ (due to } V_{S2})$$

$$= 4 \text{ mA} - 3 \text{ mA} = 1 \text{ mA}$$

This current is upward through R_3.

Review for 8–4

1. State the superposition theorem.
2. Why is the superposition theorem useful for analysis of multiple-source circuits?
3. Why is a voltage source shorted and a current source opened when the superposition theorem is applied?
4. Using the superposition theorem, find the current through R_1 in Figure 8–28.

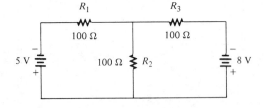

FIGURE 8–28

5. If two currents are in opposing directions through a branch of a circuit, in what direction is the net or total current?

8-5

THEVENIN'S THEOREM

Thevenin's theorem, as applied to resistive dc circuits, provides a method for reducing *any* circuit to an *equivalent* circuit consisting of *an equivalent voltage source in series with an equivalent resistance*. Although we are dealing with dc sources at this point, most circuit theorems, including Thevenin's, apply equally to ac circuits.

The form of Thevenin's equivalent circuit is shown in Figure 8–29. Regardless of how complex the original circuit is, it can be always reduced to this single equivalent form. The equivalent voltage source is designated V_{TH}, and the equivalent resistance, R_{TH}.

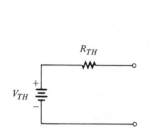

FIGURE 8–29 *Form of Thevenin's equivalent circuit.*

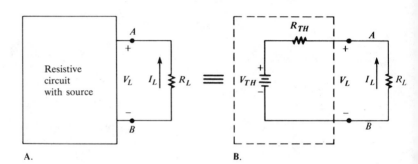

FIGURE 8–30 *Circuit equivalency.*

To apply Thevenin's theorem, you must know how to find the two quantities V_{TH} and R_{TH}. Once you have found them for a given circuit, simply connect them in series to get the complete Thevenin circuit.

Meaning of Equivalency

Figure 8–30A shows a block diagram representing a resistive circuit of any complexity. This circuit has two output terminals A and B. There is a load resistor, R_L, connected across these two terminals. The circuit inside the block produces a certain voltage, V_L, across the load, and a certain current, I_L, through the load as illustrated.

By Thevenin's theorem, the circuit in the block, regardless of how complex it is, can be reduced to an *equivalent* circuit of the form shown by the dashed lines of Figure 8–30B. The term *equivalent* means that when the *same value of load* is connected to both the original circuit (block) and Thevenin's equivalent circuit, the voltages across the loads are equal. Also, the currents through the loads are equal. Therefore, *as far as the load is concerned,* there is no difference between the original circuit and Thevenin's circuit—they are equivalent. The load resistance "sees" the same values of V_L and I_L regardless of whether it is connected to the original circuit or to Thevenin's circuit, and therefore it does not "know" the difference.

Thevenin's Equivalent Voltage (V_{TH})

As you have seen, the equivalent voltage, V_{TH}, is one part of the complete Thevenin equivalent circuit. The other part is R_{TH}. V_{TH} *is defined to be the open circuit voltage between two points in a circuit.* Any component connected between these two points effectively "sees" V_{TH} in series with R_{TH}.

To illustrate, suppose that a resistive circuit of some kind has a resistor connected between two points, as shown in Figure 8–31A. We wish to find the Thevenin circuit that is equivalent to the one shown as "seen" by R. V_{TH} is the voltage between points A and B *with R removed,* as shown in Figure 8–31B. In other words, we view the rest of the circuit from the viewpoint of the open terminals AB. R is considered external to the portion of the circuit to which Thevenin's theorem is applied. Some examples will show how to find V_{TH}.

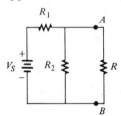

A. Original circuit

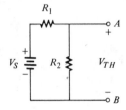

B. Remove R to open the terminals AB to get V_{TH}.

FIGURE 8–31

Example 8–10

Determine V_{TH} for the circuit within the dashed lines in Figure 8–32.

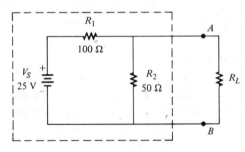

FIGURE 8–32

Solution:

Remove R_L and determine the voltage from A to B, which is V_{TH}. In this case, the voltage from A to B is the same as the voltage across R_2. We determine the voltage across R_2 using the voltage divider formula:

$$V_2 = \left(\frac{R_2}{R_1 + R_2}\right)V_S = \left(\frac{50 \ \Omega}{150 \ \Omega}\right)25 \text{ V}$$

THEVENIN'S THEOREM

$$= 8.33 \text{ V}$$
$$V_{TH} = V_{AB} = V_2 = 8.33 \text{ V}$$

Example 8–11

For the circuit in Figure 8–33, determine the Thevenin voltage V_{TH} as seen by R_L.

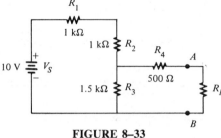

FIGURE 8–33

Solution:

Thevenin's voltage for the circuit between terminals A and B is the voltage that appears across A and B with R_L removed.

There is no voltage drop across R_4 because the open terminals AB prevent current through it. Thus, V_{AB} is the same as V_3 and can be found by the voltage divider formula:

$$V_{AB} = V_3 = \left(\frac{R_3}{R_1 + R_2 + R_3}\right) V_S$$
$$= \left(\frac{1.5 \text{ k}\Omega}{3.5 \text{ k}\Omega}\right) 10 \text{ V} = 4.29 \text{ V}$$

V_{TH} is the open terminal voltage from A to B. Therefore,

$$V_{TH} = V_{AB} = 4.29 \text{ V}$$

Example 8–12

For the circuit in Figure 8–34, find V_{TH} as seen by R_L.

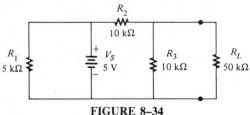

FIGURE 8–34

Example 8–12 (continued)

Solution:

First remove R_L and determine the voltage across the resulting open terminals, which is V_{TH}. We find V_{TH} by applying the voltage divider formula to R_2 and R_3:

$$V_{TH} = V_{R_3} = \left(\frac{R_3}{R_2 + R_3}\right) V_S$$

$$= \left(\frac{10 \text{ k}\Omega}{20 \text{ k}\Omega}\right) 5 \text{ V} = 2.5 \text{ V}$$

Notice that R_1 has no effect on the result since 5 V still appear across the R_2 and R_3 combination.

Thevenin's Equivalent Resistance (R_{TH})

The previous examples showed how to find only one part of a Thevenin equivalent circuit: the equivalent voltage, V_{TH}. Now we will illustrate how to find the equivalent resistance, R_{TH}. As defined by Thevenin's theorem, R_{TH} *is the total resistance appearing between two terminals in a given circuit with all sources replaced by their internal resistances.* Thus, if we wish to find R_{TH} between any two terminals in a circuit, first we *short* all voltage sources and *open* all current sources, leaving only their internal resistances, if any. Then we determine the *total* resistance between those two terminals. Some examples will illustrate how to find R_{TH}.

Example 8–13

Find R_{TH} for the circuit within the dashed lines of Figure 8–32 (Example 8–10).

Solution:

First reduce V_S to zero by shorting it, as shown in Figure 8–35.

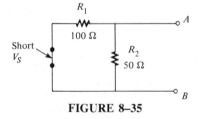

FIGURE 8–35

Looking in at terminals A and B, we see that R_1 and R_2 are in parallel. Thus,

THEVENIN'S THEOREM

$$R_{TH} = \frac{R_1 R_2}{R_1 + R_2} = \frac{(50\ \Omega)(100\ \Omega)}{150\ \Omega}$$

$$= 33.33\ \Omega$$

Example 8–14

For the circuit in Figure 8–33 (Example 8–11), determine R_{TH} as seen by R_L.

Solution:

First short the voltage source as shown in Figure 8–36.

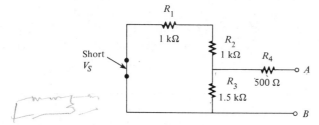

FIGURE 8–36

Looking in at the terminals A and B, we note that R_3 is in parallel with the series combination of R_1 and R_2, and this combination is in series with R_4. The calculation for R_{TH} is as follows:

$$R_{TH} = R_4 + \frac{R_3(R_1 + R_2)}{R_1 + R_2 + R_3}$$

$$= 500\ \Omega + \frac{(1.5\ k\Omega)(2\ k\Omega)}{3.5\ k\Omega} = 1.357\ k\Omega$$

Example 8–15

For the circuit in Figure 8–34 (Example 8–12), determine R_{TH} as seen by R_L.

Solution:

With the voltage source shorted, R_1 is effectively out of the circuit. R_2 and R_3 appear in parallel, as indicated in Figure 8–37. R_{TH} is calculated as follows:

Example 8–15 (continued)

$$R_{TH} = \frac{10 \text{ k}\Omega}{2} = 5 \text{ k}\Omega$$

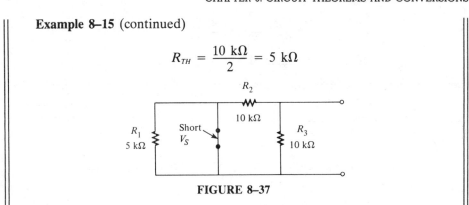

FIGURE 8–37

The previous examples have shown you how to find the two equivalent components of a Thevenin circuit, V_{TH} and R_{TH}. Keep in mind that V_{TH} and R_{TH} can be found for any circuit. Once these equivalent values are determined, they must be connected in *series* to form the Thevenin equivalent circuit. The following examples illustrate this final step.

Example 8–16

Draw the complete Thevenin circuit for the original circuit within the dashed lines of Figure 8–32 (Example 8–10).

Solution:

We found in Examples 8–10 and 8–13 that V_{TH} = 8.33 V and R_{TH} = 33.33 Ω. The Thevenin equivalent circuit is shown in Figure 8–38.

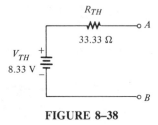

FIGURE 8–38

Example 8–17

For the original circuit in Figure 8–33 (Examples 8–11), draw the Thevenin equivalent circuit as seen by R_L.

THEVENIN'S THEOREM

Solution:

We found in Examples 8–11 and 8–14 that $V_{TH} = 4.29$ V and $R_{TH} = 1.375$ kΩ. Figure 8–39 shows these values combined to form the Thevenin equivalent circuit.

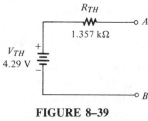

FIGURE 8–39

Example 8–18

For the original circuit in Figure 8–34 (Example 8–12), determine the Thevenin equivalent circuit as seen by R_L.

Solution:

From Examples 8–12 and 8–15, we know that V_{TH} is 2.5 V and R_{TH} is 5 kΩ. Connected in series, they produce the Thevenin equivalent circuit as shown in Figure 8–40.

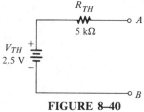

FIGURE 8–40

Summary of Thevenin's Theorem

Remember, the Thevenin equivalent circuit is *always* of the series form regardless of the original circuit that it replaces. The significance of Thevenin's theorem is that the equivalent circuit can replace the original circuit as far as any external load is concerned. Any load resistor connected between the terminals of a Thevenin equivalent circuit will have the same current through it and the same voltage across it as if it were connected to the terminals of the original circuit.

A summary of steps for applying Thevenin's theorem is as follows:

1. Open the two terminals (remove any load) between which you want to find the Thevenin equivalent circuit.

2. Determine the voltage (V_{TH}) across the two open terminals.

3. Determine the resistance (R_{TH}) between the two terminals with all voltage sources shorted and all current sources opened.

4. Connect V_{TH} and R_{TH} in series to produce the complete Thevenin equivalent for the original circuit.

Determining V_{TH} and R_{TH} by Measurement

Thevenin's theorem is largely an analytical tool that is applied theoretically in order to simplify circuit analysis. However, in many cases, Thevenin's equivalent can be found for an actual circuit by the following general measurement methods. These steps are illustrated in Figure 8–41.

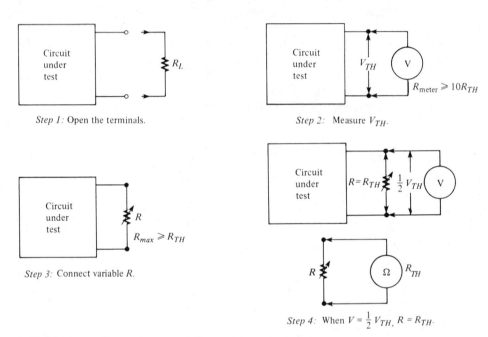

FIGURE 8–41 *Determination of Thevenin's equivalent by measurement.*

1. Remove any load from the output terminals of the circuit.

2. Measure the open terminal voltage. The voltmeter used must have an internal resistance much greater (at least 10 times greater) than the R_{TH} of the circuit. (V_{TH} is the open terminal voltage.)

3. Connect a variable resistor (rheostat) across the output terminals. Its maximum value must be greater than R_{TH}.

4. Adjust the rheostat and measure the terminal voltage. When the terminal voltage equals ½ V_{TH}, the resistance of the rheostat is equal to R_{TH}. It should be disconnected from the terminals and measured with an ohmmeter.

NORTON'S THEOREM

This procedure for determining R_{TH} differs from the theoretical procedure because it is impractical to short voltage sources or open current sources in an actual circuit.

Also, when measuring R_{TH}, be certain that the circuit is capable of providing the required current to the variable resistor load and that the variable resistor can handle the required power. These considerations may make the procedure impractical in some cases.

Review for 8-5

1. State Thevenin's theorem.
2. What are the two components of a Thevenin equivalent circuit?
3. Draw the general form of a Thevenin equivalent circuit.
4. How is V_{TH} determined?
5. How is R_{TH} determined theoretically?
6. For the original circuit in Figure 8–42, find the Thevenin equivalent circuit as seen by R_L.

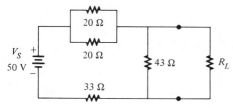

FIGURE 8–42

8-6

NORTON'S THEOREM

Like Thevenin's theorem, Norton's theorem provides a method of reducing a more complex circuit to a simpler form. The basic difference is that Norton's theorem gives *an equivalent current source in parallel with an equivalent resistance*. The form of Norton's equivalent circuit is shown in Figure 8–43. Regardless of how complex the original circuit is, it can always be reduced to this equivalent form. The equivalent current source is designated I_N, and the equivalent resistance, R_N.

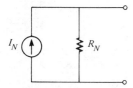

FIGURE 8–43 *Form of Norton's equivalent circuit.*

CHAPTER 8: CIRCUIT THEOREMS AND CONVERSIONS

To apply Norton's theorem, you must know how to find the two quantities I_N and R_N. Once you know them for a given circuit, simply connect them in parallel to get the complete Norton circuit.

Norton's Equivalent Current (I_N)

As stated, I_N is one part of the complete Norton equivalent circuit; R_N is the other part. I_N is defined to be the *short circuit current* between two points in a circuit. Any component connected between these two points effectively "sees" a current source of value I_N in parallel with R_N.

To illustrate, suppose that a resistive circuit of some kind has a resistor connected between two points in the circuit, as shown in Figure 8–44A. We wish to find the Norton circuit that is equivalent to the one shown as "seen" by R. To find I_N, calculate the current between points A and B with these two points *shorted*, as shown in Figure 8–44B. An example will demonstrate how to find I_N.

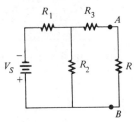
A. Original circuit

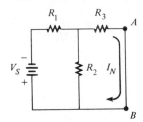
B. Short the terminals to get I_N.

FIGURE 8–44

Example 8–19

Determine I_N for the circuit within the dashed lines in Figure 8–45A.

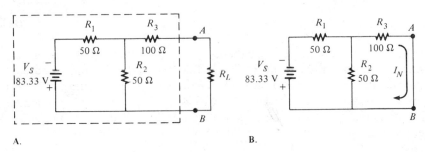

A.

B.

FIGURE 8–45

Solution:

Short terminals A and B as shown in Figure 8–45B. I_N is the current through the short and is calculated as follows: First, the total resistance seen by the voltage source is

NORTON'S THEOREM

$$R_T = R_1 + \frac{R_2 R_3}{R_2 + R_3}$$

$$= 50 \, \Omega + \frac{(50 \, \Omega)(100 \, \Omega)}{150 \, \Omega} = 83.33 \, \Omega$$

The total current from the source is

$$I_T = \frac{V_S}{R_T} = \frac{83.33 \text{ V}}{83.33 \, \Omega}$$

$$= 1 \text{ A}$$

Now apply the current divider formula to find I_N (the current through the short):

$$I_N = \left(\frac{R_2}{R_2 + R_3}\right) I_T = \left(\frac{50 \, \Omega}{150 \, \Omega}\right) 1 \text{ A}$$

$$= 0.33 \text{ A}$$

This is the value for the equivalent Norton current source.

Norton's Equivalent Resistance (R_N)

We define R_N in the same way as R_{TH}: It is the total resistance appearing between two terminals in a given circuit with all sources replaced by their internal resistances.

Example 8–20

Find R_N for the circuit within the dashed lines of Figure 8–45 (Example 8–19).

Solution:

First reduce V_S to zero by shorting it, as shown in Figure 8–46.

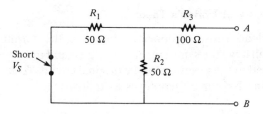

FIGURE 8–46

Looking in at terminals A and B, we see that the parallel combination of R_1 and R_2 is in series with R_3. Thus,

Example 8–20 (continued)

$$R_N = R_3 + \frac{R_1}{2} = 100 \, \Omega + \frac{50 \, \Omega}{2}$$
$$= 125 \, \Omega$$

The last two examples have shown how to find the two equivalent components of a Norton equivalent circuit, I_N and R_N. Keep in mind that these values can be found for any circuit. Once these are known, they must be connected in *parallel* to form the Norton equivalent circuit. The following example will illustrate.

Example 8–21

Draw the complete Norton circuit for the original circuit in Figure 8–45 (Example 8–19).

Solution:

We found in Examples 8–19 and 8–20 that $I_N = 0.33$ A and $R_N = 125 \, \Omega$. The Norton equivalent circuit is shown in Figure 8–47.

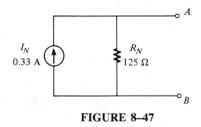

FIGURE 8–47

Summary of Norton's Theorem

Any load resistor connected between the terminals of a Norton equivalent circuit will have the same current through it and the same voltage across it as if it were connected to the terminals of the original circuit. A summary of steps for theoretically applying Norton's theorem is as follows:

1. Short the two terminals between which you want to find the Norton equivalent circuit.
2. Determine the current (I_N) through the shorted terminals.
3. Determine the resistance (R_N) between the two terminals (opened)

with all voltage sources shorted and all current sources opened ($R_N = R_{TH}$).

4. Connect I_N and R_N in parallel to produce the complete Norton equivalent for the original circuit.

Norton's equivalent circuit can also be derived from Thevenin's equivalent circuit by use of the source conversion method discussed in Section 8–3.

Review for 8–6

1. State Norton's theorem.
2. What are the two components of a Norton equivalent circuit?
3. Draw the general form of a Norton equivalent circuit.
4. How is I_N determined?
5. How is R_N determined?
6. Find the Norton circuit as seen by R in Figure 8–48.

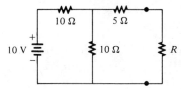

FIGURE 8–48

8–7 MILLMAN'S THEOREM

Millman's theorem allows us to reduce *any number of parallel voltage sources to a single equivalent voltage source*. It simplifies finding the voltage across or current through a load. Millman's theorem gives the same results as Thevenin's theorem for the special case of parallel voltage sources. A conversion by Millman's theorem is illustrated in Figure 8–49.

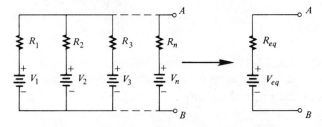

FIGURE 8–49 *Reduction of parallel voltage sources to a single equivalent voltage source.*

Millman's Equivalent Voltage (V_{eq}) and Equivalent Resistance (R_{eq})

Millman's theorem gives us a formula for calculating the equivalent voltage, V_{eq}. To find V_{eq}, convert each of the parallel voltage sources into current sources, as shown in Figure 8–50.

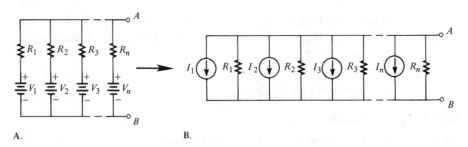

FIGURE 8–50 *Parallel voltage sources converted to current sources.*

In Figure 8–50B, the total current from the parallel current sources is

$$I_T = I_1 + I_2 + I_3 + \cdots + I_n$$

The total conductance between terminals A and B is

$$G_T = G_1 + G_2 + G_3 + \cdots + G_n$$

where $G_T = 1/R_T$, $G_1 = 1/R_1$, and so on. Remember, the current sources are effectively open. Therefore, by Millman's theorem, the *equivalent resistance* is the total resistance R_T:

$$R_{eq} = \frac{1}{G_T} = \frac{1}{(1/R_1) + (1/R_2) + (1/R_3) + \cdots + (1/R_n)} \qquad (8\text{--}3)$$

By Millman's theorem, the *equivalent voltage* is $I_T R_{eq}$, where I_T is expressed as follows:

$$I_T = I_1 + I_2 + I_3 + \cdots + I_n$$
$$= \frac{V_1}{R_1} + \frac{V_2}{R_2} + \frac{V_3}{R_3} + \cdots + \frac{V_n}{R_n}$$

The following is the formula for the equivalent voltage:

$$V_{eq} = \frac{(V_1/R_1) + (V_2/R_2) + (V_3/R_3) + \cdots + (V_n/R_n)}{(1/R_1) + (1/R_2) + (1/R_3) + \cdots + (1/R_n)} \qquad (8\text{--}4)$$

Equations (8–3) and (8–4) are the two Millman formulas. The equivalent voltage source has a polarity such that the total current through a load will be in the same direction as in the original circuit.

MILLMAN'S THEOREM

Example 8–22

Use Millman's theorem to find the voltage across R_L and the current through R_L in Figure 8–51.

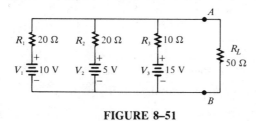

FIGURE 8–51

Solution:

Apply Millman's theorem as follows:

$$R_{eq} = \frac{1}{(1/R_1) + (1/R_2) + (1/R_3)}$$

$$= \frac{1}{(1/20\ \Omega) + (1/20\ \Omega) + (1/10\ \Omega)}$$

$$= \frac{1}{0.2} = 5\ \Omega$$

$$V_{eq} = \frac{(V_1/R_1) + (V_2/R_2) + (V_3/R_3)}{(1/R_1) + (1/R_2) + (1/R_3)}$$

$$= \frac{(10\ \text{V}/20\ \Omega) + (5\ \text{V}/20\ \Omega) + (15\ \text{V}/10\ \Omega)}{(1/20\ \Omega) + (1/20\ \Omega) + (1/10\ \Omega)}$$

$$= \frac{2.25\ \text{A}}{0.2\ \text{S}} = 11.25\ \text{V}$$

The single equivalent voltage source is shown in Figure 8–52.

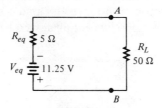

FIGURE 8–52

Now we calculate I_L and V_L for the load resistor.

$$I_L = \frac{V_{eq}}{R_{eq} + R_L} = \frac{11.25\ \text{V}}{55\ \Omega}$$

$$= 0.205\ \text{A}$$

> **Example 8–22** (continued)
> $$V_L = I_L R_L = (0.205 \text{ A})(50 \text{ }\Omega)$$
> $$= 10.25 \text{ V}$$

Review for 8–7

1. To what type of circuit does Millman's theorem apply?
2. Write the Millman theorem formula for R_{eq}.
3. Write the Millman theorem formula for V_{eq}.
4. Find the load current and the load voltage in Figure 8–53.

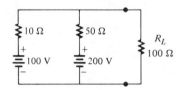

FIGURE 8–53

8–8

MAXIMUM POWER TRANSFER THEOREM

The maximum power transfer theorem states that when a circuit is connected to a load, *maximum power is delivered to the load when the load resistance is equal to the source resistance of the circuit*. The source resistance of a circuit is the equivalent resistance as viewed from the output terminals and using Thevenin's theorem. An equivalent circuit with its source resistance and load is shown in Figure 8–54. When $R_L = R_S$, the maximum power possible is transferred from the voltage source to R_L.

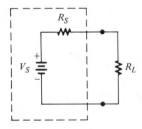

FIGURE 8–54

Practical applications of this theorem include audio systems such as stereo, radio, public address, and so on. In these systems the resistance of the speaker is the load. The circuit that drives the speaker is a power amplifier. The

MAXIMUM POWER TRANSFER THEOREM

systems are sometimes optimized for maximum power to the speakers. Thus, the resistance of the speaker must equal the source resistance of the amplifier (sometimes called the *output resistance*).

Example 8–23 will show that maximum power occurs at the *matched* condition (that is, when $R_L = R_s$).

Example 8–23

The circuit to the left of terminals A and B in Figure 8–55 provides power to the load, R_L. It is the Thevenin equivalent of a more complex circuit. Calculate the power delivered to R_L for each following value of R_L: 1 Ω, 5 Ω, 10 Ω, 15 Ω, and 20 Ω.

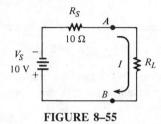

FIGURE 8–55

Solution:

For $R_L = 1\ \Omega$:

$$I = \frac{10\text{ V}}{11\ \Omega}$$
$$= 0.91\text{ A}$$
$$P_L = I^2 R_L = (0.91\text{ A})^2(1\ \Omega)$$
$$= 0.83\text{ W}$$

For $R_L = 5\ \Omega$:

$$I = \frac{10\text{ V}}{15\ \Omega}$$
$$= 0.67\text{ A}$$
$$P_L = I^2 R_L = (0.67\text{ A})^2(5\ \Omega)$$
$$= 2.24\text{ W}$$

For $R_L = 10\ \Omega$:

$$I = \frac{10\text{ V}}{20\ \Omega}$$
$$= 0.5\text{ A}$$
$$P_L = I^2 R_L = (0.5\text{ A})^2(10\ \Omega)$$
$$= 2.50\text{ W}$$

Example 8–23 (continued)

For $R_L = 15\ \Omega$:

$$I = \frac{10\ \text{V}}{25\ \Omega}$$
$$= 0.4\ \text{A}$$
$$P_L = I^2R_L = (0.4\ \text{A})^2(15\ \Omega)$$
$$= 2.40\ \text{W}$$

For $R_L = 20\ \Omega$:

$$I = \frac{10\ \text{V}}{30\ \Omega}$$
$$= 0.33\ \text{A}$$
$$P_L = I^2R_L = (0.33\ \text{A})^2(20\ \Omega)$$
$$= 2.18\ \text{W}$$

Notice in Example 8–23 that the maximum load power occurs when $R_L = R_S = 10\ \Omega$. If R_L is decreased below this value or increased above it, the power falls off, as shown in the graph in Figure 8–56.

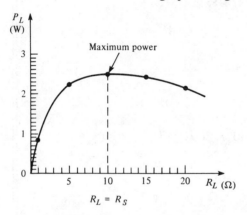

FIGURE 8–56 *Load power versus load resistance.*

Review for 8–8

1. State the maximum power transfer theorem.
2. When is maximum power delivered from a source to a load?
3. A given circuit has a source resistance of 50 Ω. What will be the value of the load to which the maximum power is delivered?

8–9 DELTA-WYE (Δ-Y) AND WYE-DELTA (Y-Δ) NETWORK CONVERSIONS

A resistive delta (Δ) network has the form shown in Figure 8–57A. A wye (Y) network is shown in Figure 8–57B. Notice that letter subscripts are used to designate resistors in the delta network, and numerical subscripts are used to designate resistors in the wye.

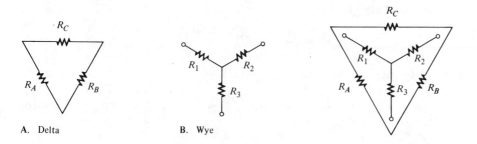

FIGURE 8–57 *Delta and wye networks.*

FIGURE 8–58 *"Y within Δ" aid for conversion formulas.*

Conversion between these two forms of circuits is sometimes helpful. In this section, the conversion formulas and rules for remembering them are given.

Δ-to-Y Conversion

It is convenient to think of the wye positioned within the delta, as shown in Figure 8–58. To convert from delta to wye, we need R_1, R_2, and R_3 in terms of R_A, R_B, and R_C. The conversion rule is as follows: *Each resistor in the wye is equal to the product of the resistors in two adjacent delta branches, divided by the sum of all three delta resistors.*

In Figure 8–58, R_A and R_C are "adjacent" to R_1:

$$R_1 = \frac{R_A R_C}{R_A + R_B + R_C} \tag{8-5}$$

Also, R_B and R_C are "adjacent" to R_2:

$$R_2 = \frac{R_B R_C}{R_A + R_B + R_C} \tag{8-6}$$

and R_A and R_B are "adjacent" to R_3:

$$R_3 = \frac{R_A R_B}{R_A + R_B + R_C} \tag{8-7}$$

Y-to-Δ Conversion

To convert from wye to delta, we need R_A, R_B, and R_C in terms of R_1, R_2, and R_3. The conversion rule is as follows: *Each resistor in the delta is equal to the sum of all possible products of wye resistors taken two at a time, divided by the opposite wye resistor.*

In Figure 8–58, R_2 is "opposite" to R_A:

$$R_A = \frac{R_1 R_2 + R_1 R_3 + R_2 R_3}{R_2} \qquad (8\text{–}8)$$

Also, R_1 is "opposite" to R_B:

$$R_B = \frac{R_1 R_2 + R_1 R_3 + R_2 R_3}{R_1} \qquad (8\text{–}9)$$

and R_3 is "opposite" to R_C:

$$R_C = \frac{R_1 R_2 + R_1 R_3 + R_2 R_3}{R_3} \qquad (8\text{–}10)$$

Example 8–24

Convert the delta network in Figure 8–59 to a wye network.

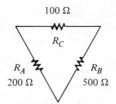

FIGURE 8–59

Solution:

Use Equations (8–5), (8–6), and (8–7):

$$R_1 = \frac{R_A R_C}{R_A + R_B + R_C} = \frac{(200\ \Omega)(100\ \Omega)}{200\ \Omega + 500\ \Omega + 100\ \Omega}$$
$$= 25\ \Omega$$

$$R_2 = \frac{R_B R_C}{R_A + R_B + R_C} = \frac{(500\ \Omega)(100\ \Omega)}{800\ \Omega}$$
$$= 62.5\ \Omega$$

$$R_3 = \frac{R_A R_B}{R_A + R_B + R_C} = \frac{(200\ \Omega)(500\ \Omega)}{800\ \Omega}$$
$$= 125\ \Omega$$

DELTA-WYE (Δ-Y) AND WYE-DELTA (Y-Δ) NETWORK CONVERSIONS

The resulting wye network is shown in Figure 8–60.

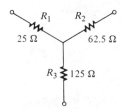

FIGURE 8–60

Example 8–25

Convert the wye network in Figure 8–61 to a delta network.

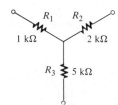

FIGURE 8–61

Solution:

Use Equations (8–8), (8–9), and (8–10):

$$R_A = \frac{R_1 R_2 + R_1 R_3 + R_2 R_3}{R_2}$$

$$= \frac{(1 \text{ k}\Omega)(2 \text{ k}\Omega) + (1 \text{ k}\Omega)(5 \text{ k}\Omega) + (2 \text{ k}\Omega)(5 \text{ k}\Omega)}{2 \text{ k}\Omega} = 8.5 \text{ k}\Omega$$

$$R_B = \frac{R_1 R_2 + R_1 R_3 + R_2 R_3}{R_1}$$

$$= \frac{(1 \text{ k}\Omega)(2 \text{ k}\Omega) + (1 \text{ k}\Omega)(5 \text{ k}\Omega) + (2 \text{ k}\Omega)(5 \text{ k}\Omega)}{1 \text{ k}\Omega} = 17 \text{ k}\Omega$$

$$R_C = \frac{R_1 R_2 + R_1 R_3 + R_2 R_3}{R_3}$$

$$= \frac{(1 \text{ k}\Omega)(2 \text{ k}\Omega) + (1 \text{ k}\Omega)(5 \text{ k}\Omega) + (2 \text{ k}\Omega)(5 \text{ k}\Omega)}{5 \text{ k}\Omega} = 3.4 \text{ k}\Omega$$

The resulting delta network is shown in Figure 8–62 on page 250.

Example 8–25 (continued)

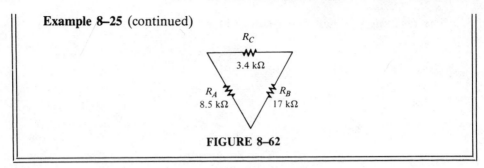

FIGURE 8–62

Review for 8–9

1. Sketch a delta network.
2. Sketch a wye network.
3. Write the formulas for delta-to-wye conversion.
4. Write the formulas for wye-to-delta conversion.

Formulas

$$I_S = \frac{V_S}{R_S} \tag{8-1}$$

$$V_S = I_S R_S \tag{8-2}$$

$$R_{eq} = \frac{1}{G_T} = \frac{1}{(1/R_1) + (1/R_2) + (1/R_3) + \cdots + (1/R_n)} \tag{8-3}$$

$$V_{eq} = \frac{(V_1/R_1) + (V_2/R_2) + (V_3/R_3) + \cdots + (V_n/R_n)}{(1/R_1) + (1/R_2) + (1/R_3) + \cdots + (1/R_n)} \tag{8-4}$$

Δ-to-Y conversions:

$$R_1 = \frac{R_A R_C}{R_A + R_B + R_C} \tag{8-5}$$

$$R_2 = \frac{R_B R_C}{R_A + R_B + R_C} \tag{8-6}$$

$$R_3 = \frac{R_A R_B}{R_A + R_B + R_C} \tag{8-7}$$

Y-to-Δ conversions:

$$R_A = \frac{R_1 R_2 + R_1 R_3 + R_2 R_3}{R_2} \tag{8-8}$$

$$R_B = \frac{R_1 R_2 + R_1 R_3 + R_2 R_3}{R_1} \tag{8-9}$$

$$R_C = \frac{R_1 R_2 + R_1 R_3 + R_2 R_3}{R_3} \tag{8-10}$$

Summary

1. An ideal voltage source has zero internal resistance. It provides a constant voltage across its terminals regardless of the load resistance.
2. A practical voltage source has a nonzero internal resistance. Its terminal voltage is essentially constant when $R_L \geq 10R_s$ (rule of thumb).
3. An ideal current source has infinite internal resistance. It provides a constant current regardless of the load resistance.
4. A practical current source has a finite internal resistance. Its current is essentially constant when $10R_L \leq R_S$.
5. The superposition theorem is useful for multiple-source circuits.
6. Thevenin's theorem provides for the reduction of any resistive circuit to an equivalent form consisting of an equivalent voltage source in series with an equivalent resistance.
7. The term *equivalency*, as used in Thevenin's and Norton's theorems, means that when a given load resistance is connected to the equivalent circuit, it will have the same voltage across it and the same current through it as when it was connected to the original circuit.
8. Norton's theorem provides for the reduction of any resistive circuit to an equivalent form consisting of an equivalent current source in parallel with an equivalent resistance.
9. Millman's theorem provides for the reduction of parallel voltage sources to a single equivalent voltage source consisting of an equivalent voltage and an equivalent series resistance.
10. Maximum power is transferred to a load from a source when the load resistance equals the source resistance.

Self-Test

1. A voltage source has the values $V_s = 25$ V and $R_s = 5\,\Omega$. What are the values for the equivalent current source?
2. Convert the voltage source in Figure 8–63 to an equivalent current source.

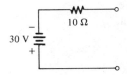
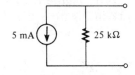

FIGURE 8–63   **FIGURE 8–64**

3. Convert the current source in Figure 8–64 to an equivalent voltage source.

4. In Figure 8–65, use the superposition theorem to find the total current in R_3.

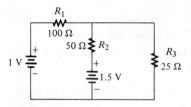

FIGURE 8–65

5. In Figure 8–65, what is the total current through R_2?
6. For Figure 8–66, determine the Thevenin equivalent circuit as seen by R.

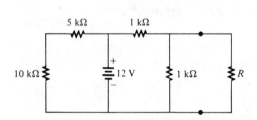

FIGURE 8–66

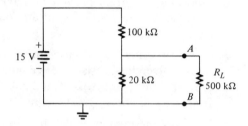

FIGURE 8–67

7. Using Thevenin's theorem, find the current in R_L for the circuit in Figure 8–67.
8. Reduce the circuit in Figure 8–67 to its Norton equivalent as seen by R_L.
9. Use Millman's theorem to simplify the circuit in Figure 8–68 to a single voltage source.

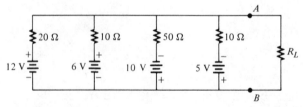

FIGURE 8–68

10. What value of R_L in Figure 8–68 is required for maximum power transfer?
11. (a) Convert the delta network in Figure 8–69A to a wye.
    (b) Convert the wye network in Figure 8–69B to a delta.

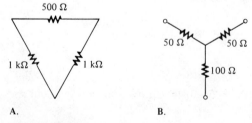

A.   B.

FIGURE 8–69

Problems

8–1. A voltage source has the values $V_s = 300$ V and $R_s = 50\ \Omega$. Convert it to an equivalent current source.

8–2. Convert the practical voltage sources in Figure 8–70 to equivalent current sources.

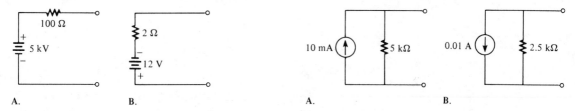

FIGURE 8–70   FIGURE 8–71

8–3. A current source has an I_s of 600 mA and an R_s of 1.2 kΩ. Convert it to an equivalent voltage source.

8–4. Convert the practical current sources in Figure 8–71 to equivalent voltage sources.

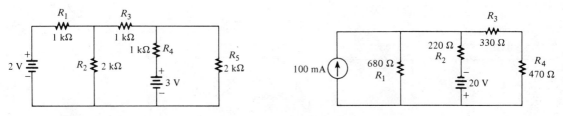

FIGURE 8–72   FIGURE 8–73

8–5. Using the superposition method, calculate the current in the right-most branch of Figure 8–72.

8–6. Use the superposition theorem to find the current in and the voltage across the R_2 branch of Figure 8–72.

8–7. Using the superposition theorem, solve for the current through R_3 in Figure 8–73.

8–8. Using the superposition theorem, find the load current in each circuit of Figure 8–74.

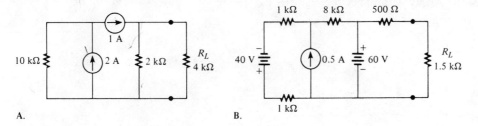

FIGURE 8–74

8–9. For each circuit in Figure 8–75, determine the Thevenin equivalent as seen by R_L.

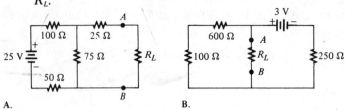

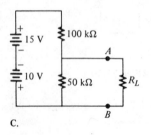

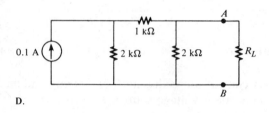

FIGURE 8–75

8–10. Using Thevenin's theorem, determine the current through the load R_L in Figure 8–76.

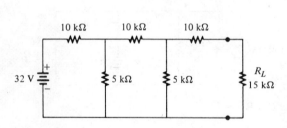

FIGURE 8–76

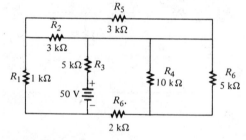

FIGURE 8–77

8–11. Using Thevenin's theorem, find the voltage across R_4 in Figure 8–77.

8–12. For each circuit in Figure 8–75, determine the Norton equivalent as seen by R_L.

8–13. Using Norton's theorem, find the current through the load resistor R_L in Figure 8–76.

8–14. Using Norton's theorem, find the voltage across R_4 in Figure 8–77.

8–15. Apply Millman's theorem to the circuit of Figure 8–78.

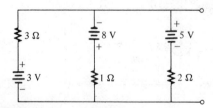

FIGURE 8–78

PROBLEMS

8–16. Use Millman's theorem and source conversions to reduce the circuit in Figure 8–79 to a single voltage source.

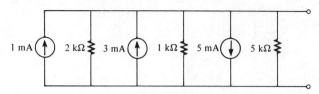

FIGURE 8–79

8–17. For each circuit in Figure 8–80, maximum power is to be transferred to the load R_L. Determine the appropriate value for R_L in each case.

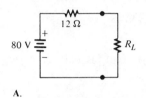

A.

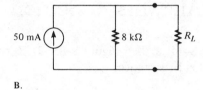

B.

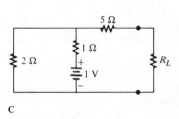

C.

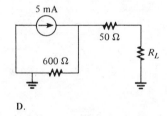

D.

FIGURE 8–80

8–18. Determine R_L for maximum power in Figure 8–81.

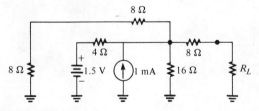

FIGURE 8–81

8–19. In Figure 8–82, convert each delta network to a wye network.

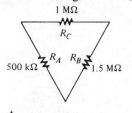

A.

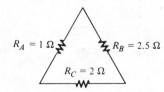

B.

FIGURE 8–82

8–20. In Figure 8–83, convert each wye network to a delta network.

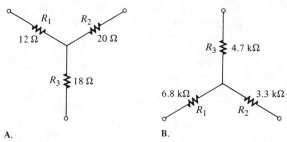

A. **B.**

FIGURE 8–83

Answers to Section Reviews

Section 8–1:
1. See Figure 8–84. **2.** See Figure 8–85. **3.** Zero ohms. **4.** Output voltage varies directly with load resistance.

FIGURE 8–84 **FIGURE 8–85**

Section 8–2:
1. See Figure 8–86. **2.** See Figure 8–87. **3.** Infinite. **4.** Load current varies inversely with load resistance.

FIGURE 8–86 **FIGURE 8–87**

Section 8–3:
1. $I_S = V_S/R_S$. **2.** $V_S = I_S R_S$. **3.** See Figure 8–88. **4.** See Figure 8–89.

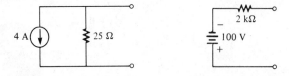

FIGURE 8–88 **FIGURE 8–89**

Section 8–4:
1. The total current in any branch of a multiple-source circuit is equal to the algebraic sum of the currents due to the individual sources acting alone, with the

ANSWERS TO SECTION REVIEWS

other sources replaced by their internal resistances. **2.** Because it allows each source to be treated independently. **3.** A short simulates the internal resistance of an ideal voltage source. An open simulates the internal resistance of an ideal current source. **4.** 6.67 mA. **5.** In the direction of the larger current.

Section 8–5:
1. Any resistive network can be replaced by an equivalent circuit consisting of an equivalent voltage source and an equivalent series resistance. **2.** V_{TH} and R_{TH}. **3.** See Figure 8–90. **4.** V_{TH} is the open circuit voltage between two terminals in a circuit. **5.** R_{TH} is the resistance as viewed from two terminals in a circuit, with all sources replaced by their internal resistances. **6.** See Figure 8–91.

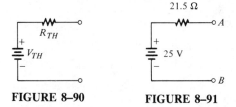

FIGURE 8–90 **FIGURE 8–91**

Section 8–6:
1. Any resistive network can be replaced by an equivalent circuit consisting of an equivalent current source and an equivalent parallel resistance. **2.** I_N and R_N. **3.** See Figure 8–92. **4.** I_N is the short circuit current between two terminals in a circuit. **5.** R_N is the resistance as viewed from the two open terminals in a circuit. **6.** See Figure 8–93.

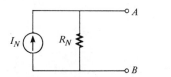 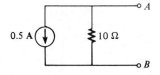

FIGURE 8–92 **FIGURE 8–93**

Section 8–7:
1. Parallel voltage sources. **2.** The equation is as follows:

$$R_{eq} = \frac{1}{(1/R_1) + (1/R_2) + (1/R_3) + \cdots + (1/R_n)}$$

3. The equation is as follows:

$$V_{eq} = \frac{(V_1/R_1) + (V_2/R_2) + (V_3/R_3) + \cdots + (V_n/R_n)}{(1/R_1) + (1/R_2) + (1/R_3) + \cdots + (1/R_n)}$$

4. $I_L = 1.1$ A; $V_L = 110$ V.

Section 8–8:
1. Maximum power is transferred from a source to a load when the load resistance is equal to the source resistance. **2.** When $R_L = R_S$. **3.** 50 Ω.

Section 8–9:
1. See Figure 8–94. **2.** See Figure 8–95. **3.** The equations are as follows:

$$R_1 = \frac{R_A R_C}{R_A + R_B + R_C}$$

$$R_2 = \frac{R_B R_C}{R_A + R_B + R_C}$$

$$R_3 = \frac{R_A R_B}{R_A + R_B + R_C}$$

4. The equations are as follows:

$$R_A = \frac{R_1 R_2 + R_1 R_3 + R_2 R_3}{R_2}$$

$$R_B = \frac{R_1 R_2 + R_1 R_3 + R_2 R_3}{R_1}$$

$$R_C = \frac{R_1 R_2 + R_1 R_3 + R_2 R_3}{R_3}$$

FIGURE 8–94

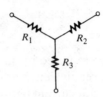

FIGURE 8–95

In this chapter we will discuss three circuit anaylsis methods based on Kirchhoff's voltage and current laws. These methods are particularly useful in the analysis of circuits with two or more voltage or current sources. The methods can be used alone or in conjunction with the circuit theorems and other techniques covered in previous chapters. Each of these methods can be used to solve the same circuit problem. With experience, you will learn which method is best for a particular problem, or you may develop a preference for one of them.

The solution of simultaneous equations by determinants is also discussed in this chapter.

9–1 Branch Current Method
9–2 Determinants
9–3 Mesh Current Method
9–4 Node Voltage Method

CIRCUIT ANALYSIS METHODS

9-1

BRANCH CURRENT METHOD

In the branch current method, we use Kirchhoff's voltage law and Kirchhoff's current law to solve for the current in each branch of a circuit. Once we know the currents, we can find the voltages.

Loops and Nodes

Figure 9-1 shows a circuit with two voltage sources. It will be used as the basic model throughout the chapter to illustrate each of the circuit analysis methods. In this circuit, there are two *closed loops*, as indicated by arrows 1 and 2. A loop is a complete current path within a circuit. Also, there are four *nodes* in this circuit, as indicated by the letters A, B, C, and D. A node is a junction where two or more current paths come together.

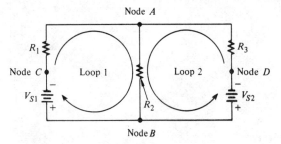

FIGURE 9-1 *Basic multiple-source circuit showing loops and nodes.*

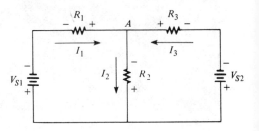

FIGURE 9-2 *Circuit for demonstrating branch current analysis.*

The following are the general steps used in applying the *branch current method*. These steps will be demonstrated with aid of Figure 9-2.

1. Assign a current in each circuit branch in an *arbitrary* direction.
2. Show the polarities of the resistor voltages according to the assigned branch current directions.
3. Apply Kirchhoff's voltage law around each closed loop (sum of voltages is equal to zero).
4. Apply Kirchhoff's current law at the *minimum* number of nodes so that *all* branch currents are included (sum of currents at a node equals zero).
5. Solve the equations resulting from Steps 3 and 4 for the branch current values.

First, the branch currents are assigned in the direction shown. Do not worry about the *actual* current directions at this point.

BRANCH CURRENT METHOD

Second, the polarities of the voltage drops across R_1, R_2, and R_3 are indicated in the figure according to the current directions.

Third, Kirchhoff's voltage law applied to the two loops gives the following equations:

$$R_1I_1 + R_2I_2 - V_{S1} = 0 \quad \text{for loop 1}$$
$$R_2I_2 + R_3I_3 - V_{S2} = 0 \quad \text{for loop 2}$$

Fourth, Kirchhoff's current law is applied to node A, including all branch currents as follows:

$$I_1 - I_2 + I_3 = 0$$

The negative sign indicates that I_2 is out of the junction.

Fifth and last, the equations must be solved for I_1, I_2, and I_3. The three equations in the above steps are called *simultaneous equations* and can be solved in two ways: by *substitution* or by *determinants*. You will see how to solve equations by substitution in Example 9–1. In the next section we will study the use of determinants and apply it to the later methods.

Example 9–1

Use the branch current method to find each branch current in Figure 9–3.

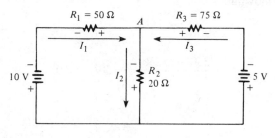

FIGURE 9–3

Solution:

Step 1: Assign branch currents as shown in Figure 9–3. Keep in mind that you can assume any current direction at this point, and the final solution will have a negative sign if the *actual* current is opposite to the assigned current.

Step 2: Mark the polarities of the resistor voltage drops as shown in the figure.

Step 3: Kirchhoff's voltage law around the left loop gives

$$50I_1 + 20I_2 = 10$$

Example 9–1 (continued)

Around the right loop we get

$$20I_2 + 75I_3 = 5$$

Step 4: At node A, the current equation is

$$I_1 - I_2 + I_3 = 0$$

Step 5: The equations are solved by substitution as follows. First find I_1 in terms of I_2 and I_3:

$$I_1 = I_2 - I_3$$

Now substitute $I_2 - I_3$ for I_1 in the first loop equation:

$$50(I_2 - I_3) + 20I_2 = 10$$
$$50I_2 - 50I_3 + 20I_2 = 10$$
$$70I_2 - 50I_3 = 10$$

Next, take the second loop equation and solve for I_2 in terms of I_3:

$$20I_2 = 5 - 75I_3$$
$$I_2 = \frac{5 - 75I_3}{20}$$

Substituting this expression for I_2 into $70I_2 - 50I_3 = 10$, we get the following:

$$70\left(\frac{5 - 75I_3}{20}\right) - 50I_3 = 10$$
$$\frac{350 - 5250I_3}{20} - 50I_3 = 10$$
$$17.5 - 262.5I_3 - 50I_3 = 10$$
$$-312.5I_3 = -7.5$$
$$I_3 = \frac{7.5}{312.5}$$
$$I_3 = 0.024 \text{ A}$$

Now, substitute this value of I_3 into the second loop equation:

$$20I_2 + 75(0.024) = 5$$

Solve for I_2:

$$I_2 = \frac{5 - 75(0.024)}{20}$$
$$= \frac{3.2}{20} = 0.16 \text{ A}$$

DETERMINANTS

Substituting I_2 and I_3 values into the current equation at node A, we obtain

$$I_1 - 0.16 + 0.024 = 0$$
$$I_1 = 0.16 - 0.024$$
$$= 0.136 \text{ A}$$

Review for 9-1

1. What basic circuit laws are used in the branch current method?
2. When assigning branch currents, you should be careful of the directions (T or F).
3. What is a loop?
4. What is a node?

9-2

DETERMINANTS

When there are several unknown quantities to be found, such as the three currents in the last example, you must have a number of equations equal to the number of unknowns. In this section you will learn how to solve for two unknowns using a systematic method known as *determinants*. This method is an alternate to the substitution method, which we used in the previous section.

Solving Two Simultaneous Equations

To illustrate the method of *second-order determinants*, we will assume two loop equations as follows:

$$10I_1 + 5I_2 = 15$$
$$2I_1 + 4I_2 = 8$$

We want to find the value of I_1 and I_2. To do so, we form a *determinant* with the coefficients of the unknown currents. A *coefficient* is the number associated with an unknown. For example, 10 is the coefficient for I_1 in the first equation.

The first column in the determinant consists of the coefficients of I_1, and the second column consists of the coefficients of I_2. The resulting determinant appears as follows:

$$\begin{vmatrix} 10 & 5 \\ 2 & 4 \end{vmatrix}$$

This is called the *characteristic determinant* for the set of equations.

Next, we form another determinant and use it in conjunction with the characteristic determinant to solve for I_1. We form the determinant for our example by replacing the coefficients of I_1 in the characteristic determinant with the *constants* on the right side of the equations. Doing this, we get the following determinant:

$$\begin{vmatrix} 15 & 5 \\ 8 & 4 \end{vmatrix}$$

We can now solve for I_1 by *evaluating* each determinant and then dividing by the characteristic determinant. To evaluate the determinants, we cross-multiply with appropriate signs and sum the resulting products:

$$\begin{vmatrix} 10 & 5 \\ 2 & 4 \end{vmatrix} = (10)(4) - (2)(5)$$
$$= 40 - 10 = 30$$

The value of this characteristic determinant is 30. Next we cross-multiply the second determinant:

$$\begin{vmatrix} 15 & 5 \\ 8 & 4 \end{vmatrix} = (15)(4) - (8)(5)$$
$$= 60 - 40 = 20$$

The value of this determinant is 20. Now we can solve for I_1 as follows:

$$I_1 = \frac{\begin{vmatrix} 15 & 5 \\ 8 & 4 \end{vmatrix}}{\begin{vmatrix} 10 & 5 \\ 2 & 4 \end{vmatrix}}$$

$$= \frac{20}{30} = 0.667 \text{ A}$$

To find I_2, we form another determinant by substituting the *constants* on the right side of the equations for the coefficients of I_2:

$$\begin{vmatrix} 10 & 15 \\ 2 & 8 \end{vmatrix}$$

We solve for I_2 by dividing this determinant by the characteristic determinant already evaluated:

$$I_2 = \frac{\begin{vmatrix} 10 & 15 \\ 2 & 8 \end{vmatrix}}{30} = \frac{(10)(8) - (2)(15)}{30}$$

$$= \frac{80 - 30}{30} = \frac{50}{30} = 1.67 \text{ A}$$

Three equations with three unknowns can be solved with *third-order determinants*. Since we will limit most of our coverage of analysis methods in this book to those producing only two unknowns, third-order determinants are not covered.

MESH CURRENT METHOD

Example 9–2

Solve the following set of equations for the unknown currents:

$$2I_1 - 5I_2 = 10$$
$$6I_1 + 10I_2 = 20$$

Solution:

The characteristic determinant is

$$\begin{vmatrix} 2 & -5 \\ 6 & 10 \end{vmatrix} = (2)(10) - (-5)(6)$$
$$= 20 - (-30) = 20 + 30 = 50$$

Solving for I_1 yields

$$I_1 = \frac{\begin{vmatrix} 10 & -5 \\ 20 & 10 \end{vmatrix}}{50} = \frac{(10)(10) - (-5)(20)}{50}$$
$$= \frac{100 - (-100)}{50} = \frac{200}{50} = 4 \text{ A}$$

Solving for I_2 yields

$$I_2 = \frac{\begin{vmatrix} 2 & 10 \\ 6 & 20 \end{vmatrix}}{50} = \frac{(2)(20) - (6)(10)}{50}$$
$$= \frac{40 - 60}{50} = -0.4 \text{ A}$$

In a circuit problem, the negative sign would indicate that the direction of actual current is opposite to the assigned direction.

Review for 9–2

1. What is a determinant?
2. For what are second-order determinants used?

9–3

MESH CURRENT METHOD

In the mesh current method, we will work with *loop currents* rather than branch currents. As you perhaps realize, a branch current is the *actual* current through a branch. An ammeter in that branch will measure the value of the branch current. Loop currents are abstract or fictitious quanities that are used to make circuit analysis somewhat easier than it is with the branch current method. Keep this in mind as we proceed through this section. The term *mesh* comes from the fact that a multiple-loop circuit, when drawn out, resembles a wire mesh.

A *systematic* method of mesh analysis is listed in the following steps and is illustrated in Figure 9–4, which is the same circuit configuration used in the branch current section. It demonstrates the basic principles well.

1. Assign a current in the *clockwise* (CW) direction around each closed loop. This may not be the actual current direction, but it does not matter. The number of current assignments must be sufficient to include current through all components in the circuit. No *redundant* current assignments should be made.
2. Indicate the voltage drop polarities in each loop based on the *assigned* current directions.
3. Apply Kirchhoff's voltage law around each closed loop. When more than one loop current passes through a component, include its voltage drop.
4. Using substitution or determinants, solve the resulting equations simultaneously for the loop currents.

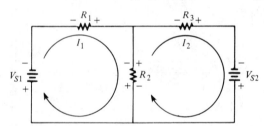

FIGURE 9–4 *Circuit for mesh analysis.*

First, the loop currents I_1 and I_2 are assigned in the CW direction as shown in the figure. A loop current could be assigned around the outer perimter of the circuit, but this information would be redundant since I_1 and I_2 already pass through all of the components.

Second, the polarities of the voltage drops across R_1, R_2, and R_3 are shown based on the loop current directions. Notice that I_1 and I_2 flow in opposite directions through R_2 because R_2 is common to both loops. Therefore, two *fictitious* voltage polarities are indicated. In reality, R_2 currents cannot be separated into two parts, but remember that the loop currents are basically abstract quantities used for analysis purposes. The polarities of the voltage sources are fixed and are not affected by the current assignments.

Third, Kirchhoff's voltage law applied to the two loops results in the following two equations:

$$R_1 I_1 + R_2(I_1 - I_2) = V_{S1} \quad \text{for loop 1}$$
$$R_3 I_2 + R_2(I_2 - I_1) = V_{S2} \quad \text{for loop 2}$$

MESH CURRENT METHOD

Fourth, the like terms in the equations are combined and rearranged for convenient solution. The equations are rearranged into the following form. Once the loop currents are evaluated, all of the branch currents can be determined.

$$(R_1 + R_2)I_1 - R_2I_2 = V_{S1} \quad \text{for loop 1}$$
$$-R_2I_1 + (R_2 + R_3)I_2 = V_{S2} \quad \text{for loop 2}$$

Notice that only *two* equations are required for the same circuit that required *three* equations in the branch current method. The last two equations follow a certain form which can be used as a *format* to make mesh analysis easier. Referring to these last two equations, notice that for loop 1, the total resistance in the loop, $R_1 + R_2$, is multiplied by I_1 (its loop current). Also in the loop 1 equation, the common resistance R_2 is multiplied by the other loop current I_2 and subtracted from the first term. The same general form is seen in the loop 2 equation. From these observations, a set of rules can be established for applying the same format repeatedly to each loop:

1. Sum the resistances around the loop, and multiply by the loop current.
2. Subtract the common resistance(s) times the adjacent loop current(s).
3. Set the terms in Steps 1 and 2 equal to the total source voltage in the loop. The sign of the source voltage is positive if the assigned loop current is *out of* its negative terminal. The sign is negative if the loop current is *into* its negative terminal.

Example 9–3 will illustrate the application of these rules to the mesh current analysis of a circuit.

Example 9–3

Using the mesh current method, find the branch currents in Figure 9–5, which is the same circuit as in Example 9–1.

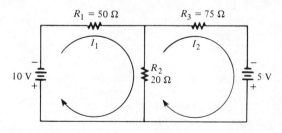

FIGURE 9–5

Solution:

The loop currents are assigned as shown. The format rules are followed for setting up the two equations.

Example 9-3 (continued)

$$(50 + 20)I_1 - 20I_2 = 10 \text{ for loop 1}$$
$$70I_1 - 20I_2 = 10$$
$$-20I_1 + (20 + 75)I_2 = -5 \text{ for loop 2}$$
$$-20I_1 + 95I_2 = -5$$

Using determinants to find I_1, we obtain

$$I_1 = \frac{\begin{vmatrix} 10 & -20 \\ -5 & 95 \end{vmatrix}}{\begin{vmatrix} 70 & -20 \\ -20 & 95 \end{vmatrix}} = \frac{(10)(95) - (-5)(-20)}{(70)(95) - (-20)(-20)}$$

$$= \frac{950 - 100}{6650 - 400} = 0.136 \text{ A}$$

Solving for I_2 yields

$$I_2 = \frac{\begin{vmatrix} 70 & 10 \\ -20 & -5 \end{vmatrix}}{6250} = \frac{(70)(-5) - (-20)(10)}{6250}$$

$$= \frac{-350 - (-200)}{6250} = -0.024 \text{ A}$$

The negative sign on I_2 means that its direction must be reversed.

Now we find the actual *branch* currents. Since I_1 is the *only* current through R_1, it is also the branch current I_{R1}:

$$I_{R1} = I_1 = 0.136 \text{ A}$$

Since I_2 is the *only* current through R_3, it is also the branch current I_{R3}:

$$I_{R3} = I_2 = 0.024 \text{ A} \quad \text{(opposite direction to } I_2\text{)}$$

Both loop currents I_1 and I_2 are through R_2 in the *same* direction. Remember, the negative I_2 value told us to reverse its assigned direction.

$$I_{R2} = I_1 - I_2 = 0.136 \text{ A} - (-0.024 \text{ A})$$
$$= 0.16 \text{ A}$$

Keep in mind that once we know the branch currents, we can find the voltages by using Ohm's law.

MESH CURRENT METHOD

Circuits with More than Two Loops

The mesh method also can be systematically applied to circuits with any number of loops. Of course, the more loops there are, the more difficult is the solution. However, the basic rules still apply. For example, for a three-loop circuit, three simultaneous equations are required. It is beyond the scope of this book to solve more than two simultaneous equations, but we will use Example 9–4 to *set up* the equations for a solution to a three-loop circuit.

Example 9–4

Set up the loop equations for Figure 9–6.

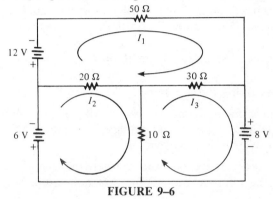

FIGURE 9–6

Solution:

Assign three CW loop currents as shown in the figure. Then use the format rules to write the loop equations. A concise statement of these rules is as follows:

(Sum of resistors in loop) times (loop current) minus (each common resistor) times (associated adjacent loop current) equals (source voltage in the loop).

$$100I_1 - 20I_2 - 30I_3 = 12 \quad \text{for loop 1}$$
$$-20I_1 + 30I_2 - 10I_3 = 6 \quad \text{for loop 2}$$
$$-30I_1 - 10I_2 + 40I_3 = 8 \quad \text{for loop 3}$$

These three equations can be solved for the currents by substitution or more easily with *third-order determinants*.

Appendix C presents a computer program for solving three equations using the BASIC language. The program can also be used to solve for loop currents given the resistor and voltage source values.

Review for 9-3

1. Do the loop currents necessarily represent the actual currents in the branches?
2. When you solve for a loop current and get a negative value, what does it mean?
3. What circuit law is used in the mesh current method?

9-4

NODE VOLTAGE METHOD

Another alternate method of analysis of multiple-source circuits is called the *node voltage method*. It is based on finding the voltages at each node in the circuit using *Kirchhoff's current law*. Remember that a *node* is the junction of two or more current paths.

The general steps for this method are as follows:

1. Determine the number of nodes.
2. Select one node as a *reference*. All voltages will be relative to the reference node. Assign voltage designations to each node where the voltage is unknown.
3. Assign currents at each node where the voltage is unknown, except at the reference node. The directions are arbitrary.
4. Apply Kirchhoff's current law to each node where currents are assigned.
5. Express the current equations in terms of voltages, and solve the equations for the unknown node voltages.

We will use Figure 9-7 to illustrate the general approach to node voltage analysis.

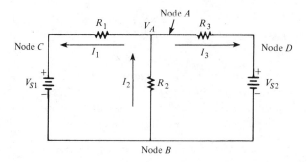

FIGURE 9-7 *Circuit for node voltage analysis.*

NODE VOLTAGE METHOD

First, establish the nodes. In this case there are *four,* as indicated in the figure.

Second, let us use node B as reference. Think of it as circuit ground. Node voltages C and D are already known to be the source voltages. The voltage at node A is the only unknown in this case. It is designated as V_A.

Third, arbitrarily assign the currents at node A as indicated in the figure.

Fourth, the Kirchhoff current equation at node A is

$$I_1 - I_2 + I_3 = 0$$

Fifth, express the currents in terms of circuit voltages using Ohm's law as follows:

$$I_1 = \frac{V_1}{R_1} = \frac{V_{S1} - V_A}{R_1}$$

$$I_2 = \frac{V_2}{R_2} = \frac{V_A}{R_2}$$

$$I_3 = \frac{V_3}{R_3} = \frac{V_{S2} - V_A}{R_3}$$

Substituting these into the current equation, we get

$$\frac{V_{S1} - V_A}{R_1} - \frac{V_A}{R_2} + \frac{V_{S2} - V_A}{R_3} = 0$$

The only unknown is V_A; so we can solve the single equation by combining and rearranging terms. Once the voltage is known, all branch currents can be calculated. An example will illustrate this method further.

Example 9–5

Find the node voltages in Figure 9–8.

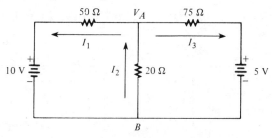

FIGURE 9–8

Solution:

The reference node is chosen at B. The unknown node voltage is V_A, as indicated in the figure. This is the only unknown voltage. Currents are assigned at node A as shown. The current equation is

$$I_1 - I_2 + I_3 = 0$$

Example 9–5 (continued)

Substitution for currents using Ohm's law gives the equation in terms of voltages:

$$\frac{10 - V_A}{50} - \frac{V_A}{20} + \frac{5 - V_A}{75} = 0$$

Solving for V_A yields

$$\frac{10}{50} - \frac{V_A}{50} - \frac{V_A}{20} + \frac{5}{75} - \frac{V_A}{75} = 0$$

$$-\frac{V_A}{50} - \frac{V_A}{20} - \frac{V_A}{75} = -\frac{10}{50} - \frac{5}{75}$$

$$\frac{6V_A + 15V_A + 4V_A}{300} = \frac{30 + 10}{150}$$

$$\frac{25V_A}{300} = \frac{40}{150}$$

$$V_A = \frac{(40)(300)}{(25)(150)}$$

$$= 3.2 \text{ V}$$

Using the same basic procedure, we can solve circuits with more than one unknown node voltage. Example 9–6 illustrates this calculation for two unknown node voltages. Also, the computer program in Appendix C can be used to find two or three unknown node voltages given the values of the resistors and voltage sources in the circuit.

Example 9–6

Using the node analysis method, solve for V_1 and V_2 in the circuit of Figure 9–9.

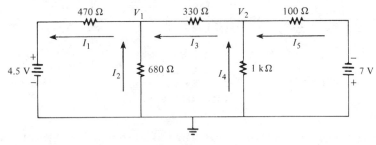

FIGURE 9–9

Solution:

First, the branch currents are assigned as shown in the diagram. Next, Kirchhoff's current law is applied at each node. At node 1,

NODE VOLTAGE METHOD

$$I_1 - I_2 - I_3 = 0$$

Using Ohm's law substitution for the currents, we get

$$\left(\frac{4.5 - V_1}{470}\right) - \left(\frac{V_1}{680}\right) - \left(\frac{V_1 - V_2}{330}\right) = 0$$

$$\frac{4.5}{470} - \frac{V_1}{470} - \frac{V_1}{680} - \frac{V_1}{330} + \frac{V_2}{330} = 0$$

$$\left(\frac{1}{470} + \frac{1}{680} + \frac{1}{330}\right)V_1 - \left(\frac{1}{330}\right)V_2 = \frac{4.5}{470}$$

Now, using the $1/x$ key of the calculator, evaluate the coefficients and constant. The resulting equation for node 1 is

$$0.00663 V_1 - 0.00303 V_2 = 0.00957$$

At node 2,

$$I_3 - I_4 - I_5 = 0$$

Again using Ohm's law substitution, we get

$$\left(\frac{V_1 - V_2}{330}\right) - \left(\frac{V_2}{1000}\right) - \left(\frac{V_2 - (-7)}{100}\right) = 0$$

$$\frac{V_1}{330} - \frac{V_2}{330} - \frac{V_2}{1000} - \frac{V_2}{100} - \frac{7}{100} = 0$$

$$\left(\frac{1}{330}\right)V_1 - \left(\frac{1}{330} + \frac{1}{1000} + \frac{1}{100}\right)V_2 = \frac{7}{100}$$

Evaluating the coefficients and constant, we obtain the equation for node 2:

$$0.00303 V_1 - 0.01403 V_2 = 0.07$$

Now, these two node equations must be solved for V_1 and V_2. Using determinants, we get the following solutions:

$$V_1 = \frac{\begin{vmatrix} 0.00957 & -0.00303 \\ 0.07 & -0.01403 \end{vmatrix}}{\begin{vmatrix} 0.00663 & -0.00303 \\ 0.00303 & -0.01403 \end{vmatrix}}$$

$$= \frac{(0.00957)(-0.01403) - (0.07)(-0.00303)}{(0.00663)(-0.01403) - (0.00303)(-0.00303)}$$

$$= -0.928 \text{ V}$$

$$V_2 = \frac{\begin{vmatrix} 0.00663 & 0.00957 \\ 0.00303 & 0.07 \end{vmatrix}}{\begin{vmatrix} 0.00663 & -0.00303 \\ 0.00303 & -0.01403 \end{vmatrix}}$$

$$= \frac{(0.00663)(0.07) - (0.00303)(0.00957)}{(0.00663)(-0.01403) - (0.00303)(-0.00303)}$$

$$= -5.19 \text{ V}$$

Review for 9-4

1. What circuit law is the basis for the node voltage method?
2. What is the reference node?

Summary

1. A loop is a closed current path in a circuit.
2. A node is the junction of two or more current paths.
3. The branch current method is based on Kirchhoff's voltage law and Kirchhoff's current law.
4. A branch current is an *actual* current in a branch.
5. Determinants are used to solve simultaneous equations.
6. The mesh current method is based on Kirchhoff's voltage law.
7. A loop current is not necessarily the actual current in a branch.
8. The node voltage method is based on Kirchhoff's current law.

Self-Test

1. How many loops and nodes are there in the circuit of Figure 9-10?

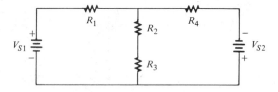

FIGURE 9-10

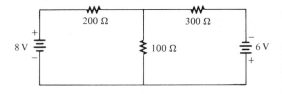

FIGURE 9-11

2. Use the branch current method to find the currents in Figure 9-11.
3. Solve the following two equations by substitution:

$$2I_1 + 5I_2 = 15$$
$$6I_1 - 8I_2 = 10$$

4. Using determinants, solve the following two equations for I_1:

$$15I_1 - 12I_2 = 25$$
$$9I_1 + 3I_2 = 18$$

5. Use the mesh current method to find the current through R_3 in Figure 9-12.

PROBLEMS

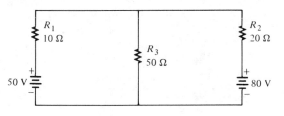

FIGURE 9–12

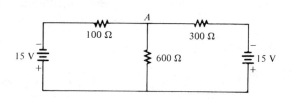

FIGURE 9–13

6. Use the node voltage method to find the voltage at node A in Figure 9–13.

Problems

9–1. Identify all *possible* loops in Figure 9–14.

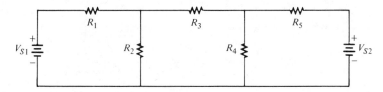

FIGURE 9–14

9–2. Identify all nodes in Figure 9–14. Which ones have a *known* voltage?

9–3. Write the Kirchhoff current equation for the current assignment shown at node A in Figure 9–15.

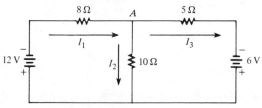

FIGURE 9–15

9–4. Solve for each of the branch currents in Figure 9–15.

9–5. Find the voltage drop across each resistor in Figure 9–15, and indicate its actual polarity.

9–6. Using the substitution method, solve the following set of equations for I_1 and I_2:

$$100I_1 + 50I_2 = 30$$
$$75I_1 + 90I_2 = 15$$

9-7. Using the substitution method, solve the following set of three equations for all currents:

$$5I_1 - 2I_2 + 8I_3 = 1$$
$$2I_1 + 4I_2 - 12I_3 = 5$$
$$10I_1 + 6I_2 + 9I_3 = 0$$

9-8. Evaluate each determinant.

(a) $\begin{vmatrix} 4 & 6 \\ 2 & 3 \end{vmatrix}$ (b) $\begin{vmatrix} 9 & -1 \\ 0 & 5 \end{vmatrix}$

(c) $\begin{vmatrix} 12 & 15 \\ -2 & -1 \end{vmatrix}$ (d) $\begin{vmatrix} 100 & 50 \\ 30 & -20 \end{vmatrix}$

9-9. Using determinants, solve the following set of equations for both currents:

$$-I_1 + 2I_2 = 4$$
$$7I_1 + 3I_2 = 6$$

9-10. Using the mesh current method, find the loop currents in Figure 9-16.

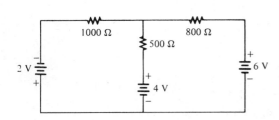

FIGURE 9-16

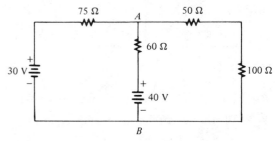

FIGURE 9-17

9-11. Find the branch currents in Figure 9-16.

9-12. Determine the voltages and their proper polarities for each resistor in Figure 9-16.

9-13. In Figure 9-17, use the node voltage method to find the voltage at point A with respect to point B.

9-14. What are the branch current values in Figure 9-17? Show the actual direction of current flow in each branch.

9-15. Write the loop equations for the circuit in Figure 9-18.

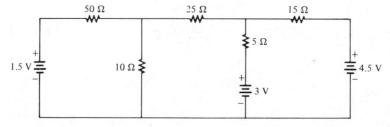

FIGURE 9-18

ANSWERS TO SECTION REVIEWS

9–16. Write the node voltage equations for Figure 9–18.

9–17. Using the computer program in Appendix C, solve for the loop currents in Problem 9–15.

9–18. Using the computer program in Appendix C, solve for the node voltages in Problem 9–16.

Answers to Section Reviews

Section 9–1:
1. Kirchhoff's voltage law and Kirchhoff's current law. **2.** False, but write the equations so that they are consistent with your assigned directions. **3.** A closed path within a circuit. **4.** A junction of two or more current paths.

Section 9–2:
1. The value of a matrix of the coefficients of the unknowns in a set of equations. **2.** To solve two equations for two unknowns.

Section 9–3:
1. No. **2.** The direction should be reversed. **3.** Kirchhoff's voltage law.

Section 9–4:
1. Kirchhoff's current law. **2.** The junction to which all circuit voltages are referenced.

In this chapter you will learn about electrical signals. An *electrical signal*, for our purposes, is a voltage or a current that changes in some manner with time. In other words, the voltage and current values fluctuate according to a certain pattern.

One of the most common electrical signals is represented by a wave form called the *sine wave*. The sine wave is encountered throughout the electrical and electronics field. Therefore, a large portion of this chapter is devoted to this important form of electrical signal.

Pulse signals are also very important in electronics. They are found in digital computers and many other types of electronic equipment. Triangular and sawtooth wave forms are also discussed in this chapter.

10–1 The Sine Wave
10–2 Period and Frequency
10–3 Amplitude Values of a Sine Wave
10–4 Phase Relationships
10–5 Equation for a Sine Wave
10–6 Phasors
10–7 Pulse Wave Forms
10–8 Triangular and Sawtooth Wave Forms
10–9 Harmonics

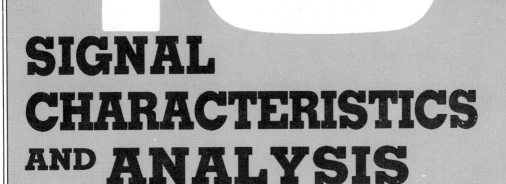

SIGNAL CHARACTERISTICS AND ANALYSIS

10-1

THE SINE WAVE

The sine wave represents the most common type of ac voltage. It is also called a *sinusoidal wave form* or *sinusoid*. The electrical voltage at your wall outlet is sinusoidal, as are radio frequency carrier signals. Other types of electrical signals are made up of many individual sine waves called *harmonics*.

The graph in Figure 10–1 shows the general shape of a sine wave. This graph shows how a sinusoidal voltage (or current) changes with *time*. Notice that the curve increases to a maximum and then decreases to a minimum.

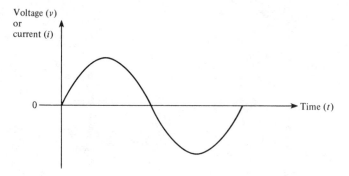

FIGURE 10–1 *Sine wave.*

Changes in Polarity

A sine wave voltage changes polarity at its zero value; that is, it alternates between positive and negative voltage values. If the sine wave voltage is applied to a resistive circuit, as shown in Figure 10–2, there is an alternating current flow. That is, the current reverses direction following a sine wave pattern when the voltage changes polarity. Notice the symbol for the sine wave voltage source in Figure 10–2.

During a *positive alternation* of the voltage, there is current through R in the direction shown in Figure 10–2A. During a *negative alternation* of the voltage, the current is in the opposite direction, as shown in Part B. A positive and a negative alternation make up *one cycle* of a sine wave.

Review for 10–1

1. Sketch a sine wave.
2. What is the zero crossing of a sine wave?
3. Describe one cycle of a sine wave.

PERIOD AND FREQUENCY

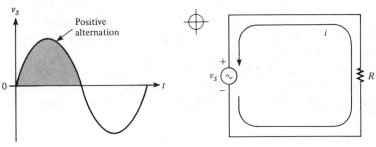

A. Positive voltage: current flow as shown.

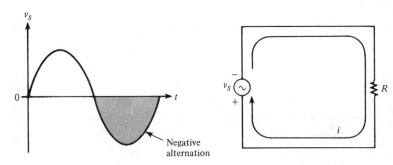

B. Negative voltage: current reverses direction.

FIGURE 10–2 *Alternating current and voltage (lower-case v and i represent instantaneous quantities).*

10–2

PERIOD AND FREQUENCY

Period

As you have seen, a sine wave varies in a certain manner *with time (t)*. In Figure 10–3A, starting at t_0, the voltage increases to a positive maximum and then decreases back to its zero value at t_1. It continues to decrease to a negative maximum and then increases back to its zero value at t_2. It then repeats. The *time interval* from t_0

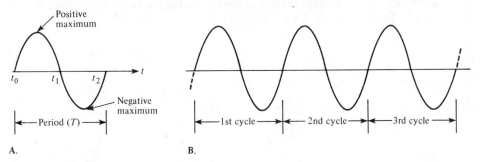

FIGURE 10-3 *Period of a sine wave.*

to t_2 is called the *period* of the sine wave. *The period, designated by the letter T, is the time for completion of one full cycle.* During each successive period the cycle is repeated, as shown in Figure 10–3B.

Each period is the same as all other periods for a given sine wave and is therefore a definable characteristic of that sine wave. The period can be measured from any point during a given cycle to the *corresponding* point in the next cycle.

Example 10–1

What is the period of the sine wave in Figure 10–4?

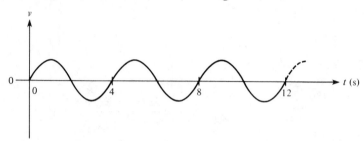

FIGURE 10–4

Solution:

Notice that each successive cycle takes 4 seconds (4 s) to complete. This is the period; that is,

$$T = 4 \text{ s}$$

Example 10–2

Show three possible ways to measure the period of the sine wave in Figure 10–5A. How many cycles are shown?

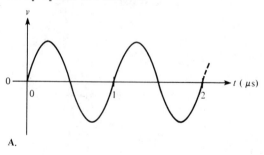

A.

FIGURE 10–5A

PERIOD AND FREQUENCY

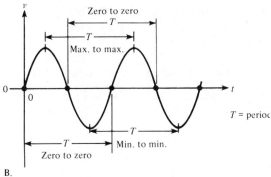

B.

FIGURE 10-5B

Solution:

Method 1: The period can be measured from one zero crossing to the corresponding zero crossing in the next cycle.

Method 2: The period can be measured from the positive peak, or maximum point, in one cycle to the positive peak in the next cycle.

Method 3: The period can be measured from the negative peak, or minumum point, in one cycle to the negative peak in the next cycle.

These period measurements are shown in Figure 10-5B. You obtain the same value for the period no matter what points on the wave form you use.

Two cycles are shown in Figure 10-5A.

Frequency

Frequency defines how fast a sine wave is changing. That is, *frequency is the number of cycles that the signal completes in one second.* The more cycles the signal completes in a second, the higher the frequency is.

Figure 10-6 shows two sine waves. The sine wave in Part A goes through two complete cycles in one second. The sine wave in Part B goes through four complete cycles in one second. Therefore, wave form B has twice the frequency of wave form A.

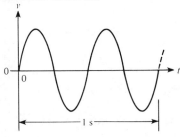

A.

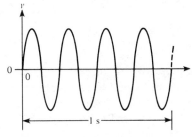

B.

FIGURE 10-6

The unit of frequency is the hertz, abbreviated Hz. One hertz is one cycle per second; 60 Hz is 60 cycles per second; and so on. The symbol used for frequency is f.

Relationship of Frequency and Period

Frequency and period are, of course, related to each other. The following equations give the formulas for this relationship:

$$f = \frac{1}{T} \qquad (10\text{--}1)$$

$$T = \frac{1}{f} \qquad (10\text{--}2)$$

As you can see, there is a *reciprocal* relationship between f and T. If we know the value of one quantity, we can calculate the value of the other with Equation (10–1) or (10–2).

Since the period T is the length of time for completion of one cycle, if the period increases, the frequency decreases, and vice versa. This relationship makes sense because a sine wave with a larger period has fewer cycles per second, and a sine wave with a shorter period has more cycles per second.

Example 10–3

How many cycles are shown in Figure 10–7A and B? What is the period in each case?

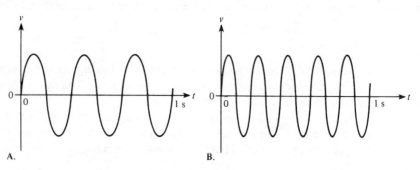

A.　　　　　　　　　　　　B.

FIGURE 10–7

Solution:

In Figure 10–7A, three cycles are shown. The period is 0.333 s because it takes one second to complete *three* cycles. So one cycle is completed in one-third of a second.

In Part B, five cycles are shown. The period is 0.2 s, which is one-fifth of a second.

PERIOD AND FREQUENCY

Example 10–4

If the period of a certain sine wave is 10 milliseconds (10 ms), what is the frequency?

Solution:

Using Equation (10–1), we obtain

$$f = \frac{1}{T} = \frac{1}{10 \text{ ms}} = \frac{1}{10 \times 10^{-3} \text{ s}}$$
$$= 0.1 \times 10^3 \text{ Hz} = 100 \text{ Hz}$$

Example 10–5

If the frequency of a sine wave is 60 Hz, what is the period?

Solution:

Using Equation (10–2), we obtain

$$T = \frac{1}{f} = \frac{1}{60 \text{ Hz}}$$
$$= 0.01667 \text{ s} = 16.67 \text{ ms}$$

Rate of Change of a Sine Wave

The average rate of change of a sine wave determines the frequency. The faster the rate of change is, the higher the frequency is. Now let us see what the *instantaneous rate of change* means and how it varies at different points on a sine wave. *The instantaneous rate of change is the rate at which the sine wave is increasing or decreasing at any instant of time.*

In Figure 10–8A, an interval of time between t_1 and t_2 along the horizontal axis is projected up to points on the sine wave. These points are then projected across to the points v_1 and v_2 on the vertical axis. As you can see, the voltage changes from a value v_1 to a value v_2 in the time interval from t_1 to t_2. If Δv is the change in voltage and Δt is the change in time, the *average rate of change* in this interval is $\Delta v/\Delta t$. If this interval is made extremely small, we approach a single point and thus an instantaneous rate of change called dv/dt. This rate is expressed in V/s and is illustrated in Figure 10–8B. A similar discussion applies to current sine waves where the instantaneous rate of change is called di/dt and its units are A/s.

Notice on the sine wave in Figure 10–9A that the maximum rate of change occurs at the zero crossing and a zero rate of change occurs at the peaks. The

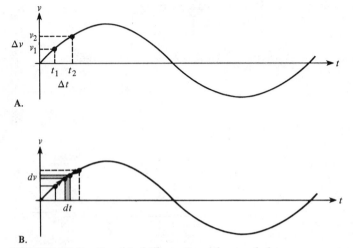

A.

B.

FIGURE 10–8 *Graphical illustration of rate of change.*

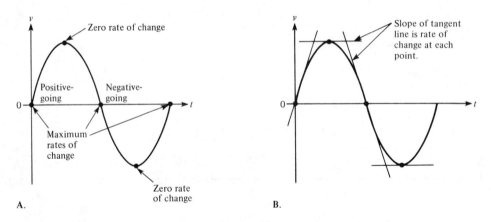

A.

B.

FIGURE 10–9 *Variation in the rate of change.*

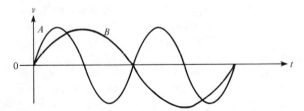

FIGURE 10–10

instantaneous rate of change is actually the slope of the tangent line at any point, as shown in Part B of the figure. The rate of change varies from one point to the next along the sine wave curve.

For equal value sine waves, a higher frequency corresponds to a greater *maximum* rate of change, as Figure 10–10 shows. Sine wave *A* has a higher frequency and is changing at a faster rate at the zero crossing than is sine wave *B*. The maximum rate of change of a sine wave always occurs at the zero crossing and is indicated by the slope of the curve at that point.

AMPLITUDE VALUES OF A SINE WAVE

Example 10-6

What is the *average* rate of change of the sine wave between 1 μs and 2 μs in Figure 10-11?

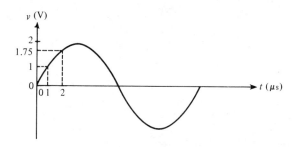

FIGURE 10-11

Solution:

Project the 1-μs and 2-μs points up and across as shown. We find that v_1 is 1 V and v_2 is 1.75 V. The change in voltage divided by the change in time is expressed as follows:

$$\frac{\Delta v}{\Delta t} = \frac{1.75 \text{ V} - 1 \text{ V}}{2 \text{ μs} - 1 \text{ μs}} = \frac{0.75 \text{ V}}{1 \text{ μs}}$$
$$= 0.75 \text{ V/μs}$$

Review for 10-2

1. What is one cycle of a sine wave?
2. What is the period of a sine wave?
3. What is the frequency of a sine wave?
4. Define *rate of change*.
5. If $T = 5$ μs, what is f?
6. If $f = 120$ Hz, what is T?

10-3
AMPLITUDE VALUES OF A SINE WAVE

There are several ways to express the value of a sine wave in terms of its voltage or current magnitude. The *amplitude, A,* is a general term referring to the amount of voltage or current for a given sine wave. The amplitude of a sine wave can be described in terms of its voltage or current in several ways: *instantaneous, peak, peak-to-peak, rms (effective),* and *average* values.

Instantaneous Value

The instantaneous value of a sine wave varies from one instant to the next. Instantaneous values of voltage and current are expressed as lower-case v and i, respectively. (Constant values such as the peak, peak-to-peak, rms, and average are described using capital letters.)

Figure 10–12 illustrates that at any point in time along the sine curve, the voltage has a unique value at that instant.

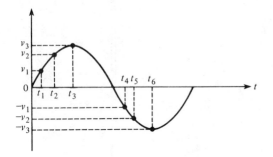
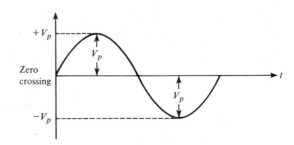

FIGURE 10–12 *Instantaneous values.*   **FIGURE 10–13** *Peak values.*

Peak Value

The peak value of a sine wave is the amount of voltage or current at the positive and negative maximum points *referenced to the zero crossing*. The positive peak value occurs at the positive maximum. The negative peak value occurs at the negative maximum. This concept is illustrated in Figure 10–13. Note that V_p designates the peak voltage. A capital letter is used because for a particular sine wave, the peak value is a constant.

Peak-to-Peak Value

The peak-to-peak value is the voltage (V_{pp}) or current (I_{pp}) from the positive peak to the negative peak (see Figure 10–14). It is always *twice* the peak value. These values are given in the following equations:

$$V_{pp} = 2V_p \tag{10-3}$$

$$I_{pp} = 2I_p \tag{10-4}$$

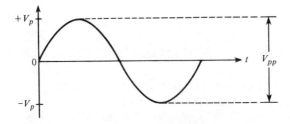

FIGURE 10–14 *Peak-to-peak value.*

AMPLITUDE VALUES OF A SINE WAVE

Root Mean Square (rms) Value

The term rms stands for *root mean square*. It refers to the mathematical process whereby this value is obtained. (See Appendix D for the derivation.) The rms value is also referred to as the *effective value*. Most ac voltmeters read voltage in rms. The 115 volts at your wall outlet is an rms value.

The rms value is actually a measure of the *heating effect* of the sine wave. If a resistor is connected to an ac voltage source as shown in Figure 10–15A, a certain amount of heat is generated by power in the resistor. A dc voltage source can be connected to the *same* resistor, as shown in Part B of the figure. The value of the dc voltage can be adjusted so that the resistor gives off the same amount of heat as it does with ac source. *The dc value that produces the same amount of heat as the sine wave is the rms value of that sine wave.*

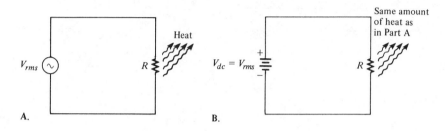

FIGURE 10–15 *rms value.*

The rms value is expressed in terms of the peak value by the following equations:

$$V_{rms} = 0.707 V_p \qquad (10\text{–}5)$$

$$I_{rms} = 0.707 I_p \qquad (10\text{–}6)$$

Note that these equations are identical except that Equation (10–5) is for voltage and (10–6) is for current.

Using these formulas, we can also solve for the peak values from the rms values as follows:

$$V_p = \frac{V_{rms}}{0.707} = \left(\frac{1}{0.707}\right) V_{rms}$$

$$V_p = 1.414 V_{rms} \qquad (10\text{–}7)$$

$$I_p = 1.414 I_{rms} \qquad (10\text{–}8)$$

The peak values are doubled to give the peak-to-peak values:

$$V_{pp} = 2.828 V_{rms} \qquad (10\text{–}9)$$

$$I_{pp} = 2.828 I_{rms} \qquad (10\text{–}10)$$

Average Value

If the sine wave is *averaged* around the zero crossing for *one full cycle*, the average value is always *zero*, because the positive values (above the zero crossing) offset the negative values (below the zero crossing).

For the average value to be useful for comparison purposes, it is defined over a *half-cycle* rather than a full cycle. We find the average value of a sine wave by adding each instantaneous value over one half-cycle and dividing the sum by the number of values. To get an exact value, we must use integral calculus. This derivation is shown in Appendix E. The result is expressed in terms of the peak value as follows:

$$V_{avg} = 0.637V_p \qquad (10\text{--}11)$$

$$I_{avg} = 0.637I_p \qquad (10\text{--}12)$$

Figure 10–16 illustrates the average value of a sine wave.

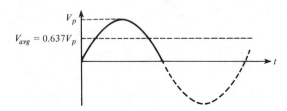

FIGURE 10–16 *Average value.*

Example 10–7

Determine V_p, V_{pp}, V_{rms}, and V_{avg} for the sine wave graphed in Figure 10–17.

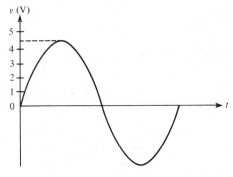

FIGURE 10–17

Solution:

$$V_p = 4.5 \text{ V} \quad \text{as taken from the graph}$$

$$V_{pp} = 2V_p = 2(4.5 \text{ V})$$
$$= 9 \text{ V}$$

AMPLITUDE VALUES OF A SINE WAVE

$$V_{rms} = 0.707V_p = 0.707(4.5 \text{ V})$$
$$= 3.182 \text{ V}$$
$$V_{avg} = 0.637V_p = 0.637(4.5 \text{ V})$$
$$= 2.867 \text{ V}$$

Ohm's Law in ac Circuits

When a sine wave voltage is applied to a resistive circuit, as shown in Figure 10–18, a *sine wave current* results. *Ohm's law can be used in ac circuits such as this.* For example, when the rms value of voltage is used in Ohm's law, the current value is also an rms value. Equations (10–13) through (10–17) express Ohm's law in terms of the sine wave values:

$$i = \frac{v}{R} \qquad (10\text{–}13)$$

$$I_p = \frac{V_p}{R} \qquad (10\text{–}14)$$

$$I_{pp} = \frac{V_{pp}}{R} \qquad (10\text{–}15)$$

$$I_{rms} = \frac{V_{rms}}{R} \qquad (10\text{–}16)$$

$$I_{avg} = \frac{V_{avg}}{R} \qquad (10\text{–}17)$$

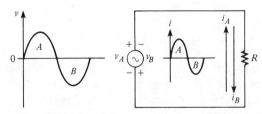

FIGURE 10–18

Review for 10–3

1. If V_p is 1 V, what is V_{pp}?
2. If V_p is 1 V, what is V_{rms}?
3. If V_p is 1 V, what is V_{avg}?
4. A sine wave voltage with an rms value of 5 V is applied to a circuit with a resistance of 10 Ω. What is the peak value of the current?

10-4

PHASE RELATIONSHIPS

As you have seen, when the frequency of a sine wave changes, the time required for completion of one cycle (period) also changes. For this reason, it is useful to express points on the sine wave in terms of an *angular measurement* in degrees or in radians. The angular or phase measurement of a sine wave is *independent of frequency*.

One way of producing a sine wave voltage is by rotating electromechanical machines (ac generators). When the rotor of the ac generator goes through a *full 360° of rotation,* the resulting voltage output is *one full cycle* of a sine wave. Thus, the angular measurement of a sine wave can be related to the angular rotation of a generator.

Angular Measurement

A *radian* (rad) is the angular distance along the circumference of a circle, equal to the radius of the circle. In terms of degrees, *one radian (1 rad) is equivalent to 57.3°,* as illustrated in Figure 10–19.

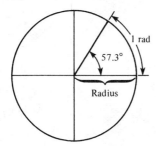

FIGURE 10–19

In a 360° revolution there are 2π radians; π *is the ratio of the circumference of any circle to its diameter and has a constant value of 3.1416.* Most calculators have a π key so that you do not have to enter the actual value of π.

Table 10–1 lists several values of degrees and the corresponding radian values. These angular measurements are illustrated in Figure 10–20.

TABLE 10–1

Degrees	Radians (rad)
0°	0
45°	$\pi/4$
90°	$\pi/2$
135°	$3\pi/4$
180°	π
225°	$5\pi/4$
270°	$3\pi/2$
315°	$7\pi/4$
360°	2π

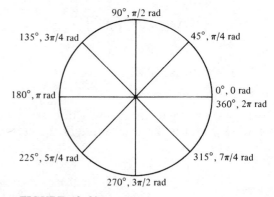

FIGURE 10–20

PHASE RELATIONSHIPS

We can convert degrees to radians and radians to degrees by using Equations (10–18) and (10–19). These conversions are illustrated in Example 10–8.

$$\text{rad} = \left(\frac{\pi \text{ rad}}{180°}\right)(\text{degrees}) \qquad (10\text{--}18)$$

$$\text{Degrees} = \left(\frac{180°}{\pi \text{ rad}}\right)(\text{rad}) \qquad (10\text{--}19)$$

Example 10–8

(a) Convert 60° to radians.   (b) Convert π/6 rad to degrees.

Solution:

(a) $\text{Rad} = \left(\dfrac{\pi \text{ rad}}{180°}\right) 60° = \dfrac{\pi}{3} \text{ rad} = 1.047 \text{ rad}$

(b) $\text{Degrees} = \left(\dfrac{180°}{\pi \text{ rad}}\right)\left(\dfrac{\pi}{6} \text{ rad}\right) = 30°$

Sine Wave Angles

The angular measurement of a sine wave is based on 360° or 2π rad for a complete cycle. A half-cycle is 180° or π rad; a quarter cycle is 90° or π/2 rad; and so on. Figure 10–21A shows angles in degrees on the sine wave curve, and Part B shows the angles in radian measure.

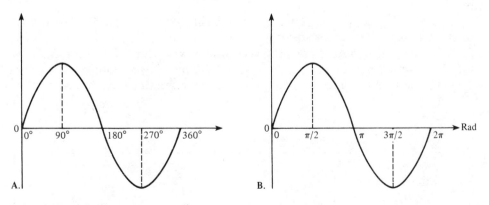

FIGURE 10–21 *Sine wave angles.*

Phase Shift

The phase of a sine wave is relative. Figure 10–21 shows one cycle of a *reference* sine wave. The positive-going zero crossing is at 0°; the positive peak is at

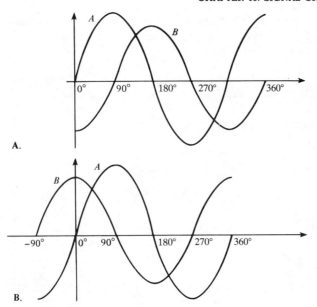

FIGURE 10–22 *Phase shifts.*

90°. The negative-going zero crossing is at 180°; the negative peak is at 270°. The cycle is completed at 360°. If the sine wave is moved left or right with respect to this reference, there is a *phase shift*. That is, the sine wave is shifted left or right by a certain number of degrees (or radians) *with respect to a reference*.

Figure 10–22 illustrates phase shifts of a sine wave. In Part A of the figure, wave form *B* is shifted to the right by 90°. Thus, there is a *phase angle* of 90° between wave form *A* and wave form *B*. In terms of time, the positive peak of wave form *B* occurs *later in time* than the positive peak of wave form *A*, because time *increases* to the right along the horizontal axis. Wave form *B* is said to *lag* wave form *A*. In this case, *B* lags by 90° or π/2 radians. Stated another way, *wave form A leads B by 90°*.

In Figure 10–22B, wave form *B* is shifted to the left by 90°; so there is a phase angle of 90° between wave form *A* and wave form *B*. The positive peak of wave form *B* now occurs *sooner* than that of wave form *A*. Wave form *B* is said to *lead* wave form *A*, or wave form *A* lags *B*. In this case, *B* leads by 90° or π/2 radians. Wave form *B* in Figure 10–22B is called a *cosine* wave. It is actually a sine wave shifted by 90° to the left.

Example 10–9

What are the phase angles between the two wave forms in Figure 10–23A and B?

EQUATION FOR A SINE WAVE

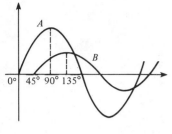

A.

B.

FIGURE 10-23

Solution:

In Part A of the figure, the phase angle is 45°, Wave form A leads wave form B (and B lags A).

In Part B, the phase angle is 30°. Wave form A lags wave form B (and B leads A).

Review for 10-4

1. If the positive-going zero crossing of a sine wave occurs at 0°, at what angles does each of the following points occur?
   (a) positive peak
   (b) negative-going zero crossing
   (c) negative peak
   (d) end of complete cycle

2. A half-cycle consists of _____ degrees or _____ rad.

3. A full cycle consists of _____ degrees or _____ rad.

4. What is the phase angle between the two wave forms in Figure 10-24?

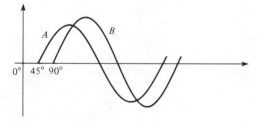

FIGURE 10-24

10-5

EQUATION FOR A SINE WAVE

As you have seen, a sine wave can be graphed on the basis of two quantities: amplitude *(A)* and angle (θ, the Greek letter theta). A general graph of a sine wave is

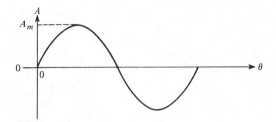

FIGURE 10–25

shown in Figure 10–25. The values of amplitude run along the vertical axis, and the values of angle run along the horizontal axis.

A sine curve follows a specific mathematical formula. The general expression for a sine wave that starts at the zero reference is as follows:

$$A_m \sin \theta \quad (10\text{–}20)$$

where A_m is the peak amplitude. This expression states that any point on the curve, which is an instantaneous value, is equal to the peak value of the sine wave times the sine (sin) of the angle at that point. For example, suppose that a sine wave voltage has a peak value of 10 V. The instantaneous voltage, v, at 45° can be calculated as follows:

$$v = V_p \sin \theta = 10 \sin 45° = 10(0.707) = 7.07 \text{ V}$$

Figure 10–26 shows this particular instantaneous value. You can find the sine of any angle on your calculator by first entering the value of the angle and then pressing the *sin* key. You can also use the trigonometric tables in Appendix G.

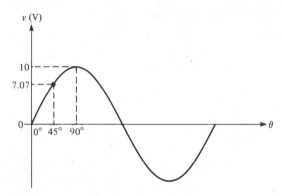

FIGURE 10–26 *Illustration of the instantaneous value at $\theta = 45°$.*

Equations for Shifted Sine Waves

When a sine wave is shifted to the right of the reference by a certain phase angle, the amount of phase shift is called ϕ (the Greek letter phi). This phase shift is illustrated in Figure 10–27A and is expressed by the following equation:

$$A_m \sin(\theta - \phi) \quad (10\text{–}21)$$

EQUATION FOR A SINE WAVE

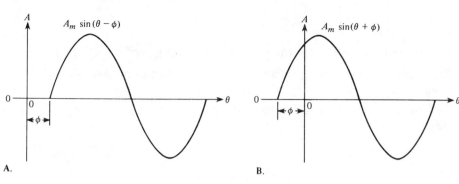

FIGURE 10–27

If the sine wave is shifted left by a phase angle φ (see Figure 10–27B), its instantaneous value is

$$A_m \sin(\theta + \phi) \tag{10-22}$$

Example 10–10

Determine the instantaneous value of each sine wave voltage in Figure 10–28 at the 90° point on the horizontal axis.

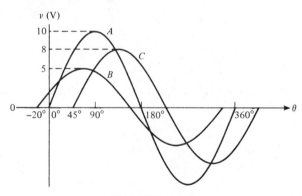

FIGURE 10–28

Solution:

Sine wave A is the reference. Sine wave B is shifted left 20° with respect to A; so it leads. Sine wave C is shifted right 45° with respect to A; so it lags.

Sine wave A: $v = 10 \sin 90° = 10(1) = 10$ V

Sine wave B: $v = 5 \sin(90° + 20°) = 5 \sin 110° = 5(0.9397) = 4.7$ V

Sine wave C: $v = 8 \sin(90° - 45°) = 8 \sin 45° = 8(0.7071) = 5.66$ V

Review for 10-5

1. Calculate the instantaneous value at 60° on the sine wave shown in Figure 10-26.
2. Determine the instantaneous value at the 45° point of a sine wave shifted 10° left from the zero reference ($V_p = 10$ V).
3. Determine the instantaneous value at the 90° point of a sine wave shifted 25° to the right from the zero reference ($V_p = 5$ V).

10-6

PHASORS

A *phasor* is used to represent a time-varying quantity in terms of both magnitude and direction. Phasors are used to represent the amplitude and phase angle of sine waves. This section is an introduction to using phasors to represent sine waves.

Examples of phasors are shown in Figure 10-29. The length of the phasor arrow is the magnitude, and the angle θ relative to 0° is the direction. The phasor in Part A of the figure has a magnitude of *two* and a phase angle of 45°. The phasor in Part B has a magnitude of *three* and a phase angle of 180°. The phasor in Part C has a magnitude of *one* and a phase angle of −45°.

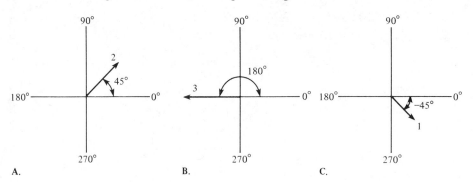

A.   B.   C.

FIGURE 10-29   *Examples of phasors.*

Phasor Representation of a Sine Wave

A full cycle of a sine wave can be represented by rotation of a phasor through 360° *The instantaneous value of the sine wave at any point is equal to the vertical distance from the tip of the phasor to the horizontal axis.* Figure 10-30 illustrates how the phasor "traces out" the sine wave as it goes from 0° to 360°. You can relate this concept to the rotation in an ac generator.

Notice that in Figure 10-30, the length of the phasor is the *peak* value of the sine wave (observe the 90° and 270° points). The angle of the phasor measured from 0° is the corresponding phase of the sine wave.

PHASORS

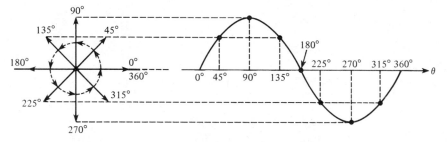

FIGURE 10-30 *Sine wave represented by rotational phasor motion.*

Phasors and the Sine Wave Formula

Let us examine a phasor at one specific angle. Figure 10-31 shows a voltage phasor at the 45° point and also the corresponding point on the sine curve. The instantaneous value of the sine wave at this point is related to the *position and length* of the phasor. As mentioned before, the vertical distance from the phasor tip down to the horizontal axis is the instantaneous value of the sine wave. Notice that when a vertical line is drawn from the tip of the phasor, a *right triangle* is formed, as shown in the figure. The length of the phasor is the *hypotenuse* of the triangle, and the vertical projection is the *opposite* side. From trigonometry, *the opposite side of a right triangle is equal to the hypotenuse times the sine of the angle* θ. In this case, the length of the phasor is the peak value of the sine wave voltage, V_p. Thus, the opposite side of the triangle, which is the instantaneous voltage, is $v = V_p \sin \theta$. The same values would hold true if this were a current wave form. Recall that this formula is the one that you learned earlier for calculating instantaneous sine wave values.

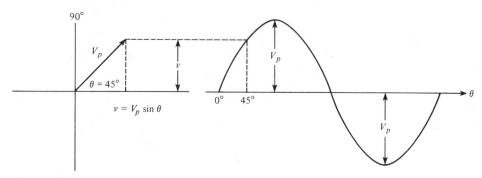

FIGURE 10-31 *Right triangle derivation of sine wave formula.*

Positive and Negative Phasor Angles

The angle of a phasor at any point can be expressed as a positive angle or an *equivalent* negative angle. Positive angles are measured counterclockwise from 0°. Negative angles are measured clockwise from 0°. For a given positive angle θ, the corresponding negative angle is $\theta - 360°$, as illustrated in Figure 10-32A. Part B of the figure illustrates a specific example. The angle of the phasor in this instance can be expressed as a positive 225° or, equivalently, as a negative 135°.

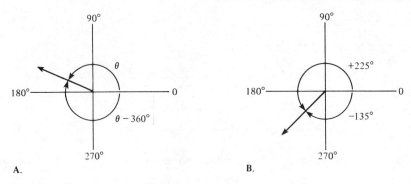

FIGURE 10–32 *Positive and negative phasor angles.*

Example 10–11

For each phasor A–F in Figure 10–33, determine the instantaneous sine wave value. Also express each positive angle shown as an equivalent negative angle. The length of the phasor in each case is the peak value of the sine wave that it represents.

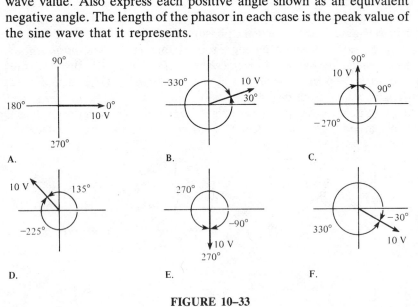

FIGURE 10–33

Solution:

Part A: $\theta = 0°$
$v = 10 \sin 0° = 10(0) = 0$ V
Part B: $\theta = 30° = -330°$
$v = 10 \sin 30° = 10(0.5) = 5$ V
Part C: $\theta = 90° = -270°$
$v = 10 \sin 90° = 10(1) = 10$ V

Part D: $\theta = 135° = -225°$
$v = 10 \sin 135° = 10(0.707) = 7.07 \text{ V}$
Part E: $\theta = 270° = -90°$
$v = 10 \sin 270° = 10(-1) = -10 \text{ V}$
Part F: $\theta = 330° = -30°$
$v = 10 \sin 330° = 10(-0.5) = -5 \text{ V}$

The equivalent negative angles are shown in the figure.

Phasor Diagrams

Several sine waves of different angles and amplitudes can be represented by a *phasor diagram*. All of the sine waves represented must be of the same frequency. When comparing phase angles, you must use a reference phasor and draw all other phasors relative to it. A phasor in a *fixed* position is used to represent the *complete* sine wave, because once the phase angles between two or more sine waves are established, they remain constant throughout the cycles. For example, the two sine waves in Figure 10–34A can be represented by a phasor diagram as shown in Part B of the figure. Sine wave A is the reference. Sine wave B is leading A by 30°.

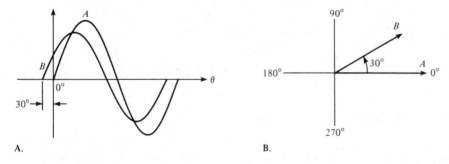

FIGURE 10–34 *Example of a phasor diagram.*

The phasor diagram shows the *relative relationship* of these two waves. In these diagrams, the length of the phasors can represent peak, rms, or average values, as long as a *consistent* value is used.

Example 10–12

Represent the sine waves in Figure 10–35 with a phasor diagram.

Example 10–12 (continued)

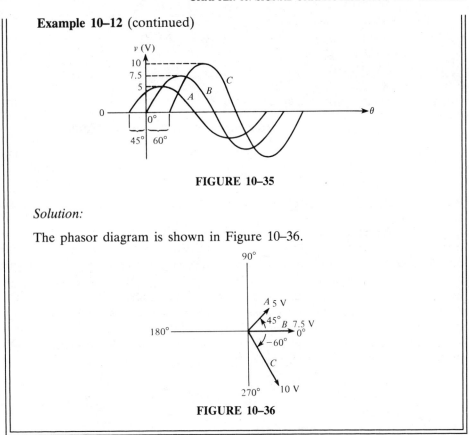

FIGURE 10–35

Solution:

The phasor diagram is shown in Figure 10–36.

FIGURE 10–36

Angular Velocity of a Phasor

As you have seen, a sine wave is traced out when a phasor is rotated. The faster it is rotated, the faster the cycle of the sine wave is traced out. Thus, the period and frequency are related to the velocity of rotation of the phasor. The velocity of rotation is called the *angular velocity* and is symbolized by ω (the Greek letter omega).

We will now establish the relationship between angular velocity and frequency. When a phasor rotates through 360° or 2π rad, one complete cycle is traced out. Therefore, the time required for the phasor to go through 2π rad is the *period* of the sine wave. The angular velocity, ω, is $2\pi/T$ because the phasor rotates through 2π rad in a time equal to the period T. Since $f = 1/T$, then $\omega = 2\pi(1/T) = 2\pi f$. The following equation expresses this important relationship:

$$\omega = 2\pi f \quad\quad\quad (10\text{–}23)$$

If the phasor is rotating at a velocity ω, then ω*t* is the angular distance at any instant in time through which the phasor has passed; so ω*t* is the same as θ. The basic sine wave formula can therefore be expressed as $A_m \sin \omega t$ as well as $A_m \sin \theta$.

PULSE WAVE FORMS **305**

Review for 10-6

1. Define *phasor*.
2. What two quantities of a sine wave does a phasor represent?
3. Draw a phasor representing a cosine wave.
4. Sketch a phasor diagram to represent the two sine waves in Figure 10-37.

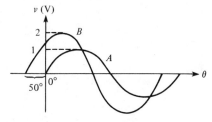

FIGURE 10-37

5. The angular velocity of a given phasor is 628 rad/s. To what frequency does this value correspond?

10-7

PULSE WAVE FORMS

Sine waves are important, but they are by no means the only type of wave form. Two other categories will be discussed next: the *pulse* wave form and the *triangular* wave form.

Ideal Steps

An *electrical step* is an abrupt change in voltage or current from one value to another. Figure 10-38A shows an ideal positive-going step. This step goes from a lower value to a higher value instantaneously. The difference between the lower and the higher value is the *step amplitude*. Part B of the figure shows an ideal negative-going step. This step goes from a higher value to a lower value.

A. Positive-going step

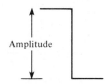

B. Negative-going step

FIGURE 10-38 *Positive- and negative-going steps.*

Ideal Pulses

An *ideal pulse* consists of two opposite steps of equal amplitude. Figure 10–39A shows an ideal positive-going pulse. It consists of a positive-going step followed by a negative-going step. The time that the pulse remains at the higher level is called the *pulse width* (PW). Part B of the figure shows an ideal negative-going pulse which consists of a negative-going step followed by a positive-going step. In this case, the pulse width is the time that the pulse is at its lower level.

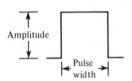
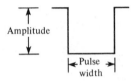

A. Positive-going pulse B. Negative-going pulse

FIGURE 10–39 *Ideal pulses.*

Nonideal Pulses

In many applications, analysis is simplified by treating all pulses as ideal. Actual pulses, however, are not ideal. They possess certain characteristics that cause them to be different from the ideal rectangular shape shown in Figure 10–39.

In practice, pulses cannot change from one value to another instantaneously. Time is always required for a transition (step) to occur, as illustrated in Figure 10–40A. As you can see, there is an interval of time for the pulse to rise from its lower value to its higher value. This interval is called the *rise time*, t_r. The accepted definition of rise time is *the time required for the pulse to go from 10% of its full amplitude to 90% of its full amplitude.*

The interval of time for the pulse to fall from its higher value to its lower value is called the *fall time* t_f. The accepted definition of fall time is *the time required for the pulse to go from 90% of its full amplitude to 10% of its full amplitude.*

Pulse width also requires a precise definition for the nonideal pulse because the rising and falling edges are not vertical. The generally accepted definition of pulse width (PW) is *the time between the point on the rising edge where the value is 50% of full amplitude to the point on the falling edge where the value is 50% of full amplitude.* Pulse width is shown in Figure 10-40B.

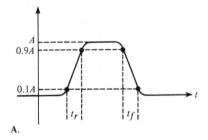
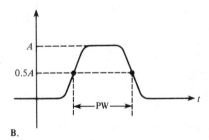

A. B.

FIGURE 10–40 *Nonideal pulse.*

PULSE WAVE FORMS

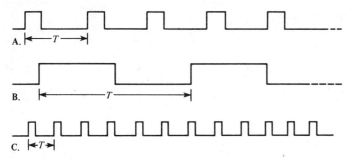

FIGURE 10-41 *Repetitive pulse wave forms.*

Repetitive Pulses

When pulses repeat at fixed intervals, the wave form is *periodic*. Examples of periodic pulse wave forms are shown in Figure 10-41. Notice that in each case the pulses repeat at regular intervals. The rate at which the pulses repeat is called the *pulse repetition* rate (PRR) or *pulse repetition frequency* (PRF), which is the *fundamental frequency* of the wave form. The frequency can be expressed in hertz or in *pulses per second* (p/s). The time from one pulse to the next is the period T of the wave form. The relationship between PRF and T is the same as the frequency-period relationship of a sine wave. The following equations state this relationship:

$$\text{PRF} = \frac{1}{T} \quad \quad (10\text{-}24)$$

$$T = \frac{1}{\text{PRF}} \quad \quad (10\text{-}25)$$

Another important relationship in repetitive pulse wave forms is the *duty cycle*. The duty cycle is *the ratio of the pulse width (PW) to the period (T)* and is usually exprssed as a percentage:

$$\% \text{ duty cycle} = \left(\frac{\text{PW}}{T}\right) 100 \quad \quad (10\text{-}26)$$

Example 10-13

Determine the period, PRF, and % duty cycle for the pulse wave form in Figure 10-42.

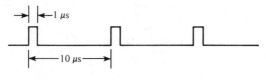

FIGURE 10-42

Example 10–13 (continued)

Solution:

$$T = 10 \text{ μs}$$

$$\text{PRF} = \frac{1}{T} = \frac{1}{10 \text{ μs}} = 0.1 \text{ MHz}$$

$$\% \text{ duty cycle} = \left(\frac{1 \text{ μs}}{10 \text{ μs}}\right)100 = 10\%$$

Square Wave

A *square wave* is a special case of a pulse wave form with a duty cycle of 50%. Thus, the pulse width is equal to one-half of the period. The amplitude can be any value. A square wave is shown in Figure 10–43.

FIGURE 10–43 *Square wave.*

Average Value of a Pulse Wave Form

The average value (V_{avg}) of a pulse wave form is equal to its baseline value plus its duty cycle times its amplitude. We will define the *lower level* of the wave form to be its *baseline*. The formula for average value is

$$V_{avg} = \text{baseline} + (\text{duty cycle})(\text{amplitude}) \tag{10–27}$$

Some examples will illustrate calculation of average values for several different pulse wave forms.

Example 10–14

Determine the average value of the wave form in Figure 10–44.

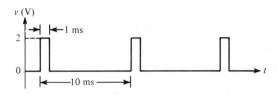

FIGURE 10–44

Solution:

Here the baseline value is 0 V; the amplitude is 2 V; and the duty cycle is 10%. The average value is determined as follows:

PULSE WAVE FORMS

$$V_{avg} = \text{baseline} + (\text{duty cycle})(\text{amplitude})$$
$$= 0 \text{ V} + (0.1)(2 \text{ V}) = 0.2 \text{ V}$$

Example 10–15

Determine the average value of the wave form in Figure 10–45.

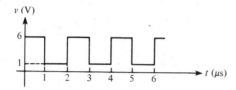

FIGURE 10–45

Solution:

This wave form has a baseline value of $+1$ V, an amplitude of 5 V, and a duty cycle of 50%. The average value is as follows:

$$V_{avg} = \text{baseline} + (\text{duty cycle})(\text{amplitude})$$
$$= 1 \text{ V} + (0.5)(5 \text{ V}) = 1 \text{ V} + 2.5 \text{ V} = 3.5 \text{ V}$$

Example 10–16

What is the average value of the pulse wave form in Figure 10–46?

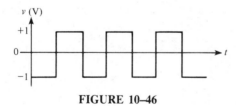

FIGURE 10–46

Solution:

This is a square wave with a baseline of -1 V and an amplitude of 2 V. The average value is determined as follows:

$$V_{avg} = \text{baseline} + (\text{duty cycle})(\text{amplitude})$$
$$= -1 \text{ V} + (0.5)(2 \text{ V}) = -1 \text{ V} + 1 \text{ V} = 0 \text{ V}$$

This is an *alternating* square wave, and, like an alternating sine wave, it has an average of zero.

Review for 10-7

1. Define *ideal pulse*.
2. What distinguishes an actual pulse from an ideal one?
3. Define the following:
   (a) rise time (b) fall time (c) pulse width
4. In a given repetitive pulse wave form, the pulses occur once every millisecond. What is the PRF of this wave form?
5. For the wave form in Figure 10-47, determine the duty cycle, amplitude, and average value.

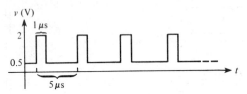

FIGURE 10-47

10-8

TRIANGULAR AND SAWTOOTH WAVE FORMS

Ramps

A voltage or a current ramp is a *linear* increase or decrease in the voltage or current with time. Figure 10-48 shows both positive- and negative-going ramps. In Part A of the figure, the ramp has a positive *slope*, and in Part B, the ramp has a negative slope. The slope of a voltage ramp is $\pm \Delta v / \Delta t$ and is expressed in V/s. The slope of a current ramp is $\pm \Delta i / \Delta t$ and is expressed in A/s.

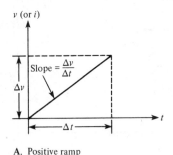

A. Positive ramp

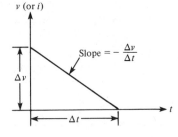

B. Negative ramp

FIGURE 10-48 *Ramps.*

TRIANGULAR AND SAWTOOTH WAVE FORMS

Example 10–17

What are the slopes of the voltage ramps in Figure 10–49?

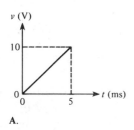
A.

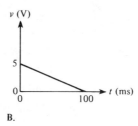
B.

FIGURE 10–49

Solution:

In Part A, the voltage increases from 0 V to +10 V in 5 ms. Thus, $\Delta v = 10$ V and $\Delta t = 5$ ms. The slope is

$$\frac{\Delta v}{\Delta t} = \frac{10 \text{ V}}{5 \text{ ms}} = 2 \text{ V/ms}$$
$$= 2 \text{ kV/s}$$

In Part B, the voltage decreases from +5 V to 0 V in 100 ms. Thus, $\Delta v = -5$ V and $\Delta t = 100$ ms. The slope is

$$\frac{\Delta v}{\Delta t} = \frac{-5 \text{ V}}{100 \text{ ms}} = -0.05 \text{ V/ms}$$
$$= -50 \text{ V/s}$$

Triangular Waves

A triangular wave form is shown in Figure 10–50. It consists of positive and negative ramps with equal-value slopes. The period of this wave form is measured from one peak to the next corresponding peak as illustrated. This triangular wave is alternating with an average value of zero.

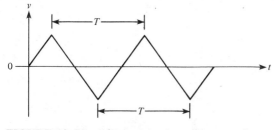

FIGURE 10–50 *Alternating triangular wave form.*

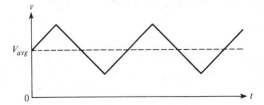

FIGURE 10–51 *Triangular wave form with a nonzero average value.*

Figure 10–51 shows a triangular wave with a nonzero average value. The frequency calculation for a triangle is the same as for a sine wave or a pulse wave form, that is, $f = 1/T$.

Sawtooth Waves

The sawtooth is actually a special case of triangular wave. Sawtooth wave forms are commonly found in many electronic systems. For example, the electron beam that sweeps across the screen of your TV set and recreates the picture is controlled basically by sawtooth voltages and currents. One sawtooth wave produces the horizontal beam movement, and the other produces the vertical beam movement. The sawtooth is sometimes called a *sweep wave form*.

Figure 10–52 is an example of a sawtooth wave. Notice that it consists of a positive ramp of relatively long duration, followed by a negative ramp of relatively short duration. The longer ramp typically has a much smaller slope than the short ramp. The name comes from the waveform's resemblance to the teeth of a saw blade.

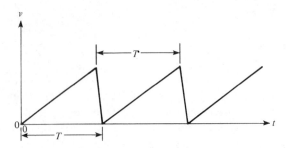

FIGURE 10–52 *Sawtooth wave form.*

Review for 10–8

1. What is the period of the sawtooth wave in Figure 10–53?

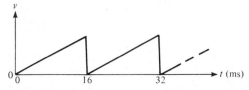

FIGURE 10–53

HARMONICS

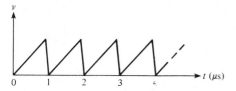

FIGURE 10-54

2. What is the frequency of the sawtooth in Figure 10-54?

10-9
HARMONICS

A pulse wave form or any other nonsinusoidal wave form is composed of a *fundamental* frequency and *harmonic* frequencies. The fundamental frequency is the repetition rate of the wave form. *Harmonics* are higher-frequency sine waves that are *exact multiples* of the fundamental frequency.

A nonsinusoidal wave is the composite (sum) of the fundamental and certain harmonics.

Odd Harmonics

Odd harmonics are frequencies that are *odd multiples* of the fundamental frequency. For example, a 1-kHz square wave consists of a fundamental of 1 kHz and odd harmnics of 3 kHz, 5 kHz, 7 kHz, and so on. The 3-kHz quantity in this case is called the *third harmonic;* the 5-kHz is the *fifth harmonic;* and so on.

Even Harmonics

Even harmonics are frequencies that are *even multiples* of the fundamental frequency. For example, if a certain wave has a fundamental of 200 Hz, the second harmonic is 400 Hz, the fourth harmonic is 800 Hz, the sixth harmonic is 1200 Hz, and so on. These are even harmonics.

Composite Wave Form

Any variation from a *pure* sine wave produces harmonics. The nonsinusoidal wave is a composite of the fundamental and the harmonics. Some waves have only odd harmonics. Some have only even harmonics. Some consist of both odd and even harmonics. The *shape* of the wave is determined by its *harmonic content.* Higher harmonics have less amplitude than lower harmonics. Usually, only the fundamental and the first few harmonics need be considered.

A *square wave* is an example of a wave form that consists of a fundamental and only odd harmonics. If the instantaneous values of the fundamental and each odd harmonic are added *algebraically* at each point, the resulting curve will

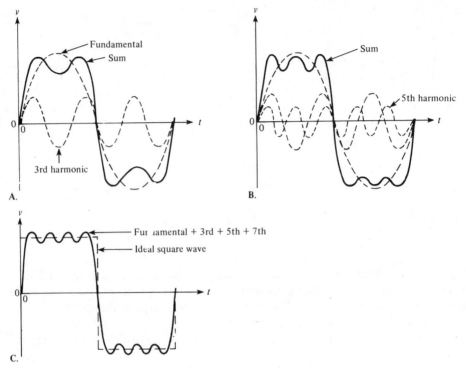

FIGURE 10–55 *Odd harmonics produce a square wave.*

have the shape of a square wave, as illustrated in Figure 10–55. In Part A of the figure, the fundamental and the third harmonic produce a wave shape that begins to resemble a square wave. In Part B, the fundamental, third, and fifth harmonics produce a closer resemblance. When the seventh harmonic is included, as in Part C, the resulting wave shape becomes even more like a square wave. As more harmonics are included, a square wave is approached.

Review for 10–9

1. Define *fundamental frequency*.
2. What are odd harmonics?
3. What are even harmonics?
4. Does a pure sine wave have harmonics?

Formulas

$$f = \frac{1}{T} \tag{10-1}$$

FORMULAS

$$T = \frac{1}{f} \tag{10-2}$$

$$V_{pp} = 2V_p \tag{10-3}$$

$$I_{pp} = 2I_p \tag{10-4}$$

$$V_{rms} = 0.707V_p \tag{10-5}$$

$$I_{rms} = 0.707I_p \tag{10-6}$$

$$V_p = 1.414V_{rms} \tag{10-7}$$

$$I_p = 1.414I_{rms} \tag{10-8}$$

$$V_{pp} = 2.828V_{rms} \tag{10-9}$$

$$I_{pp} = 2.828I_{rms} \tag{10-10}$$

$$V_{avg} = 0.637V_p \tag{10-11}$$

$$I_{avg} = 0.637I_p \tag{10-12}$$

$$i = \frac{V}{R} \tag{10-13}$$

$$I_p = \frac{V_p}{R} \tag{10-14}$$

$$I_{pp} = \frac{V_{pp}}{R} \tag{10-15}$$

$$I_{rms} = \frac{V_{rms}}{R} \tag{10-16}$$

$$I_{avg} = \frac{V_{avg}}{R} \tag{10-17}$$

$$\text{rad} = \left(\frac{\pi \text{ rad}}{180°}\right)(\text{degrees}) \tag{10-18}$$

$$\text{Degrees} = \left(\frac{180°}{\pi \text{ rad}}\right)(\text{rad}) \tag{10-19}$$

$$A_m \sin \theta \tag{10-20}$$

$$v = V_p \sin \theta$$

$$i = I_p \sin \theta$$

$$A_m \sin(\theta - \phi) \tag{10-21}$$

$$A_m \sin(\theta + \phi) \tag{10-22}$$

$$\omega = 2\pi f \tag{10-23}$$

$$\text{PRF} = \frac{1}{T} \qquad (10\text{--}24)$$

$$T = \frac{1}{\text{PRF}} \qquad (10\text{--}25)$$

$$\% \text{ duty cycle} = \left(\frac{\text{PW}}{T}\right)100 \qquad (10\text{--}26)$$

$$V_{avg} = \text{baseline} + (\text{duty cycle})(\text{amplitude}) \qquad (10\text{--}27)$$

Summary

1. The sine wave is a time-varying electrical signal. Current or voltage can vary sinusoidally.
2. The sine wave is the most common form of ac (alternating current).
3. Alternating current reverses direction in response to changes in the voltage polarity.
4. A sine wave is periodic; that is, it repeats at fixed intervals.
5. The time required for a sine wave to repeat is the period.
6. It takes one period to complete a cycle.
7. Frequency is the rate of change of the sine wave in cycles per second, or hertz (Hz).
8. There is a reciprocal relationship between frequency and period.
9. The instantaneous rate of change for a sine wave is maximum at the zero crossings and minimum at the peaks.
10. The amplitude of a sine wave is its current or voltage value expressed as instantaneous, peak, peak-to-peak, rms, or average value.
11. The instantaneous value of a sine wave is the value of the voltage or current at any point (instant) in time.
12. The peak value of a sine wave is its maximum positive value or its maximum negative value measured from the zero crossing.
13. The peak-to-peak value of a sine wave is measured from the positive peak to the negative peak and is twice the peak value.
14. The term *rms* stands for *root mean square*. The rms value of a sine wave, also known as the *effective* value, is a measure of the heating effect of a sine wave. It is 0.707 times the peak value.
15. The average value of a sine wave is defined over a half-cycle. It is 0.637 times the peak value. The average value over a full sine wave cycle is zero.

16. A full cycle of a sine wave is 360° or 2π radians (2π rad). A half-cycle is 180° or π rad. A quarter-cycle is 90° or $\pi/2$ rad.
17. A phase angle is the difference in degrees or radians between two sine waves.
18. A phasor represents a time-varying quantity in terms of both magnitude and direction.
19. A phasor can be used to represent a sine wave for analytical purposes.
20. The angular position of a phasor represents the angle of the sine wave, and the length of a phasor represents the amplitdue.
21. Angular velocity equals 2π times the frequency of the sine wave.
22. A step is a voltage or a current that rapidly changes from one value or level to a second value.
23. A pulse consists of two equal and opposite-going steps.
24. The amplitude of a pulse is the distance from one level to the other level expressed in voltage or current values.
25. The rise time of a pulse is the time required for the pulse to change from the 10% point to the 90% point on its rising edge.
26. The fall time of a pulse is the time required for the pulse to change from the 90% point to the 10% point on its falling edge.
27. Pulse width (PW) is the time between the 50% points on the rising and falling edges. It is the time duration of the pulse.
28. The frequency of a pulse wave form, sometimes called the PRF, can be expressed in pulses per second (p/s).
29. The duty cycle is the ratio of the pulse width to the period.
30. All nonsinusoidal wave forms are composed of a fundamental frequency and certain harmonic frequencies.
31. The fundamental frequency is the repetition rate of the wave.
32. Harmonics are the odd or even multiples of the fundamental frequency.

Self-Test

1. A sine wave has a period of 200 ms. What is the frequency?
2. A sine wave has a frequency of 25 kHz. What is its period?
3. A sine wave takes 5 μs to complete one cycle. Determine its frequency.
4. During a given 1-μs interval, a sine wave changes from 1 V to 1.5 V. What is the average rate of change during this time interval?
5. What does dv/dt stand for?

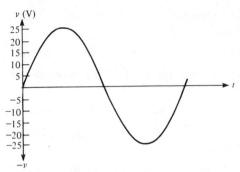

FIGURE 10–56

6. For the sine wave in Figure 10–56, determine the peak, peak-to-peak, rms, and average values.

7. The rms value of a given sine wave is 115 V. What is its peak value?

8. A 10-V rms sine wave is applied to a 100-Ω resistive circuit. Determine the rms value of the current.

9. Make a sketch of two sine waves as follows: Sine wave A is the reference, and sine wave B lags A by 90°. Both have the same amplitude.

10. A certain sine wave has a peak value of 20 V and begins at 0° reference. Calculate its instantaneous value at each of the following points: 10°, 25°, 30°, 90°, 180°, 210°, and 300°.

11. Calculate the instantaneous value of the sine wave in Figure 10–57 at 30°, 60°, 100°, and 225° points on the horizontal axis.

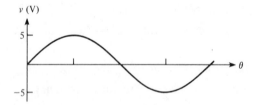

FIGURE 10–57

12. Determine the instantaneous value represented by each of the phasors in Figure 10–58.

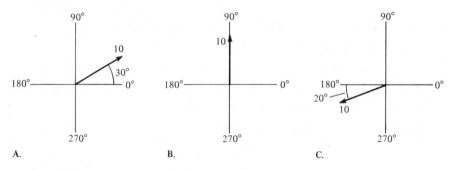

FIGURE 10–58

PROBLEMS

13. Express each of the following angles as an equivalent negative angle: 20°, 60°, 135°, 200°, 315°, and 330°.
14. If ω is 1000 rad/s, what is the frequency?
15. Determine the approximate rise time, fall time, pulse width, and amplitude of the pulse in Figure 10–59.

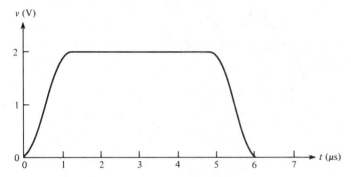

FIGURE 10–59

16. The period of a pulse wave form is 25 μs. What is the PRF?
17. The frequency of a pulse wave form is 2 kp/s, and the pulse width is 1 μs. What is the duty cycle?
18. Calculate the average value of the pulse wave form in Figure 10–60.

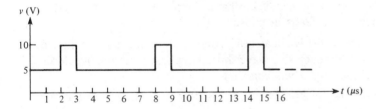

FIGURE 10–60

19. What is the fundamental frequency of the pulse wave form in Figure 10–60?
20. What is the second harmonic of 25 kHz?

Problems

10–1. Calculate the frequency for each following value of period:
    (a) 1 s   (b) 0.2 s   (c) 50 ms
    (d) 1 ms   (e) 500 μs   (f) 10 μs

10–2. Calculate the period for each following value of frequency:
    (a) 1 Hz   (b) 60 Hz   (c) 500 Hz
    (d) 1 kHz   (e) 200 kHz   (f) 5 MHz

10–3. A sine wave goes through 5 cycles in 10 μs. What is the period?

10–4. A sine wave has a frequency of 50 kHz. How many cycles does it complete in 10 ms?

10–5. A sine wave has a peak value of 12 V. Determine the following values:
(a) rms (b) Peak-to-peak (c) Average

10–6. A sine wave current has an rms value of 5 mA. Determine the following values:
(a) Peak (b) Average (c) Peak-to-peak

10–7. A sine wave voltage is applied to the resistive circuit in Figure 10–61. Determine the following:
(a) I_{rms} (b) I_{avg} (c) I_p
(d) I_{pp} (e) i at the positive voltage peak

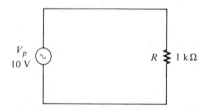

FIGURE 10–61

10–8. Sketch a graph of the voltage and current in Problem 10–7.

10–9. Sine wave A has a positive-going zero crossing at 30°. Sine wave B has a positive-going zero crossing at 45°. What is the phase angle between the two signals? Which signal leads in phase?

10–10. One sine wave has a positive peak at 75°, and another one has a positive peak at 100°. How much is each sine wave shifted in phase from the 0° reference? What is the phase angle between them?

10–11. For a certain 0° reference sine wave, V_{rms} is 20 V. Calculate its instantaneous value at each following angle:
(a) 15° (b) 33° (c) 50° (d) 110°
(e) 70° (f) 145° (g) 250° (h) 325°

10–12. For a particular 0° reference sine wave current, the peak value is 100 mA. Determine the instantaneous value at each of the following points:
(a) 35° (b) 95° (c) 190°
(d) 215° (e) 275° (f) 360°

10–13. For a 0° reference sine wave with an average value of 6.37 V, determine its instantaneous value at each following radian point:
(a) $\pi/8$ (b) $\pi/4$ (c) $\pi/2$ (d) $3\pi/4$
(e) π (f) $3\pi/2$ (g) 2π

10–14. Sine wave A lags sine wave B by 30°. Both have peak values of 15 V. Assume that sine wave A is the reference. Determine the instantaneous value of sine wave B at 30°, 45°, 90°, 180°, 200°, and 300°.

10–15. Repeat Problem 10–14 for the case when sine wave A leads sine wave B by 30°.

PROBLEMS

10–16. Draw a phasor diagram to represent the sine waves in Figure 10–62.

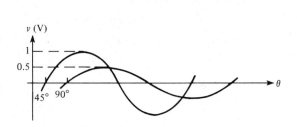

FIGURE 10–62

FIGURE 10–63

10–17. Sketch a graph of the wave forms represented by the phasor diagram in Figure 10–63. The phasor length is peak value.

10–18. Determine the frequency for each following angular velocity:
  (a) 60 rad/s   **(b)** 360 rad/s
  (c) 2 rad/s   **(d)** 1256 rad/s

10–19. From the graph in Figure 10–64, determine the approximate values of t_r, t_f, PW, and A.

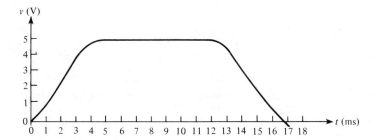

FIGURE 10–64

10–20. Calculate the PRF for each of the following values of period:
  (a) 1 ms   **(b)** 5 ms   **(c)** 0.250 µs   **(d)** 500 µs

10–21. Calculate the period for each of the following pulse frequencies:
  (a) 60 p/s   **(b)** 1 kp/s   **(c)** 2.5 kp/s   **(d)** 2 Mp/s

10–22. Determine the duty cycle for each wave form in Figure 10–65.

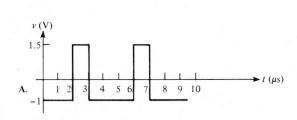 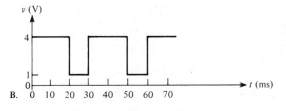

FIGURE 10–65

10-23. Find the average value of each pulse wave form in Figure 10-65.

10-24. What is the frequency of each wave form in Figure 10-65?

10-25. A square wave has a period of 40 μs. List the first six odd harmonics.

10-26. What is the frequency of each sawtooth wave form in Figure 10-66?

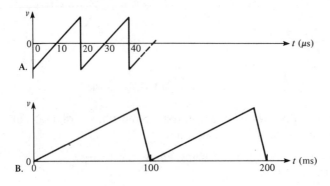

FIGURE 10-66

Answers to Section Reviews

Section 10-1:
1. See Figure 10-67. **2.** The points of maximum rate of change. The sine wave is equal above and below this line. **3.** One cycle consists of a positive and a negative alternation.

FIGURE 10-67

Section 10-2:
1. From the zero crossing through a positive peak, then through a zero to a negative peak, and back to the zero crossing. **2.** The time required to complete one cycle. **3.** The number of cycles completed in one second. **4.** The measure of how fast a quantity is increasing or decreasing. **5.** 200 kHz. **6.** 8.33 ms.

Section 10-3:
1. 2 V. **2.** 0.707 V. **3.** 0.637 V. **4.** 0.707 A.

Section 10-4:
1. (a) 90° (b) 180° (c) 270° (d) 360°. **2.** 180°, π. **3.** 360°, 2π. **4.** 45°.

Section 10-5:
1. 8.66 V. **2.** 8.19 V. **3.** 4.53 V.

ANSWERS TO SECTION REVIEWS

Section 10–6:
1. A representation of magnitude and phase angle of a time-varying function.
2. Amplitude and phase. 3. See Figure 10–68. 4. See Figure 10–69. 5. 100 Hz.

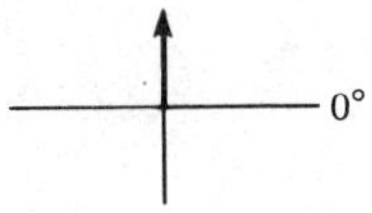

FIGURE 10–68

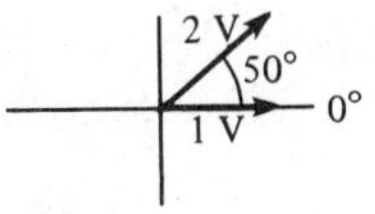

FIGURE 10–69

Section 10–7:
1. A pulse having instantaneous rise and fall times. 2. An actual pulse always has nonzero rise and fall times. 3. (a) The time interval from 10% to 90% of the rising pulse edge. (b) The time interval from 90% to 10% of the falling edge. (c) The time interval from 50% of the leading pulse edge to 50% of the trailing pulse edge.
4. 1 kp/s, or 1 kHz. 5. 20%, 1.5 V, 0.8 V.

Section 10–8:
1. 16 ms. 2. 1 MHz.

Section 10–9:
1. The repetition rate of the wave form. 2. Odd multiple frequencies of the fundamental. 3. Even multiple frequencies of the fundamental. 4. No.

The ability to store electrical charge is called *capacitance*, and the electrical component designed to have capacitance is called a *capacitor*. In this chapter you will study the construction and characteristics of capacitors. You will learn the basic functions of a capacitor in circuit applications, for example, its reactance and its charging and discharging abilities.

The effects of series and parallel combinations of capacitors are presented, along with their energy-storage properties and their power characteristics. The method of testing capacitors is also discussed.

11–1 The Basic Capacitor
11–2 Charging and Discharging a Capacitor
11–3 Unit of Capacitance
11–4 Characteristics of Capacitors
11–5 Types of Capacitors
11–6 Capacitor Labeling
11–7 Series Capacitors
11–8 Parallel Capacitors
11–9 Relationship of Capactive Current and Voltage
11–10 Capacitive Reactance
11–11 Energy and Power in a Capacitor
11–12 Testing a Capacitor

11

CAPACITANCE

11-1

THE BASIC CAPACITOR

A capacitor is a device that stores electrical charge. In its simplest form, a capacitor consists of *two parallel conductive plates* separated by an *insulating material* called the *dielectric*. A basic capacitor is shown in Figure 11–1.

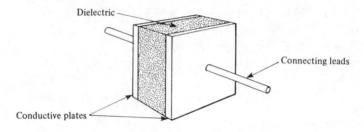

FIGURE 11–1 *Basic capacitor.*

Capacitance

Capacitance is a measure of a capacitor's ability to store electrical charge. Charge is stored when more electrons accumulate on one plate than on the other so that one plate has a negative charge with respect to the other plate. To charge a capacitor, you must connect a voltage source to the plates. This connection causes one plate to have a negative charge and the other a positive charge, as shown in Figure 11–2A.

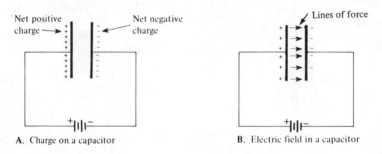

A. Charge on a capacitor   B. Electric field in a capacitor

FIGURE 11–2

Electric Field

When the plates of a capacitor are oppositely charged (one plate positive and one plate negative), an *electric field* is set up in the dielectric, as shown in Figure 11–2B. The electric field is made up of *lines of force* between the plates, caused by the attraction of opposite charges. These lines of force are concentrated within the dielectric. Energy is stored in the electric field within the dielectric of a capacitor.

Review for 11-1

1. What is a capacitor?
2. What are the basic parts of a capacitor?
3. Define *capacitance*.
4. What is an electric field?

11-2

CHARGING AND DISCHARGING A CAPACITOR

Charging

A capacitor will charge if it is connected to a voltage source, as shown in Figure 11-3. Notice that the capacitor in Part A of the figure is uncharged. That is, plate A and plate B have the same number of free electrons. When the switch is closed, as shown in Part B, electrons move away from plate A around the circuit to plate B. As plate A loses electrons and plate B gains electrons, plate A becomes *positive with respect to plate B*. As this charging process continues, the voltage builds up across the capacitor until it is equal and opposite in polarity to the applied voltage, V_S, as shown in Figure 11-3C. *When the capacitor is fully charged, there is no additional current.* The capacitor *blocks* the nonchanging dc.

Ideally, if the charged capacitor is disconnected from the source, as shown in Figure 11-3D, it will remain charged indefinitely. Actually the charge will gradually *leak off* due to imperfections in the dielectric material.

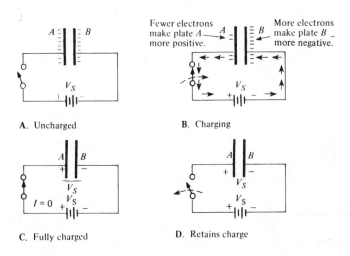

FIGURE 11-3 *Charging a capacitor.*

Discharging

If a conductor is connected across a charged capacitor, as shown in Figure 11–4, the capacitor will *discharge*. In this case a low-resistance path (wire) is connected across the capacitor with a switch. Before the switch is closed, the capacitor is charged to 50 V, as indicated in Part A of the figure. When the switch is closed, the extra electrons on plate *B* move through the circuit to plate *A*, as illustrated in Figure 11–4B. The current continues until the charge on the capacitor is neutralized. Neutrality occurs when the number of free electrons on plate *A* becomes equal to the number of free electrons on plate *B*, as shown in Figure 11–4C. At this time, the voltage across the capacitor has decreased to zero, and the capacitor is completely *discharged*.

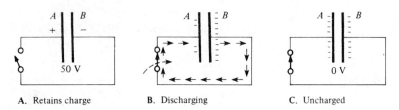

A. Retains charge   B. Discharging   C. Uncharged

FIGURE 11–4 *Discharging a capacitor.*

Charging and Discharging Current

Notice in Figures 11–3 and 11–4 that the direction of the current during discharge is opposite to that of the charging current. It is also important to understand that *there is no current from one plate to the other through the dielectric,* because the dielectric material is an insulator. There is current from one plate to the other only in the external circuit when the capacitor is charging or discharging, as illustrated in Figure 11–5.

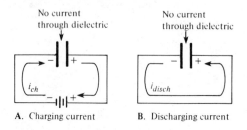

A. Charging current   B. Discharging current

FIGURE 11–5

Review for 11–2

1. How is a capacitor charged?
2. Does the voltage across the plates increase or decrease as the capacitor charges?

UNIT OF CAPACITANCE

3. How is a capacitor discharged?
4. If a capacitor is connected to a 120V dc source, what is the voltage across it when it reaches full charge?
5. When a capacitor is completely discharged, what is the voltage across its plates?

11-3
UNIT OF CAPACITANCE

The *farad*, abbreviated F, is the basic unit of capacitance. A capacitor has *one farad* of capacitance if *one coulomb* of charge is stored when *one volt* is applied to its plates.

The formula for capacitance in terms of *charge* and *voltage* is

$$C = \frac{Q}{V} \quad (11\text{–}1)$$

where C is capacitance, Q is charge, and V is voltage. When charge is in *coulombs* (C) and voltage in *volts* (V), capacitance is in *farads* (F).

By rearranging Equation (11–1), we obtain the following:

$$Q = CV \quad (11\text{–}2)$$

$$V = \frac{Q}{C} \quad (11\text{–}3)$$

Example 11–1

A capacitor stores 50 C and has 10 V across its plates. What is its capacitance in farads?

Solution:

$$C = \frac{Q}{V} = \frac{50 \text{ C}}{10 \text{ V}} = 5 \text{ F}$$

Example 11–2

A 2-F capacitor has 100 V across it. How much charge does it store?

Solution:

$$Q = CV = (2 \text{ F})(100 \text{ V}) = 200 \text{ C}$$

Example 11-3

What is the voltage across a 1-F capacitor that has stored 20 coulombs of charge?

Solution:

$$V = \frac{Q}{C} = \frac{20 \text{ C}}{1 \text{ F}} = 20 \text{ V}$$

Smaller Units of Capacitance

The farad is a relatively large amount of capacitance. In practice, most capacitance values used in electronic and electrical circuit work are much smaller. Typically, the *microfarad* (μF) and the *picofarad* (pF) are the practical units of capacitance. One microfarad is 10^{-6} farad, and one picofarad is 10^{-12} farad. A picofarad is sometimes referred to as a micro-microfarad ($\mu\mu$F), which is an older designation. Conversion formulas for farads, microfarads, and picofarads are given in Table 11-1.

TABLE 11-1

1 F = 1 × 10^6 μF
1 F = 1 × 10^{12} pF
1 μF = 1 × 10^{-6} F
1 μF = 1 × 10^6 pF
1 pF = 1 × 10^{-12} F
1 pF = 1 × 10^{-6} μF

Example 11-4

Convert the following values to microfarads:
(a) 0.00001 F   (b) 0.005 F   (c) 1000 pF   (d) 200 pF

Solution:

(a) 0.00001 F × 10^6 = 10 μF   (b) 0.005 F × 10^6 = 5000 μF
(c) 1000 pF × 10^{-6} = 0.001 μF   (d) 200 pF × 10^{-6} = 0.0002 μF

Example 11-5

Convert the following values to picofarads:
(a) 0.1 × 10^{-8} F   (b) 0.000025 F
(c) 0.01 μF   (d) 0.005 μF

CHARACTERISTICS OF CAPACITORS

Solution:

(a) 0.1×10^{-8} F $\times 10^{12} = 1000$ pF
(b) 0.000025 F $\times 10^{12} = 25 \times 10^{6}$ pF
(c) 0.01 μF $\times 10^{6} = 10,000$ pF
(d) 0.005 μF $\times 10^{6} = 5000$ pF

Review for 11-3

1. What is the basic unit of capacitance?
2. How many microfarads are in a farad?
3. How many picofarads are in a farad?
4. How many picofarads are in a microfarad?
5. Convert 0.0015 μF to picofarads.

11-4

CHARACTERISTICS OF CAPACITORS

Capacitance

As defined earlier, capacitance is the measure of the ability to store charge. For a fixed voltage, the greater the capacitance, the greater is the amount of charge that can be stored.

The capacitance value of a capacitor is normally indicated by a color code or by a typographical label on the body or case of the capacitor. For most capacitors used in electronics, the units are in microfarads or picofarads. The amount of capacitance is dependent on the physical construction of and the material used in the capacitor.

Voltage Rating

Every capacitor has a voltage rating which specifies the maximum voltage that can be applied across the capacitor without risk of damage to the device. This maximum voltage is commonly called the *breakdown voltage*. If it is exceeded, permanent damage can result. Like capacitance, the voltage rating is dependent on construction and material.

Both capacitance and voltage rating must be taken into account before a capacitor is used in a circuit application. The choice of capacitance is based on the particular circuit requirement and on factors that we will study later. The voltage rating should always be well above the actual voltage expected in the particular application.

We will now discuss the factors that determine the capacitance and breakdown voltage.

Plate Area

Capacitance is directly proportional to the plate area A. A larger plate area produces a larger capacitance, and vice versa. In Figure 11–6, the plate area A is the surface area of one of the plates of a basic parallel plate capacitor. If the plates are moved in relation to each other, the *overlapping area* determines the effective plate area. This variation in effective plate area is the basis for a type of variable capacitor used in radio receivers.

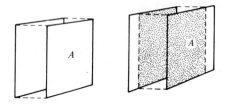

FIGURE 11–6 *Plate area.*

Plate Separation

Plate separation is the distance between the plates, designated d. It is also the thickness of the dielectric. *Capacitance is inversely proportional to the distance between the plates.* A greater separation of the plates produces a smaller capacitance, as illustrated in Figure 11–7.

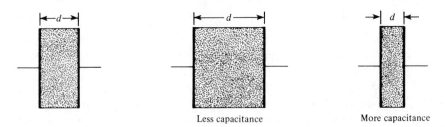

FIGURE 11–7 *Plate separation.*

Dielectric Constant

As you know, the insulating material between the plates of a capacitor is called the *dielectric*. Every dielectric material is able to concentrate the lines of force of an electric field and thus increase the capacity for energy storage. A measure of the ability of a material to establish an electric field is called the *dielectric constant* or *relative permittivity,* symbolized ϵ_r (the Greek letter epsilon).

Capacitance is directly proportional to the dielectric constant, ϵ_r. A larger value of dielectric constant produces a larger capacitance. The dielectric constant of a vacuum is 1, and that of air is very close to 1. These values are used as a reference; that is, all other materials have values of dielectric constant with respect to

CHARACTERISTICS OF CAPACITORS

those of a vacuum or air. For example, a material with $\epsilon_r = 8$ can provide an electric field intensity eight times greater than that of air, and, as a result, it can produce a capacitance eight times that of air.

Table 11–2 lists some common dielectric materials and a typical dielectric constant of each. Values can vary because they depend upon the specific composition of the material.

TABLE 11–2 *Dielectric constants.*

Material	Typical Dielectric Constant (ϵ_r)
Vacuum	1.0
Air	1.006
Teflon®	2.0
Paper	2.5
Oil	4.0
Mica	5.0
Glass	7.5
Ceramic	1200

The dielectric constant (relative permittivity) is dimensionless. It has no units because it is a relative measure and is a ratio of the absolute permittivity, ϵ, of a material to the absolute permittivity, ϵ_0, of a vacuum:

$$\epsilon_r = \frac{\epsilon}{\epsilon_0} \qquad (11\text{--}4)$$

The absolute permittivity for any material can therefore be calculated as $\epsilon = \epsilon_r \epsilon_0$. The value of ϵ_0 is 8.85×10^{-12} F/m.

Example 11–6

Determine the absolute permittivity of mica.

Solution:

$$\epsilon_r = 5.0 \quad \text{for mica}$$
$$\epsilon_0 = 8.85 \times 10^{-12} \text{ F/m}$$
$$\epsilon = 5.0 \times 8.85 \times 10^{-12} \text{ F/m}$$
$$= 44.25 \times 10^{-12} \text{ F/m}$$

Dielectric Strength

The dielectric strength describes the breakdown voltage of a capacitor. It is expressed in volts per mil (1 mil = 0.001 in.). Table 11–3 lists typical dielectric strength values for several materials.

TABLE 11-3 *Dielectric strengths.*

Material	Dielectric Strength (V/mil)
Air	20
Oil	375
Ceramic	1000
Paper	1250
Teflon®	1500
Mica	1500
Glass	2000

Dielectric strength can be explained by an example. Assume that a certain capacitor has a plate separation of 1 mil, and the dielectric is ceramic. This capacitor can withstand a maximum voltage of 1000 V because its dielectric strength is 1000 V/mil. If the maximum voltage is exceeded, the dielectric will break down and conduct current, causing permanent damage to the capacitor. Similarly, if the capacitor has a plate separation of 2 mils, its breakdown voltage is 2000 V.

A Formula for Capacitance in Terms of Physical Parameters

You have already seen how capacitance is directly related to plate area A and the dielectric constant ϵ_r, and inversely related to plate separation d. The formula for capacitance in terms of these three quantities is

$$C = \frac{A\epsilon_r(8.85 \times 10^{-12} \text{ F/m})}{d} \tag{11-5}$$

where A is in square meters (m²), d is in meters (m), and C is in farads (F). Recall that 8.85×10^{-12} F/m is the permittivity of a vacuum and that $\epsilon_r \times 8.85 \times 10^{-12}$ F/m is the absolute permittivity of the dielectric.

Example 11-7

Determine the capacitance of a parallel plate capacitor having a plate area of 0.01 m² and a plate separation of 0.02 m. The dielectric is mica.

Solution:

Use Equation (11-5):

$$C = \frac{(0.01 \text{ m}^2)(5.0)(8.85 \times 10^{-12} \text{ F/m})}{0.02 \text{ m}}$$

$$= 22.13 \text{ pF}$$

CHARACTERISTICS OF CAPACITORS

Temperature Coefficient

The temperature coefficient indicates the amount and direction of the change in capacitance with temperature. A positive temperature coefficient means that capacitance increases as temperature increases. A negative coefficient means that capacitance decreases as temperature increases.

The temperature coefficient is normally specified in parts per million per degree Celsius (ppm/°C). For example, a negative temperature coefficient of 150 ppm/°C for a 1-μF capacitor means that for every degree rise in temperature, the capacitance decreases 150 pF, since there are one million picofarads in a microfarad.

Leakage

No insulating material is perfect. The dielectric of all capacitors will conduct some current (although the current is usually so small that it can be neglected). Thus, the charge on any capacitor will eventually leak off. Some capacitors, particularly electrolytic ones, have a higher leakage than others. An equivalent circuit for a nonideal capacitor is shown in Figure 11–8. The parallel resistor represents the resistance of the dielectric through which there is leakage current.

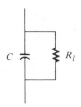

FIGURE 11–8 *Equivalent circuit for a nonideal capacitor.*

Review for 11–4

1. What is the significance of the voltage rating of a capacitor?
2. What precaution should be taken regarding the voltage rating before a capacitor is used in a circuit?
3. If the plate area is increased, will the capacitance increase or decrease?
4. If the distance between the plates is increased, will the capacitance increase or decrease?
5. If the dielectric constant is increased, will the capacitance be greater or smaller?
6. The plates of a ceramic capacitor are separated by 10 mils. What is the typical breakdown voltage?
7. A ceramic capacitor has a plate area of 0.2 m². The thickness of the dielectric is 0.005 m. What is the capacitance?

8. A 2-µF capacitor has a positive temperature coefficient of 50 ppm/°C. What is the capacitance when the temperature increases by 100°C?

11-5

TYPES OF CAPACITORS

There are many types of capacitors available. We will discuss some of the most common ones in this section.

Capacitors can be either *fixed* or *variable* and are generally classified according to their dielectric material. The standard capacitor symbols for circuit schematics are shown in Figure 11–9.

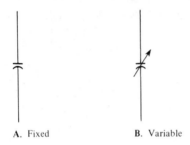

A. Fixed   B. Variable

FIGURE 11–9 *Schematic symbols for fixed and variable capacitors.*

Fixed Capacitors

A fixed capacitor has a value that cannot be altered. Let us examine several kinds of fixed capacitors which, as mentioned, are classified by their dielectric material.

Mica Capacitors: Figure 11–10 shows the typical construction of a mica capacitor. Thin sheets of mica form the dielectric. They are stacked alternately with sheets of metal foil, which form the plates. Alternate foil sections are connected together to form a single plate. More sections are used to increase the plate area, thus increasing capacitance. A reduction in the number of foil sections will reduce the capacitance. The mica and foil assembly is encapsulated in Bakelite® or other protective materials.

Mica capacitors are most commonly used for low capacitance values ranging from a few picofarads to several hundred picofarads. Some typical mica capacitors are shown in Figure 11–11.

Ceramic Capacitors: Ceramic dielectrics provide very high dielectric constants. As a result, comparatively high capacitance values can be achieved in a small physical size. Some typical ceramic capacitors are shown in Figure 11–12.

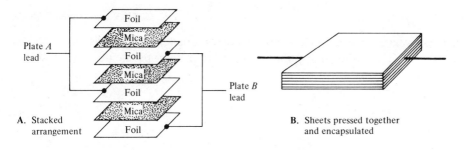

FIGURE 11–10 Construction of a typical mica capacitor.

FIGURE 11–11 Typical mica capacitors.

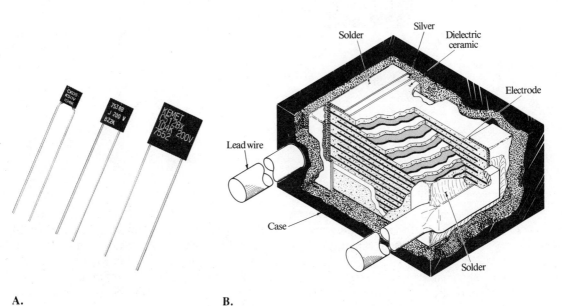

FIGURE 11–12 Ceramic capacitors. **A.** Typical capacitors. **B.** Construction view. (**A** and **B**, *courtesy of Union Carbide Corporation, Electronics Division*)

Paper Capacitors: The typical construction of a paper capacitor is illustrated in Figure 11–13. Commonly, the paper dielectric has been permeated by paraffin.

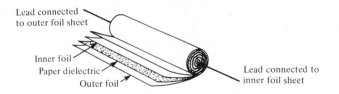

FIGURE 11–13   *Construction of a paper tubular capacitor.*

The tubular construction consists of a long strip of paper dielectric between two strips of foil. A wire lead is connected to each foil sheet. The foil sheets form the plates of the capacitor. The assembly is then rolled into a tubular shape and encapsulated in a molded case. Some paper capacitors also have a flat type of construction.

The end of the capacitor from which the lead is attached to the *outer* foil is often marked with a band around the case; other markings are also used. When the capacitor is connected in a circuit, the banded end should be grounded to take advantage of the shielding effect of the outer foil. Figure 11–14 shows some typical paper capacitors.

FIGURE 11–14   *Typical paper capacitors.*

Electrolytic Capacitors: Electrolytic capacitors are polarized; that is, one plate is positive and the other negative. These capacitors are used for comparatively high capacitance values ranging from a few microfarads to several thousand microfarads. The basic construction of an electrolytic capacitor is illustrated in Figure 11–15. As shown in Part A, the capacitor consists of a roll of foil with a layer of aluminum oxide deposited on its inner surface by electrolytic action, with the paper or gauze separator saturated with an electrolyte. This foil is the *positive plate,* and the oxide layer acts as the dielectric. The electrolyte in the paper and an inner foil layer act as the *negative plate.*

TYPES OF CAPACITORS

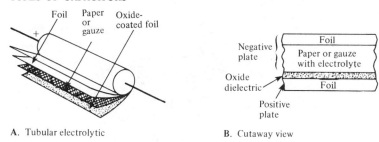

A. Tubular electrolytic   B. Cutaway view

FIGURE 11-15 *Construction of an electrolytic capacitor.*

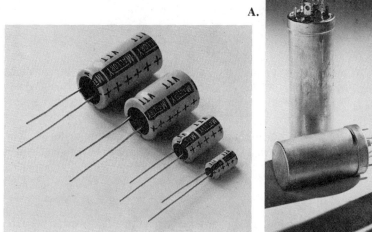

FIGURE 11-16 *Electrolytic capacitors.* **A.** *Typical examples. (Courtesy of Mallory Capacitor Division)* **B.** *Symbol.*

Since the capacitor is polarized, *the positive plate must be connected to the more positive side of a circuit.* Always be sure that this connection is correct.

Another type of electrolytic capacitor, called a *tantalum capacitor,* uses tantalum rather than aluminum. These capacitors are commonly used in transitor circuit applications. Figure 11-16 shows several typical electrolytic capacitors.

Film Capacitors: There are several basic types of film capacitors. Various types of materials are used as a dielectric, including Teflon®, Mylar®, Parylene®, and metalized film. They exhibit very high insulation resistance and are available in capacitance ranges from 0.001 µF to 1 µF. Figure 11-17 illustrates the construction of typical film capacitors.

Integrated Circuit Capacitors: An integrated circuit is constructed on a single tiny silicon chip. All of the circuit components, including transistors, diodes, resistors,

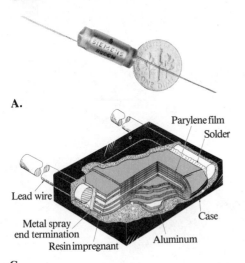
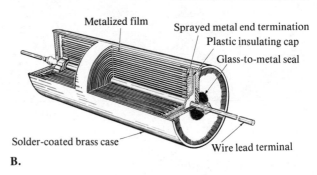

FIGURE 11-17 *Film capacitors.* **A.** *Typical example. (Courtesy of Siemens Corporation)* **B.** *Construction view of metalized film capacitor.* **C.** *Construction view of molded film capacitor. (Courtesy of Union Carbide Corporation, Electronics Division)*

and capacitors, are *integrated* onto this one piece of semiconductor material. You will study the total integrated circuit in a later course and examine the construction of this type of capacitor at that time.

Variable Capacitors

Variable capacitors are used in circuits when there is a need to vary the capacitance either manually or automatically.

Air Capacitors: Variable air capacitors, such as the one shown in Figure 11-18, are commonly used in radio receivers as tuning capacitors to provide frequency selection. This type of variable capacitor is constructed of several plates that mesh together. One set of plates can be moved relative to the other, thus changing the effective plate area and the capacitance. The plates are linked together mechanically so that all the plates are moved by rotation of a shaft.

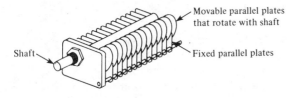
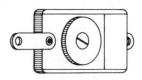
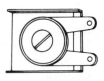

FIGURE 11-18 *Typical variable air capacitor.*   **FIGURE 11-19** *Mica trimmer capacitors.*

Trimmer Capacitors: Trimmer capacitors are small, variable devices commonly having mica dielectrics and a mechanical adjustment, as shown in Figure 11-19. Low-value air capacitors are also referred to as *trimmers*.

CAPACITOR LABELING

Review for 11-5

1. How are capacitors commonly classified?
2. What is the difference between a fixed and a variable capacitor?
3. What type of capacitor is normally polarized?
4. What precaution must be taken when a polarized capacitor is installed in a circuit?

11-6
CAPACITOR LABELING

Capacitor values are indicated on the body of the capacitor either by *typographical labels* or by *color codes*. Typographical labels consist of letters and numbers that indicate various parameters such as capacitance, voltage rating, tolerance, and others. Figure 11-20 shows some typically marked tantalum capacitors.

FIGURE 11-20 *Solid tantalum capacitors showing typical labeling. (Courtesy of Union Carbide Corporation, Electronics Division)*

Some capacitors carry no unit designation for capacitance. In these cases, the units are implied by the value indicated. For example, a ceramic capacitor marked .001 or .01 has units of microfarads because picofarad values that small are not available. As another example, a ceramic capacitor labeled 50 or 330 has units of picofarads because microfarad units that large normally are not available in this type.

In some instances, the units are labeled as pF or μF; often the microfarad unit is labeled as MF or MFD. Voltage rating appears on some types of capacitors and is omitted on others. When it is omitted, the voltage rating can be determined from information supplied by the manufacturer. The tolerance of the capacitor is sometimes labeled as a percentage, such as ±10%, and sometimes by a *parts per million* marking. The latter type of label consists of a *P* or an *N* followed by a number. For example, N750 means a negative temperature coefficient of 750 ppm/°C, and P30 means a positive temperature coefficient of 30 ppm/°C.

Color Coding

Some capacitors have color-coded designations. The color code used for capacitors is basically the same as that used for resistors. Some variations occur in tolerance designation. The basic color codes are shown in Table 11–4, and some typical color-coded capacitors are illustrated in Figure 11–21.

TABLE 11–4 *Typical composite color codes for capacitors (picofarads).*

Color	Digit	Multiplier	Tolerance
Black	0	1	20%
Brown	1	10	1%
Red	2	100	2%
Orange	3	1000	3%
Yellow	4	10000	
Green	5	100000	5%(EIA)
Blue	6	1000000	
Violet	7		
Gray	8		
White	9		
Gold		0.1	5%(JAN)
Silver		0.01	10%
No color			20%

NOTE: EIA stands for Electronic Industries Association, and JAN stands for Joint Army-Navy, a military standard.

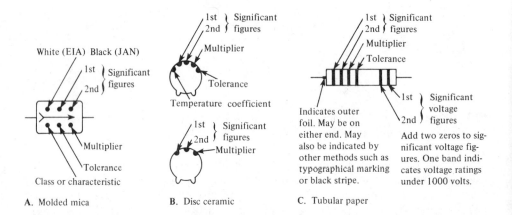

A. Molded mica   B. Disc ceramic   C. Tubular paper

FIGURE 11–21 *Typical color-coded capacitors.*

Example 11–8

Determine the values of the two capacitors in Figure 11–22.

SERIES CAPACITORS

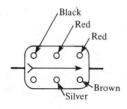
A. Molded mica

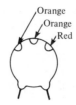
B. Disc ceramic (3-dot)

FIGURE 11-22

Solution:

The values are shown in Figure 11-23.

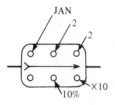
A. $C = 220$ pF $\pm 10\%$

B. $C = 3300$ pF

FIGURE 11-23

Review for 11-6

1. Determine the capacitance in Figure 11-24.

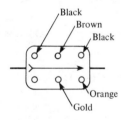

FIGURE 11-24

2. What does the designation N330 stand for?

11-7
SERIES CAPACITORS

Total Capacitance

When capacitors are connected in series, the *effective* plate separation increases, and *the total capacitance is less than that of the smallest capacitor.* The reason is as follows: Consider the generalized circuit in Figure 11–25A which has n capacitors in series with a voltage source and a switch. When the switch is closed, the capacitors will charge as current is established through the circuit. Since this is a series circuit, the current must be the same at all points, as illustrated. Since current is the rate of flow of charge, *the amount of charge stored by each capacitor is equal to the total charge,* expressed as follows:

$$Q_T = Q_1 = Q_2 = Q_3 = \cdots = Q_n \qquad (11\text{–}6)$$

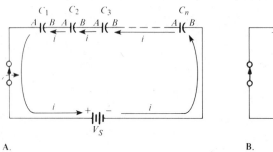
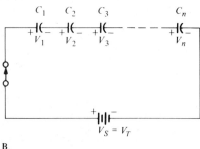

A.   B.

FIGURE 11–25 *Series capacitive circuit.*

Next, by *Kirchhoff's voltage law,* the *sum* of the voltages across the charged capacitors (voltage drop) must equal the total voltage, V_S, as shown in Figure 11–25B. This is expressed in equation form as

$$V_T = V_1 + V_2 + V_3 + \cdots + V_n$$

From Equation (11–3), we know that $V = Q/C$. Substituting this relationship into each term of the voltage equation, we obtain the following result:

$$\frac{Q_T}{C_T} = \frac{Q_1}{C_1} + \frac{Q_2}{C_2} + \frac{Q_3}{C_3} + \cdots + \frac{Q_n}{C_n} \qquad (11\text{–}7)$$

By Equation (11–6), $Q_T = Q_1 = Q_2 = Q_3 = \cdots = Q_n$. Thus, the Q terms can be factored from Equation (11–7) and canceled, resulting in

$$\frac{1}{C_T} = \frac{1}{C_1} + \frac{1}{C_2} + \frac{1}{C_3} + \cdots + \frac{1}{C_n} \qquad (11\text{–}8)$$

We can obtain a formula for the total series capacitance C_T by taking the reciprocal of both sides of Equation (11–8). The result is

$$C_T = \frac{1}{(1/C_1) + (1/C_2) + (1/C_3) + \cdots + (1/C_n)} \qquad (11\text{–}9)$$

Special Case of Two Capacitors in Series

When only two capacitors are in series, a special form of Equation (11–9) can be used. It is developed as follows:

SERIES CAPACITORS

$$C_T = \frac{1}{(1/C_1) + (1/C_2)}$$
$$= \frac{1}{(C_1 + C_2)/C_1 C_2}$$
$$C_T = \frac{C_1 C_2}{C_1 + C_2} \qquad (11\text{--}10)$$

Capacitors of Equal Value in Series

This special case is another in which a formula can be developed from Equation (11–9). If all of the values are the same and equal to C, Equation (11–9) can be written as

$$C_T = \frac{1}{(1/C) + (1/C) + (1/C) + \cdots + (1/C)}$$

Adding all of the terms in the denominator, we get

$$C_T = \frac{C}{n} \qquad (11\text{--}11)$$

The capacitance value divided by the number of capacitors in series gives the total capacitance.

Notice that in all cases the total series capacitance is calculated in the same manner as *total parallel resistance*.

Example 11–9

Determine the total capacitance C_T in Figure 11–26.

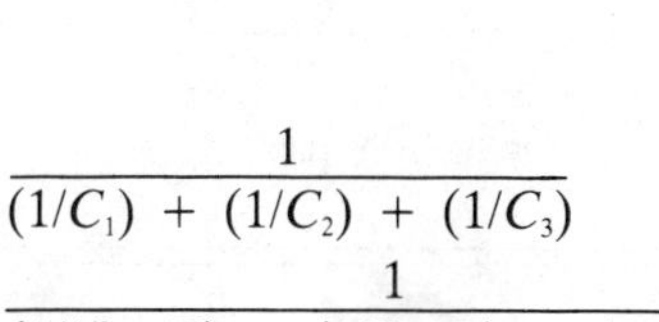

C_1 10 μF, C_2 5 μF, C_3 8 μF

FIGURE 11–26

Solution:

$$C_T = \frac{1}{(1/C_1) + (1/C_2) + (1/C_3)}$$
$$= \frac{1}{(1/10 \ \mu F) + (1/5 \ \mu F) + (1/8 \ \mu F)}$$
$$= \frac{1}{0.425} \ \mu F = 2.35 \ \mu F$$

Example 11–10

Find C_T in Figure 11–27.

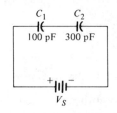

FIGURE 11–27

Solution:

$$C_T = \frac{C_1 C_2}{C_1 + C_2} = \frac{(100 \text{ pF})(300 \text{ pF})}{400 \text{ pF}}$$
$$= 75 \text{ pF}$$

Example 11–11

Determine C_T for the series capacitors in Figure 11–28.

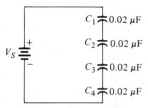

FIGURE 11–28

Solution:

$$C_1 = C_2 = C_3 = C_4 = C$$
$$C_T = \frac{C}{n} = \frac{0.02 \text{ }\mu\text{F}}{4}$$
$$= 0.005 \text{ }\mu\text{F}$$

Capacitor Voltages

A series connection of charged capacitors acts as a *voltage divider*. The voltage across each capacitor in series is inversely proportional to its capacitance value because $V = Q/C$.

SERIES CAPACITORS

The voltage across any capacitor in series can be calculated as follows:

$$V_x = \left(\frac{C_T}{C_x}\right)V_T \tag{11-12}$$

where C_x is C_1, or C_2, or C_3, and so on. The derivation of Equation (11-12) is as follows: Since the charge on any capacitor in series is the same as the total charge ($Q_x = Q_T$), and since $Q_x = V_x C_x$ and $Q_T = V_T C_T$, then

$$V_x C_x = V_T C_T$$

Solving the last equation for V_x, we get

$$V_x = \frac{C_T V_T}{C_x}$$

The largest capacitor in series will have the smallest voltage, and the smallest will have the largest voltage.

Example 11-12

Find the voltage across each capacitor in Figure 11-29.

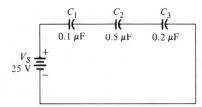

FIGURE 11-29

Solution:

$$C_T = \frac{1}{(1/C_1) + (1/C_2) + (1/C_3)} = \frac{1}{(1/0.1\ \mu F) + (1/0.5\ \mu F) + (1/0.2\ \mu F)}$$

$$= 1/17\ \mu F = 0.0588\ \mu F$$

$$V_S = V_T = 25\ V$$

$$V_1 = \left(\frac{C_T}{C_1}\right)V_T = \left(\frac{0.0588\ \mu F}{0.1\ \mu F}\right)25\ V$$

$$= 14.71\ V$$

$$V_2 = \left(\frac{C_T}{C_2}\right)V_T = \left(\frac{0.0588\ \mu F}{0.5\ \mu F}\right)25\ V$$

$$= 2.94\ V$$

$$V_3 = \left(\frac{C_T}{C_3}\right)V_T = \left(\frac{0.0588\ \mu F}{0.2\ \mu F}\right)25\ V$$

$$= 7.35\ V$$

Check: $V_1 + V_2 + V_3 = 14.71\ V + 2.94\ V + 7.35\ V = 25\ V$.

Review for 11-7

1. Do capacitors in series produce a total capacitance that is less or greater than that of the smallest capacitor?
2. The following capacitors are in series: 100 pF, 250 pF, and 500 pF. What is the total capacitance?
3. A 0.01-μF and a 0.015-μF capacitor are in series. What is C_T?
4. Five 100-pF capacitors are connected in series. What is C_T?
5. Determine the voltage across C_1 in Figure 11–30.

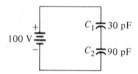

FIGURE 11–30

11-8

PARALLEL CAPACITORS

When capacitors are connected in parallel, the *effective* plate area increases, and *the total capacitance is the sum of the individual capacitances.* You can understand the reason by considering what happens when the switch is closed in Figure 11–31.

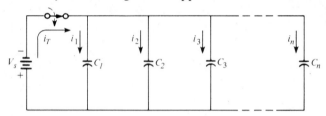

FIGURE 11–31 *Capacitors in parallel.*

The total charging current from the source divides at the junction of the parallel branches. There is a different charging current through each branch so that a different charge can be stored by each capacitor. By Kirchhoff's current law, the *sum* of all of the charging currents is equal to the total current. Therefore, the *sum* of the charges on the capacitors is equal to the total charge. Also, the voltages across all of the parallel branches are the same. These observations are used to develop the formula for total parallel capacitance as follows:

$$Q_T = Q_1 + Q_2 + Q_3 + \cdots + Q_n$$

Since $Q = CV$,

$$C_T V_T = C_1 V_1 + C_2 V_2 + C_3 V_3 + \cdots + C_n V_n$$

PARALLEL CAPACITORS

Since $V_T = V_1 = V_2 = V_3 = \cdots = V_n$, the voltages can be factored and canceled, giving

$$C_T = C_1 + C_2 + C_3 + \cdots + C_n \tag{11-13}$$

Equation (11–13) is the generalized formula for total parallel capacitance where n is the number of capacitors. Remember that *capacitors add in parallel*.

For the special case when all capacitors have the same value, multiply the value by the number of capacitors in parallel:

$$C_T = nC \tag{11-14}$$

Notice that in all cases the total parallel capacitance is calculated in the same manner as *total series resistance*.

Example 11–13

What is the total capacitance in Figure 11–32? What is the voltage across each capacitor?

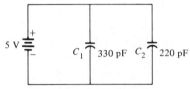

FIGURE 11–32

Solution:

$$C_T = C_1 + C_2 = 330 \text{ pF} + 220 \text{ pF}$$
$$= 550 \text{ pF}$$
$$V_1 = V_2 = 5 \text{ V}$$

Example 11–14

Determine C_T in Figure 11–33.

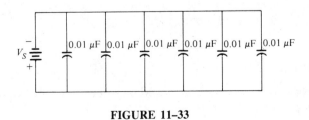

FIGURE 11–33

Example 11–14 (continued)

Solution:

There are six capacitors in parallel; so $n = 6$.

$$C_T = nC = (6)(0.01 \; \mu F)$$
$$= 0.06 \; \mu F$$

Review for 11–8

1. When capacitors are in parallel, how is the total capacitance determined?
2. In a certain application, you need 0.05 µF. The only values available are 0.01 µF. How can you get the total capacitance that you need?
3. The following capacitors are in parallel: 10 pF, 5 pF, 33 pF, and 50 pF. What is C_T?

11–9

RELATIONSHIP OF CAPACITIVE CURRENT AND VOLTAGE

In Chapter 10, we introduced the concept of a derivative in relation to the rate of change of a sine wave. We learned that *a derivative is the instantaneous rate of change of a quantity*. Recall that current is the *rate of flow of charge (electrons)*. Therefore, instantaneous current, i, can be expressed as the instantaneous rate of change of charge, q, with respect to time, t:

$$i = \frac{dq}{dt} \qquad (11\text{--}15)$$

Also, we know that in terms of instantaneous quantities, $q = Cv$. Therefore, from a basic rule of calculus, the derivative of q is $dq/dt = C(dv/dt)$. Since $i = dq/dt$, we get the following relationship:

$$i = C \frac{dv}{dt} \qquad (11\text{--}16)$$

This equation says that *the instantaneous capacitor current is equal to the capacitance times the instantaneous rate of change of the voltage.* From this you can see that the faster the voltage across a capacitor changes, the greater the current is. For example, if the rate of change of voltage is zero, the current is zero $[i = C(dv/dt) = C(0) = 0]$.

Phase Relationship

Now consider what happens when a sine wave voltage is applied across a capacitor, as shown in Figure 11–34.

CAPACITIVE RESISTANCE

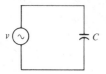

FIGURE 11–34 *Sine wave voltage applied to a capacitor.*

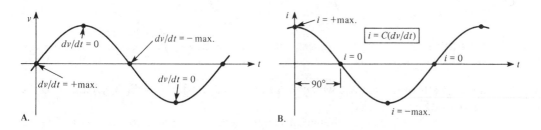

FIGURE 11–35 *Phase relation of v and i in a capacitor. Current leads voltage by 90°.*

The voltage wave form, one cycle of which is shown in Figure 11–35A, has a maximum rate of change (dv/dt = max.) at the zero crossings. It has a zero rate of change (dv/dt = 0) at the peaks.

Using Equation (11–16), we can establish the phase relationship between the current and the voltage for the capacitor. When dv/dt = 0, i is also zero because $i = C(dv/dt) = C(0) = 0$. When dv/dt is a positive-going maximum, i is a positive maximum (peak). When dv/dt is a negative-going maximum, i is a negative maximum (peak). *A sine wave voltage produces a cosine wave current in a capacitive circuit.* Therefore, we can plot the current in relation to the voltage by knowing the points on the voltage curve at which the current is zero and maximum. This plotting is shown in Figure 11–35B. Notice that *the current leads the voltage by 90°,* as is always true in a purely capacitive circuit.

Review for 11–9

1. What does *derivative* mean?
2. Write the formula for the current-voltage relationship in a capacitor.
3. State the phase relationship between current and voltage in a capacitor.

11–10

CAPACITIVE REACTANCE

Capacitive reactance is the opposition to sinusoidal current, expressed in ohms and symbolized by X_C. It is *inversely* dependent on *capacitance* and *frequency*. To

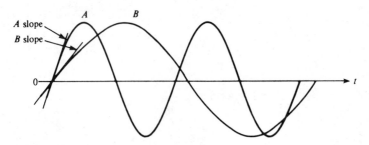

FIGURE 11–36 *A higher-frequency wave has a greater slope at its zero crossings, corresponding to a higher rate of change.*

develop a formula for X_C, we again use the relationship $i = C(dv/dt)$ and also the curves in Figure 11–36.

The rate of change of voltage is directly related to *frequency*. That is, the faster the voltage changes, the higher the frequency is. For example, you can see that in Figure 11–36, the slope of sine wave A at the zero crossings is greater than that of sine wave B. The *slope* of a curve at a point indicates its rate of change at that point. Since sine wave A has a higher frequency than sine wave B, it also has *a greater dv/dt* at the zero crossings. Its *average* rate of change, of course, is also greater.

If frequency increases, *dv/dt* increases, and thus *i* increases; if frequency decreases, *dv/dt* decreases, and thus *i decreases*:

$$\overset{\uparrow}{i} = C\overset{\uparrow}{\frac{dv}{dt}} \qquad \underset{\downarrow}{i} = C\underset{\downarrow}{\frac{dv}{dt}}$$

An increase in *i* means less opposition (X_C is less), and a decrease in *i* means more opposition (X_C is greater). Thus, *X_C is inversely proportional to i and therefore inversely proportional to frequency*:

$$X_C \text{ is proportional to } \frac{1}{f}$$

If *dv/dt* is constant and the capacitance is varied, an increase in *C* produces an increase in *i*, and a decrease in *C* produces a decrease in *i*:

$$\overset{\uparrow}{i} = \overset{\uparrow}{C}\frac{dv}{dt} \qquad \underset{\downarrow}{i} = \underset{\downarrow}{C}\frac{dv}{dt}$$

Again, an increase in *i* means less opposition (X_C is less), and a decrease in *i* means more opposition (X_C is greater). Thus, *X_C is inversely proportional to i and therefore inversely proportional to capacitance in addition to frequency*:

$$X_C \text{ is proportional to } \frac{1}{fC}$$

CAPACITIVE REACTANCE

So far we have determined that X_c is *proportional* to $1/fC$. What we need is a formula that tells us what X_c is *equal to* so that we can calculate X_c. This formula is as follows:

$$X_C = \frac{1}{2\pi fC} \quad (11\text{--}17)$$

where X_c is in *ohms* if f is in *hertz* and C is in *farads*. Notice that 2π appears in the denominator as a constant factor. This term is derived from the relationship of a sine wave to rotational motion. A mathematical derivation of Equation (11–17) is given in Appendix F.

Example 11–15

A sine wave voltage is applied to a capacitor as shown in Figure 11–37. The frequency of the sine wave is 1 kHz. Determine the reactance of the capacitor.

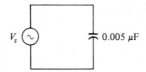

FIGURE 11–37

Solution:

$$X_C = \frac{1}{2\pi fC} = \frac{1}{2\pi (1 \times 10^3 \text{ Hz})(0.005 \times 10^{-6} \text{ F})}$$
$$= 31.83 \text{ k}\Omega$$

Ohm's Law Applies to X_C

Ohm's law applies to reactance as well as to resistance. It is stated in the following three forms:

$$V = IX_C \quad (11\text{--}18)$$

$$I = \frac{V}{X_C} \quad (11\text{--}19)$$

$$X_C = \frac{V}{I} \quad (11\text{--}20)$$

where X_C is used in place of R. Both current and voltage must be expressed in the same units of rms, peak, peak-to-peak, or average values.

Example 11-16

Determine the rms current in Figure 11-38.

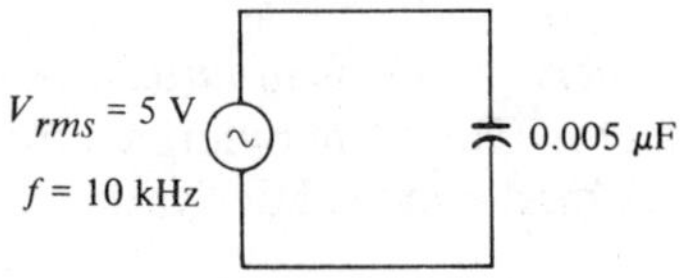

FIGURE 11-38

Solution:

$$X_C = \frac{1}{2\pi f C} = \frac{1}{2\pi(10 \times 10^3 \text{ Hz})(0.005 \times 10^{-6} \text{ F})}$$
$$= 3.18 \text{ k}\Omega$$

$$I_{rms} = \frac{V_{rms}}{X_C} = \frac{5 \text{ V}}{3.18 \text{ k}\Omega}$$
$$= 1.57 \text{ mA}$$

Review for 11-10

1. Define *capacitive reactance*.
2. Calculate X_C for $f = 5$ kHz and $C = 50$ pF.
3. To obtain a reactance of 2 kΩ from a 0.1-μF capacitor, what frequency is required?
4. Calculate the current in Figure 11-39.

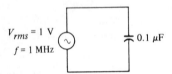

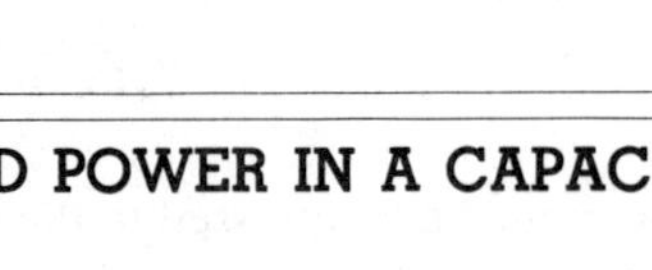

FIGURE 11-39

11-11

ENERGY AND POWER IN A CAPACITOR

Energy

As mentioned in an earlier section, a charged capacitor stores energy in the electric field existing within the dielectric. It can be shown that the energy, $\mathscr{E}$, stored by a charged capacitor is

ENERGY AND POWER IN A CAPACITOR

$$\mathcal{E} = \tfrac{1}{2}CV^2 \tag{11-21}$$

where V is the voltage across the capacitor and, if C is in farads and V is in volts, $\mathcal{E}$ is in joules.

Power

An ideal capacitor does not dissipate energy; it only stores it. When an ac voltage is applied to a capacitor, energy is stored by the capacitor during a portion of the voltage cycle and then is *returned* to the source during another portion. *There is no net energy loss.* Figure 11–40 shows one cycle of capacitor voltage and current.

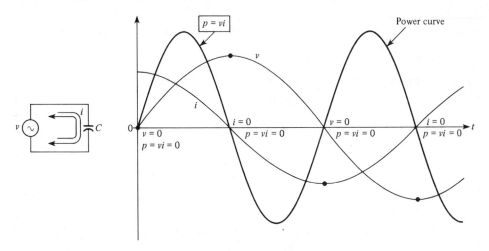

FIGURE 11–40 *Power curve.*

Instantaneous Power (p): The product of v and i gives instantaneous power, p. At points where v or i is zero, p is zero. When both v and i are positive, p is positive. When either v or i is positive and the other negative, p is negative. When both v and i are negative, p is positive. Between the zero points, the power follows a sine wave curve as indicated in Figure 11–40. *Positive* values of power indicate that energy is *stored* by the capacitor. *Negative* values of power indicate that energy is *returned* from the capacitor to the source. Note that the power fluctuates at a frequency *twice* that of the voltage or current as energy is alternately stored and returned to the source.

Average Power (P_{avg}): Ideally, all of the energy stored by a capacitor during the positive portion of the power cycle is returned to the source during the negative portion. Thus, no net energy is consumed in the capacitor; so the *average power is zero*.

Actually, because of the leakage resistance in a practical capacitor, a very small percentage of the total energy is dissipated.

Reactive Power (P_r): The rate at which a capacitor stores or returns energy is called its *reactive power*, P_r. The reactive power is a nonzero quantity, because at any instant, the capacitor is actually taking energy from or returning energy to the source. The following equations express the reactive power in a capacitor:

$$P_r = V_{rms} I_{rms} \tag{11-22}$$

$$P_r = \frac{V_{rms}^2}{X_C} \tag{11-23}$$

$$P_r = I_{rms}^2 X_C \tag{11-24}$$

Notice that these equations are of the same *form* as those for average power in a resistor. The voltage and current are rms values. The unit of reactive power is *volt-amperes reactive.* (VAR).

Example 11–17

Determine the energy stored by the capacitor in Figure 11–41.

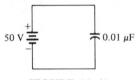

FIGURE 11–41

Solution:

$$\mathscr{E} = \tfrac{1}{2}CV^2 = \tfrac{1}{2}(0.01 \times 10^{-6} \text{ F})(50 \text{ V})^2$$
$$= 12.5 \times 10^{-6} \text{ J} = 12.5 \text{ }\mu\text{J}$$

Example 11–18

Determine the average power and the reactive power in Figure 11–42.

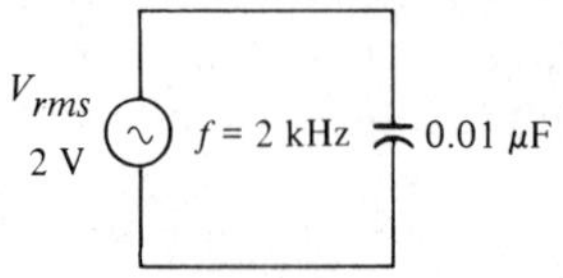

FIGURE 11–42

Solution:

P_{avg} *is always zero for a capacitor.* The reactive power is determined as follows:

$$X_C = \frac{1}{2\pi f C} = \frac{1}{2\pi (2 \times 10^3 \text{ Hz})(0.01 \times 10^{-6} \text{ F})}$$

TESTING A CAPACITOR

$$= 7.958 \text{ k}\Omega$$

$$P_r = \frac{V_{rms}^2}{X_C} = \frac{(2 \text{ V})^2}{7.958 \text{ k}\Omega} = 0.503 \times 10^{-3} \text{ VAR}$$

$$= 0.503 \text{ mVAR}$$

Review for 11–11

1. A 100-pF capacitor is connected to a 10-V dc source. When the capacitor is fully charged, how much energy does it store?
2. What is the average power in a capacitor connected to an ac source?

11–12

TESTING A CAPACITOR

One common failure in a capacitor occurs when the dielectric breaks down and a short results. The short can be a permanent one or an intermittent one that reoccurs only at a certain voltage level. Also, a capacitor can become very leaky and develop a much lower dielectric resistance than normal. In some cases, an open can occur in a capacitor. All of these types of failures destroy or reduce the capacitor's ability to store charge.

Ohmmeter Check

When there is a suspected problem, you should remove the capacitor from the circuit to check it with an ohmmeter. First, to be sure that the capacitor is discharged, short its leads as indicated in Figure 11–43A. Connect the meter, set on a high ohms range such as RX1M, to the capacitor as shown in Figure 11–43B, and observe the needle. It should initially indicate near-zero ohms. Then it should begin to move toward the high-resistance end of the scale as the capacitor charges, as shown in Figure 11–43C. When the capacitor is fully charged, the meter will indicate an extremely high resistance.

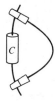
A. Discharging

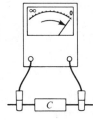
B. Initially ($i = 0$)

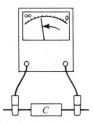
C. Charging

FIGURE 11–43 *Checking a capacitor with an ohmmeter.*

The capacitor is charged from the internal battery of the ohmmeter, and the meter responds to charging current. The larger the capacitance value, the more slowly the capacitor will charge, as indicated by the needle movement. For very small pF values, the meter response may be too slow to indicate the fast charging action.

If the capacitor is shorted, the meter will go to zero and stay there. If it is leaky, the final meter reading will be much less than normal. Most capacitors have a resistance of several hundred megohms. The exception is the electrolytic, which may normally have less than one megohm.

If the capacitor is open, no charging action will be observed, and the meter will indicate infinite resistance.

Review for 11–12

1. How can a capacitor be discharged out of the circuit?
2. Describe how the needle of an ohmmeter behaves when a good capacitor is checked.

Formulas

$$C = \frac{Q}{V} \tag{11-1}$$

$$Q = CV \tag{11-2}$$

$$V = \frac{Q}{C} \tag{11-3}$$

$$\epsilon_r = \frac{\epsilon}{\epsilon_0} \tag{11-4}$$

$$C = \frac{A\epsilon_r(8.85 \times 10^{-12} \text{ F/m})}{d} \tag{11-5}$$

$$Q_T = Q_1 = Q_2 = Q_3 = \cdots = Q_n \tag{11-6}$$

$$C_T = \frac{1}{(1/C_1) + (1/C_2) + (1/C_3) + \cdots + (1/C_n)} \tag{11-9}$$

$$C_T = \frac{C_1 C_2}{C_1 + C_2} \tag{11-10}$$

$$C_T = \frac{C}{n} \tag{11-11}$$

$$V_X = \left(\frac{C_T}{C_X}\right) V_T \tag{11-12}$$

$$C_T = C_1 + C_2 + C_3 + \cdots + C_n \tag{11-13}$$

$$C_T = nC \tag{11-14}$$

SUMMARY

$$i = \frac{dq}{dt} \qquad (11\text{--}15)$$

$$i = C\frac{dv}{dt} \qquad (11\text{--}16)$$

$$X_C = \frac{1}{2\pi fC} \qquad (11\text{--}17)$$

$$V = IX_C \qquad (11\text{--}18)$$

$$I = \frac{V}{X_C} \qquad (11\text{--}19)$$

$$X_C = \frac{V}{I} \qquad (11\text{--}20)$$

$$\mathscr{E} = \tfrac{1}{2}CV^2 \qquad (11\text{--}21)$$

$$P_r = V_{rms}I_{rms} \qquad (11\text{--}22)$$

$$P_r = \frac{V_{rms}^2}{X_C} \qquad (11\text{--}23)$$

$$P_r = I_{rms}^2 X_C \qquad (11\text{--}24)$$

Summary

1. A capacitor is composed of two parallel conducting plates separated by a dielectric insulator.
2. Capacitance is a measure of a capacitor's ability to store electrical charge.
3. Energy is stored by a capacitor in its electric field.
4. When a capacitor charges, the voltage across its plates builds up.
5. When a capacitor discharges, the voltage across its plates decreases.
6. A capacitor will charge to the maximum value of the applied dc voltage. At this point it is fully charged.
7. A capacitor blocks nonchanging direct current.
8. A completely discharged capacitor has zero volts across it.
9. Capacitance is the ratio of charge per volt ($C = Q/V$).
10. The farad is the unit of capacitance.
11. *One farad* is the amount of capacitance when *one coulomb* of charge is stored and *one volt* is across the plates.
12. The dielectric constant is an indication of the ability of a material to establish an electric field.
13. The dielectric strength determines the breakdown voltage of a capacitor.
14. Electrolytic capacitors are generally polarized.
15. Capacitors are usually classified according to their dielectric material.
16. Total series capacitance is less than that of the smallest capacitor.

17. Capacitance adds in parallel.
18. Instantaneous current in a capacitor equals the capacitance times the rate of change of voltage: $i = C(dv/dt)$.
19. A sine wave voltage produces a cosine wave current in a capacitor.
20. Current leads voltage by 90° in a capacitor.
21. Capacitive reactance, X_C, is the opposition to sinusoidal current flow. It is expressed in ohms.
22. X_C is inversely proportional to frequency and capacitance.
23. Ohm's law applies to capacitive reactance in the same way that it applies to resistance.
24. The average power in a capacitor is zero.

Self-Test

1. Indicate true or false for each of the following statements:
   (a) The plates of a capacitor are conductive.
   (b) The dielectric is the insulating material separating the plates.
   (c) A fully charged capacitor allows a constant current.
   (d) A practical capacitor stores charge indefinitely when it is disconnected from the source.
2. Indicate true or false for each of the following statements:
   (a) There is charging current through the dielectric of a capacitor.
   (b) When a capacitor is connected to a dc source, it will charge to the value of the source.
   (c) You can discharge an ideal capacitor by disconnecting it from the voltage source.
   (d) When a capacitor is completely discharged, the voltage across its plates is zero.
3. A 0.006-µF capacitor is charged to 10 V. How many coulombs of charge does it store?
4. A given capacitor stores 50×10^{-6} coulomb of charge when its voltage is 5 V. Determine its capacitance.
5. A 0.01-µF capacitor stores 5×10^{-8} C. What is the voltage across its plates?
6. Convert 0.005 µF to picofarads.
7. Convert 2000 pF to microfarads.
8. Calculate the absolute permittivity, ϵ, for ceramic.
9. Calculate the capacitance for the following physical parameters: $A = 0.009$ m², $d = 0.0015$ m, and the dielectric is Teflon®.
10. A 0.1-µF capacitor has a temperature coefficient designated N750. By what amount will a 25°C rise in temperature change the capacitance?

SELF-TEST

11. Indicate true or false for each of the following statements:
    (a) Capacitors are classified according to their shape.
    (b) A polarized capacitor has one positive plate and one negative plate.
    (c) The positive lead of an electrolytic capacitor must be connected to the positive side of a circuit.
    (d) Air capacitors are normally variable.
12. What is the value of the molded mica capacitor in Figure 11–44?

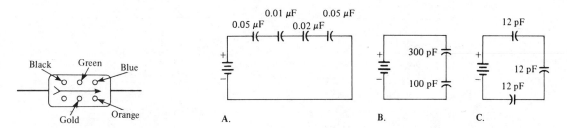

FIGURE 11–44   **FIGURE 11–45**

13. What is the total capacitance in each of the circuits in Figure 11–45?
14. Determine the voltage across each capacitor in Figure 11–46.

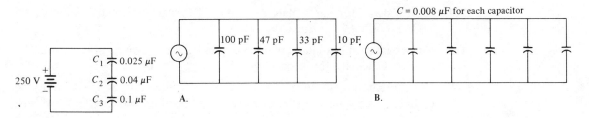

FIGURE 11–46   **FIGURE 11–47**

15. Find C_T in each circuit of Figure 11–47.
16. A 2-MHz sine wave voltage is applied to a 1-μF capacitor. What is the reactance?
17. The frequency of the source in Figure 11–48 can be adjusted. To get 50 milliamperes of rms current, to what frequency must you adjust the source?

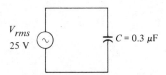

FIGURE 11–48

18. A 10-μF capacitor is charged to 100 V. How many joules of energy are stored?

Problems

11-1. What is the charge, in coulombs, of each capacitor in Figure 11-49?

11-2. How much voltage is there across each capacitor in Figure 11-49?

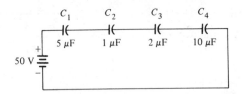
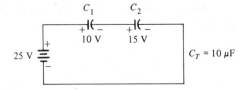

FIGURE 11-49   **FIGURE 11-50**

11-3. Determine each capacitance value in Figure 11-50.

11-4. Make the following conversions:
 (a) 0.1 µF to picofarads   (b) 0.0025 µF to picofarads
 (c) 5 µF to picofarads

11-5. Make the following conversions:
 (a) 1000 pF to microfarads   (b) 3500 pF to microfarads
 (c) 250 pF to microfarads

11-6. Calculate the absolute permittivity ϵ for each of the following materials: air, oil, glass, and Teflon®.

11-7. A mica capacitor has a plate area of 0.04 m² and a dielectric thickness of 0.008 m. What is its capacitance?

11-8. An air capacitor has square plates 0.1 m by 0.1 m. The plates are separated by 0.01 m. Calculate the capacitance.

11-9. At ambient temperature (25°C), a certain capacitor is specified to be 1000 pF. It has a negative temperature coefficient of 200 ppm/°C. What is its capacitance at 75°C?

11-10. A 0.001-µF capacitor has a temperature coefficient of positive 500 ppm/°C. How much change in capacitance will a 25°C increase in temperature cause?

11-11. Determine the value of the typographically labeled ceramic disk capacitors in Figure 11-51.

A.   B.   C.   D.

FIGURE 11-51

11-12. Determine the values of the color-coded capacitors in Figure 11-52.

11-13. Find the total capacitance C_T of each circuit in Figure 11-53.

11-14. For each circuit in Figure 11-53, determine the voltage across each capacitor.

PROBLEMS

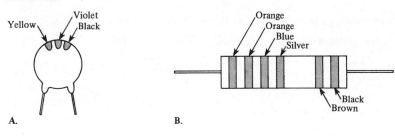

FIGURE 11-52

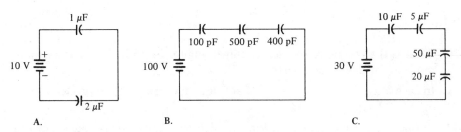

FIGURE 11-53

11-15. Two capacitors (one 1-μF, the other of unknown value) are charged from a 12-V source. The 1-μF capacitor is charged to 8 V, and the other to 4 V. What is the value of the other capacitor?

11-16. Determine C_T for each circuit in Figure 11-54.

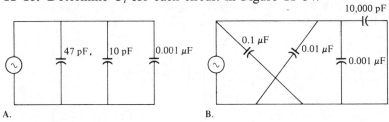

FIGURE 11-54

11-17. Determine C_T for each circuit in Figure 11-55.

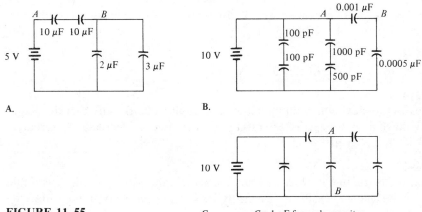

FIGURE 11-55

11–18. What is the voltage between points A and B in each circuit in Figure 11–55?

11–19. What is the reactance in each circuit in Figure 11–56?

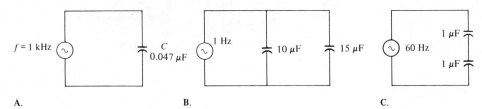

A.   B.   C.

FIGURE 11–56

11–20. What is the reactance of each circuit in Figure 11–55 if the dc sources are replaced by 2-kHz ac sources?

11–21. In each circuit of Figure 11–56, what frequency is required to produce an X_c of 100 Ω? An X_c of 1 kΩ?

11–22. A sine wave voltage of 20 V rms applied to a capacitor produces an rms current of 100 mA. What is the reactance?

11–23. A 10-kHz voltage source is applied to a 0.0047-μF capacitor. One milliampere of rms current is measured. What is the value of the voltage?

11–24. A 4.7-μF capacitor is charged to 30 V dc. How many joules of energy does it store?

11–25. A certain capacitor stores 50 millijoules of energy when it is charged to 12 V. What is the capacitance value?

Answers to Section Reviews

Section 11–1:
1. Two conductive plates separated by a dielectric material. **2.** Two plates, a dielectric, and connecting leads. **3.** The ability to store electrical charge. **4.** Lines of force between opposite charges.

Section 11–2:
1. By connecting it to an energy source. **2.** Increase. **3.** By connecting a conductive path across its plates. **4.** 120 V. **5.** 0 V.

Section 11–3:
1. Farad. **2.** 10^6. **3.** 10^{12}. **4.** 10^6. **5.** 1500 pF.

Section 11–4:
1. It indicates the maximum voltage that can be applied. **2.** Be sure that the actual in-circuit voltage does not exceed the rating. **3.** Increase. **4.** Decrease. **5.** Greater. **6.** 10,000 V. **7.** 0.425 μF. **8.** 2.01 μF.

Section 11–5:
1. By their dielectric material. **2.** A fixed capacitor cannot be altered. A variable capacitor can be changed easily. **3.** Electrolytic. **4.** Connect the positive lead to the positive side of the circuit.

ANSWERS TO SECTION REVIEWS

Section 11–6:
1. 10,000 pF ± 5%. **2.** Negative temperature coefficient of 330 ppm/°C.

Section 11–7:
1. Less. **2.** 62.5 pF. **3.** 0.006 μF. **4.** 20 pF. **5.** 75 V.

Section 11–8:
1. The individual capacitances are added. **2.** By using five 0.01-μF capacitors in parallel. **3.** 98 pF.

Section 11–9:
1. Instantaneous rate of change. **2.** $i = C(dv/dt)$. **3.** Current leads voltage by 90°.

Section 11–10:
1. Opposition to ac presented by a capacitor. **2.** 637 kΩ. **3.** 796 Hz. **4.** 628 mA.

Section 11–11:
1. 5000 μJ. **2.** 0 watts.

Section 11–12:
1. By shorting its leads. **2.** Initially the needle jumps to zero; then it slowly moves to the high-resistance end of scale.

An RC circuit contains both *resistance* and *capacitance*. It is one of the basic types of reactive circuits that you will study. In this chapter the *basic series RC circuit* is covered. As you will learn in a later chapter, more complex RC circuits can be converted to this basic series form.

The *basic parallel RC circuit* is discussed in the last section of this chapter. Both the series and the parallel RC circuits are of fundamental importance in electronics. Applications of the RC circuit include filters, amplifier coupling, oscillators, wave-shaping circuits, and many others.

Two fundamental aspects of RC circuit operation are the *frequency response* and the *time response.* Frequency response analysis is based on the use of sine waves. As you learned in Chapter 10, any electrical signal can be broken down into harmonics, which are the sinusoidal components of the signal. Any electrical signal can be treated as a composite of sine waves of different frequencies and amplitudes. As a result, the response of a circuit to sine waves is of fundamental importance in many areas of electronics. In other areas, particularly in pulse and digital electronics, the emphasis is often on the circuit response to pulsed signals on a *time basis* rather than on a frequency basis. Pulse response will be studied in Chapter 13.

12–1 Sine Wave Response
12–2 Basic Phasor Algebra
12–3 Impedance of a Series RC Circuit
12–4 Phase Angle
12–5 Voltage and Current Magnitudes
12–6 The RC Lag Network
12–7 The RC Lead Network
12–8 The Series RC Circuit as a Filter
12–9 Power in an RC Circuit
12–10 The Parallel RC Circuit

12
FREQUENCY RESPONSE OF RC CIRCUITS

12-1

SINE WAVE RESPONSE

When a sinusoidal voltage is applied to an RC circuit, *each resulting voltage drop and current in the circuit will be a sine wave with the same frequency as that of the applied voltage.* As shown in Figure 12–1, the resistor voltage, the capacitor voltage, and the current are *all* sine waves with the same frequency as that of the source.

The amplitudes and the phase relationships of the voltages and currents depend on the ohmic values of both the resistance and the capacitive reactance.

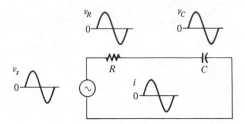

FIGURE 12–1 *Illustration of sine wave response.*

Review for 12-1

1. When a sine wave voltage is applied to an RC circuit, are the voltage drops and the currents also sine waves?
2. There is current in a series RC circuit as a result of an applied sine wave voltage of 60 Hz. What is the frequency of the current?

12-2

BASIC PHASOR ALGEBRA

Phasors were introduced in Chapter 10. In this chapter we will use phasors in working with voltages and currents that are not in phase with each other. You learned in the last chapter that in a capacitor, the voltage and current are *always 90° out of phase* with each other, and the current *leads*.

When a 90° (right angle) relationship exists between two quantities, the quantities can be represented by a *phasor diagram,* as shown in Figure 12–2. Both leading and lagging conditions are shown. Recall that the length of the phasor represents the magnitude of the quantities, such as V_{rms} or I_{rms}. In this section we are interested not in a particular quantity but in a *general* representation of *any* quantity.

BASIC PHASOR ALGEBRA

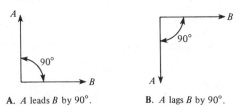

A. *A* leads *B* by 90°.   **B.** *A* lags *B* by 90°.

FIGURE 12–2 *Phasor diagrams for 90° relationships.*

Example 12–1

The voltage across a capacitor is 10 V rms. The current is 5 A rms. Sketch the phasor diagram.

Solution:

The phasor diagram is shown in Figure 12–3.

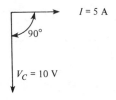

FIGURE 12–3

The Resultant Phasor

Two phasor quantities can be combined into a *resultant phasor*, as shown in Figure 12–4. The resultant phasor can be thought of as the sum of the two phasors. However, the *phasor sum* is not the same as the algebraic sum. We must use a relationship from trigonometry to find the phasor sum when two phasor quantities are at a right angle (90°) to each other.

  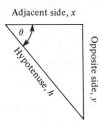 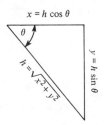

FIGURE 12–4 *Resultant phasor.*   **FIGURE 12–5** *Relationships in a right triangle.*

The *Pythagorean theorem* states that for a right triangle, the *hypotenuse (h) is the square root of the sum of the squares of the other two sides.* This theorem is illustrated in Figure 12–5 and is stated as follows:

$$h = \sqrt{x^2 + y^2} \tag{12-1}$$

Also, the two sides, x and y, can be expressed in terms of the hypotenuse, h, and the angle θ:

$$y = h \sin \theta \tag{12-2}$$
$$x = h \cos \theta \tag{12-3}$$

We need one more formula to express the angle θ in terms of the two sides. This formula is as follows:

$$\theta = \arctan\left(\frac{y}{x}\right) \tag{12-4}$$

The term "arctan" is the *arc tangent* or *inverse tangent* and is sometimes symbolized as $\tan^{-1}$. Equation (12-4) is read, "θ is the angle whose tangent is y over x." It gives the absolute value of the angle. Tangent is a trigonometric function defined as the ratio of the opposite side to the adjacent side of a right triangle. Scientific calculators have the trigonometric functions of sin, cos, and tan, in addition to their inverse functions, thereby making the calculations relatively simple. Also, a trigonometric table appears in Appendix G.

Example 12–2

Determine the hypotenuse h and angle θ of the right triangle in Figure 12–6.

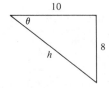

FIGURE 12–6

Solution:

$$h = \sqrt{10^2 + 8^2} = 12.8$$
$$\theta = \arctan\left(\frac{8}{10}\right) = 38.66°$$

Example 12–3

Determine the x and y sides of the triangle in Figure 12–7.

FIGURE 12–7

BASIC PHASOR ALGEBRA

Solution:

$$x = 25 \cos 60° = 12.50$$
$$y = 25 \sin 60° = 21.65$$

Phasor Diagram Forms Right Triangle

A phasor diagram can be visualized as forming a right triangle, as shown in Figure 12–8. Phasor A is the x side, phasor B is the y side, and the resultant phasor is the hypotenuse.

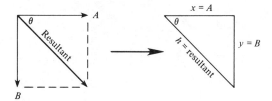

FIGURE 12–8

Review for 12–2

1. Name the three sides of a right triangle.
2. How do you find the resultant of two phasors 90° apart?
3. How do you find the adjacent side of a right triangle if the hypotenuse and the angle are known?
4. How do you find the opposite side of a right triangle if the hypotenuse and the angle are known?
5. How do you find the angle if the two sides are known?
6. Determine the hypotenuse and the angle in Figure 12–9A and the opposite and adjacent sides in Figure 12–9B.

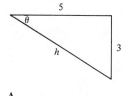
A.

B.

FIGURE 12–9

7. How is a phasor diagram related to a right triangle?

12-3

IMPEDANCE OF A SERIES RC CIRCUIT

Impedance is the total opposition to sinusoidal current and is expressed in *ohms*. In a purely resistive circuit, the impedance is simply equal to the total resistance. In a purely capacitive circuit, the impedance is the total capacitive reactance. The impedance of a series RC circuit is determined by *both the resistance and the capacitive reactance*. These cases are illustrated in Figure 12–10. The magnitude of the impedance is symbolized by Z.

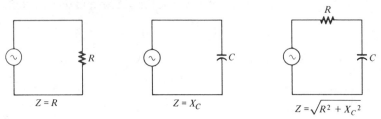

FIGURE 12–10 *Impedance.*

The magnitude of the impedance can be calculated as follows:

$$Z = \sqrt{R^2 + X_C^2} \qquad (12\text{–}5)$$

Why cannot the resistance and capacitive reactance be added algebraically in the same way that resistors in series are added? The last section provides a clue. The reason is as follows: In Chapter 11 it was shown that the capacitance caused a 90° phase difference between the capacitor voltage and the current. In a series circuit the *same* current flows through both the resistor and the capacitor. Thus, *the resistor voltage is in phase with the current, and the capacitor voltage is lagging the current by 90°*. Therefore, there is a phase difference of 90° between the resistor voltage, V_R, and the capacitor voltage, V_C, as illustrated in Figure 12–11.

We know from Kirchhoff's voltage law that the sum of the voltage drops must equal the applied voltage. However, since V_R and V_C are not in phase with

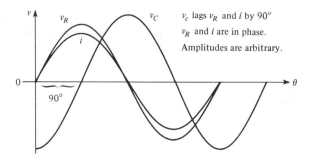

FIGURE 12–11 *Phase relation of voltages and current in an RC circuit.*

IMPEDANCE OF A SERIES RC CIRCUIT

each other, they cannot be added algebraically. They must be added as *phasor* quantities with V_C lagging V_R by 90°, as shown in Figure 12–12A. As shown in Part B, V_s is the *phasor sum* (resultant) of V_R and V_C. Therefore, there is a *right triangle* relationship among V_s, V_R, and V_C; V_s is represented by the hypotenuse, V_R by the adjacent side, and V_C by the opposite side, as shown in Figure 12–12C.

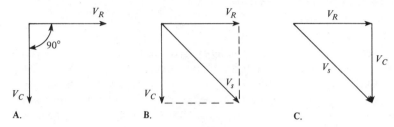

FIGURE 12–12 *Voltage phasor diagram for a series RC circuit.*

As you learned in the previous section, the length of the hypotenuse of a right triangle can be calculated from the other two sides by the Pythagorean theorem. In our case, the rms values of V_R and V_C are represented by the adjacent and opposite sides of the right triangle. This relationship is expressed as

$$V_s = \sqrt{V_R^2 + V_C^2}$$

By Ohm's law, which also applies to impedance, $Z = V_s/I$. By substitution of

$$\sqrt{V_R^2 + V_C^2}$$

for V_s, and by some algebraic manipulation, we get the following result:

$$Z = V_s/I = \sqrt{V_R^2 + V_C^2}/I$$
$$= \sqrt{V_R^2 + V_C^2} / \sqrt{I^2}$$
$$= \sqrt{(V_R^2 + V_C^2)/I^2}$$
$$= \sqrt{V_R^2/I^2 + V_C^2/I^2}$$

Now, if R^2 is substituted for V_R^2/I^2, and X_C^2 is substituted for V_C^2/I^2, we get

$$Z = \sqrt{R^2 + X_C^2}$$

which is Equation (12–5). It is a relationship that can be represented by a right triangle with Z the hypotenuse, R the adjacent side, and X_C the opposite side, as shown in Figure 12–13. This triangle is called an *impedance triangle*.

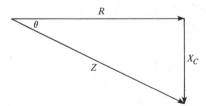

FIGURE 12–13 *Impedance triangle.*

The impedance triangle is useful in helping us to visualize how the frequency of a sine wave affects the RC circuit. Recall from Chapter 11 that the reactance of a capacitor is *inversely* dependent on the frequency of the applied signal. If X_C increases, Z increases; if X_C decreases, Z decreases:

$$\overset{\uparrow}{Z} = \sqrt{R^2 + \overset{\uparrow}{X_C^2}} \qquad \underset{\downarrow}{Z} = \sqrt{R^2 + \underset{\downarrow}{X_C^2}}$$

Therefore, Z is *inversely dependent* on frequency also. We can see this change in Z by sketching the impedance triangle for different values of frequency in Figure 12–14. The key point is that *because X_C varies inversely as the frequency, so also does the magnitude of the total impedance.* Example 12–4 illustrates this point.

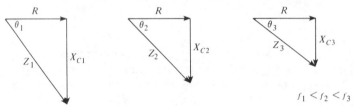

FIGURE 12–14 *Impedance varies inversely with frequency.*

Example 12–4

For the series RC circuit in Figure 12–15, determine the magnitude of the impedance for input frequency values of 10 kHz, 20 kHz, and 30 kHz.

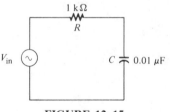

FIGURE 12–15

Solution:

For $f = 10$ kHz:

$$X_C = \frac{1}{2\pi f C}$$

$$\cong 1592 \ \Omega$$

$$Z = \sqrt{R^2 + X_C^2} = \sqrt{(1000 \ \Omega)^2 + (1592 \ \Omega)^2}$$

$$\cong 1880 \ \Omega$$

PHASE ANGLE

For $f = 20$ kHz:

$$X_C = \frac{1}{2\pi f C}$$
$$\cong 796 \; \Omega$$
$$Z = \sqrt{R^2 + X_C^2} = \sqrt{(1000 \; \Omega)^2 + (796 \; \Omega)^2}$$
$$\cong 1278 \; \Omega$$

For $f = 30$ kHz:

$$X_C = \frac{1}{2\pi f C}$$
$$\cong 531 \; \Omega$$
$$Z = \sqrt{R^2 + X_C^2} = \sqrt{(1000 \; \Omega)^2 + (531 \; \Omega)^2}$$
$$\cong 1132 \; \Omega$$

Notice that as the frequency increases, X_C and Z decrease.

Review for 12–3

1. Define *impedance*.
2. Does the applied voltage lead or lag the current in an RC circuit?
3. How many degrees of phase difference are there between the capacitor voltage and the resistor voltage in a series RC circuit?
4. Sketch a generalized impedance triangle.
5. If the frequency of the applied voltage in an RC circuit increases, what happens to the capacitive reactance? What happens to the impedance magnitude?
6. In a certain series RC circuit, $R = 1 \; k\Omega$ and $X_C = 1 \; k\Omega$. What is the magnitude of the impedance?

12–4 PHASE ANGLE

As you saw in the previous section, *a change in X_C causes a change in the magnitude (ohmic value) of Z*. A change in the phase angle also occurs, as you will learn in this section.

The *phase angle* of an RC circuit is *the angle between the source voltage and the total current* and is designated θ (the Greek letter theta). *The voltage always lags the current,* as shown in Figure 12–16A. In the corresponding impedance triangle in Part B of the figure, θ appears between R and Z.

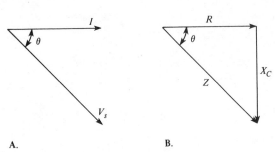

FIGURE 12–16 *Phase angle of an RC circuit.*

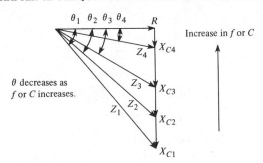

FIGURE 12–17 *Effect of frequency and capacitance on phase angle.*

Figure 12–17 shows how θ changes with frequency. If the frequency is decreased, X_c increases, creating a larger phase angle. If frequency is increased, X_c decreases, causing a smaller phase angle. This relationship is obvious, because the more reactive the circuit is (the larger the reactance is compared to the resistance), the more phase difference there is between the voltage and the current. When the resistance is insignificant compared to the capacitive reactance, θ is close to 90°, because the circuit is almost purely capacitive.

θ Varies Inversely with Capacitance

For a fixed frequency, θ will change with the capacitance value. If C is decreased, X_c increases, creating a larger phase angle. If C is increased, X_c decreases, causing a smaller phase angle. Figure 12–17 illustrates this relationship also.

θ Varies Inversely with Resistance

If capacitance and frequency are held constant and the resistance is varied, the phase angle will change. This relationship is illustrated in Figure 12–18, where a larger resistance produces a smaller phase angle and a smaller resistance results in a larger phase angle. Notice also that Z varies directly with resistance, as we would expect.

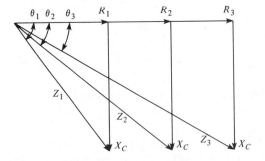

FIGURE 12–18 *Effect of resistance on phase angle.*

PHASE ANGLE

Calculation of Phase Angle

If the frequency, resistance, and capacitance of a series RC circuit are known, the phase angle can be determined. We will apply Equation (12-4) to the impedance triangle of Figure 12-13. R is the adjacent side, and X_C is the opposite side; so Equation (12-4) becomes

$$\theta = \arctan\left(\frac{X_C}{R}\right) \tag{12-6}$$

We want to know "the angle whose tangent is X_C/R." This, of course, is the phase angle of the circuit. Examples 12-5, 12-6, and 12-7 will clarify use of this formula.

Example 12-5

In a series RC circuit, R is 100 Ω and X_C is 100 Ω at a given frequency. Determine the phase angle between the applied voltage and current. Sketch the wave forms.

Solution:

$$\theta = \arctan\left(\frac{X_C}{R}\right) = \arctan\left(\frac{100\ \Omega}{100\ \Omega}\right)$$
$$= \arctan 1 = 45°$$

This result points out that the phase angle is always 45° when the resistance and capacitive reactance are equal.

A sketch of the wave forms of the current and voltage are shown in Figure 12-19. The actual amplitudes are not known from the information given. The purpose of this wave-form diagram is to show the phase relationship.

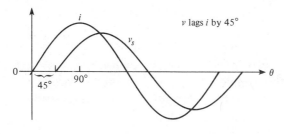

FIGURE 12-19

Example 12-6

Repeat Example 12-5 for $X_C = 50\ \Omega$.

Example 12-6 (continued)

Solution:

$$\theta = \arctan\left(\frac{50\ \Omega}{100\ \Omega}\right) = \arctan(0.5)$$
$$= 26.57°$$

Notice that θ is less than 45° when X_C is less than R. *A smaller phase angle corresponds to a less reactive circuit.*

The wave-form diagram is shown in Figure 12-20, and again the relative amplitudes are arbitrarily assigned.

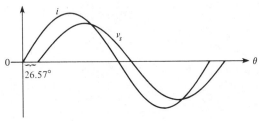

FIGURE 12-20

Example 12-7

Repeat Example 12-5 for $X_C = 150\ \Omega$.

Solution:

$$\theta = \arctan\left(\frac{150\ \Omega}{100\ \Omega}\right) = \arctan(1.5)$$
$$= 56.31°$$

Notice that θ is greater than 45° when X_C is greater than R. *A greater phase angle corresponds to a more reactive circuit.*

The wave forms are shown in Figure 12-21.

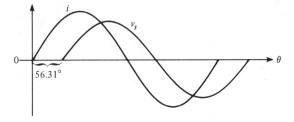

FIGURE 12-21

Review for 12-4

1. What produces the phase difference between the applied voltage and the current in an RC circuit?
2. Is the current in an RC circuit in phase or out of phase with the resistor voltage?
3. Does the current in an RC circuit lead or lag the capacitor voltage?
4. Does the current in an RC circuit lead or lag the applied voltage?
5. If the capacitive reactance in an RC circuit is increased, does the phase angle increase or decrease?
6. For a given frequency and capacitance in an RC circuit, if the resistance is increased, what happens to the phase angle?
7. When does a 45° phase angle occur?
8. Does a phase angle greater than 45° mean that the reactance is greater or less than the resistance?

12-5

VOLTAGE AND CURRENT MAGNITUDES

In this section, formulas for the voltage and current magnitudes in a series RC circuit are developed. You have already seen that Ohm's law can be extended to reactive circuits. Thus, we develop Equations (12–7), (12–8), and (12–9) using Ohm's law, and we can use these equations for calculating current and voltages in a series RC circuit.

We know that $Z = \sqrt{R^2 + X_C^2}$ and $I = V_s/Z$; thus,

$$I = \frac{V_s}{\sqrt{R^2 + X_C^2}} \tag{12-7}$$

Using Equation (12–7), we derive the voltage divider equations as applied to a series RC circuit as follows:

$$V_R = IR$$

$$= \left(\frac{V_s}{\sqrt{R^2 + X_C^2}}\right) R$$

$$V_R = \left(\frac{R}{\sqrt{R^2 + X_C^2}}\right) V_s \tag{12-8}$$

$$V_C = IX_C$$

$$= \left(\frac{V_s}{\sqrt{R^2 + X_C^2}}\right) X_C$$

$$V_C = \left(\frac{X_C}{\sqrt{R^2 + X_C^2}}\right) V_s \tag{12-9}$$

The voltage drops are directly proportional to the resistance or reactance across which they are taken. If R is larger than X_C, it will drop more of the total voltage, and vice versa. The total voltage is the phasor sum of V_R and V_C:

$$V_s = \sqrt{V_R^2 + V_C^2} \qquad (12\text{--}10)$$

Example 12–8 illustrates voltage current magnitude calculations using these formulas.

Example 12–8

For the circuit of Figure 12–22, determine the rms values of the voltages across R and C and the current. The input has an rms value of 5 V.

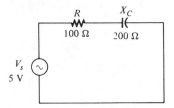

FIGURE 12–22

Solution:

First calculate the impedance of the circuit:

$$Z = \sqrt{R^2 + X_C^2} = \sqrt{(100\ \Omega)^2 + (200\ \Omega)^2}$$
$$= 223.6\ \Omega$$

Now we can find the current and voltages:

$$I = \left(\frac{V_s}{Z}\right) = \frac{5\ \text{V}}{223.6\ \Omega} = 0.02236\ \text{A}$$
$$= 22.36\ \text{mA}$$

$$V_R = \left(\frac{R}{Z}\right)V_s = \left(\frac{100\ \Omega}{223.6\ \Omega}\right)5\ \text{V}$$
$$= 2.24\ \text{V}$$

$$V_C = \left(\frac{X_C}{Z}\right)V_s = \left(\frac{200\ \Omega}{223.6\ \Omega}\right)5\ \text{V}$$
$$= 4.47\ \text{V}$$

We could also determine the voltage drops using Ohm's law directly, once we know the current. The results are the same:

$$V_R = IR = (22.36\ \text{mA})(100\ \Omega)$$
$$\cong 2.24\ \text{V}$$
$$V_C = IX_C = (22.36\ \text{mA})(200\ \Omega)$$
$$\cong 4.47\ \text{V}$$

THE RC LAG NETWORK

Check: $V_s = \sqrt{V_R^2 + V_C^2} = \sqrt{(2.24 \text{ V})^2 + (4.47 \text{ V})^2} \cong 5 \text{ V}.$

Review for 12–5

1. Define *magnitude* of voltage or current.
2. Explain how the series RC circuit acts as a voltage divider.
3. If X_C is larger than R, where does the greater voltage drop occur?
4. If you are measuring the voltage across the resistor in a series RC circuit, what would you expect to see as the frequency of the source voltage is increased?
5. Calculate the voltages across the resistor and the capacitive reactance in Figure 12–23.

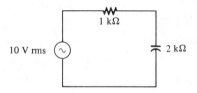

FIGURE 12–23

12–6

THE RC LAG NETWORK

Figure 12–24A shows a series RC circuit with *the output voltage taken across the capacitor*. The source voltage is the *input voltage*, V_{in}. As you know, θ, the phase angle between the current and the applied voltage, is also the phase angle between the resistor voltage and the applied voltage, because V_R and I are in phase. Since V_C lags V_R by 90°, the phase angle between the capacitor voltage and the applied voltage is the difference of 90° and θ (90° − θ), as shown in Figure 12–24B. The capacitor voltage is taken as the *output* voltage, and it *lags the input voltage* by 90° − θ. Output lags input, thus creating a basic lag network.

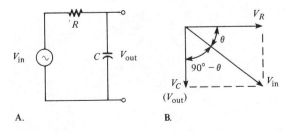

FIGURE 12–24 *RC lag network.*

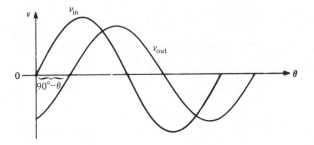

FIGURE 12–25 *Input and output voltages of an RC lag network.*

If the input and output wave forms of the lag network are displayed on an oscilloscope, you will see a display similar to that shown in Figure 12–25. The amount of phase difference between the input and the output is dependent on the relative sizes of the capacitive reactance and the resistance, as is the magnitude of the output voltage.

Phase Difference between Input and Output

As mentioned, θ is the phase angle between V_R and V_{in}. We will call the angle between V_{out} and V_{in}, ϕ (phi), as indicated in Figure 12–26A. Since we already have an expression for θ in Equation (12–6), ϕ can be written as $90° - \theta$:

$$\phi = 90° - \arctan\left(\frac{X_C}{R}\right) \quad \text{(12–11)}$$

Equation (12–11) can be stated another way, using the voltage phasor diagram with V_{out} as the adjacent side and V_R as the side opposite to the angle ϕ, as shown in Figure 12–26B. Because the current through R and C is the same, $V_R/V_C = R/X_C$, and Equation (12–11) can be stated in the following form:

$$\phi = \arctan\left(\frac{R}{X_C}\right) \quad \text{(12–12)}$$

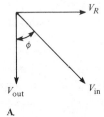
A.

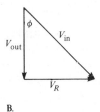
B.

FIGURE 12–26

Example 12–9

Determine the phase angle between the input and the output voltages in the lag network of Figure 12–27.

THE RC LAG NETWORK

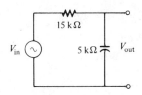

FIGURE 12–27

Solution:

$$\phi = 90° - \arctan\left(\frac{X_C}{R}\right) = 90° - \arctan\left(\frac{5 \text{ k}\Omega}{15 \text{ k}\Omega}\right)$$
$$= 90° - 18.43° = 71.57°$$

or

$$\phi = \arctan\left(\frac{R}{X_C}\right) = \arctan\left(\frac{15 \text{ k}\Omega}{5 \text{ k}\Omega}\right)$$
$$= 71.57°$$

Example 12–10

Determine the phase angle between the input and the output voltages for the lag network in Figure 12–28.

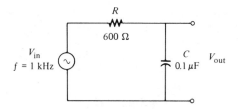

FIGURE 12–28

Solution:

$$X_C = \frac{1}{2\pi fC} = \frac{1}{2\pi(1 \text{ kHz})(0.1 \text{ }\mu\text{F})}$$
$$= 1592 \text{ }\Omega$$

$$\phi = \arctan\left(\frac{R}{X_C}\right) = \arctan\left(\frac{600 \text{ }\Omega}{1592 \text{ }\Omega}\right)$$
$$\phi = 20.65°$$

Magnitude of Output Voltage

To evaluate the output voltage in terms of its magnitude, visualize the RC circuit as a voltage divider. A portion of the total input voltage is dropped across R and a portion across X_C. Since the output voltage *is* the voltage across C for a lag network, we can calculate it by using the voltage divider formula of Equation (12–9).

Example 12–11

For the lag network in Figure 12–28, determine the magnitude of the output voltage if V_{in} has an rms value of 10 V. Then use the value of the output phase angle found in Example 12–10 to sketch the input and output wave forms in proper relationship.

Solution:

$$V_{out} = \left(\frac{X_C}{\sqrt{R^2 + X_C^2}}\right)V_{in} = \left(\frac{1592\ \Omega}{\sqrt{(600\ \Omega)^2 + (1592\ \Omega)^2}}\right)10\ V$$

$$= 9.36\ V$$

$$\phi = 20.65° \quad \text{from Example 12–10}$$

The wave forms are shown in Figure 12–29.

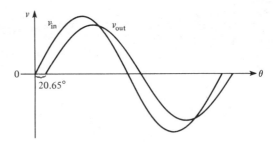

FIGURE 12–29

Review for 12–6

1. In an RC lag network, does the input voltage lag or lead the output voltage?
2. In a series RC lag network, across which component is the output taken?
3. If the resistance and capacitive reactance are equal, what is the phase angle between the input and the output?
4. If the input frequency is increased, will the output phase angle, ϕ, increase or decrease?
5. If the input frequency is decreased, will the output magnitude increase or decrease?

12-7

THE RC LEAD NETWORK

If we use an RC series circuit and take the output voltage across the resistor rather than across the capacitor, we have a basic RC *lead* network, as shown in Figure 12–30A. The output leads the input.

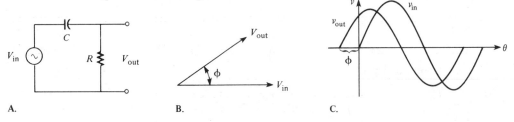

FIGURE 12–30 *RC lead network.*

Phase Difference between Input and Output

In a series RC circuit, the total current leads the input voltage. Also, the resistor voltage is in phase with the current. Since the output voltage is taken across the resistor, the output *leads* the input, as indicated by the phasor diagram in Figure 12–30B. The wave-form diagrams are shown in Figure 12–30C.

Just as in the lag network, in the lead network the amount of phase difference between input and output, and also the magnitude of the output voltage, are dependent on the relative sizes of the resistance and the capacitive reactance. If the input voltage is given a *reference angle* of 0°, then the phase angle of the output is the same as that previously expressed for θ, because the resistor voltage and the current have the same phase angle:

$$\phi = \arctan\left(\frac{X_C}{R}\right) \tag{12-13}$$

The following examples will illustrate calculations of output phase angles for lead networks.

Example 12–12

Calculate the output phase angle for the lead network in Figure 12–31.

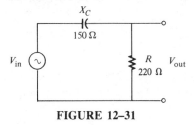

FIGURE 12–31

Example 12-12 (continued)

Solution:

$$\phi = \arctan\left(\frac{X_C}{R}\right) = \arctan\left(\frac{150\ \Omega}{220\ \Omega}\right)$$

$$= 34.29°$$

The output leads the input by 34.29°.

Example 12-13

Calculate the output phase angle for the lead network in Figure 12-32.

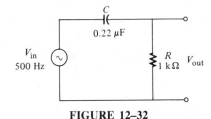

FIGURE 12-32

Solution:

First calculate X_C:

$$X_C = \frac{1}{2\pi f C} = \frac{1}{2\pi(500\ \text{Hz})(0.22\ \mu\text{F})}$$

$$= 1446.86\ \Omega$$

$$\phi = \arctan\left(\frac{X_C}{R}\right) = \arctan\left(\frac{1446.86\ \Omega}{1000\ \Omega}\right)$$

$$= 55.35°$$

The output leads the input by 55.35°

Magnitude of Output Voltage

We can determine the magnitude of the output voltage by using the voltage divider formula of Equation (12-8), as Example 12-14 illustrates.

Example 12-14

The input voltage in Figure 12-32 has an rms value of 10 V. Determine the rms value of the output voltage, and sketch the input and output voltage wave forms.

THE SERIES RC CIRCUIT AS A FILTER

Solution:

$$V_{out} = \left(\frac{R}{\sqrt{R^2 + X_C^2}}\right)V_{in} = \left(\frac{1000\ \Omega}{1759\ \Omega}\right)10\ V$$
$$= 5.69\ V$$

The peak value of the input voltage is 1.414 × (10 V) = 14.14 V. The peak value of the output voltage is 1.414 × (5.69 V) = 8.05 V.

The wave forms are drawn in Figure 12–33.

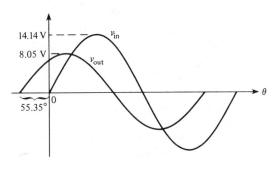

FIGURE 12–33

Review for 12–7

1. In an RC lead network, does the output voltage lag or lead the input voltage?
2. In a series RC lead network, across which component is the output taken?
3. If the resistance and capacitive reactance are equal, what is the phase angle between the input and the output?
4. If the input frequency is decreased, does the output phase angle, θ, increase or decrease?
5. If the input frequency is increased, does the output magnitude increase or decrease?

12–8

THE SERIES RC CIRCUIT AS A FILTER

An electrical filter, as the name implies, is a circuit that permits electrical signals having certain frequencies to pass from the input to the output while blocking all others. That is, it *filters out* all frequencies but the *selected* ones. In this section you will see how the RC circuit acts as a filter.

There are two types of series RC circuit filters. The first that we will examine, called a *low-pass filter,* occurs when the output is taken across the capacitor, as in a lag network. The second, called a high-pass filter, occurs when the output is taken across the resistor, as in a lead network.

Low-Pass Filter

You have already seen what happens to the output phase angle and the output voltage magnitude in the lag network. In terms of filtering action, we are interested primarily in what happens to the output magnitude as the input frequency is varied.

Figure 12-34 is an example of the filtering action of an RC lag network. In Part A of the figure, the input is *zero* frequency (dc). Since the capacitor blocks the flow of direct current, the output voltage is equal to the full value of the input voltage, because no voltage is dropped across *R*. Therefore, the circuit passes *all* of the input voltage through to the output.

In Figure 12-34B, the frequency of the input voltage has been increased to 1 kHz, causing the capacitive reactance to *decrease* to 159 Ω. For an input voltage of 10 V rms, the output voltage is approximately 8.5 V, which we calculate using the voltage divider formula.

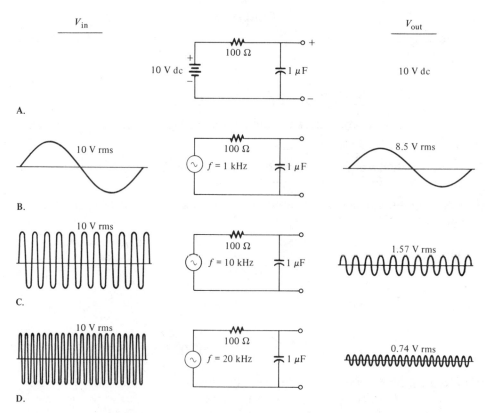

FIGURE 12-34 *Low-pass filter action.*

THE SERIES RC CIRCUIT AS A FILTER

In Figure 12–34C, the input frequency is increased to 10 kHz, causing the capacitive reactance to decrease further to 15.9 Ω. For a constant input voltage of 10 V rms, the output voltage is now 1.57 V.

As the input frequency is increased further, the output voltage approaches zero, as Figure 12–34D shows. The action is as follows: As the frequency of the input increases, the capacitive reactance decreases. Because the resistance is constant and the capacitive reactance decreases, the voltage across the capacitor also decreases, exhibiting the familiar voltage divider principle. The input frequency can be increased until it reaches a value at which the capacitive reactance is so small compared to the resistance that the output voltage can be neglected because it is very small compared to the input voltage. At this point, the circuit is essentially blocking all input signals.

As you can see in Figure 12–34, the circuit passes dc (zero frequency) completely. As the frequency of the input signal increases, less of the input voltage is passed through to the output. That is, the output voltage decreases as the frequency increases. It is apparent that the lower frequencies pass through the circuit better than the higher frequencies. This lag network is therefore a basic form of *low-pass filter*.

Figure 12–35 shows a graph of output voltage versus frequency for the low-pass circuit. This graph, called a *response curve,* shows that the output decreases as the frequency increases.

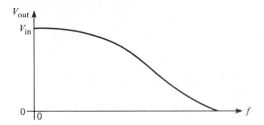

FIGURE 12–35 *Frequency response curve for a low-pass filter.*

High-Pass Filter

Next, refer to Figure 12–36A, where the output is taken across the resistor, as in a lead network. If the input voltage is dc (zero frequency), the output is zero volts, because the capacitor blocks direct current and therefore no voltage is dropped across R.

In Figure 12–36B, the frequency of the input signal has been increased to 100 Hz with an rms value of 10 V. The output voltage is 0.63 V. Thus, only a small percentage of the input voltage is getting through to the output at this frequency.

In Figure 12–36C, the input frequency is increased further to 1 kHz, causing more voltage to be dropped across the resistor because the capacitive reactance has decreased further. The output voltage at this frequency is 5.32 V. As you can see, the output voltage increases as the input frequency increases. A value of frequency will be reached at which the X_c is negligible compared to the resistance, and most of the input voltage will appear across the output as shown in Figure 12–36D.

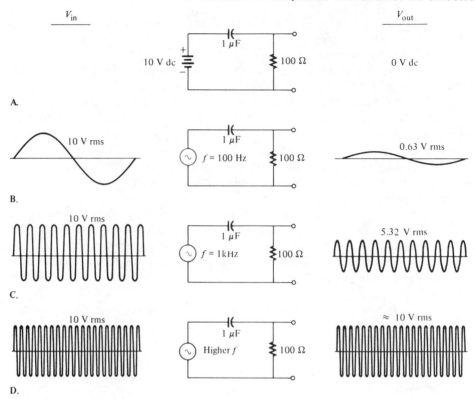

FIGURE 12-36 *High-pass filter action.*

This circuit tends to prevent lower frequencies from passing through to the output and to allow the higher frequencies to pass through. This lead network is therefore a basic form of *high-pass filter*.

Figure 12-37 shows a plot of output voltage versus frequency for the high-pass circuit. This is a response curve, and it shows that the output increases as the frequency increases and then levels off, approaching the value of the input voltage.

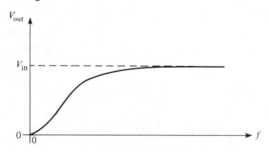

FIGURE 12-37 *Frequency response curve for a high-pass filter.*

RC Coupling: A common application of the RC high-pass filter is *amplifier coupling*. The RC circuit as a coupling network is shown in Figure 12-38A. Its purpose is to completely pass or *couple* the input signal to the output and block any dc

POWER IN AN RC CIRCUIT

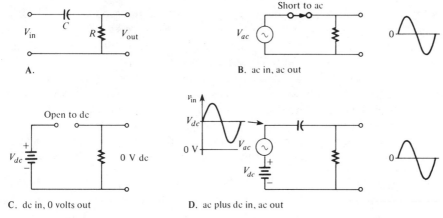

FIGURE 12-38 *RC coupling.*

level. The capacitance is chosen large enough so that X_C is practically a *short to the ac signal*. The capacitor appears as *an open to dc*, as indicated in Figure 12-38B and C.

When both ac and dc are superimposed, only the ac is coupled to the output, as shown in Figure 12-38D. This situation occurs in transistor amplifiers in which a dc bias voltage and an ac signal voltage are superimposed. The signal is then passed to the next amplifier stage, and the dc is blocked by an RC coupling network.

Review for 12-8

1. In terms of an electrical circuit, what does *filter* mean?
2. When an RC circuit is used as a low-pass filter, across which component is the output taken?
3. Does a high-pass filter tend to discriminate against signals with higher frequencies or those with lower frequencies?
4. Does the output magnitude increase or decrease in a high-pass filter when the frequency is decreased?
5. What does a filter response curve show?
6. In an RC coupling network, X_C must be very small compared to the resistance (T or F).

12-9

POWER IN AN RC CIRCUIT

In a purely resistive circuit, all of the energy delivered by the source is dissipated by the resistance in the form of heat. In a purely capacitive circuit, all of the energy delivered by the source is stored by the capacitor and then returned to the source; so there is no energy loss. What happens when both resistance *and* capacitance exist in a circuit? *Some of the energy is stored and returned to the source by the capacitor, and some is dissipated by the resistor.* The amount of energy loss is determined by the relative sizes of the resistance and the capacitance reactance.

It is reasonable to assume that if the resistance is greater than the reactance, *more* of the total energy delivered by the source is dissipated than is stored by the capacitor. Likewise, if the reactance is greater than the resistance, more of the total energy is stored and returned than is lost.

The power in a resistor, sometimes called *average power* (P_{avg}), and the power in a capacitor, called *reactive power* (P_r), are restated as follows:

$$P_{avg} = I^2 R \tag{12-14}$$

$$P_r = I^2 X_C \tag{12-15}$$

The Power Triangle

If the phasor relationship is as shown in Figure 12–39A for R, X_C, and Z, then the resistive and reactive powers also can be represented by the same phasor relationship, because *the respective magnitudes of the powers differ from R and X_C by a factor of I^2*, as shown in Figure 12–39B. The resultant phasor represents the *apparent power*, P_a, because at any instant this is the total power that *appears* to be delivered to the load. The unit of apparent power is the volt-ampere, abbreviated VA.

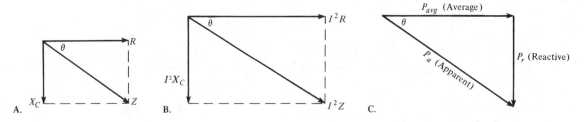

FIGURE 12–39 *Development of the power triangle for an RC circuit.*

The equation for apparent power is

$$P_a = I^2 Z \tag{12-16}$$

The power phasor diagram of Figure 12–39B can be drawn in the form of a right triangle, as shown in Part C of the figure. This is called the *power triangle*. According to the rules for a right triangle, the adjacent side, P_{avg}, is equal to the hypotenuse, P_a, times the *cosine* of the angle θ. This relationship is stated in equation form as follows:

$$P_{avg} = P_a \cos \theta$$

Since P_a is equal to $I^2 Z$ or $V_T I$, we can write the equation for the average power loss, which is the power in the resistor, as follows:

$$P_{avg} = V_T I \cos \theta \tag{12-17}$$

where V_T is total voltage and I is total current.

POWER IN AN RC CIRCUIT

For the special case of a purely resistive circuit, $\theta = 0°$ and $\cos 0° = 1$; so P_{avg} is just $V_T I$. For a purely capacitive circuit, $\theta = 90°$ and $\cos 90° = 0$; so P_{avg} is 0 W. This value indicates that there is no average power in a capacitor, as we already know.

The term $\cos \theta$ is called the *power factor* (PF), stated as follows:

$$PF = \cos \theta \tag{12–18}$$

Example 12–15

Determine the average power in the circuit of Figure 12–40.

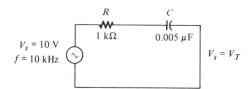

FIGURE 12–40

Solution:

$$X_C = \frac{1}{2\pi f C} = \frac{1}{2\pi(10 \text{ kHz})(0.005 \text{ μF})}$$
$$= 3183 \text{ Ω}$$

$$Z = \sqrt{R^2 + X_C^2}$$
$$= 3336.4 \text{ Ω}$$

$$I = \frac{V_s}{Z} = \frac{10 \text{ V}}{3336.4 \text{ Ω}} = 0.003 \text{ A}$$
$$= 3 \text{ mA}$$

$$\theta = \arctan\left(\frac{X_C}{R}\right) = \arctan\left(\frac{3183 \text{ Ω}}{1000 \text{ Ω}}\right)$$
$$= 72.56°$$

$$P_{avg} = V_s I \cos \theta = (10 \text{ V})(0.003 \text{ A}) \cos 72.56°$$
$$= 8.99 \text{ mW}$$

Example 12–16

For the circuit of Figure 12–41, find the power in the resistor given that $X_C = 2 \text{ kΩ}$.

Example 12-16 (continued)

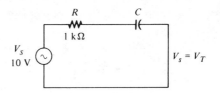

FIGURE 12-41

Solution:
$$Z = \sqrt{R^2 + X_C^2} = \sqrt{(1000\ \Omega)^2 + (2000\ \Omega)^2}$$
$$= 2236\ \Omega$$

$$I = \frac{V_s}{Z} = \frac{10\ V}{2236\ \Omega} = 0.00447\ A$$
$$= 4.47\ mA$$

$$\theta = \arctan\left(\frac{X_C}{R}\right) = \arctan\left(\frac{2000\ \Omega}{1000\ \Omega}\right)$$
$$= 63.43°$$

$$P_{avg} = V_s I \cos\theta = (10\ V)(4.47\ mA)\cos 63.43°$$
$$= 20\ mW$$

Review for 12-9

1. To which component in a circuit is the energy loss due?
2. How do average power and reactive power differ?
3. Write the equation for average power in a pure resistance.
4. Write the equation for reactive power.
5. Write the equation for apparent power.
6. What does the power factor (PF) equal?
7. Write the equation for average power in an RC circuit.
8. In a series RC circuit, what happens to the average power as the input frequency is increased?

12-10

THE PARALLEL RC CIRCUIT

A *basic parallel RC circuit* is shown in Figure 12-42A. Because R and C are in parallel, the voltage across each component has the same phase and magnitude as does the source voltage, V_s.

THE PARALLEL RC CIRCUIT

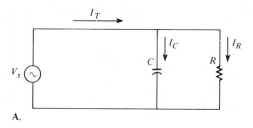
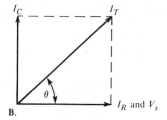

FIGURE 12–42 *Parallel RC circuit.*

Total Current

In a parallel RC circuit, the current through the resistor is in phase with the voltage. The current through the capacitor leads the voltage by 90°. The total current is the *phasor* sum of the two branch currents, as shown in Figure 12–42B. The magnitude of the total current is expressed as

$$I_T = \sqrt{I_R^2 + I_C^2} \qquad (12\text{–}19)$$

By Ohm's law, the resistor current is V_s/R, and the capacitor current is V_s/X_C.

Phase Angle

The angle by which the total current leads the source voltage is θ, as shown in Figure 12–42B. The formula for this angle is developed as follows: $\theta = \arctan(I_C/I_R)$, $I_C = V_s/X_C$, and $I_R = V_s/R$. By substitution, we get

$$\theta = \arctan\left(\frac{R}{X_C}\right) \qquad (12\text{–}20)$$

Impedance

The magnitude of the impedance of a parallel RC circuit is the total voltage divided by the total current, V_s/I_T. The following steps show how the impedance can be expressed in terms of R and X_C:

$$Z = \frac{V_s}{I_T} = \frac{V_s}{\sqrt{I_R^2 + I_C^2}}$$

$$= \frac{V_s}{\sqrt{(V_s^2/R^2) + (V_s^2/X_C^2)}} = \frac{V_s}{V_s\sqrt{(1/R^2) + (1/X_C^2)}}$$

$$= \frac{1}{\sqrt{(1/R^2) + (1/X_C^2)}} = \frac{1}{\sqrt{(R^2 + X_C^2)/R^2 X_C^2}}$$

$$= \frac{\sqrt{R^2 X_C^2}}{\sqrt{R^2 + X_C^2}}$$

$$Z = \frac{R X_C}{\sqrt{R^2 + X_C^2}} \qquad (12\text{–}21)$$

Example 12–17

Determine the total current in the circuit of Figure 12–43.

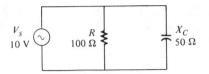

FIGURE 12–43

Solution:

$$I_R = \frac{V_s}{R} = \frac{10 \text{ V}}{100 \text{ }\Omega}$$
$$= 0.1 \text{ A}$$

$$I_C = \frac{V_s}{X_C} = \frac{10 \text{ V}}{50 \text{ }\Omega}$$
$$= 0.2 \text{ A}$$

$$I_T = \sqrt{I_R^2 + I_C^2} = \sqrt{(0.1 \text{ A})^2 + (0.2 \text{ A})^2}$$
$$\cong 0.224 \text{ A}$$

Example 12–18

Determine the total impedance and the phase angle for Figure 12–44.

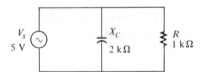

FIGURE 12–44

Solution:

$$Z = \frac{RX_C}{\sqrt{R^2 + X_C^2}} = \frac{(1 \text{ k}\Omega)(2 \text{ k}\Omega)}{\sqrt{(1 \text{ k}\Omega)^2 + (2 \text{ k}\Omega)^2}}$$
$$= 894.4 \text{ }\Omega$$

$$\theta = \arctan\left(\frac{R}{X_C}\right) = \arctan\left(\frac{1 \text{ k}\Omega}{2 \text{ k}\Omega}\right)$$
$$= 26.57°$$

I_T leads V_s by 26.57°.

FORMULAS

Review for 12–10

1. In a parallel RC circuit, if X_C is 100 Ω and R is 50 Ω, which component has the largest current?
2. In a parallel RC circuit, we can directly add the capacitive current and the resistive current to obtain the total current (T or F).
3. If the frequency of the voltage applied across a parallel RC circuit is increased, does the phase angle increase or decrease?

Formulas

$$h = \sqrt{x^2 + y^2} \quad (12\text{–}1)$$

$$y = h \sin \theta \quad (12\text{–}2)$$

$$x = h \cos \theta \quad (12\text{–}3)$$

$$\theta = \arctan\left(\frac{y}{x}\right) \quad (12\text{–}4)$$

For series RC circuits:

$$Z = \sqrt{R^2 + X_C^2} \quad (12\text{–}5)$$

$$\theta = \arctan\left(\frac{X_C}{R}\right) \quad (12\text{–}6)$$

$$I = \frac{V_s}{\sqrt{R^2 + X_C^2}} \quad (12\text{–}7)$$

$$V_R = \left(\frac{R}{\sqrt{R^2 + X_C^2}}\right) V_s \quad (12\text{–}8)$$

$$V_C = \left(\frac{X_C}{\sqrt{R^2 + X_C^2}}\right) V_s \quad (12\text{–}9)$$

$$V_s = \sqrt{V_R^2 + V_C^2} \quad (12\text{–}10)$$

$$\phi = 90° - \arctan\left(\frac{X_C}{R}\right) \quad (12\text{–}11)$$

$$\phi = \arctan\left(\frac{R}{X_C}\right) \quad (12\text{–}12)$$

$$\phi = \arctan\left(\frac{X_C}{R}\right) \quad (12\text{–}13)$$

$$P_{avg} = I^2 R \quad (12\text{–}14)$$

$$P_r = I^2 X_C \quad (12\text{–}15)$$

$$P_a = I^2 Z \quad (12\text{–}16)$$

$$P_{avg} = V_T I \cos \theta \quad (12\text{–}17)$$

$$PF = \cos \theta \quad (12\text{–}18)$$

For parallel RC circuits:

$$I_T = \sqrt{I_R^2 + I_C^2} \qquad (12\text{--}19)$$

$$\theta = \arctan\left(\frac{R}{X_C}\right) \qquad (12\text{--}20)$$

$$Z = \frac{RX_C}{\sqrt{R^2 + X_C^2}} \qquad (12\text{--}21)$$

Summary

1. A sine wave voltage applied to an RC circuit produces a sine wave current.
2. Current leads the voltage in an RC circuit.
3. Impedance is the total opposition to sinusoidal current flow.
4. The unit of impedance is the ohm.
5. For a series RC circuit, the impedance is the phasor sum of the resistance and the capacitive reactance.
6. Resistor voltage is in phase with the current.
7. Capacitor voltage lags the current by 90°.
8. The magnitude of the impedance in an RC circuit varies inversely with frequency.
9. The phase angle is dependent on the relative values of R and X_C.
10. If X_C is negligible with respect to R, the phase angle is approximately zero.
11. For a given value of R, the phase angle increases as X_C increases.
12. When R and X_C are equal, the phase angle is 45°.
13. The phase angle is the phase difference between the total voltage and the total current.
14. The term *arctan* means "the angle whose tangent is."
15. In an RC lag network, the output voltage is taken across the capacitor.
16. In an RC lag network, the output voltage lags the input voltage in phase.
17. In an RC lead network, the output voltage is taken across the resistor.
18. In an RC lead network, the output voltage leads the input voltage in phase.
19. An RC lag network acts as a low-pass filter.
20. A low-pass filter passes low frequencies and tends to reject high frequencies.
21. An RC lead network acts as a high-pass filter.
22. A high-pass filter passes high frequencies and tends to reject low frequencies.

SELF-TEST

23. An RC coupling circuit passes the signal voltage (ac) from one point to another but blocks dc voltage.
24. The average power in a capacitor is zero.
25. A capacitor stores energy on the positive half of the ac cycle and then returns that stored energy to the source on the negative half of the ac cycle.
26. When a sine wave source drives an RC circuit, part of the total power delivered by the source is resistive power (average power), and part of it is reactive power.
27. Reactive power is the rate at which power is stored by a capacitor.
28. The unit of reactive power is the volt-ampere reactive, abbreviated VAR.
29. The total power being delivered by the source is the combination of average power and reactive power, and its unit is the volt-ampere, abbreviated VA. It is called the *apparent power.*
30. The power factor is the cosine of the phase angle.

Self-Test

1. If a voltage sine wave is applied to an RC circuit, what is the shape of the voltage wave form that is observed across the capacitor?
2. In a given series RC circuit, $R = 2.5$ kΩ and $C = 0.005$ μF. What is the magnitude of the impedance at a frequency of 10 kHz?
3. In Problem 2, if the frequency is doubled, what will the impedance be?
4. In Figure 12–45, at what frequency does a 45° phase angle between the source voltage and the current occur?

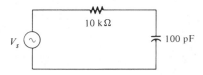

FIGURE 12–45

5. In a given series RC circuit the capacitance is 0.01 μF. At 5 kHz, what value of resistance will produce a 60° phase angle?
6. In the circuit of Figure 12–45, what is the phase difference between the capacitor voltage and the resistor voltage for $f = 100$ kHz?
7. In a given circuit, R is 500 Ω and X_c is 1 kΩ. What is the phase angle?
8. A circuit has a resistance that is twice the capacitive reactance at a given frequency. By how many degrees does the voltage lag the current?

9. Determine the magnitudes of the capacitor voltage and the resistor voltage in Figure 12–46.

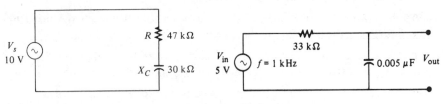

FIGURE 12–46   **FIGURE 12–47**

10. How much current is in the circuit of Figure 12–46?
11. For the lag network shown in Figure 12–47, determine the angle by which the output voltage lags the input voltage.
12. What is the magnitude of the output voltage in Figure 12-47?
13. For the lead network shown in Figure 12–48, determine the angle by which the output voltage leads the input voltage.

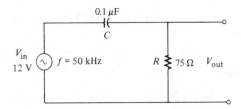

FIGURE 12–48

14. What is the magnitude of the output voltage for Figure 12–48?
15. For Figure 12–47, is the output voltage greater at 2 kHz than it is at 200 Hz?
16. For Figure 12–48, is the output voltage greater at 5 kHz or at 10 Hz?
17. Calculate the average power, the reactive power, and the apparent power for the circuits of Figures 12–47 and 12–48.
18. Calculate the power factors in Problem 17.
19. For the parallel circuit in Figure 12–49, determine each branch current and the total current.

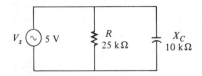

FIGURE 12–49

20. What is the phase angle between the applied voltage and the total current in Figure 12–49?

Problems

12–1. Calculate the impedance magnitude of the circuits in Figure 12–50.

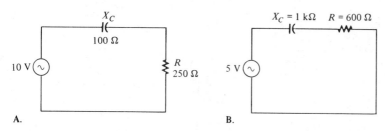

FIGURE 12–50

12–2. Determine the impedance magnitude of each circuit in Figure 12–51.

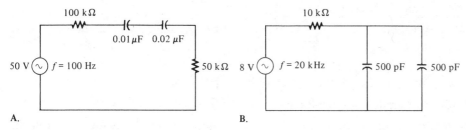

FIGURE 12–51

12–3. Calculate the current in each circuit of Figure 12–50.

12–4. Calculate the total current in each circuit of Figure 12–51.

12–5. For the circuit of Figure 12–52, determine the impedance magnitude for each of the following frequencies:
 (a) 100 Hz
 (b) 500 Hz
 (c) 1 kHz
 (d) 2.5 kHz

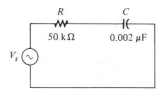

FIGURE 12–52

12–6. Repeat Problem 12–5 for $C = 0.005 \, \mu F$.

12–7. Calculate the phase angle between the applied voltage and the total current for the circuits in Figure 12–50.

12–8. Calculate the phase angle between the applied voltage and the current for the circuits in Figure 12–51.

12–9. Find the phase angle for each frequency in Problem 12–5.

12-10. For the circuit in Figure 12–53, draw the phasor diagram showing all voltages and the current. Indicate phase angles.

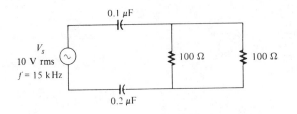

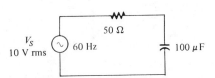

FIGURE 12–53

FIGURE 12–54

12-11. Calculate θ for the circuit in Figure 12–54.

12-12. If the capacitance in Figure 12–54 is doubled, by how many degrees does θ change? Does it increase or decrease?

12-13. In Problem 12-12, if R is halved and C is doubled, what is the new phase angle?

12-14. On a single graph, sketch the wave forms for V_s, V_R, V_C, and i for the circuit in Figure 12–54. Assume that V_s has a peak value of 10 V.

12-15. Determine the magnitude of the voltages across R and C in Figure 12–55.

12-16. For the circuit in Figure 12–55, determine the rms values for V_R and V_C for each of the following frequencies:
(a) 60 Hz
(b) 200 Hz
(c) 500 Hz
(d) 1 kHz

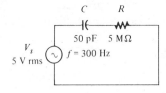

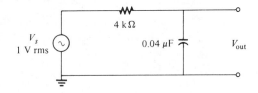

FIGURE 12–55

FIGURE 12–56

12-17. For the lag network in Figure 12–56, determine the phase lag of the output voltage with respect to the input for the following frequencies:
(a) 1 Hz
(b) 100 Hz
(c) 1 kHz
(d) 10 kHz

12-18. The lag network in Figure 12–56 acts as a low-pass filter. Draw a response curve for this circuit by plotting the output voltage versus frequency for 0 Hz to 10 kHz in 1-kHz increments.

12-19. Repeat Problem 12–17 for the lead network in Figure 12–57.

PROBLEMS

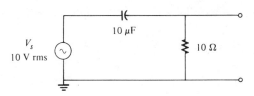

FIGURE 12–57

12–20. Sketch the frequency response curve for the lead network in Figure 12–57 for a frequency range of 0 kHz to 10 kHz in 1-kHz increments.

12–21. Draw the voltage phasor diagram for each circuit in Figures 12–56 and 12–57 for a frequency of 5 kHz and $V_s = 1$ V.

12–22. Determine the phase angle between the input and the output for the lag network in Figure 12–58 given that the frequency is 5 kHz.

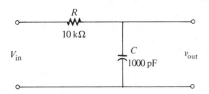 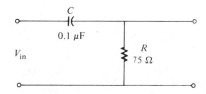

FIGURE 12–58 **FIGURE 12–59**

12–23. Calculate the output phase angle for the lead network in Figure 12–59 given that the frequency is 20 kHz.

12–24. Calculate the average power, reactive power, and apparent power in Figure 12–58 for $V_{in} = 10$ V_{rms} and $f = 5$ kHz.

12–25. Calculate the average power, reactive power, and apparent power in Figure 12–59 for $V_{in} = 10$ V_{rms} and $f = 20$ kHz.

12–26. What is the power factor in Figure 12–58? In Figure 12–59?

12–27. (a) Sketch the power triangle for Figure 12–58.
(b) Sketch the power triangle for Figure 12–59.

12–28. What value of coupling capacitor is required in Figure 12–60 so that the signal voltage at the input of amplifier 2 is at least 70.7% of the signal voltage at the output of amplifier 1 when the frequency is 20 Hz? Neglect the input resistance of the amplifier.

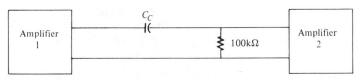

FIGURE 12–60

12–29. Determine the total current for the parallel RC circuit in Figure 12–61.

FIGURE 12–61

12–30. In Figure 12–61, what is the phase angle between the total current and the source voltage?

12–31. For the circuit in Figure 12–62, determine the following values:
(a) Z_T
(b) I_R
(c) I_C
(d) I_T
(e) θ

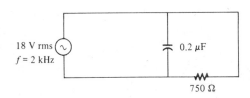

FIGURE 12–62

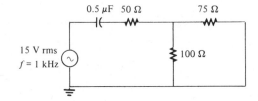

FIGURE 12–63

12–32. Repeat Problem 12–31 for $R = 5$ kΩ, $C = 0.05$ μF, and $f = 500$ Hz.

12–33. For the circuit in Figure 12–63, determine the following:
(a) I_T
(b) θ
(c) P_{avg}
(d) PF

Answers to Section Reviews

Section 12–1:
1. Yes. **2.** 60 Hz.

Section 12–2:
1. Adjacent, opposite, and hypotenuse. **2.** Square each magnitude; add the squares; and then take the square root of the sum. **3.** Adjacent side equals the hypotenuse times cos θ. **4.** Opposite side equals the hypotenuse times sin θ. **5.** Angle equals the arc tangent of the opposite over the adjacent. **6.** Part A, 5.83, 30.96°; Part B, $x = 6.43$, $y = 7.66$. **7.** Two phasors 90° apart are the adjacent and opposite sides, and their resultant is the hypotenuse.

Section 12–3:
1. The opposition to sinusoidal current. **2.** Lag. **3.** 90°. **4.** See Figure 12-64. **5.** X_C decreases; Z decreases. **6.** 1414 Ω.

FIGURE 12-64

Section 12-4:
1. The capacitive reactance. **2.** In phase. **3.** Lead. **4.** Lead. **5.** Increase. **6.** It decreases. **7.** When R equals X_C. **8.** Greater.

Section 12-5:
1. The value in volts or amperes, respectively. **2.** A portion of the total applied voltage is dropped across X_C, and a portion across R. The relative values of X_C and R determine how the total voltage is divided. **3.** Across X_C. **4.** The voltage across R would increase, because X_C would be decreasing. **5.** $V_R = 4.47$ V; $V_C = 8.94$ V.

Section 12-6:
1. Lead. **2.** Capacitor. **3.** 45°. **4.** Increase. **5.** Increase.

Section 12-7:
1. Lead. **2.** Resistor. **3.** 45°. **4.** Increase. **5.** Increase.

Section 12-8:
1. To block certain frequencies and pass others. **2.** Capacitor. **3.** Lower. **4.** Decrease. **5.** It shows how the output of the filter changes with frequency for a fixed input magnitude. **6.** T.

Section 12-9:
1. Resistance. **2.** Average power is heat loss in the resistance. Reactive power is stored by the capacitor and returned to the source. **3.** $P_{avg} = I^2 R$. **4.** $P_r = I^2 X_C$. **5.** $P_a = I^2 Z$. **6.** $\cos \theta$. **7.** $P_{avg} = V_T I \cos \theta$. **8.** It increases.

Section 12-10:
1. Resistive branch. **2.** F. **3.** Increase.

In the area of pulse and digital circuits, the technician is often concerned with how a circuit responds over an interval of time to rapid changes in voltage or current magnitudes. As you know, pulses are made up of fast transitions from one level to another. The *pulse response*, often referred to as *time response*, is discussed in this chapter.

13–1 Pulses
13–2 Time Constant
13–3 Exponential Curves
13–4 The RC Integrator
13–5 Integrator Response to a Rectangular Pulse
13–6 Integrator Response to Periodic Pulse Wave Forms
13–7 The RC Differentiator
13–8 Differentiator Response to a Rectangular Pulse
13–9 Differentiator Response to Periodic Pulse Wave Forms
13–10 Current in the Integrator and Differentiator
13–11 Relationship of Time Response to Frequency Response

13
PULSE RESPONSE
OF RC CIRCUITS

13-1
PULSES

An introduction to pulses was presented in Chapter 10. In this section we will discuss some additional pulse characteristics.

Leading and Trailing Edges

Recall that a pulse is composed basically of two steps or transitions: one transition from the low to the high level, and another transition from the high to the low level. The first transition to occur in time is called the *leading edge* of the pulse; the second transition to occur in time is called the *trailing edge,* as illustrated in Figure 13-1. When the pulse is positive-going, as shown in Part A of the figure, the leading edge is a positive-going transition or *rising edge*, and the trailing edge is a negative-going transition or *falling edge.* When the pulse is negative-going, as shown in Figure 13-1B, the leading edge is a negative-going transition, and the trailing edge is a positive-going transition.

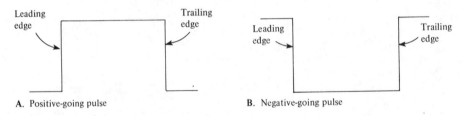

FIGURE 13-1 *Pulse edges.*

Overshoot and Undershoot

Certain circuit conditions can misshape a pulse and cause short spikes of voltage or current to occur on the leading and trailing edges. Such spikes exceed the normal levels of the pulse. These *overshoot* and *undershoot* conditions are illustrated in Figure 13-2.

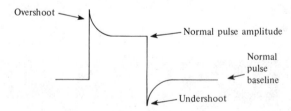

FIGURE 13-2 *Overshoot and undershoot of a pulse.*

PULSES

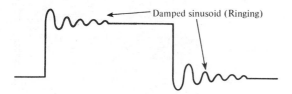

FIGURE 13-3 *Pulse with ringing.*

Ringing

Ringing is another condition often observed on a pulse wave form. In an RC circuit this situation may occur due to unwanted or stray *inductance* (inductance will be covered in Chapter 14).

Ringing appears as a *damped sine wave* occurring immediately after the leading and trailing edges of a pulse. A damped sine wave is one in which the amplitude decreases with each cycle and eventually dies out. Figure 13-3 illustrates a pulse with ringing.

Tilt

The *tilt* of a pulse is a slope in the normally flat portion, as shown in Figure 13-4. It is usually expressed as the percentage change in voltage or current between the leading and trailing edges, compared to the full amplitude of the pulse.

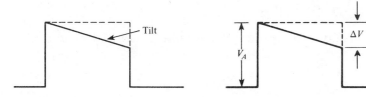

FIGURE 13-4 *Pulse tilt.*   **FIGURE 13-5**

Tilt of a pulse of voltage is illustrated in Figure 13-5, and the formula is expressed as follows:

$$\% \text{ tilt} = \left(\frac{\Delta v}{V_A}\right)100 \qquad (13\text{-}1)$$

Example 13-1

For the pulse in Figure 13-6A, determine the amount of overshoot and undershoot, and identify the leading and trailing edges. For the pulse shown in Figure 13-6B, determine the percentage of tilt.

Example 13–1 (continued)

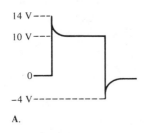
A.

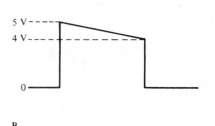
B.

FIGURE 13–6

Solution:

In Figure 13–6A, the overshoot and undershoot are both 4 V. The leading edge is the first edge to occur, and the trailing edge is the second edge.
    The percentage of tilt in Figure 13–6B is determined as follows:

$$\% \text{ tilt} = \left(\frac{\Delta v}{V_A}\right)100 = \left(\frac{1 \text{ V}}{5 \text{ V}}\right)100$$
$$= 20\%$$

Review for 13–1

1. In Figure 13–7, identify each portion of the pulse indicated by the circled numbers.

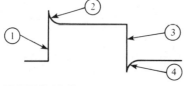

FIGURE 13–7

FIGURE 13–8

2. Define *ringing*.
3. Calculate the tilt of the pulse in Figure 13–8.

13–2

TIME CONSTANT

When a capacitor *charges* or *discharges* through a resistance, a certain time is required for the capacitor to fully charge or fully discharge, because the voltage across a capacitor cannot change instantaneously. The rate at which the capacitor

TIME CONSTANT

charges or discharges is determined by the *time constant* of the circuit. *The time constant of a series RC circuit is a time interval that equals the product of the resistance and the capacitance.*

The symbol for time constant is τ (the Greek letter tau). The formula for time constant is

$$\tau = RC \tag{13-2}$$

where τ is expressed in seconds.

Example 13–2

A series RC circuit has a resistance of 1 MΩ and a capacitance of 5 μF. What is the time constant?

Solution:

$$\tau = RC = (1 \times 10^6 \, \Omega)(5 \times 10^{-6} \, F)$$
$$= 5 \text{ s}$$

During a time interval equal to one time constant, the charge on a capacitor will change approximately 63%. Therefore, an uncharged capacitor will *charge* to 63% of the full charge voltage in one time constant. For the case of discharging, the voltage across a capacitor drops to approximately 37% of its initial voltage in one time constant. This change also corresponds to a 63% change.

Charging Time Constants

First Time Constant of Charging: In Figure 13–9, a 10-V source is connected to an RC circuit through a switch. The time constant for this circuit is $RC = (1 \text{ k}\Omega)(10 \text{ }\mu\text{F}) = 10 \text{ ms}$. If there is no charge on the capacitor when the switch is closed, the capacitor will charge to 6.3 V (63% of 10 V) during the first 10 ms (one time constant), as illustrated in Figure 13–9B.

Second Time Constant of Charging: During the second 10 ms, as illustrated in Figure 13–9C, the capacitor will charge 63% of the *remaining* amount of voltage required to reach the full charge of 10 V. In this example, the remaining voltage after one time constant is 10 V − 6.3 V = 3.7 V. Sixty-three percent of 3.7 V is approximately 2.3 V. Therefore, the capacitor will charge an additional 2.3 V during the second time constant. It will be charged to 6.3 V + 2.3 V = 8.6 V after two time constants, which is 86% of its full charge voltage.

Third Time Constant of Charging: During the third 10 ms, as illustrated in Figure 13–9D, the capacitor will again charge 63% of the *remaining* amount of voltage required to reach the full charge voltage. The remaining voltage after two time constants is 10 V − 8.6 V = 1.4 V. Sixty-three percent of 1.4 V is approximately

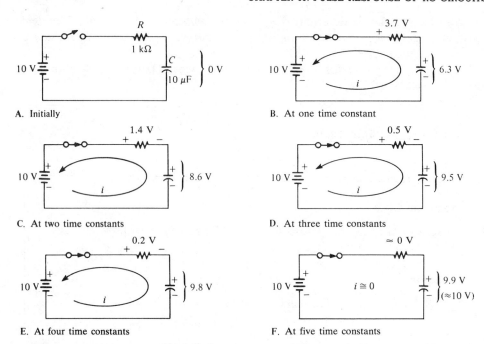

FIGURE 13–9 *Time constants in the charging of an RC circuit. Note that the voltages across the resistor and capacitor always add up to 10 V.*

0.9 V. Therefore, the capacitor will charge an additional 0.9 V during the third time constant. It will be charged to 8.6 V + 0.9 V = 9.5 V after three time constants, which is 95% of the full charge voltage.

Fourth Time Constant of Charging: During the fourth 10 ms, as shown in Figure 13–9E, the capacitor will once again charge 63% of the *remaining* amount of voltage required to reach full charge. The remaining voltage after three time constants is 10 V − 9.5 V = 0.5 V. Sixty-three percent of 0.5 V is approximately 0.3 V. Therefore, the capacitor will charge an additional 0.3 V during the fourth time constant. It will be charged to 9.5 V + 0.3 V = 9.8 V after four time constants, which is 98% of the full charge voltage.

Fifth Time Constant of Charging: During the fifth 10-ms interval, as shown in Figure 13–9F, the capacitor will charge 63% of the remaining amount of voltage required to reach full charge. The remaining voltage after four time constants is 10 V − 9.8 V = 0.2 V. Sixty-three percent of 0.2 V is approximately 0.1 V. Therefore, the capacitor will charge an additional 0.1 V during the fifth time-constant interval. It will be charged to 9.8 V + 0.1 V = 9.9 V after five time constants, which is 99% of the full charge voltage. This is so close to full charge that *it is accepted practice to consider the capacitor fully charged after five time constants.* A five-time-constant interval (5τ) is called the *transient time*.

It is important to keep in mind that the 63% charge is a constant *regardless* of the time-constant value or the value of the full charge voltage. Table 13–1 summarizes the *percentage of full charge* that a capacitor reaches after each time-constant interval of charging.

TIME CONSTANT

TABLE 13–1 *RC time-constant charging.*

τ	% full charge
1	63
2	86
3	95
4	98
5	99 (considered 100% for all practical purposes)

Discharging Time Constants

When a charged capacitor *discharges,* it loses 63% of its voltage during each time-constant interval. Therefore, after each time-constant interval, its voltage is 37% of its voltage at the beginning of the interval. For example, assume that the capacitor in Figure 13–10 is fully charged to 10 V. When the switch is closed, the capacitor discharges through the resistor with a time constant of $RC = (10\ k\Omega) \times (10\ \mu F) = 100$ ms.

First Time Constant of Discharging: In Figure 13–10A, the capacitor starts with a charge of 10 V and drops to 37% of that value after 100 ms (one time constant). Therefore, the voltage after one time constant is 3.7 V, as illustrated in Figure 13–10B.

Second Time Constant of Discharging: At the beginning of the second time-constant interval, the capacitor voltage is 3.7 V. It will drop to 37% of 3.7 V by the end of the second time-constant interval. Therefore, as illustrated in Figure 13–10C,

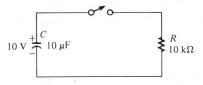

A. Initially

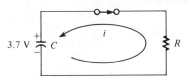

B. At one time constant

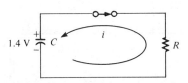

C. At two time constants

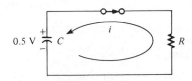

D. At three time constants

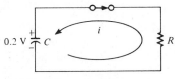

E. At four time constants

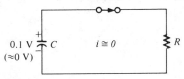

F. At five time constants

FIGURE 13–10 *Time constants in the discharging of an RC circuit.*

the voltage after two time constants is approximately 1.4 V, which is 14% of its original full charge voltage.

Third Time Constant of Discharging: At the beginning of the third time-constant interval, the capacitor voltage is 1.4 V. It will drop to 37% of 1.4 V by the end of the third time-constant interval. Therefore, as illustrated in Figure 13–10D, the voltage after three times constants is approximately 0.5 V, which is 5% of its original full charge voltage.

Fourth Time Constant of Discharging: At the beginning of the fourth time-constant interval, the capacitor voltage is 0.5 V. It will drop to 37% of 0.5 V by the end of the fourth time-constant interval. Therefore, as illustrated in Figure 13–10E, the voltage after four time constants is approximately 0.2 V, which is 2% of its original full charge voltage.

Fifth Time Constant of Discharging: At the beginning of the fifth time-constant interval, the capacitor voltage is 0.2 V. It will drop to 37% of 0.2 V by the end of the fifth time-constant interval. Therefore, as illustrated in Figure 13–10F, the voltage after five time constants is approximately 0.1 V, which is 1% of its original full charge voltage. This is so close to being completely discharged (0 V) that *it is accepted practice to consider the capacitor fully discharged after five time constants.*

In summary, *five time constants are required for a capacitor to fully charge or discharge.* Table 13–2 summarizes the percentage of full charge that a capacitor reaches at each time constant of discharging.

TABLE 13–2 *RC time-constant discharging.*

τ	% full charge
1	37
2	14
3	5
4	2
5	1 (considered 0 for all practical purposes)

Example 13–3

Calculate the time constant for the circuit in Figure 13–11. Then determine the capacitor voltage and the time at each time-constant interval, measured from the instant the switch is closed.

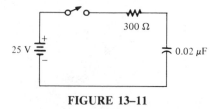

FIGURE 13–11

TIME CONSTANT

Solution:

$$\tau = RC = (300 \ \Omega)(0.02 \ \mu F)$$
$$= 6 \ \mu s$$

At 6 μs:
$$v_C = 0.63(25 \ V) = 15.75 \ V$$

At 12 μs:
$$v_C = 0.86(25 \ V) = 21.5 \ V$$

At 18 μs:
$$v_C = 0.95(25 \ V) = 23.75 \ V$$

At 24 μs:
$$v_C = 0.98(25 \ V) = 24.5 \ V$$

At 30 μs:
$$v_C = 0.99(25 \ V) = 24.75 \ V \quad \text{(approximately fully charged)}$$

Example 13–4

Calculate the time constant for the circuit in Figure 13–12. Then determine the capacitor voltage and the time at each time-constant interval, measured from the time the switch is closed.

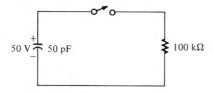

FIGURE 13–12

Solution:

$$\tau = RC = (100 \ k\Omega)(50 \ pF)$$
$$= 5 \ \mu s$$

At 5 μs:
$$v_C = 0.37(50 \ V) = 18.5 \ V$$

At 10 μs:
$$v_C = 0.14(50 \ V) = 7 \ V$$

At 15 μs:
$$v_C = 0.05(50 \ V) = 2.5 \ V$$

At 20 μs:
$$v_C = 0.02(50 \ V) = 1 \ V$$

> **Example 13–4** (continued)
>
> At 25 μs:
>
> $v_C = 0.01(50 \text{ V}) = 0.5 \text{ V}$ (approximately fully discharged)

Review for 13–2

1. In a given series RC circuit, $R = 2.2 \text{ k}\Omega$ and $C = 0.001 \text{ μF}$. What is the time constant for this circuit?
2. If the circuit mentioned in Problem 1 is charged with a 5-V source, how long will it take the capacitor to reach full charge? At full charge, what is the capacitor voltage?
3. A certain circuit has a time constant of 1 ms. If it is charged with a 10-V battery, what will the capacitor voltage be after each of the following intervals: 2 ms, 3 ms, 4 ms, and 5 ms?
4. A capacitor is charged to 100 V. If it is discharged through a resistor, what is the capacitor voltage after one time constant?
5. Define *transient time*.

13–3

EXPONENTIAL CURVES

A capacitor charges and discharges following a nonlinear curve, as shown in Figure 13–13. In these graphs, the percentage of full charge is shown at each time-constant interval, as discussed in the last section, and a smooth curve is drawn between these points. This type of curve follows a precise mathematical formula and is called an *exponential curve*. The charging curve is an *increasing exponential,* and the discharging curve is a *decreasing exponential*.

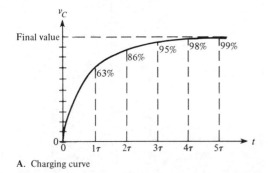
A. Charging curve

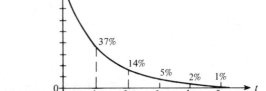
B. Discharging curve

FIGURE 13–13 *Charging and discharging exponential curves for an RC circuit.*

EXPONENTIAL CURVES

General Formula for Exponential Curves

The general expressions for either increasing or decreasing exponential curves are given in the following equations. They are identical except that (13–3) is for voltage and (13–4) for current:

$$v = V_F + (V_i - V_F)e^{-t/\tau} \quad (13\text{–}3)$$

$$i = I_F + (I_i - I_F)e^{-t/\tau} \quad (13\text{–}4)$$

where V_F and I_F are the *final* values, and V_i and I_i are the *initial* values. For example, V_F can be the voltage to which a capacitor will charge or discharge after 5τ, and I_i can be the current at the instant the capacitor begins to charge. Also, in these equations, v is the instantaneous value of the capacitor voltage at time t, and e is the base of natural logarithms and has a constant value of 2.718. The e^x key on your calculator provides an easy means of evaluating this exponential term.

Formula for RC Charging Curve

The formula for the special case in which an increasing exponential voltage curve begins at zero is given in Equation (13–5). It is developed as follows, starting with the general formula (where $V_i = 0$ V):

$$v = V_F + (V_i - V_F)e^{-t/\tau}$$
$$= V_F + (0 - V_F)e^{-t/RC}$$
$$v = V_F(1 - e^{-t/RC}) \quad (13\text{–}5)$$

Using Equation (13–5), we can calculate the value of the charging voltage in an RC circuit at any instant of time. The same is true, of course, for an increasing current.

Example 13–5

For Figure 13–14, determine the capacitor voltage 50 μs after the switch is closed. Sketch the charging curve.

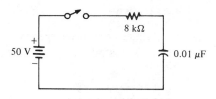

FIGURE 13–14

Solution:

The time constant is $RC = (8 \text{ k}\Omega)(0.01 \text{ μF}) = 80$ μs. The voltage to which the capacitor will charge is 50 V, and the initial voltage is zero. Notice that 50 μs is less than one time-constant interval; so the capacitor will charge less than 63% of the full voltage in that time. Using Equation (13–5), we calculate as follows:

Example 13–5 (continued)

$$v_C = V_F(1 - e^{-t/RC}) = 50 \text{ V}(1 - e^{-50\ \mu s/80\ \mu s})$$
$$= 50 \text{ V}(1 - e^{-0.625}) = 50 \text{ V}(1 - 0.535)$$
$$= 23.2 \text{ V}$$

We determine the value of $e^{-0.625}$ on the calculator by entering -0.625 and then pressing the e^x key.

The charging curve for the capacitor is shown in Figure 13–15.

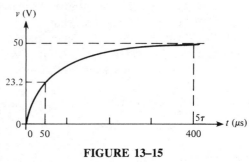

FIGURE 13–15

It should be noted that if *exact* values of time-constant intervals are substituted for t in Equation (13–5), voltages that are the appropriate percentages of the final voltage as listed in Table 13–1 will result.

Formula for RC Discharging Curve

The formula for the special case in which a decreasing exponential voltage curve ends at zero is derived from the general formula as follows (with $V_F = 0 \text{ V}$):

$$v = V_F + (V_i - V_F)e^{-t/\tau}$$
$$= 0 + (V_i - 0)e^{-t/RC}$$
$$v = V_i e^{-t/RC} \tag{13-6}$$

where V_i is the voltage at the beginning of the discharge. We can use this formula to calculate the discharging voltage at any instant of time, as Example 13–6 illustrates.

Example 13–6

Determine the capacitor voltage in Figure 13–16 at a point in time 6 ms after the switch is closed. Sketch the discharging curve.

EXPONENTIAL CURVES

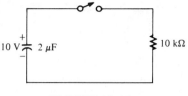

FIGURE 13–16

Solution:

The discharge time constant is $RC = (10\ \text{k}\Omega)(2\ \mu\text{F}) = 20\ \text{ms}$. The initial capacitor voltage is 10 V (this is V_i). Notice that 6 ms is less than one time constant; so the capacitor will discharge less than 63%. Therefore, it will have a voltage greater than 37% of the initial voltage after 6 ms:

$$v_C = V_i e^{-t/RC} = 10 e^{-6\ \text{ms}/20\ \text{ms}}$$
$$= 10 e^{-0.3} = 10(0.741)$$
$$= 7.41\ \text{V}$$

Again, the value of $e^{-0.3}$ can be determined with a calculator.
    The discharging curve for the capacitor voltage is shown in Figure 13–17.

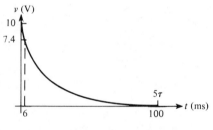

FIGURE 13–17

Again, it should be noted that if *exact* values of time-constant intervals are substituted for t in Equation (13–6), voltages that are the appropriate percentages of the full voltage as shown in Table 13–2 will result.

Solving for Time

Occasionally, we may want to determine how long it will take a capacitor to charge or discharge to a specified voltage. Equation (13–5) or (13–6) can be solved for t if v is specified. The natural logarithm (abbreviated "ln") of $e^{-t/RC}$ is the exponent $-t/RC$. Therefore, taking the natural logarithm of both sides of the equation will allow us to solve for time. This calculation is done as follows for Equation (13–6):

$$v = V_i e^{-t/RC}$$

$$\frac{v}{V_i} = e^{-t/RC}$$

$$\ln\left(\frac{v}{V_i}\right) = \ln e^{-t/RC}$$

$$\ln\left(\frac{v}{V_i}\right) = -t/RC$$

$$t = -RC \ln\left(\frac{v}{V_i}\right)$$

The same procedure can be used for the increasing exponential formula in Equation (13–5).

Example 13–7

In Figure 13–18, how long will it take the capacitor to discharge to 25 V when the switch is closed?

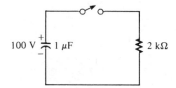

FIGURE 13–18

Solution:

$$t = -RC \ln\left(\frac{v}{V_i}\right) = -(2 \text{ ms}) \ln\left(\frac{25 \text{ V}}{100 \text{ V}}\right)$$
$$= -(2 \text{ ms}) \ln(0.25) = -2 \text{ ms} \, (-1.39)$$
$$= 2.77 \text{ ms}$$

We can determine ln(0.25) with a calculator by first entering 0.25 and then pressing the ln *x* key.

Universal Exponential Curves

The universal curves in Figure 13–19 provide a graphical solution of the charge and discharge of capacitors. Example 13–8 will illustrate this method of solution.

EXPONENTIAL CURVES

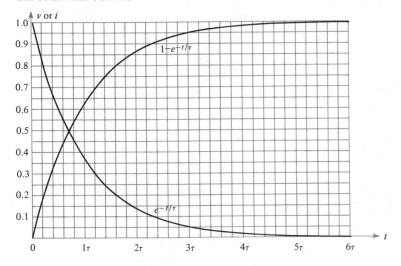

FIGURE 13–19 *Universal exponential curves.*

Example 13–8

How long will it take the capacitor in Figure 13–20 to charge to 75 V? What is the capacitor voltage 2 ms after the switch is closed? Use the universal exponential curves in Figure 13–19 to determine the answers.

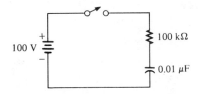

FIGURE 13–20

Solution:

The full charge voltage is 100 V, which is the 100% level on the graph. Since 75 V = 75% of maximum, you can see that this value occurs at 1.4 time constants. One time constant is 1 ms. Therefore, the capacitor voltage will reach 75 V, 1.4 ms after the switch is closed.

The capacitor is at approximately 87 V in 2 ms. These graphical solutions are shown in Figure 13–21 on page 422.

Example 13–8 (continued)

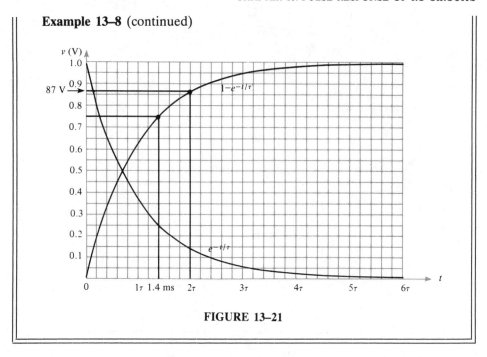

FIGURE 13–21

Review for 13-3

1. What are the two basic types of exponential curves?
2. In Figure 13–22, determine the voltage across the capacitor 2.5 time constants after the switch is closed.

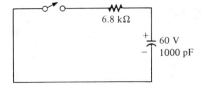

FIGURE 13–22

3. In Figure 13–22, how long will it take the capacitor to discharge to 15 V?

13–4

THE RC INTEGRATOR

Figure 13–23 shows a series RC circuit with an input and an output. When the output is taken across the capacitor, this circuit is known as an *integrator* in terms of its time response. Recall that in terms of frequency response, it is a basic *low-pass* filter. The

THE RC INTEGRATOR

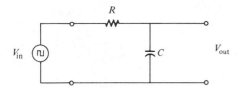

FIGURE 13–23 *RC integrator.*

term *integrator* derives from a mathematical function in calculus which this circuit approximates under certain conditions.

Capacitor Charges and Discharges with Pulse Input

When a square wave generator is connected to the input, as symbolized in Figure 13–23, the capacitor will charge and discharge in response to the pulses. When the input goes from its low level to its high level, the capacitor *charges* toward the high level of the pulse through the resistor. This action is analogous to connecting a battery through a switch to the RC network, as illustrated in Figure 13–24A. When the pulse goes from its high level back to its low level, the capacitor *discharges* back through the source. The resistance of the source is assumed to be negligible compared to R. This action is analogous to connecting a short across the source, as illustrated in Figure 13–24B.

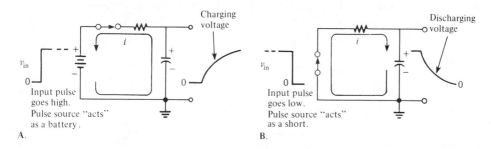

FIGURE 13–24 *Pulse source charging and discharging a capacitor.*

The capacitor will charge and discharge following an *exponential curve*. Its rate of charging and discharging, of course, depends on the RC time constant.

Capacitor Voltage

In an RC integrator, *the output is the capacitor voltage*. The capacitor charges during the time that the pulse is high. If the pulse is at its high level long enough, the capacitor will fully charge to the voltage amplitude of the pulse, as illustrated in Figure 13–25.

The capacitor discharges during the time that the pulse is low. If the low time between pulses is long enough, the capacitor will fully discharge to zero, as shown in Figure 13–25. Then when the next pulse occurs, it will begin to charge again.

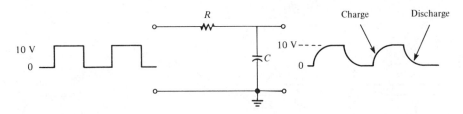

FIGURE 13-25 *Capacitor fully charging and discharging in response to a square wave.*

Review for 13-4

1. Define the term *integrator* in relation to an RC circuit.
2. What causes a capacitor in an RC circuit to charge and discharge?

13-5

INTEGRATOR RESPONSE TO A RECTANGULAR PULSE

Figure 13-26 shows an RC integrator with a rectangular pulse input. The capacitor charges for *the duration of the pulse* and then discharges when the pulse ends. Two conditions must be considered: (1) PW $\geq$ 5τ and (2) PW < 5τ.

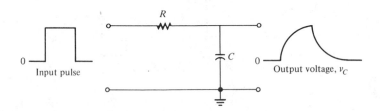

FIGURE 13-26 *RC integrator with rectangular pulse input.*

Pulse Width Equal to or Greater than Five Time Constants (PW $\geq$ 5τ)

The capacitor will *fully charge* if the pulse width is equal to or greater than five time constants (5τ). This condition is expressed as PW $\geq$ 5τ. At the end of the pulse, the capacitor fully discharges back through the source.

Figure 13-27 illustrates the output wave forms for various values of time constant and a fixed input pulse width. Notice that the shape of the output pulse approaches that of the input as the time constant is made small compared to the pulse width. In each case the output reaches the full amplitude of the input.

INTEGRATOR RESPONSE TO A RECTANGULAR PULSE

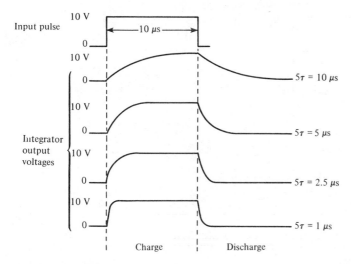

FIGURE 13-27 *Variation in output pulse shape with time constant.*

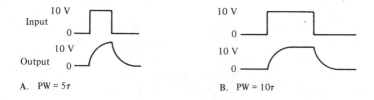

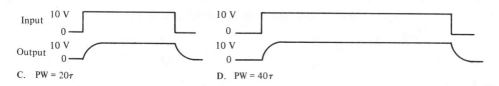

FIGURE 13-28 *Variation in output pulse shape with input pulse width (time constant fixed).*

Figure 13-28 shows how a fixed time constant and a variable input pulse width affect the integrator output. Notice that as the pulse width is increased, the shape of the output pulse approaches that of the input. Again, this means that the time constant is short compared to the pulse width.

Pulse Width Less than Five Time Constants (PW < 5τ)

Now let us examine the case in which the width of the rectangular pulse input is less than five time constants of the integrator. This condition is expressed as PW < 5τ.

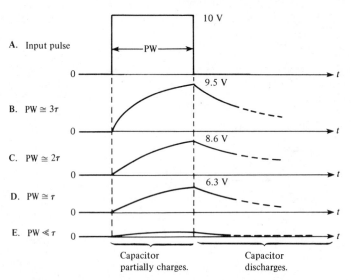

FIGURE 13–29 *Output of integrator for various time constants longer than input pulse width.*

As before, the capacitor charges for the duration of the pulse. However, because the pulse width is less than the time it takes the capacitor to fully charge (5τ), the output voltage will *not* reach the full input voltage before the end of the pulse. *The capacitor only partially charges,* as illustrated in Figure 13–29 for several values of RC time constants. Notice that for longer time constants, the output reaches a lower voltage because the capacitor does not have to charge as much. Of course, in our examples with a single pulse input, the capacitor fully discharges after the pulse ends.

When the time constant is much greater than the input pulse width, the

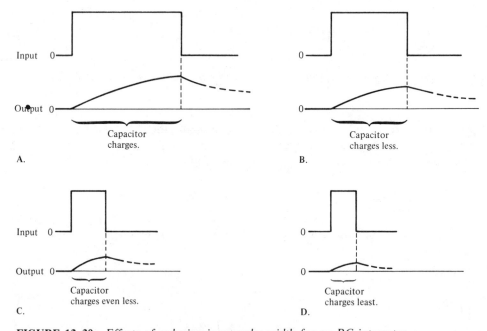

FIGURE 13–30 *Effects of reducing input pulse width for an RC integrator.*

INTEGRATOR RESPONSE TO A RECTANGULAR PULSE

capacitor charges very little, and, as a result, the output voltage becomes almost negligible, as indicated in Figure 13–29E.

Figure 13–30 illustrates the effect of reducing the input pulse width for a *fixed* time-constant value. As the width is reduced, the output voltage becomes smaller because the capacitor has less time to charge. However, it takes the capacitor the same length of time (5τ) to discharge back to zero for each condition.

Example 13–9

A single 10-V pulse with a width of 100 μs is applied to the integrator in Figure 13–31. To what voltage will the capacitor charge? How long will it take the capacitor to discharge if the internal pulse source resistance is 50 Ω? Sketch the output voltage.

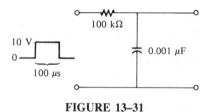

FIGURE 13–31

Solution:

Charging: The circuit time constant is

$$\tau = RC = (100 \text{ k}\Omega)(0.001 \text{ μF}) = 100 \text{ μs}$$

Notice that the pulse width is exactly equal to the time constant. Thus, the capacitor will charge 63% of the full input amplitude in one time constant; so the output will reach a maximum voltage of 6.3 V.

Discharging: The capacitor discharges back through the source when the pulse ends. We can neglect the 50-Ω source resistance in series with 100 kΩ. The total discharge time therefore is

$$5\tau = 5(100 \text{ μs}) = 500 \text{ μs}$$

The output charging and discharging curve is shown in Figure 13–32.

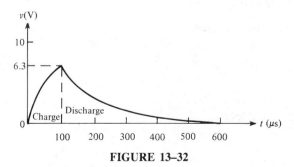

FIGURE 13–32

Example 13-10

Determine how much the capacitor will charge in Figure 13-33 when the single pulse is applied to the input.

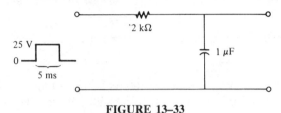

FIGURE 13-33

Solution:

Calculate the time constant:

$$\tau = RC = (2 \text{ k}\Omega)(1 \text{ }\mu\text{F}) = 2 \text{ ms}$$

Because the pulse width is 5 ms, the capacitor charges for 2.5 time constants (5 ms = 2.5 × 2 ms). The standard percentage time-constant table cannot be used in this case because an exact integer multiple of a time constant is not involved. We must use the exponential formula to find the voltage to which the capacitor will charge. The calculation is done as follows:

$$v = V_F(1 - e^{-t/RC}) \quad \text{where } V_F = 25 \text{ V} \quad \text{and} \quad t = 5 \text{ ms}$$
$$v = 25 \text{ V } (1 - e^{-5/2}) = 25 \text{ V } (1 - e^{-2.5})$$
$$= 25 \text{ V } (1 - 0.082) = 25 \text{ V } (0.918)$$
$$= 22.9 \text{ V}$$

These calculations show that the capacitor charges to 22.9 V during the 5-ms duration of the input pulse. It will discharge back to zero when the single pulse leaves.

Review for 13-5

1. When an input pulse is applied to an RC integrator, what condition must exist in order for the output voltage to reach full amplitude?
2. For the circuit in Figure 13-34, which has a single input pulse, find the maximum output voltage and determine how long the capacitor will discharge.
3. For Figure 13-34, sketch the approximate shape of the output voltage with respect to the input pulse.

INTEGRATOR RESPONSE TO PERIODIC PULSE WAVE FORMS

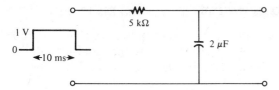

FIGURE 13-34

4. If the integrator time constant equals the input pulse width, will the capacitor fully charge?
5. Describe the condition under which the output voltage has the approximate shape of a rectangular input pulse.

13-6

INTEGRATOR RESPONSE TO PERIODIC PULSE WAVE FORMS

In the last section you learned how an RC integrator responds to a *single* pulse input. These basic ideas will be extended in this section to include the integrator response to *repetitive* pulses. In electronic systems, you will probably encounter repetitive pulse wave forms much more often than single pulses. However, an understanding of the RC integrator response to single pulses is essential to learning how these circuits act with repeated pulses.

If a periodic pulse wave form is applied to an RC integrator, as shown in Figure 13-35, *the output wave shape depends on the relationship of the circuit time constant and the frequency (period) of the input pulses.* The capacitor, of course, charges and discharges in response to a pulse input. *The amount of charge and discharge of the capacitor depends both on the circuit time constant and on the input frequency,* as mentioned.

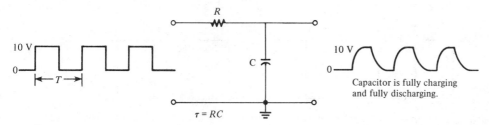

FIGURE 13-35 *RC integrator with repetitive pulse wave form.*

If the pulse width and the time between pulses are each equal to or greater than five time constants, the capacitor will fully charge and fully discharge during each period of the input wave form. This case is shown in Figure 13-35.

When the Capacitor Does Not Fully Charge and Discharge

When the pulse width and the time between pulses are *shorter than 5 time constants,* as illustrated in Figure 13–36 for a square wave, the capacitor will *not* completely charge or discharge. We will now examine the effects of this situation on the output voltage of the RC integrator.

For illustration, let us take an example of an RC integrator with a charging and discharging *time constant equal to the pulse width of a 10-V square wave input,* as in Figure 13–37. This choice will simplify the analysis and will demonstrate the basic action of the integrator under these conditions. At this point, we really do not care what the exact time-constant value is because we know that an RC circuit charges 63% during one time-constant interval.

We will assume that the capacitor in Figure 13–37 begins initially uncharged. We will examine the output voltage on a pulse-by-pulse basis. The results of this analysis are shown in Figure 13–38.

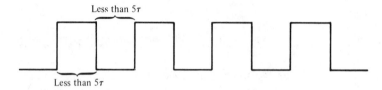

FIGURE 13–36 *Waveform that does not allow full charge or discharge of integrator capacitor.*

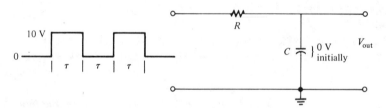

FIGURE 13–37

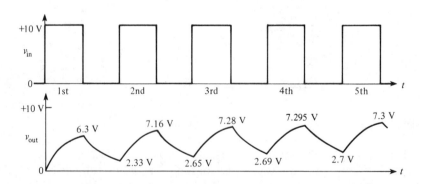

FIGURE 13–38 *Input and output for the initially uncharged integrator in Figure 13–37.*

INTEGRATOR RESPONSE TO PERIODIC PULSE WAVE FORMS

First Pulse: During the first pulse, the capacitor charges. The output voltage reaches 6.3 V (63% of 10 V), as shown in Figure 13–38.

Between First and Second Pulses: The capacitor discharges, and the voltage decreases to 37% of the voltage at the beginning of this interval: 0.37(6.3 V) = 2.33 V.

Second Pulse: The capacitor voltage begins at 2.33 V and increases 63% of the way to 10 V. This calculation is as follows: The total charging range is 10 V − 2.33 V = 7.67 V. The capacitor voltage will increase an additional 63% of 7.67 V, which is 4.83 V. Thus, at the end of the second pulse, the output voltage is 2.33 V + 4.83 V = 7.16 V, as indicated in Figure 13–38. Notice that the *average* is building up.

Between Second and Third Pulses: The capacitor discharges during this time, and therefore the voltage decreases to 37% of the voltage by the end of the second pulse: 0.37(7.16 V) = 2.65 V.

Third Pulse: At the start of the third pulse, the capacitor voltage begins at 2.65 V. The capacitor charges 63% of the way from 2.65 V to 10 V: 0.63(10 V − 2.65 V) = 4.63 V. Therefore, the voltage at the end of the third pulse is 2.65 V + 4.63 V = 7.28 V.

Between Third and Fourth Pulses: The voltage during this interval decreases due to capacitor discharge. It will decrease to 37% of its value by the end of the third pulse. The final voltage in this interval is 0.37(7.28 V) = 2.69 V.

Fourth Pulse: At the start of the fourth pulse, the capacitor voltage is 2.69 V. The voltage increases by 0.63(10 V − 2.69 V) = 4.605 V. Therefore, at the end of the fourth pulse, the capacitor voltage is 2.69 V + 4.61 V = 7.295 V. Notice that the values are leveling off as the pulses continue.

Between Fourth and Fifth Pulses: Between these pulses, the capacitor voltage drops to 0.37(7.295 V) = 2.7 V.

Fifth Pulse: During the fifth pulse, the capacitor charges 0.63(10 V − 2.7 V) = 4.6 V. Since it started at 2.7 V, the voltage at the end of the pulse is 2.7 V + 4.6 V = 7.3 V.

Steady State Response

In the preceding discussion the output voltage gradually built up and then leveled off. It takes approximately 5τ for the output voltage to build up to a constant *average* value. This value is the *transient time* of the circuit. Once the output voltage reaches *the average value of the input voltage, a steady state condition* is reached which continues as long as the periodic input continues. This state is illustrated in Figure 13–39 based on the values obtained in the preceding discussion.

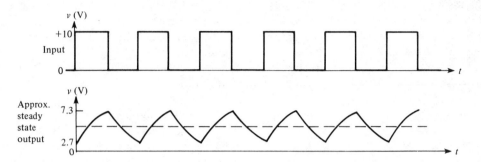

FIGURE 13–39 *Output reaches steady state after 5τ.*

The transient time for our example circuit is the time from the beginning of the first pulse to the end of the third pulse. The reason for this interval is that the capacitor voltage at the end of the third pulse is 7.28 V, which is about 99% of the final voltage.

Increase in Time Constant

What happens to the output voltage if the RC time constant of the integrator is increased with a variable resistor, as indicated in Figure 13–40? As the time constant is increased, the capacitor charges *less* during a pulse and discharges *less* between pulses. The result is a *smaller* fluctuation in the output voltage, as shown in Figure 13–40A and B for increasing values of time constant.

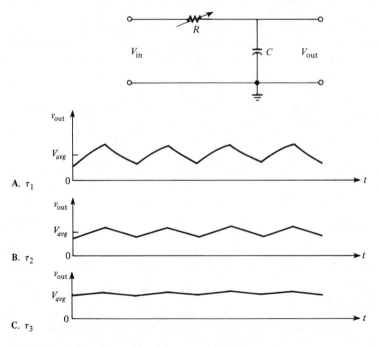

FIGURE 13–40 *Response to longer time constants ($\tau_3 > \tau_2 > \tau_1$).*

INTEGRATOR RESPONSE TO PERIODIC PULSE WAVE FORMS

As the time constant becomes extremely long compared to the pulse width, the output voltage approaches a constant dc voltage, as shown in Figure 13–40C. This value is the *average value* of the input. For a square wave, it is one-half the amplitude.

Example 13–11

Determine the output voltage wave form for the first *two* pulses applied to the integrator circuit in Figure 13–41. Assume that the capacitor is initially uncharged.

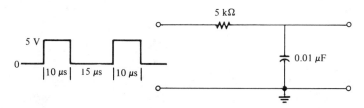

FIGURE 13–41

Solution:

First calculate the circuit time constant:

$$\tau = RC = (5 \text{ k}\Omega)(0.01 \text{ }\mu\text{F}) = 50 \text{ }\mu\text{s}$$

Obviously, the time constant is much longer than the input pulse width or the interval between pulses (notice that the input is not a square wave). As a result, the fixed percentages from the time-constant tables cannot be used. The exponential formulas must be applied, and thus the analysis is relatively difficult. Follow the solution carefully.

Calculation for first pulse: Use Equation (13–5) because C is charging. Note that V_F is 5 V, and t equals the pulse width of 10 μs. Therefore,

$$v_C = V_F(1 - e^{-t/RC}) = 5 \text{ V } (1 - e^{-10 \text{ }\mu\text{s}/50 \text{ }\mu\text{s}})$$
$$= 5 \text{ V } (1 - 0.819) = 0.906 \text{ V}$$

This result is plotted in Figure 13–42A on page 434.

Calculation for interval between first and second pulse: Use Equation (13–6) because C is discharging. Note that V_i is 0.906 V because C begins to discharge from this value at the end of the first pulse. The discharge time is 15 μs. Therefore,

$$v_C = V_i e^{-t/RC} = 0.906 e^{-15 \text{ }\mu\text{s}/50 \text{ }\mu\text{s}}$$
$$= 0.906(0.741) = 0.671 \text{ V}$$

This result is shown in Figure 13–42B.

Example 13-11 (continued)

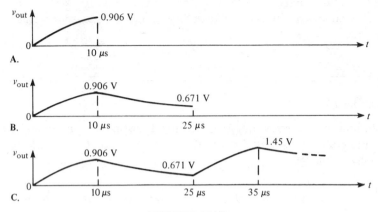

FIGURE 13-42

Calculation for second pulse: At the beginning of the second pulse, the output voltage is 0.671 V. During the second pulse, the capacitor will again charge. In this case it does not begin at zero volts. It already has 0.671 V from the previous charge and discharge. To handle this situation, we must use the general formula of Equation (13-3):

$$v = V_F + (V_i - V_F)e^{-t/\tau}$$

Using this equation, we can calculate the voltage across the capacitor at the end of the second pulse as follows:

$$\begin{aligned}
v_C &= V_F + (V_i - V_F)e^{-t/RC} \\
&= 5\text{ V} + (0.671\text{ V} - 5\text{ V})e^{-10\,\mu s/50\,\mu s} \\
&= 5\text{ V} + (-4.33\text{ V})(0.819) \\
&= 5\text{ V} - 3.55\text{ V} = 1.45\text{ V}
\end{aligned}$$

This result is shown in Figure 13-42C.

Notice that the output wave form builds up on successive input pulses. After approximately 5τ, it will reach its steady state and will fluctuate between a constant maximum and a constant minimum, with an average equal to the average value of the input. We could demonstrate this pattern by carrying the analysis in this example further.

Review for 13-6

1. What conditions allow an RC integrator capacitor to fully charge and discharge when a periodic pulse wave form is applied to the input?
2. What will the output wave form look like if the circuit time constant is extremely small compared to the pulse width of a square wave input?

DIFFERENTIATOR RESPONSE TO A RECTANGULAR PULSE

3. When 5τ is greater than the pulse width of an input square wave, the time required for the output voltage to build up to a constant average value is called _____.
4. Define *steady state response*.
5. What does the average value of the output voltage of an integrator equal during steady state?

13–7

THE RC DIFFERENTIATOR

Figure 13–43 shows a series RC circuit. Notice that the output is taken across the *resistor*, thereby distinguishing the RC differentiator from the RC integrator. This circuit is called a *differentiator* in terms of its time response. Recall that in terms of its frequency response, it is a basic *high-pass* filter.

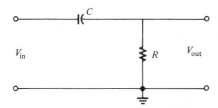

FIGURE 13–43 *RC differentiator.*

The same action occurs in the differentiator as in the integrator, except the output voltage is taken across the resistor rather than the capacitor. The capacitor charges exponentially at a rate depending on the RC time constant. The shape of the differentiator's resistor voltage is determined by the charging and discharging action of the capacitor.

Review for 13–7

1. Define *differentiator* in relation to an RC circuit.
2. What determines the shape of the differentiator's resistor voltage? Are the resistor voltage and the output voltage the same?

13–8

DIFFERENTIATOR RESPONSE TO A RECTANGULAR PULSE

To understand how the output voltage is shaped by a differentiator, we must consider its response to the rising pulse edge, the response between the rising and falling edges, and the response to the falling pulse edge.

Response to the Rising Edge of the Input Pulse

Assume that the capacitor is initially uncharged prior to the rising pulse edge. Prior to the pulse, the input is zero volts. Thus, there are zero volts across the capacitor and also zero volts across the resistor, as indicated in Figure 13–44A where point A is the input and point B is the output *with respect to ground*.

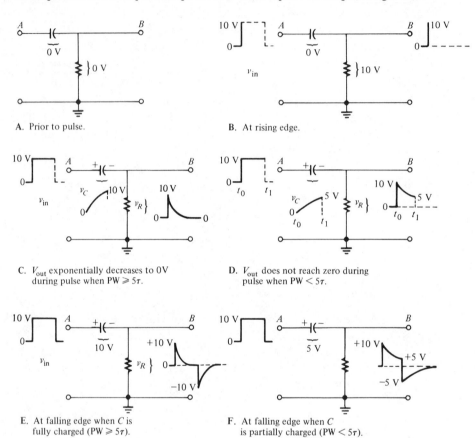

FIGURE 13–44 *Response of a differentiator to a single pulse for two conditions.*

Now assume that a 10-V pulse is applied to the input. When the rising edge occurs, point A goes to $+10$ V. Recall from Chapter 11 that *the voltage across a capacitor cannot change instantaneously.* Therefore, if point A instantly goes to $+10$ V, then point B *must* also instantly go to $+10$ V, keeping the capacitor voltage zero for the instant of the rising edge. The capacitor voltage is the voltage from point A to point B, as illustrated in Figure 13–44B.

The voltage at point B with respect to ground is the voltage across the resistor (and the output voltage). Thus, the output voltage suddenly goes to $+10$ V in response to the rising pulse edge, as indicated in Figure 13–44B.

DIFFERENTIATOR RESPONSE TO A RECTANGULAR PULSE

Response during Pulse When PW ≥ 5τ

While the pulse is at its high level between the rising edge and the falling edge, the capacitor is charging. When the pulse width is equal to or longer than five time constants (PW ≥ 5τ), the capacitor has time to *fully charge*.

As the voltage across the capacitor builds up exponentially, the voltage across the resistor *decreases* exponentially until it reaches zero volts at the time the capacitor reaches full charge (+10 V in this case). This decrease in the resistor voltage occurs because the sum of the capacitor voltage and the resistor voltage at any instant must be equal to the applied voltage, in compliance with Kirchhoff's voltage law ($v_C + v_R = v_{in}$). This response is illustrated in Figure 13–44C.

Response during Pulse When PW < 5τ

When the pulse width is less than five time constants (PW < 5τ), *the capacitor does not have time to fully charge*. Its partial charge depends on the relation of the time constant and the pulse width.

Because the capacitor does not reach the full +10 V, *the resistor voltage will not reach zero volts* by the end of the pulse. For example, if the capacitor charges to +5 V during the pulse interval, the resistor voltage will decrease to +5 V, as illustrated in Figure 13–44D.

Response to Falling Edge When PW ≥ 5τ

Let us first examine the case in which the capacitor is *fully charged* at the end of the pulse (PW ≥ 5τ). Refer to Figure 13–44E. On the falling edge, the input pulse suddenly goes from +10 V back to zero. An instant before the falling edge, the capacitor is charged to 10 V; so point A is +10 V and point B is 0 V. Since the voltage across a capacitor cannot change instantaneously, when point A makes a transition from +10 V to zero on the falling edge, point B *must* also make a 10-V transition from zero to −10 V. This keeps the voltage across the capacitor at 10 V for the instant of the falling edge.

The capacitor now begins to *discharge* exponentially. As a result, the resistor voltage goes from −10 V to zero in an exponential curve, as indicated in Figure 13–44E.

Response to Falling Edge When PW < 5τ

Next, let us examine the case in which the capacitor is only partially charged at the end of the pulse (PW < 5τ). For example, if the capacitor charges to +5 V, the resistor voltage at the instant before the falling edge is also +5 V, because the capacitor voltage plus the resistor voltage must add up to +10 V, as illustrated in Figure 13–44D.

When the falling edge occurs, point A goes from +10 V to zero. As a result, point B goes from +5 V to −5 V, as illustrated in Figure 13–44F. This decrease occurs, of course, because the capacitor voltage cannot change at the instant of the falling edge. Immediately after the falling edge, the capacitor begins to discharge to zero. As a result, the resistor voltage goes from −5 V to zero, as shown.

Summary of Differentiator Response to a Single Pulse

Perhaps a good way to summarize this section is to look at the general output wave forms of a differentiator as the time constant is varied from one extreme, when 5τ is much less than the pulse width, to the other extreme, when 5τ is much greater than the pulse width. These situations are illustrated in Figure 13–45. In Part A of the figure, the output consists of narrow positive and negative "spikes." In Part F, the output approaches the shape of the input. Various conditions between these extremes are illustrated in Parts B through E.

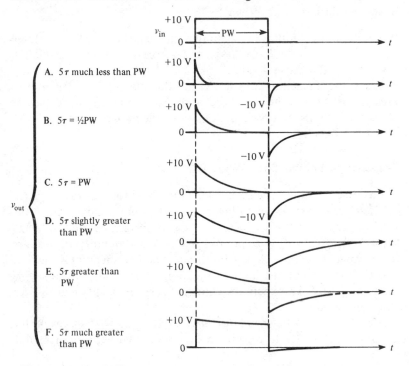

FIGURE 13–45 *Differentiator response as τ is varied.*

Example 13–12

Sketch the output voltage for the circuit in Figure 13–46.

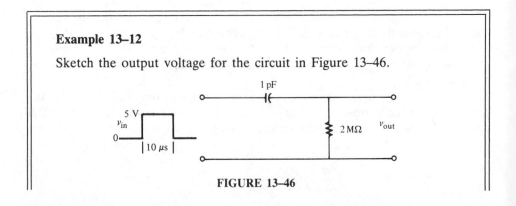

FIGURE 13–46

DIFFERENTIATOR RESPONSE TO A RECTANGULAR PULSE

Solution:

First calculate the time constant:

$$\tau = RC = (2\ \text{M}\Omega)(1\ \text{pF}) = 2\ \mu\text{s}$$

In this case, $5\tau = \text{PW}$; so the capacitor reaches full charge at the end of the pulse.

On the rising edge, the resistor voltage jumps to $+5$ V and then decreases exponentially to zero by the end of the pulse. On the falling edge, the resistor voltage jumps to -5 V and then goes back to zero exponentially. The resistor voltage is, of course, the output, and its shape is shown in Figure 13–47.

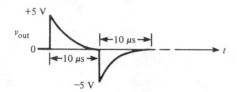

FIGURE 13–47

Example 13–13

Determine the output voltage wave form for the differentiator in Figure 13–48.

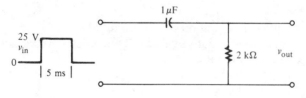

FIGURE 13–48

Solution:

First calculate the time constant:

$$\tau = (2\ \text{k}\Omega)(1\ \mu\text{F}) = 2\ \text{ms}$$

On the rising edge, the resistor voltage immediately jumps to $+25$ V. Because the pulse width is 5 ms, the capacitor charges for 2.5 time constants and therefore does not reach full charge. Thus, we must use the formula for a decreasing exponential, Equation (13–6), in order to calculate to what voltage the output decreases by the end of the pulse:

$$v_{out} = V_i e^{-t/RC} \quad \text{where } V_i = 25\ \text{V} \text{ and } t = 5\ \text{ms}$$

$$v_{out} = 25 e^{-5\ \text{ms}/2\ \text{ms}} = 25(0.082)$$

$$= 2.05\ \text{V}$$

Example 13-13 (continued)

This calculation gives us the resistor voltage at the end of the 5-ms pulse-width interval.

On the falling edge, the resistor voltage immediately jumps from +2.05 V down to −22.95 V (a 25-V transition). The resulting wave form of the output voltage is shown in Figure 13–49.

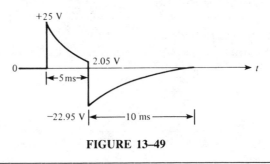

FIGURE 13–49

Review for 13-8

1. Sketch the output of a differentiator for a 10-V input pulse when $5\tau = \frac{1}{2}$PW.
2. Under what condition does the output pulse shape most closely resemble the input pulse for a differentiator?
3. What does the differentiator output look like when 5τ is much less than the pulse width of the input?
4. If the resistor voltage in a differentiating circuit is down to +5 V at the end of a 15-V input pulse, to what negative value will the resistor voltage go in response to the *falling* edge of the input?

13–9

DIFFERENTIATOR RESPONSE TO PERIODIC PULSE WAVE FORMS

The differentiator response to a single pulse, covered in the preceding section, is extended in this section to repetitive pulses.

If a periodic pulse wave form is applied to an RC differentiating circuit, two conditions again are possible: $5\tau \leq$ PW or $5\tau >$ PW. Figure 13–50 shows the output when $5\tau =$ PW. As the time constant is reduced, both the positive and the negative portions of the output become narrower. Notice that the *average* value of the output is *zero*.

DIFFERENTIATOR RESPONSE TO PERIODIC PULSE WAVE FORMS

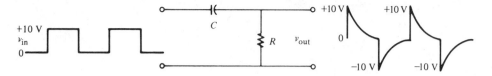

FIGURE 13–50 *RC differentiator with repetitive pulse input ($5\tau = PW$).*

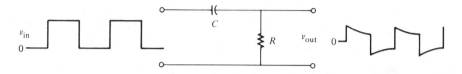

FIGURE 13–51 *RC differentiator with $5\tau > PW$.*

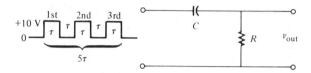

FIGURE 13–52 *RC differentiator with $\tau = PW$.*

Figure 13–51 shows the *steady state* output when $5\tau > PW$. As the time constant is increased, the positively and negatively sloping portions become flatter. For a very long time constant, the output approaches the shape of the input, but with an average value of zero. An average value of zero means that the wave form has equal positive and negative portions. Recall that the average value of a wave form is its *dc* component. Because a capacitor blocks dc, the dc component of the input is prevented from passing through to the output.

Like the integrator, the differentiator output takes time (5τ) to reach steady state. To illustrate the response, let us examine an example in which the time constant equals the input pulse width.

At this point, we do not care what the circuit time constant is, because we know that the resistor voltage will decrease to 37% of its maximum value during one pulse (1τ). We will assume that the capacitor in Figure 13–52 begins initially uncharged, and then we will examine the output voltage on a pulse-by-pulse basis. The results of the analysis to follow are shown in Figure 13–53.

First Pulse

On the rising edge, the output instantaneously jumps to $+10$ V. Then the capacitor partially charges to 63% of 10 V, which is 6.3 V. Thus, the output voltage must decrease to 3.7 V, as shown in Figure 13–53.

On the falling edge, the output instantaneously makes a negative-going 10-V transition to -6.3 V (-10 V $+ 3.7$ V $= -6.3$ V).

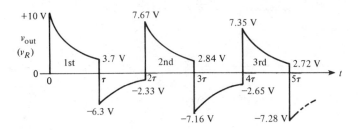

FIGURE 13–53 *Differentiator output wave form during transient time for the circuit in Figure 13–52.*

Between First and Second Pulses

The capacitor discharges to 37% of 6.3 V, which is 2.33 V. Thus, the resistor voltage, which starts at -6.3 V, must increase to -2.33 V. Why? Because at *the instant prior to* the next pulse, the input voltage is zero. Therefore, the sum of v_C and v_R must be zero ($+2.33$ V $-$ 2.33 V $=$ 0). Remember that $v_C + v_R = v_{in}$ at all times.

Second Pulse

On the rising edge, the output makes an instantaneous, positive-going, 10-V transition from -2.33 V to 7.67 V. Then the capacitor charges 0.63 $\times$ (10 V $-$ 2.33 V) $=$ 4.83 V by the end of the pulse. Thus, the capacitor voltage increases from 2.33 V to 2.33 V $+$ 4.83 V $=$ 7.16 V. The output voltage drops to 0.37 $\times$ (7.67 V) $=$ 2.84 V.

On the falling edge, the output instantaneously makes a negative-going transition from 2.84 V to -7.16 V, as shown in Figure 13–53.

Between Second and Third Pulses

The capacitor discharges to 37% of 7.16 V, which is 2.65 V. Thus, the output voltage starts at -7.16 V and increases to -2.65 V, because the capacitor voltage and the resistor voltage must add up to zero at the instant prior to the third pulse (the input is zero).

Third Pulse

On the rising edge, the output makes an instantaneous 10-V transition from -2.65 V to $+7.35$ V. Then the capacitor charges 0.63 $\times$ (10 V $-$ 2.65 V) $=$ 4.63 V to 2.65 V $+$ 4.63 V $=$ $+7.28$ V. As a result, the output voltage drops to 0.37 $\times$ 7.35 V $=$ 2.72 V.

On the falling edge, the output instantly goes from $+2.72$ V down to -7.28 V.

After the third pulse, five time constants have elapsed, and the output voltage is close to its steady state. Thus, it will continue to vary from a positive maximum of about $+7.3$ V to a negative maximum of about -7.3 V, with an average value of zero.

Review for 13-9

1. What conditions allow an RC differentiator to fully charge and discharge when a periodic pulse wave form is applied to the input?
2. What will the ouput wave form look like if the circuit time constant is extremely small compared to the pulse width of a square wave input?
3. What does the average value of the differentiator output voltage equal during steady state?

13-10

CURRENT IN THE INTEGRATOR AND DIFFERENTIATOR

So far, we have considered voltage responses. It is also useful to know what happens to the current when pulses are applied to a series RC circuit. Since both the integrator and the differentiator are *series* RC circuits, the current response is the same.

Capacitor Appears as a Short to an Instantaneous Change

On the rising edge of the pulse, the capacitor effectively "looks" like a short (its reactance is ideally zero). The *equivalent* circuit at this instant is shown in Figure 13-54A. The current at the rising edge is a maximum and equal to V_{in}/R because only R is in the circuit at this instant; V_{in} is the input pulse amplitude.

Current Decreases Exponentially as Capacitor Charges

Between the rising and falling pulse edges, the capacitor charges. As the capacitor charges, the current decreases exponentially. If the capacitor reaches

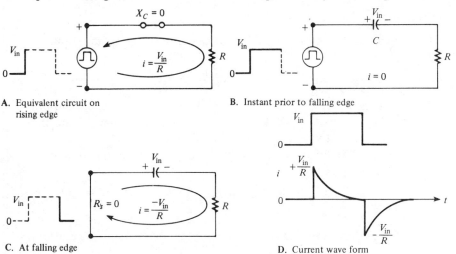

FIGURE 13-54 *Current in a pulsed series RC circuit* ($5\tau \leq PW$).

full charge, current ceases. The capacitor voltage polarity is shown in Figure 13–54B, and the current wave form is shown in Figure 13–54D.

Capacitor Acts as a Voltage Source

When the input pulse goes back to zero at the falling edge, the capacitor acts as a temporary voltage source and discharges in the direction shown in Figure 13–54C. At the instant of the falling edge, the capacitor voltage is maximum, causing a maximum instantaneous current to flow in the direction *opposite* to that of the rising-edge current. The current's maximum instantaneous value is V_C/R. If the capacitor is fully charged, $V_C = V_{in}$, as illustrated.

As the capacitor discharges, its voltage decreases exponentially, and thus the current reduces to zero exponentially, as shown in Figure 13–54D. Notice that the shape of the current wave form is the same as that of the resistor wave form studied earlier. This shape is to be expected since resistor voltage and current are directly proportional.

Example 13–14

Determine the current wave shape for the conditions indicated in Figure 13–55.

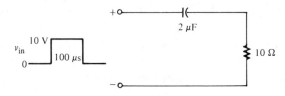

FIGURE 13–55

Solution:

The maximum value of the current at the rising edge of the input is

$$I = \frac{V_{in}}{R} = \frac{10 \text{ V}}{10 \text{ }\Omega} = 1 \text{ A}$$

The time constant is $\tau = RC = (2 \text{ }\mu\text{F})(10 \text{ }\Omega) = 20 \text{ }\mu\text{s}$. The pulse width of the input is equal to 5τ. Thus, the capacitor fully charges and the current decreases to zero by the end of the pulse.

The maximum value of the current at the falling edge is negative and equal to

$$I = \frac{-V_{in}}{R} = \frac{-10 \text{ V}}{10 \text{ }\Omega} = -1 \text{ A}$$

The resulting current wave form is shown in Figure 13–56.

RELATIONSHIP OF TIME RESPONSE TO FREQUENCY RESPONSE

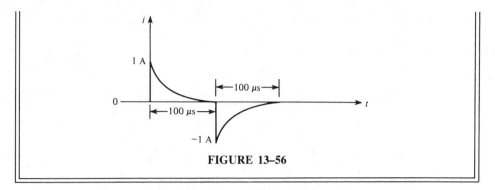

FIGURE 13-56

Review for 13-10

1. For a pulse input to an RC circuit, when is the current a positive maximum? A negative maximum?
2. Does the current have the same wave shape as that of the capacitor voltage or the resistor voltage?

13-11

RELATIONSHIP OF TIME RESPONSE TO FREQUENCY RESPONSE

There is a definite relationship between time response and frequency response: *The fast rising and falling edges of a pulse wave form contain the higher-frequency components in that wave form. The "top" or flatter portions of the pulses represent the slow changes or lower-frequency components. The average value of the pulse wave form is its dc component.* These relationships are indicated in Figure 13-57.

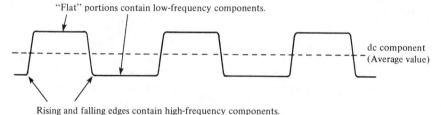

FIGURE 13-57 *General relationship of a pulse wave form to frequency content.*

Integrator Is a Low-Pass Filter

As you learned, the integrator tends to exponentially "round off" the edges of the applied pulses. This rounding off occurs to varying degrees, depending

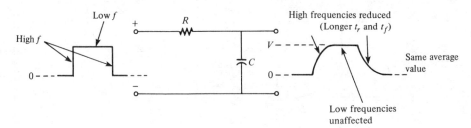

FIGURE 13–58 *Time and frequency response relationship in an integrator (one pulse in a repetitive wave form shown).*

on the relationship of the time constant to the pulse width and period. The rounding off of the edges indicates that the integrator tends to reduce the higher-frequency components of the pulse wave form, as illustrated in Figure 13–58.

Differentiator Is a High-Pass Filter

As you know, the differentiator tends to introduce *tilt* to the flat portion of a pulse. That is, it tends to reduce the lower-frequency components of a pulse wave form. Also, it completely eliminates the dc component of the input and produces a zero average-value output. This action is illustrated in Figure 13–59.

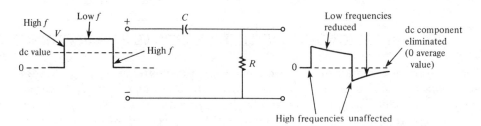

FIGURE 13–59 *Time and frequency response relationship in a differentiator (one pulse in a repetitive wave form shown).*

Formula Relating Rise Time and Fall Time to Frequency

It can be shown that the fast transitions of a pulse (rise and fall times) are related to the *highest-frequency* component, f_h, in that pulse by the following formula:

$$t_r = \frac{0.35}{f_h} \tag{13-7}$$

This formula also applies to fall time, and the fastest transition determines the highest frequency in the pulse wave form.

Equation (13–7) can be rearranged to give the highest frequency as follows:

$$f_h = \frac{0.35}{t_r} \tag{13-8}$$

Example 13–15

What is the highest frequency contained in a pulse that has rise and fall times equal to 10 nanoseconds (10 ns)?

Solution:

$$f = \frac{0.35}{t_r} = \frac{0.35}{10 \times 10^{-9} \text{ s}} = 0.035 \times 10^9 \text{ Hz}$$
$$= 35 \times 10^6 \text{ Hz} = 35 \text{ MHz}$$

Review for 13–11

1. What type of filter is an integrator?
2. What type of filter is a differentiator?
3. What is the highest-frequency component in a pulse wave form having t_r and t_f equal to 1 µs?

Formulas

$$\% \text{ tilt} = \left(\frac{\Delta v}{V_A}\right) 100 \qquad (13\text{–}1)$$

$$\tau = RC \qquad (13\text{–}2)$$

$$v = V_F + (V_i - V_F)e^{-t/\tau} \qquad (13\text{–}3)$$

$$i = I_F + (I_i - I_F)e^{-t/\tau} \qquad (13\text{–}4)$$

$$v = V_F(1 - e^{-t/RC}) \qquad (13\text{–}5)$$

$$v = V_i e^{-t/RC} \qquad (13\text{–}6)$$

$$t_r = \frac{0.35}{f_h} \qquad (13\text{–}7)$$

$$f_h = \frac{0.35}{t_r} \qquad (13\text{–}8)$$

Summary

1. The leading edge is the *first* pulse transition.
2. The trailing edge is the *second* pulse transition.
3. The leading or trailing edge is the *rising* edge if it goes to a more positive value.
4. The leading or trailing edge is the *falling* edge if it goes to a less positive value.

5. Overshoot, undershoot, ringing, tilt, and nonzero rise and fall times are distortions of the ideal pulse.
6. The time constant for a series RC circuit is the resistance times capacitance ($\tau = RC$).
7. In an RC circuit, the voltage and current in a charging or discharging capacitor make a 63% change during each time-constant interval.
8. Five time constants are required for a capacitor to reach full charge in an RC circuit.
9. In an RC circuit, the current and voltage during charge or discharge follow exponential curves.
10. An integrator is a series RC circuit in which the output voltage is across the capacitor.
11. An integrator is a basic low-pass filter.
12. A differentiator is a series RC circuit in which the output voltage is across the resistor.
13. A differentiator is a basic-high pass filter.
14. The current follows the same wave-form pattern in both the integrator and the differentiator.
15. The *transient time* is 5τ.
16. If 5τ in an integrator is much larger than the input pulse width, the output voltage approaches a constant value.
17. If 5τ in an integrator is much smaller than the input pulse width, the output voltage approaches the shape of the input pulse.
18. If 5τ in a differentiator is much larger than the input pulse width, the output voltage approaches the shape of the input, but with an average value of zero.
19. If 5τ in a differentiator is much smaller than the input pulse width, the output voltage becomes very narrow positive and negative spikes.
20. The rising and falling edges of a pulse wave form contain the higher-frequency components.
21. The flat portion of the pulse contains the lower-frequency components.
22. The average value of a pulse wave form is its dc component.

Self-Test

1. Sketch a pulse to show an example of each of the following conditions:
   (a) Overshoot and undershoot
   (b) Ringing
   (c) Tilt
2. The maximum voltage amplitude of a pulse is +10 V. Its minimum voltage amplitude is +8 V. Calculate its percentage of tilt.

PROBLEMS

3. A circuit has $R = 2\ k\Omega$ in series with $C = 0.05\ \mu F$. What is the time constant?
4. A 10-V pulse is applied to a series RC circuit. The pulse width equals *one* time constant. To what voltage does the capacitor charge during the pulse? Assume that it is initially uncharged.
5. Repeat Problem 4 for the following values of PW:
   (a) 2τ (b) 3τ (c) 4τ (d) 5τ
6. A 5-V pulse with a width of 5 μs is applied to an RC circuit with $\tau = 10$ μs. If the capacitor is initially uncharged, to how many volts does it charge by the end of the pulse?
7. The capacitor in an RC circuit is charged to 15 V. If it is allowed to discharge for an interval equal to one time constant, what is its voltage?
8. Sketch the approximate shape of an integrator output where 5τ is much less than the pulse width of a 10-V square-wave input. Repeat for the case in which 5τ is much larger than the pulse width.
9. Repeat Problem 8 for a differentiator.
10. Determine the output voltage for an integrator with a single input pulse, as shown in Figure 13–60. For repetitive pulses, how long will it take this circuit to reach steady state?

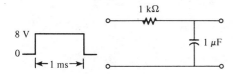

FIGURE 13–60

11. Make a differentiator from the circuit in Figure 13–60, and repeat the requirements of Problem 10.
12. What is the highest frequency for a fall time of 250 ns?

Problems

13–1. Between its rising and falling edges, a pulse decreases from 5 V to 4.8 V. What is the percentage of tilt?

13–2. If a pulse has a 20% tilt and its maximum amplitude is 8 V, what is its minimum amplitude?

13–3. Determine the time constant for each of the following series RC combinations:
   (a) $R = 100\ \Omega$, $C = 1\ \mu F$ (b) $R = 4.7\ k\Omega$, $C = 0.005\ \mu F$
   (c) $R = 10\ M\Omega$, $C = 50\ pF$ (d) $R = 1.5\ M\Omega$, $C = 0.01\ \mu F$

13–4. Determine how long it takes the capacitor to reach full charge for each of the following series RC combinations:

(a) $R = 50\ \Omega$, $C = 50\ \mu F$ (b) $R = 3300\ \Omega$, $C = 0.015\ \mu F$
(c) $R = 22\ k\Omega$, $C = 100\ pF$ (d) $R = 5\ M\Omega$, $C = 10\ pF$

13–5. In the circuit of Figure 13–61, the capacitor is initially uncharged. Determine the capacitor voltage after the following times when the switch is closed:
(a) 10 μs (b) 20 μs (c) 30 μs (d) 40 μs (e) 50 μs

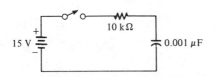
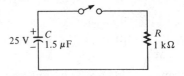

FIGURE 13–61   **FIGURE 13–62**

13–6. In Figure 13–62, the capacitor is charged to 25 V. When the switch is closed, what is the capacitor voltage after the following times?
(a) 1.5 ms (b) 4.5 ms (c) 6 ms (d) 7.5 ms

13–7. Repeat Problem 13–5 for the following time intervals:
(a) 2 μs (b) 5 μs (c) 15 μs

13–8. Repeat Problem 13–6 for the following times:
(a) 0.5 ms (b) 1 ms (c) 2 ms

13–9. Derive the formula for finding time at any point on an *increasing* exponential voltage curve. Use this formula to find the time at which the voltage in Figure 13–63 is 6 V after switch closure.

13–10. In Figure 13–62, how long does it take C to discharge to 3 V?

13–11. In Figure 13–61, how long does it take C to charge to 8 V?

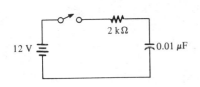
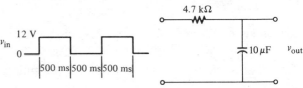

FIGURE 13–63   **FIGURE 13–64**

13–12. Sketch the integrator output in Figure 13–64, showing maximum voltages.

13–13. A 1-V, 10-kHz pulse wave form with a duty cycle of 25% is applied to an integrator with $\tau = 25\ \mu s$. Graph the output voltage for three initial pulses. C is initially uncharged.

13–14. (a) What is τ in Figure 13–65?
(b) Sketch the output voltage.

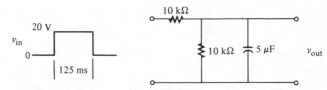

FIGURE 13–65

PROBLEMS

13–15. What is the steady state output voltage of the integrator in Figure 13–66?

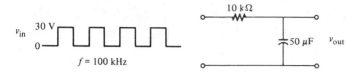

FIGURE 13–66

13–16. Sketch the differentiator output in Figure 13–67, showing maximum voltages.

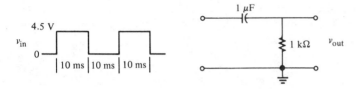

FIGURE 13–67

13–17. A 10-V, 5-kHz pulse wave form with a duty cycle of 50% is applied to a differentiator with $\tau = 50$ μs. Graph the output voltage for the three initial pulses. C is initially uncharged.

13–18. (a) What is τ in Figure 13–68?
(b) Sketch the output voltage.

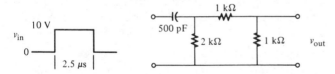

FIGURE 13–68

13–19. What is the steady state output voltage of the differentiator in Figure 13–69?

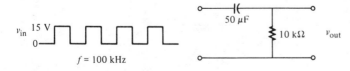

FIGURE 13–69

13–20. What is the highest-frequency component in the output of an integrator with $\tau = 10$ μs? Assume that $5\tau <$ PW.

Answers to Section Reviews

Section 13–1:
1. 1, Leading or rising edge. 2, Overshoot. 3, Trailing or falling edge. 4, Undershoot. **2.** A damped sine wave that occurs on the leading and trailing edges. **3.** 60%.

Section 13–2:
1. 2.2 μs. **2.** 11 μs, 5 V. **3.** 8.6 V, 9.5 V, 9.8 V, and 9.9 V. **4.** 37 V. **5.** The approximate time it takes a capacitor to fully charge through a resistor.

Section 13–3:
1. Increasing and decreasing. **2.** 4.93 V. **3.** 9.43 μs.

Section 13–4:
1. A series RC circuit in which the output is across the capacitor. **2.** A voltage applied to the input causes the capacitor to charge. A short across the input causes the capacitor to discharge.

Section 13–5:
1. $5\tau \leq$ PW. **2.** 0.63 V, 50 ms. **3.** See Figure 13–70. **4.** No. **5.** $5\tau \ll$ PW (5τ much less than PW).

FIGURE 13–70

Section 13–6:
1. $5\tau \leq$ PW, and $5\tau \leq$ time between pulses. **2.** Like the input. **3.** *transient time*. **4.** The response after the transient time has passed. **5.** The average value of the input voltage.

Section 13–7:
1. A series RC circuit in which the output is across the resistor. **2.** The relation of time constant to pulse width and period. Yes.

Section 13–8:
1. See Figure 13–71. **2.** 5τ much greater than PW. **3.** Positive and negative spikes. **4.** −10 V.

FIGURE 13–71

Section 13–9:
1. $5\tau \leq$ PW and $5\tau \leq$ time between pulses. **2.** Positive and negative spikes. **3.** 0 V.

ANSWERS TO SECTION REVIEWS

Section 13–10:
1. On the rising edge of the input. On the falling edge. **2.** Resistor voltage.

Section 13–11:
1. Low-pass. **2.** High-pass. **3.** 350 kHz.

Inductance is the property of a wire coil whereby a change in current produces an opposition to that change. The basis for inductance is the magnetic field that surrounds any conductor when there is current through it. The electrical component designed to have inductance is called an *inductor* or *coil*. Also, *rf choke* is a common name for a special type of inductor.

In this chapter you will learn the basic action of an inductor in circuit applications. Series and parallel inductances and inductive reactance are among the topics covered. This chapter introduces magnetism, and it also discusses self-inductance, defined in Section 14–7. Mutual inductance, which is the basis for transformer action, is covered in Chapter 17.

14–1 Magnetic Fields
14–2 Electromagnetism
14–3 Electromagnetic Properties
14–4 Electromagnetic Induction
14–5 The Basic Inductor
14–6 Faraday's Law
14–7 Self-Inductance
14–8 Characteristics of Inductors
14–9 Types of Inductors
14–10 Series Inductance
14–11 Parallel Inductance
14–12 Relationship of Inductive Current and Voltage
14–13 Inductive Reactance
14–14 Inductors in dc Circuits
14–15 Energy and Power in an Inductor
14–16 Testing an Inductor

14

14-1

MAGNETIC FIELDS

Magnetic fields are important in the study of inductance, as you will learn later in this chapter. In the first four sections, basic aspects of magnetic fields are discussed as a background for understanding inductance.

A permanent magnet, such as the bar magnet shown in Figure 14–1, has a magnetic field surrounding it. The magnetic field consists of *lines of force* that radiate from the north pole (N) to the south pole (S) and back to the north pole through the magnetic material. For clarity, only a few lines of force are shown in the figure. You can, however, imagine that many lines surround the magnet in three dimensions.

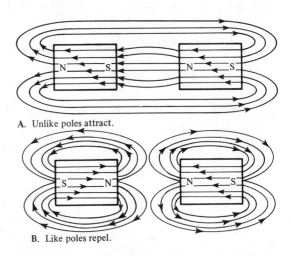

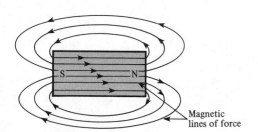

FIGURE 14–1 *Magnetic lines of force around a bar magnet.*

FIGURE 14–2 *Magnetic attraction and repulsion.*

Attraction and Repulsion of Magnetic Poles

If *unlike* poles of two permanent magnets are placed close together, an attractive force is produced by the magnetic fields, as indicated in Figure 14–2A. If two *like* poles are brought close together, they repel each other, as indicated in Figure 14–2B.

Altering a Magnetic Field

If a nonmagnetic material, such as paper, glass, wood, or plastic, is placed in a magnetic field, the lines of force are unaltered, as shown in Figure 14–3A. However, if a magnetic material such as iron is placed in the magnetic field, the lines of force will tend to change course and pass through the iron rather than through the surrounding air. They do so because the iron provides a magnetic path that is more easily traveled than that of the air. Figure 14–3B illustrates this principle.

MAGNETIC FIELDS

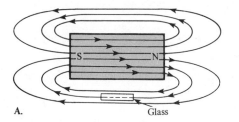
A. Glass

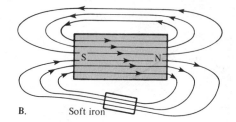
B. Soft iron

FIGURE 14–3 *Effect of nonmagnetic and magnetic materials on a magnetic field.*

Magnetic Flux

The entire group of lines of force surrounding a magnet is called the *magnetic flux*, symbolized by ϕ (phi). The number of lines of force in a magnetic field determine the value of the flux. The more lines of force there are, the greater is the flux and the stronger is the magnetic field.

The unit of magnetic flux is the weber (Wb). One weber equals 10^8 lines. In most practical situations, the weber is too large a unit, and thus the microweber (μWb) is more commonly used. One microweber equals 100 lines.

Magnetic Flux Density

The *flux density* is the amount of flux per unit area in the magnetic field. Its symbol is B, and its unit is the tesla (T). One tesla equals one weber per square meter (Wb/m²). The following formula expresses the flux density:

$$B = \frac{\phi}{A} \qquad (14\text{--}1)$$

where ϕ is the flux and A is the area.

Example 14–1

Find the flux density in a magnetic field in which the flux in 0.1 square meter is 800 μWb.

Solution:

$$B = \frac{\phi}{A} = \frac{800 \ \mu\text{Wb}}{0.1 \ \text{m}^2}$$
$$= 8000 \times 10^{-6} \ \text{T}$$

Review for 14–1

1. The magnetic field surrounding a magnet consists of lines of force that radiate from the _____ pole to the _____ pole of the magnet.

2. Two north poles attract each other (T or F).
3. Define *magnetic flux*.
4. Determine the flux density in a given magnetic field if there are 500 μWb in an area of 0.000025 m².

14–2

ELECTROMAGNETISM

With current in a conductor, a magnetic field is produced around the conductor, as illustrated in Figure 14–4. This field is called *an electromagnetic field*. The invisible lines of force of the magnetic field form a concentric circular pattern around the conductor and are continuous along its length.

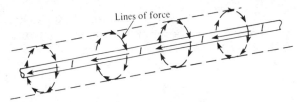

FIGURE 14–4 *Magnetic field around a current-carrying conductor.*

Although the magnetic field cannot be seen, it produces visible effects. For example, if a current-carrying wire is inserted through a sheet of paper in a perpendicular direction, iron filings on the surface of the paper arrange themselves along the magnetic lines of force in concentric rings, as illustrated in Figure 14–5A. Part B of the figure illustrates that the north pole of a compass placed in the electromagnetic field will point in the direction of the lines of force. The field is stronger closer to the conductor and becomes weaker with increasing distance from the conductor.

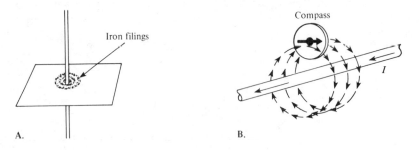

FIGURE 14–5 *Visible effects of an electromagnetic field.*

Direction of the Lines of Force

The direction of the lines of force around the conductor are indicated in Figure 14–6. When current is from right to left, as in Part A, the lines are in a clockwise direction. When the current is left to right, as in Part B, the lines are in a counterclockwise direction.

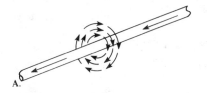 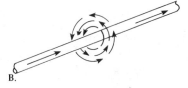

FIGURE 14–6

Left-Hand Rule

An aid to remembering the direction of the lines of force is illustrated in Figure 14–7. Imagine that you are grasping the conductor with your left hand, with your thumb pointing in the direction of electron flow. Your fingers then will be in the direction of the magnetic lines of force.

FIGURE 14–7 *Left-hand rule.*

Review for 14–2

1. A magnetic field is created by an electric current through a conductor (T or F).
2. Describe the left-hand rule.

14–3

ELECTROMAGNETIC PROPERTIES

In this section we will discuss some important properties relating to electromagnetic fields.

Magnetomotive Force (mmf)

As you learned in the last section, when there is current in a conductor, a magnetic field is produced. The force that produces the magnetic field is called the *magnetomotive force* (mmf). The unit of mmf, the ampere-turn (At), is established on the basis of the current in a single loop (turn) of wire. The formula for mmf is as follows:

$$F_m = NI \qquad (14\text{–}2)$$

where N is the number of turns of wire and I is the current.

Reluctance

Reluctance ($\mathcal{R}$) is the opposition to the establishment of a magnetic field in an electromagnetic circuit. It is the ratio of the mmf required to establish a given flux to the amount of flux, and its units are ampere-turns per weber. The formula for $\mathcal{R}$ is

$$\mathcal{R} = \frac{F_m}{\phi} \qquad (14\text{–}3)$$

This equation is sometimes known as "Ohm's law for magnetic circuits" because the reluctance is analogous to the resistance in electrical circuits.

Permeability

The ease with which a magnetic field can be established in a given material is indicated by the *permeability* of that material. The higher the permeability is, the more easily a magnetic field can be established.

The symbol for permeability is μ, and the formula is as follows:

$$\mu = \frac{l}{\mathcal{R}A} \qquad (14\text{–}4)$$

where $\mathcal{R}$ is the reluctance in ampere-turns per weber, l is the length of the material in meters, and A is the cross-sectional area in square meters.

Example 14-2

There are 2 amperes of current through a wire with 5 turns.
(a) What is the mmf?
(b) What is the reluctance of the circuit if the flux is 250 μWb?

Solution:

(a) $N = 5$ and $I = 2$ A
$$F_m = NI = (5)(2 \text{ A})$$
$$= 10 \text{ At}$$

(b)

$$\mathcal{R} = \frac{F_m}{\phi} = \frac{10 \text{ At}}{250 \text{ }\mu\text{Wb}}$$

$$= 0.04 \times 10^6 \text{ At/Wb}$$

Review for 14-3

1. What does *mmf* stand for?
2. Define *reluctance*.
3. What does the permeability of a material indicate?

14-4

ELECTROMAGNETIC INDUCTION

When a conductor is moved through a magnetic field, a voltage is produced across the conductor. This phenomenon is known as *electromagnetic induction,* and the resulting voltage is an *induced voltage.*

The principle of electromagnetic induction is used in electrical circuits, as you will learn later in this chapter and in Chapter 17 which discusses transformers. The operation of electrical motors and generators is also based on this principle.

Relative Motion

If a wire is moved across a magnetic field, there is a relative motion between the wire and the magnetic field. Likewise, if a magnetic field is moved past a stationary wire, there is still relative motion. In either case there will be an induced voltage in the wire, as Figure 14-8 indicates.

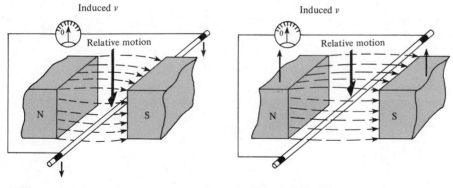

A. Wire moving

B. Magnetic field moving

FIGURE 14-8 *Relative motion between a wire and a magnetic field.*

The amount of induced voltage depends on the *rate* at which the wire and the magnetic field move with respect to each other. The faster the relative speed is, the greater the induced voltage is.

Polarity of the Induced Voltage

If the conductor in Figure 14–8 is moved first one way and then another in the magnetic field, a reversal of the polarity of the induced voltage will be observed. As the wire is moved downward, a voltage is induced with the polarity indicated in Figure 14–9A. As the wire is moved upward, the polarity is as indicated in Part B of the figure.

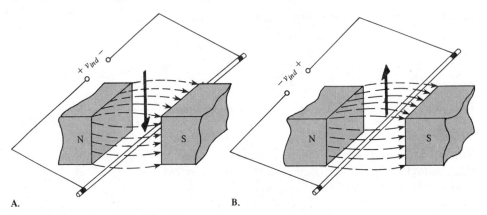

FIGURE 14–9 *Polarity of induced voltage depends on direction of motion.*

Induced Current

If a load is connected to the wire in Figure 14–9, the voltage induced by the relative motion in the magnetic field will cause a current in the load, as shown in Figure 14–10. This current is called the *induced current*.

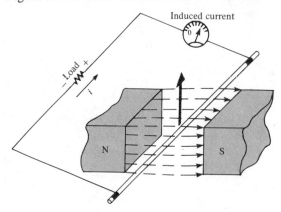

FIGURE 14–10 *Induced current in a load.*

The principle of producing a voltage and a current in a load by moving a conductor across a magnetic field is the basis for electrical generators. The concept

THE BASIC INDUCTOR

of a conductor existing in a moving magnetic field is fundamental to inductance in an electrical circuit.

Forces on a Current-Carrying Conductor in a Magnetic Field

Figure 14–11A shows current inward through a wire in a magnetic field. The electromagnetic field set up by the current interacts with the permanent magnetic field. The lines of force above the wire tend to cancel because they are in opposite directions. As a result, the flux density is reduced, and the magnetic field is weakened. The lines of force below the conductor tend to reinforce each other because they are in the same direction. Thus, the flux density is increased, and the magnetic field is strengthened. An upward force on the conductor results, and it will tend to move toward the weaker magnetic field.

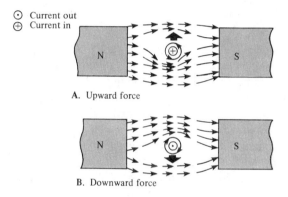

FIGURE 14–11 *Forces on a current-carrying conductor.*

Figure 14–11B shows the current outward, resulting in a force on the conductor in the downward direction. This principle is the basis for electrical motors.

Review for 14–4

1. What is the induced voltage across a stationary conductor in a stationary magnetic field?
2. If the speed with which a conductor is moving through a magnetic field is increased, does the induced voltage increase or decrease?
3. What happens to a conductor in a magnetic field if current is passed through it?

14–5

THE BASIC INDUCTOR

When a length of wire is formed into a *coil,* as shown in Figure 14–12A, it becomes a basic *inductor.* Current through the coil produces a magnetic field, as illustrated in

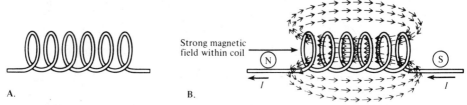

FIGURE 14–12 *Basic inductor and magnetic field.*

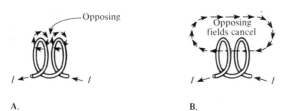

FIGURE 14–13 *Interaction of magnetic lines of force in two loops in a coil.*

Figure 14–12B. The magnetic lines of force around each loop in the coil effectively add to the lines of force around the adjoining loops, forming a strong magnetic field within the coil as shown. The net direction of the total magnetic field creates a north pole and a south pole, as shown in Part B.

To understand the formation of the total magnetic field in a coil of wire, let us discuss the interaction of the magnetic fields around two adjacent loops. The magnetic fields around each adjacent loop or turn in the wire tend to cancel *between* the loops. They tend to add in the center of the loops and on the outer edges. This effect occurs because the magnetic lines of force are in *opposing* directions between adjacent loops and in the *same* direction on the inner and outer areas of the loops. This effect is illustrated in Figure 14–13A. The total magnetic field for the two loops is illustrated in Part B of the figure. For simplicity, only single lines of force are shown. For many closely adjacent turns (loops) in a coil, this effect is additive; that is, each additional turn adds to the strength of the electromagnetic field.

Review for 14–5

1. A current creates a magnetic field around a wire (T or F).
2. Define the term *electromagnetic field*.
3. Basically, what is an inductor?
4. The terms *inductor* and *coil* mean the same thing (T or F).

14–6

FARADAY'S LAW

Michael Faraday (1791–1867), an English physicist and chemist, discovered the principle of *electromagnetic induction* in 1831. He found that moving a magnet

FARADAY'S LAW

through a coil of wire caused a current in the wire when a completed path was provided.

Faraday's law tells us that when there is relative motion between a coil of wire and a magnetic field, a voltage is created across the conductor. This voltage is called the *induced voltage*. When a complete path is provided, the induced voltage will cause an *induced current*.

The amount of induced voltage is directly proportional to the rate of change of the magnetic field with respect to the coil. This concept is illustrated in Figure 14–14, where a bar magnet is moved through a coil of wire. An induced voltage is indicated by the voltmeter connected across the coil. The *faster* the magnet is moved through the coil, the *greater* is the induced voltage.

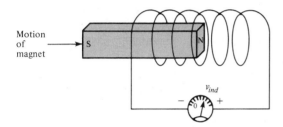

FIGURE 14–14 *Induced voltage created by a changing magnetic field.*

More about Induced Voltage

An inductor (coil) is a wire formed into a certain number of turns or loops. If it is exposed to a *changing* magnetic field, a voltage is induced across the inductor. The induced voltage is proportional to the *number of turns* of wire in the coil, N, and to the *rate* at which the magnetic field changes. The rate of change of the magnetic field is designated as $d\phi/dt$, were ϕ stands for the *magnetic flux*. As discussed in Section 14–1, flux is a measure of the number of magnetic lines of force; the weber, abbreviated Wb, is the unit of flux; $d\phi/dt$ is expressed in webers/second (Wb/s).

Faraday's law is expressed in the following equation:

$$v_{ind} = N \frac{d\phi}{dt} \qquad (14\text{--}5)$$

It states in words that the induced voltage across a coil is equal to the number of turns in the coil times the rate of flux change.

Example 14–3

Apply Faraday's law to find the induced voltage across a coil with 100 turns in a magnetic field that is changing at the rate of 5 Wb/s.

Solution:

$$v_{ind} = N \frac{d\phi}{dt} = 100(5 \text{ Wb/s}) = 500 \text{ V}$$

Review for 14-6

1. Define *induced voltage*.
2. What factors determine the amount of voltage induced across a coil in a changing magnetic field?
3. For a given rate of change of flux, if the number of turns in the coil is increased, will the induced voltage increase or decrease?

14-7

SELF-INDUCTANCE

When there is current through an inductor (coil), a magnetic field is established. When the current changes, the magnetic field changes. More current expands the magnetic field, and less current reduces it. Therefore, a changing current produces a changing magnetic field around the coil. In turn, the changing magnetic field induces a voltage across the coil, an effect called *self-inductance*.

Self-inductance is a measure of a coil's ability to establish an induced voltage as a result of a change in its current. Self-inductance is commonly referred to as simply *inductance* and is symbolized by L. The unit of inductance is the henry (H), named after the American physicist Joseph Henry (1797–1878). The schematic symbol for an inductor is shown in Figure 14–15.

FIGURE 14–15 *Standard symbol for inductor.*

Lenz's Law

The direction of induced current in a coil is such that it opposes the change in the magnetic field that produced it. This statement is Lenz's law. The effect was discovered by Heinrich F. E. Lenz, a German-born scientist working in Russia.

A change in the coil's current causes a change in the magnetic field, and the resulting induced voltage opposes that change in current. We can think of the induced voltage as producing an induced current in a direction such that *it opposes the change* in the original current. This concept of self-inductance is illustrated in Figure 14–16.

Figure 14–16A shows a current I_1 in the coil. As long as the current does not change, there is *no* induced voltage because the magnetic field does not change. At the instant the switch is closed, in Part B, the coil current tries to increase immediately to $I_1 + I_2$. This change creates an induced voltage, which produces a

SELF-INDUCTANCE

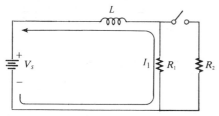

A. Initially, a constant I_1 flows. There is no induced voltage.

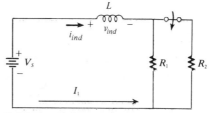

B. Switch closure attempts to increase coil current. v_{ind} is in a direction to oppose this *increase*.

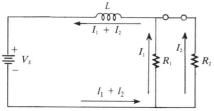

C. After current reaches constant value of $I_1 + I_2$, there is no v_{ind}.

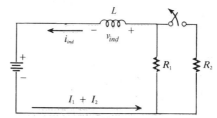

D. Switch opening attempts to decrease coil current. v_{ind} is in a direction to oppose this *decrease*.

FIGURE 14-16

current to oppose, or subtract from, the increase. After a time, the self-induced voltage dies out, and the coil current reaches a constant value of $I_1 + I_2$, as shown in Part C. In Part D, the switch is opened and the current tries to decrease back to its original I_1 value. A voltage and a current are induced in a direction to aid the existing current and thus oppose its decrease. After a time, the induced voltage dies out, and the current returns to the I_1 value, as shown in Part A.

Induced Voltage Depends on L and di/dt

The self-inductance of a coil is symbolized by L, and di/dt is the time rate of change of the current. A current change causes a change in the magnetic field, which in turn induces a voltage across the coil. The induced voltage is directly proportional to L and to di/dt:

$$v_{ind} = L \frac{di}{dt} \tag{14-6}$$

This equation indicates that the greater the inductance L is, the larger the induced voltage is. The formula also indicates that the faster the coil current changes (larger di/dt), the larger the induced voltage is. This formula is a particularly important one. Notice its similarity to Equation (11-16) for current in a capacitor:

$$i = C \frac{dv}{dt}$$

Example 14-4

How much voltage is induced across an inductor of 1 henry (1 H) when the current changes at a rate of 2 A/s?

Solution:

$$v_{ind} = L\frac{di}{dt} = (1\text{ H})(2\text{ A/s}) = 2\text{ V}$$

Review for 14-7

1. Define *self-inductance*.
2. What does Lenz's law state?
3. A nonchanging current through a coil induces a voltage across the coil (T or F).
4. What is v_{ind} when $L = 10$ H and $di/dt = 5$ A/s?

14-8

CHARACTERISTICS OF INDUCTORS

As mentioned previously, the unit of inductance is the henry, symbolized by H. Practical units commonly found in electronic circuit applications are millihenries (mH) and microhenries (μH).

Core Material

As discussed earlier, an inductor is basically a coil of wire. The material around which the coil is formed is called the *core*. If the coil is formed on a hollow tube of a material, such as plastic or paper, the core is air, as illustrated in Figure 14-17A. If a ferromagnetic material, such as iron, nickel, steel, cobalt, or an alloy of these metals, is used as the core, the inductance L is increased tremendously above that for air. A ferromagnetic core provides a better path for the magnetic lines of force and therefore allows a stronger magnetic field. A coil with a ferromagnetic core is shown in Figure 14-17B.

As you have learned, the *permeability,* symbolized by μ, of the core material is a measure of how easily a magnetic field can be established in the material. The permeability of free space or air is $4\pi \times 10^{-7}$, designated μ_0. The ratio of the permeability of a material to that of free space is called the *relative permeability,* μ_r.

Other important core characteristics are the *length* and the *cross-sectional area.*

CHARACTERISTICS OF INDUCTORS

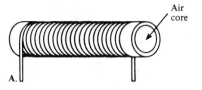
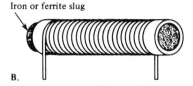

A. B.

FIGURE 14-17

Physical Factors of Inductance

Inductance depends on the square of the number of turns in the coil, N^2, and on the following physical characteristics of the core: permeability, area, and length. The relationship is

$$L = \frac{N^2 \mu A}{l} \tag{14-7}$$

where L is in henries, N is the number of turns, μ is the permeability, A is the cross-sectional area in meters squared, and l is the length of the core in meters.

Example 14-5

Determine the inductance of the coil shown in Figure 14-18. The permeability of the core is 0.25×10^{-3}.

FIGURE 14-18

Solution:

$$L = \frac{N^2 \mu A}{l} = \frac{(16)(0.25 \times 10^{-3})(0.1)}{0.01}$$

$$= 40 \text{ mH}$$

Winding Resistance (R_W)

If a coil is made from insulated copper wire, that wire has a certain *resistance* per unit length. When many turns of wire are used to construct a coil, the total resistance, sometimes called the *dc resistance* or *winding resistance,* of the wire may be significant. The resistance effectively appears in series with the inductance of the coil, as shown in Figure 14-19. In many applications the winding resistance can be

FIGURE 14–19 *Equivalent circuit diagram for a coil and its winding resistance.*

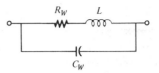

FIGURE 14–20 *Complete equivalent circuit for a nonideal inductor.*

ignored, and only the inductance is considered. In other cases the resistance must be taken into account.

Winding Capacitance (C_W)

When two conductors are placed side by side, there is always some capacitance between them. Thus, when many turns of a wire are placed close together in a coil, a certain amount of *stray* or *unwanted* capacitance (C_w) is a natural side effect. In many applications this stray capacitance is very small and has no significant effect. In other cases, particularly at higher frequencies, it may become bothersome. In some situations it can be used to advantage.

The equivalent circuit for an inductor with both its winding resistance (R_w) and its winding capacitance (C_w) is shown in Figure 14–20. The capacitance effectively acts in parallel.

Review for 14–8

1. List the physical factors that contribute to the inductance of a coil.
2. Describe what happens to L if
   (a) N is increased.
   (b) the length of the core is increased.
   (c) the area of the core is decreased.
   (d) a ferromagnetic core is replaced by an air core.
3. Why do all coils have some winding resistance?

14–9

TYPES OF INDUCTORS

Inductors are made in a variety of shapes and sizes. Basically, they fall into two general categories: fixed and variable. The standard symbols are shown in Figure 14–21.

A. Fixed   B. Variable

FIGURE 14–21 *Symbols for fixed and variable inductors.*

SERIES INDUCTANCE

Inductors are also classed according the type of core. The three classifications are air core, iron core, and ferrite core. Each has a standard symbol, as shown in Figure 14–22.

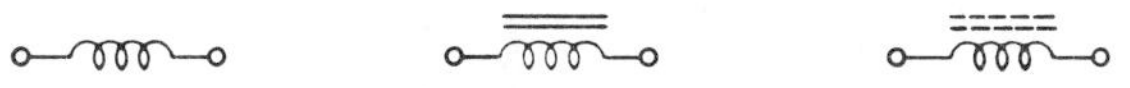

A. Air core   B. Iron core   C. Ferrite core

FIGURE 14–22 *Inductor symbols.*

Adjustable or variable inductors usually have a screw adjustment that moves a sliding core (slug) in and out. A wide variety of sizes and shapes of inductors exists, and some typical types are pictured in Figure 14–23.

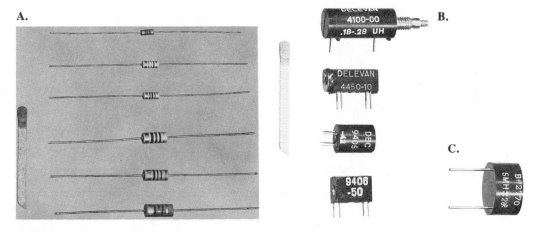

FIGURE 14–23 *Typical inductors.* **A.** *Fixed molded inductors.* **B.** *Variable coils.* **C.** *Toroid inductor.* (**A-C**, *courtesy of Delevan*)

Review for 14–9

1. Name two general categories of inductors.
2. Name three core classifications.

14–10 SERIES INDUCTANCE

When inductors are connected in series, as in Figure 14–24, *the total inductance L_T is the sum of the individual inductances*. The formula for L_T is expressed in the following equation for the general case of n inductors:

$$L_T = L_1 + L_2 + L_3 + \cdots + L_n \qquad (14\text{–}8)$$

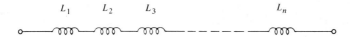

FIGURE 14–24 *Inductors in series.*

As you can see, inductance in series is similar to resistance in series.

Example 14–6

Determine the total inductance for each of the series connections in Figure 14–25.

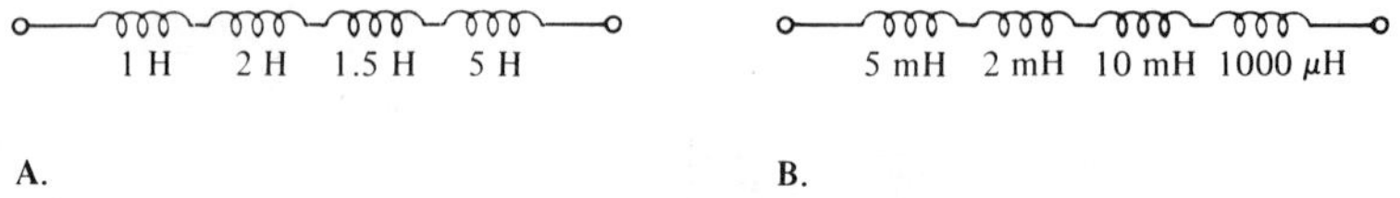

A. 1 H 2 H 1.5 H 5 H

B. 5 mH 2 mH 10 mH 1000 μH

FIGURE 14–25

Solution:

Part A: $L_T = 1\text{ H} + 2\text{ H} + 1.5\text{ H} + 5\text{ H} = 9.5\text{ H}$

Part B: $L_T = 5\text{ mH} + 2\text{ mH} + 10\text{ mH} + 1\text{ mH} = 18\text{ mH}$

(1000 μH = 1 mH)

Review for 14–10

1. What is the rule for total inductance in series?
2. What is L_T for a 100-μH, a 500-μH, and a 2-mH series connection?

14–11

PARALLEL INDUCTANCE

When inductors are connected in parallel, as in Figure 14–26, *the total inductance is less than the smallest inductance.* Total inductance is calculated in a way similar to that for R_T for parallel resistors and that for C_T for series capacitors:

$$\frac{1}{L_T} = \frac{1}{L_1} + \frac{1}{L_2} + \frac{1}{L_3} + \cdots + \frac{1}{L_n} \tag{14–9}$$

This general formula states that the reciprocal of the total is equal to the sum of the individual reciprocals. You can find L_T by taking the reciprocal of both sides of Equation (14–9) as follows:

PARALLEL INDUCTANCE

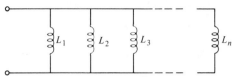

FIGURE 14–26 *Inductors in parallel.*

$$L_T = \frac{1}{(1/L_1) + (1/L_2) + (1/L_3) + \cdots + (1/L_n)} \qquad (14\text{–}10)$$

Special Case of Two Inductors in Parallel

When only two inductors are in parallel, a special "product over the sum" formula can be used. It is derived from Equation (14–9) and is stated as follows:

$$L_T = \frac{L_1 L_2}{L_1 + L_2} \qquad (14\text{–}11)$$

Equal-Value Inductors in Parallel

This is another special case in which a shortcut formula can be used. The formula is also derived from the general equation (14–9) and is stated as follows for any number, n, of equal-value inductors in parallel:

$$L_T = \frac{L}{n} \qquad (14\text{–}12)$$

Example 14–7

Determine L_T in Figure 14–27.

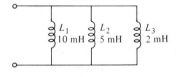

FIGURE 14–27

Solution:

Using Equation (14–10), we get

$$L_T = \frac{1}{(1/L_1) + (1/L_2) + (1/L_3)} = \frac{1}{(1/10 \text{ mH}) + (1/5 \text{ mH}) + (1/2 \text{ mH})}$$

$$= 1.25 \text{ mH}$$

Example 14-8

Find L_T for both circuits in Figure 14-28.

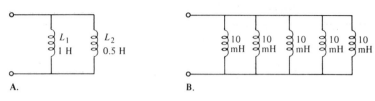

FIGURE 14-28

Solution:

Part A: Using the special formula for two inductors, we obtain

$$L_T = \frac{L_1 L_2}{L_1 + L_2} = \frac{(1 \text{ H})(0.5 \text{ H})}{1.5 \text{ H}}$$

$$= 0.33 \text{ H}$$

Part B: Using the special formula for equal inductors, we obtain

$$L_T = \frac{L}{n} = \frac{10 \text{ mH}}{5}$$

$$= 2 \text{ mH}$$

Review for 14-11

1. What is the relationship of the total inductance to the smallest inductor in parallel?
2. The calculation of parallel inductance is similar to the calculation for parallel resistance (T or F).
3. Determine L_T for each of the following parallel combinations:
   (a) A 100-mH, a 50-mH, and a 10-mH coil
   (b) A 40-μH and a 60-μH coil
   (c) Ten 1-H coils

14-12

RELATIONSHIP OF INDUCTIVE CURRENT AND VOLTAGE

As stated in Section 14-7, the induced voltage is $v_{ind} = L(di/dt)$. From this formula, you can see that the faster the current through an inductor changes, the greater the induced voltage will be. For example, if the rate of change of current is zero, the voltage is zero [$v_{ind} = L(di/dt) = L(0) = 0$].

RELATIONSHIP OF INDUCTIVE CURRENT AND VOLTAGE

Phase Relationship

Consider what happens when a sine wave of current flows through an inductor, as shown in Figure 14–29. The current wave form, one cycle of which is shown in Figure 14–30A, has a maximum rate of change (di/dt = max.) at the zero crossings. It has a zero rate of change (di/dt = 0) at the peaks.

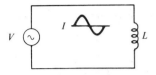

FIGURE 14–29 *Sine wave current in an inductor.*

Using the basic voltage formula of Equation (14–6), we can establish the phase relationship between the current and voltage of an inductor. When di/dt = 0, v_{ind} is also zero, because $v_{ind} = L(di/dt) = L(0) = 0$. When di/dt is a positive-going maximum, v_{ind} has a positive maximum value which is the peak. When di/dt is a negative-going maximum, v_{ind} has a negative maximum value (peak). *A sine wave current results in a cosine wave voltage across the inductor.* Therefore, we can plot the voltage in relation to the current by knowing the points at which it is zero and maximum. This plot is shown in Figure 14–30. Notice that *the current lags the voltage by 90°,* as is always true for a purely inductive circuit.

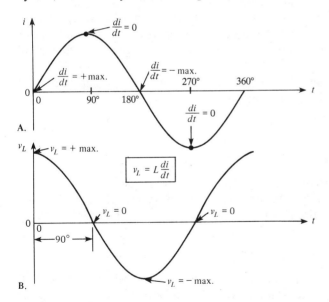

FIGURE 14–30 *Phase relation of v_{ind} and i in an inductor.*

Review for 14–12

1. Write the formula for the induced voltage in an inductor.
2. What is the phase relationship between current and voltage for an inductor?

14-13

INDUCTIVE REACTANCE

Inductive reactance (X_L) is the opposition to sinusoidal current, and it is expressed in ohms. X_L is *directly* dependent on inductance and frequency.

To develop a formula for X_L, we again use the formula $v_{ind} = L(di/dt)$ and also the curves in Figure 14-31.

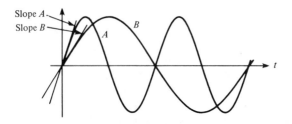

FIGURE 14-31

The rate of change of current is directly related to frequency. That is, the faster the current changes as it goes through zero, the higher the frequency is. For example, in Figure 14-31, you can see that the slope of sine wave A at the zero crossing is greater than that of sine wave B. Recall that the slope of a curve at a point indicates the rate of change at that point. Since sine wave A has a higher frequency than B, it also has a greater di/dt at the zero crossings. Its average rate of change, of course, is also greater.

If frequency increases, di/dt increases; and thus v_{ind} increases. If frequency decreases, di/dt decreases; and thus v_{ind} decreases:

$$\overset{\uparrow}{v_{ind}} = L \overset{\uparrow}{\frac{di}{dt}} \qquad \underset{\downarrow}{v_{ind}} = L \underset{\downarrow}{\frac{di}{dt}}$$

The induced voltage is therefore directly dependent on frequency.

An increase in induced voltage means more opposition (X_L is greater). Thus, X_L is directly proportional to induced voltage and, therefore, directly proportional to frequency:

X_L is proportional to f.

Now, if di/dt is constant and the inductance is varied, an increase in L produces an increase in v_{ind}, and a decrease in L produces a decrease in v_{ind}, as indicated below:

$$\overset{\uparrow}{v_{ind}} = \overset{\uparrow}{L} \frac{di}{dt} \qquad \underset{\downarrow}{v_{ind}} = \underset{\downarrow}{L} \frac{di}{dt}$$

INDUCTIVE REACTANCE

Again, an increase in v_{ind} means more opposition (greater X_L). Thus, X_L is directly proportional to induced voltage and, therefore, directly proportional to inductance in addition to frequency:

$$X_L \text{ is proportional to } fL.$$

The complete formula for X_L is given in the following equation:

$$X_L = 2\pi fL \qquad (14\text{-}13)$$

Notice that 2π appears as a constant multipler. It comes from the relationship of a sine wave to rotational motion. In this equation, X_L is in ohms if f is in hertz and L is in henries.

Example 14-9

A sine wave is applied to an inductor as shown in Figure 14-32. The frequency is 1 kHz. Determine X_L.

FIGURE 14-32

Solution:

$$1 \text{ kHz} = 1 \times 10^3 \text{ Hz} \quad \text{and} \quad 5 \text{ mH} = 5 \times 10^{-3} \text{ H}$$

$$X_L = 2\pi fL = 2\pi(1 \times 10^3 \text{ Hz})(5 \times 10^{-3} \text{ H})$$

$$= 31.4 \ \Omega$$

Ohm's Law Applies to X_L

Ohm's law applies to inductive reactance as well as to resistance and capacitive reactance, as stated in the following formulas:

$$V = IX_L \qquad (14\text{-}14)$$

$$I = \frac{V}{X_L} \qquad (14\text{-}15)$$

$$X_L = \frac{V}{I} \qquad (14\text{-}16)$$

Current and voltage must be expressed in consistent units of rms, peak, peak-to-peak, or average.

Example 14-10

Determine the rms current in Figure 14-33.

FIGURE 14-33

Solution:

$$10 \text{ kHz} = 10 \times 10^3 \text{ Hz} \quad \text{and} \quad 100 \text{ mH} = 100 \times 10^{-3} \text{ H}$$

First calculate X_L:

$$X_L = 2\pi f L = 2\pi(10 \times 10^3 \text{ Hz})(100 \times 10^{-3} \text{ H})$$

$$= 6283 \text{ }\Omega$$

Using Ohm's law, we obtain

$$I_{rms} = \frac{V_{rms}}{X_L} = \frac{5 \text{ V}}{6283 \text{ }\Omega}$$

$$= 795.8 \text{ }\mu\text{A}$$

Chokes

An inductor is often referred to as a *choke*. This term means that at a sufficiently high frequency, the coil has a very high X_L. Practically, it appears as an open circuit to the high frequencies, effectively "choking" them off and preventing them from passing from one point to another. Radio frequency (rf) chokes are commonly used to filter or block high-frequency signals in a circuit. For example, rf chokes, in conjunction with other components, are used in the tuner of a TV receiver to prevent signals from other channels from interfering with the signal from the station to which the set is tuned.

Review for 14-13

1. Define *inductive reactance*.
2. Calculate X_L for $f = 5$ kHz and $L = 500$ mH.
3. To get a reactance of 2 kΩ from a 1-H coil, what f is required?

INDUCTORS IN dc CIRCUITS

4. Calculate the current in Figure 14–34.

FIGURE 14–34

14–14

INDUCTORS IN dc CIRCUITS

An inductor presents no opposition to a *nonchanging* current other than that of its winding resistance. That is, X_L *is zero for dc.* An equivalent circuit for a dc source connected to an inductor is shown in Figure 14–35.

FIGURE 14–35 *Direct coil current.*

Example 14–11

How much current is there in a coil with a 5-Ω winding resistance if 5 V dc are connected across it?

Solution:

$$I = \frac{V}{R_W} = \frac{5 \text{ V}}{5 \text{ Ω}} = 1 \text{ A}$$

Review for 14–14

1. There are 2 amperes of dc through a coil with 1 V across it. What is its dc resistance?

14-15

ENERGY AND POWER IN AN INDUCTOR

Energy

An inductor stores energy in its magnetic field. It can be shown that the energy stored by an inductor is

$$\mathscr{E} = \tfrac{1}{2}LI^2 \qquad (14\text{-}17)$$

where I is the rms current through the inductor and, if L is in henries and I in amperes, the energy is in joules.

Power

An ideal inductor does not dissipate energy; it only stores it. When an ac voltage is applied to an inductor, energy is stored by the inductor during a portion of the voltage cycle and then is returned to the source during another portion. *There is no energy loss in the ideal inductance.* This situation is similar to that for an ideal capacitor.

Since a practical inductor has a certain winding resistance, there is an average power:

$$P_{avg} = I_{rms}^2 R_W \qquad (14\text{-}18)$$

Reactive Power

The rate at which an inductor stores or returns energy is called its *reactive power*, P_r. The following equations express the reactive power in an inductor:

$$P_r = V_{rms} I_{rms} \qquad (14\text{-}19)$$

$$P_r = \frac{V_{rms}^2}{X_L} \qquad (14\text{-}20)$$

$$P_r = I_{rms}^2 X_L \qquad (14\text{-}21)$$

Example 14-12

A 10-V rms signal with a frequency of 1 kHz is applied to a 10-mH coil with a winding resistance of 5 Ω. Determine the average power (P_{avg}) and the reactive power (P_r).

Solution:

$$X_L = 2\pi f L = 2\pi(1\text{ kHz})(10\text{ mH})$$
$$= 62.8 \text{ Ω}$$

TESTING AN INDUCTOR

$$Z = \sqrt{R^2 + X_L^2} = \sqrt{(5\ \Omega)^2 + (62.8\ \Omega)^2}$$
$$= 63\ \Omega$$
$$I = \frac{V_s}{Z} = \frac{10\ V}{63\ \Omega}$$
$$= 0.159\ A$$
$$P_{avg} = I^2 R_W = (0.159\ A)^2(5\ \Omega)$$
$$= 0.126\ W$$
$$P_r = I^2 X_L = (0.159\ A)^2(62.8\ \Omega)$$
$$= 1.59\ W$$

Review for 14-15

1. How much energy is stored by a 100-mH coil with 1 A?
2. If a coil has no winding resistance, what is the average power?

14-16

TESTING AN INDUCTOR

The most common failure of an inductor is an open. To check for an open, you should remove the coil from the circuit. If there is an open, an ohmmeter check will indicate infinite resistance, as shown in Figure 14-36A.

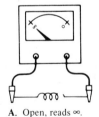
A. Open, reads ∞.

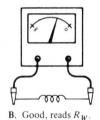
B. Good, reads R_W.

FIGURE 14-36 *Checking a coil by measuring the resistance.*

If the coil is good, the ohmmeter will show the winding resistance, R_W. The value of the winding resistance will depend on the type of coil. It can be anywhere from one ohm to several hundred ohms. This value depends on the number of turns and the size of the wire used.

Review for 14-16

1. When a coil is checked, a reading of infinity on an ohmmeter indicates a partial short (T or F).

2. An ohmmeter check of a good coil will indicate the value of the inductance (T or F).

Formulas

$$B = \frac{\phi}{A} \qquad (14\text{-}1)$$

$$F_m = NI \qquad (14\text{-}2)$$

$$\mathcal{R} = \frac{F_m}{\phi} \qquad (14\text{-}3)$$

$$\mu = \frac{l}{\mathcal{R}A} \qquad (14\text{-}4)$$

$$v_{ind} = N\frac{d\phi}{dt} \qquad (14\text{-}5)$$

$$v_{ind} = L\frac{di}{dt} \qquad (14\text{-}6)$$

$$L = \frac{N^2\mu A}{l} \qquad (14\text{-}7)$$

$$L_T = L_1 + L_2 + L_3 + \cdots + L_n \qquad (14\text{-}8)$$

$$\frac{1}{L_T} = \frac{1}{L_1} + \frac{1}{L_2} + \frac{1}{L_3} + \cdots + \frac{1}{L_n} \qquad (14\text{-}9)$$

$$L_T = \frac{1}{(1/L_1) + (1/L_2) + (1/L_3) + \cdots + (1/L_n)} \qquad (14\text{-}10)$$

$$L_T = \frac{L_1 L_2}{L_1 + L_2} \qquad (14\text{-}11)$$

$$L_T = \frac{L}{n} \qquad (14\text{-}12)$$

$$X_L = 2\pi f L \qquad (14\text{-}13)$$

$$V = IX_L \qquad (14\text{-}14)$$

$$I = \frac{V}{X_L} \qquad (14\text{-}15)$$

$$X_L = \frac{V}{I} \qquad (14\text{-}16)$$

$$\mathcal{E} = \frac{1}{2}LI^2 \qquad (14\text{-}17)$$

$$P_{avg} = I_{rms}^2 R_W \qquad (14\text{-}18)$$

$$P_r = V_{rms} I_{rms} \qquad (14\text{-}19)$$

$$P_r = \frac{V_{rms}^2}{X_L} \qquad (14\text{--}20)$$

$$P_r = I_{rms}^2 X_L \qquad (14\text{--}21)$$

Summary

1. A magnetic field exists around a conductor when there is current through it.
2. An inductor is a coil of wire formed around a core material.
3. Self-inductance is a measure of a coil's ability to establish an induced voltage as a result of a change in its current.
4. An inductor opposes a change in its current.
5. Faraday's law states that relative motion between a magnetic field and a coil produces a voltage across the coil.
6. The amount of induced voltage equals the inductance times the rate of change of current.
7. Lenz's law states that the polarity of induced voltage is such that the resulting induced current is in a direction which opposes the change in the magnetic field that produced it.
8. The henry is the unit of inductance; its symbol is H.
9. The permeability (μ) of a core material is an indication of the ability of the material to establish a magnetic field.
10. The dc resistance is the resistance of the coil windings.
11. Stray capacitance is due to the adjacent conductors (turns) in a coil.
12. Inductance adds when inductors are in series.
13. Total parallel inductance is less than the smallest inductance.
14. Voltage leads current by 90° in an inductor. That is, current lags voltage by 90°.
15. Inductive reactance (X_L) is the opposition to sinusoidal current. It is expressed in ohms.
16. X_L is directly proportional to frequency and inductance.
17. Ohm's law applies to inductive reactance in the same way that it applies to resistance and capacitive reactance.

Self-Test

1. What is the flux density in webers per square meter if the flux in one square millimeter is 1200 μWb?
2. If there is 100 mA through a wire with 50 turns, what is the mmf?
3. A coil of wire with 50 turns is placed in a magnetic field that is changing at a rate of 25 Wb/s. What is the induced voltage?

4. Convert 1500 µH to millihenries. Convert 20 mH to microhenries.
5. The current through a 100-mH coil is changing at a rate of 200 mA/s. How much voltage is induced across the coil?
6. How many turns are required to produce 30 mH in a coil wound on a cylindrical core with a cross-sectional area of 10×10^{-5} m² and a length of 0.05 m? The core has a permeability of 1.2×10^{-6}.
7. If a 12-V dc battery is connected across a coil with a winding resistance of 12 Ω, how much current flows through the coil?
8. Three 25-µH coils and two 10-µH coils are connected in series. What is the total inductance?
9. The following coils are connected in parallel: two 1-H, two 0.5-H, and one 2-H. What is L_T?
10. A 2-MHz sine wave voltage is applied across a 15-µH coil. What is the frequency of the resulting current? What is the reactance?
11. The frequency of the source in Figure 14–37 is variable. To get 50 mA of rms current, to what frequency must you adjust the source?

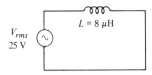

FIGURE 14–37

12. What inductance is required for an rf choke to produce an X_L of 10 MΩ at 50 MHz?

Problems

14–1. How many webers do one thousand lines of force in a magnetic field equal?

14–2. What is the mmf in 10 turns of wire when 25 mA flow through it?

14–3. How many turns of wire are required to produce an mmf of 1000 At if the current is 2.5 A?

14–4. What is the mmf necessary to produce a flux of 500 µWb if the reluctance of a given magnetic material is 3000 At/Wb?

14–5. Convert the following to millihenries:
   (a) 1 H   (b) 250 µH
   (c) 10 µH   (d) 0.0005 H

14–6. Convert the following to microhenries:
   (a) 300 mH   (b) 0.08 H
   (c) 5 mH   (d) 0.00045 mH

14–7. What is the voltage across a coil when $di/dt = 10$ mA/µs and $L = 5$ µH?

PROBLEMS

14–8. Fifty volts are induced across a coil of 25 mH. At what rate is the current changing?

14–9. A core has the following characteristics: $A = 0.005 \text{ m}^2$, $l = 0.02 \text{ m}$, and $\mu = 50 \times 10^{-6}$. How many turns are needed to produce 100 mH?

14–10. Five inductors are connected in series. The lowest value is 5 μH. If the value of each inductor is twice that of the preceding one, and if the inductors are connected in order of ascending values, what is the total inductance?

14–11. Suppose that you require a total inductance of 50 mH. You have available a 10-mH coil and a 22-mH coil. How much additional inductance do you need?

14–12. Determine the total parallel inductance for the following coils in parallel: 75 μH, 50 μH, 25 μH, and 15 μH.

14–13. You have a 12-mH inductor, which is your smallest value. You need an inductance of 8 mH. What value would you add in parallel with the 12 mH to obtain 8 mH?

14–14. Determine the inductance of each circuit in Figure 14–38.

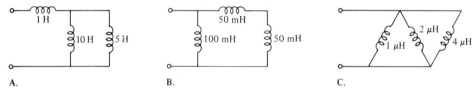

FIGURE 14–38

14–15. Determine the inductance of each circuit in Figure 14–39.

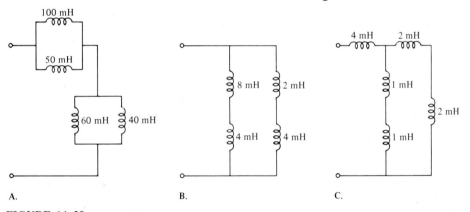

FIGURE 14–39

14–16. Find the reactance of each circuit in Figure 14–38 if a frequency of 5 kHz is applied.

14–17. Find the reactance of each circuit in Figure 14–39 if a frequency of 400 Hz is used.

14-18. Determine the total rms current in Figure 14-40. How much current is there through L_2 and L_3?

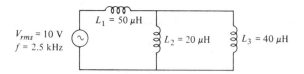

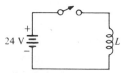

FIGURE 14-40

FIGURE 14-41

14-19. What frequency is needed to produce 500 mA total current in Figure 14-40?

14-20. What is the polarity of the induced voltage in Figure 14-41 when the switch is closed? How much final current is there if $R_W = 10\ \Omega$?

14-21. How much energy is stored by an inductor carrying 24 mA rms and having an inductance of 400 μH?

14-22. Determine the average power and the reactive power in Figure 14-42.

FIGURE 14-42

Answers to Section Reviews

Section 14-1:
1. North; south. **2.** F. **3.** All of the lines of force in a given magnetic field. **4.** 20 Wb/m².

Section 14-2:
1. T. **2.** Grasp a conductor with your left hand, with your thumb in the direction of current. Your fingers will be in the direction of the lines of force.

Section 14-3:
1. Magnetomotive force. **2.** The opposition to the establishment of a magnetic field in a magnetic material. **3.** The ease with which a magnetic field can be established in the material.

Section 14-4:
1. 0 V. **2.** It increases. **3.** A force is exerted upon it.

Section 14-5:
1. T. **2.** A magnetic field created by a current through a wire. **3.** A coil of wire. **4.** T.

Section 14-6:
1. The voltage created across a coil when the current changes. **2.** Number of turns, N, and rate of change of field, $d\phi/dt$. **3.** Increase.

ANSWERS TO SECTION REVIEWS

Section 14–7:
1. A measure of the ability of a coil to establish an induced voltage. **2.** The polarity of the induced voltage is in a direction that opposes the change in the current that produced it. **3.** F. **4.** 50 V.

Section 14–8:
1. Number of turns, area of core, length of core, and permeability of core. **2. (a)** Increases **(b)** Decreases **(c)** Decreases **(d)** Decreases. **3.** Because wire itself has resistance.

Section 14–9:
1. Fixed and variable. **2.** Air, iron, and ferrite.

Section 14–10:
1. Inductances add in series. **2.** 2600 μH.

Section 14–11:
1. Total L is smaller than smallest L. **2.** T. **3. (a)** 7.69 mH **(b)** 24 μH **(c)** 0.1 H.

Section 14–12:
1. $v = L(di/dt)$. **2.** Current lags by 90°.

Section 14–13:
1. Opposition to sinusoidal current in an inductor. **2.** 15.7 kΩ. **3.** 318.31 Hz. **4.** 15.92 mA.

Section 14–14:
1. 0.5 Ω.

Section 14–15:
1. 50 mJ. **2.** 0 W.

Section 14–16:
1. F. **2.** F.

An RL circuit, the second type of reactive circuit that we will study, contains both resistance and inductance. In this chapter we discuss the *basic series RL circuit*. More complex combinations of RL circuits can be converted to the basic series form. The *parallel RL circuit* is discussed in the last section.

Both series and parallel circuits are important in electronics. Applications include filters, oscillators, and wave shaping.

15–1 Sine Wave Response
15–2 Impedance of a Series RL Circuit
15–3 Phase Angle
15–4 Voltage and Current Magnitudes
15–5 The RL Lag Network
15–6 The RL Lead Network
15–7 The Series RL Circuit as a Filter
15–8 Power in an RL Circuit
15–9 The Parallel RL Circuit

15
FREQUENCY RESPONSE OF RL CIRCUITS

15-1

SINE WAVE RESPONSE

When a sine wave of voltage is applied to an RL circuit, *each resulting voltage drop and current in the circuit will be a sine wave with the same frequency as that of the applied voltage.* (Recall that this was also true for the RC circuit.) The amplitudes and the phase relationships of the voltages and currents depend on the ohmic values of the resistances and the inductive reactances.

Review for 15-1

1. There is current in a series RL circuit as a result of an applied sine wave voltage of 1 kHz. What is the frequency of the current?

15-2

IMPEDANCE OF A SERIES RL CIRCUIT

Impedance is the total opposition to sinusoidal current and is expressed in ohms. The impedance of an RL circuit is determined by both the resistance and the inductive reactance. The impedance magnitude is symbolized by Z, just as in the RC circuit. It is equal to the phasor sum of R and X_L, as stated in the following equation:

$$Z = \sqrt{R^2 + X_L^2} \tag{15-1}$$

Notice that this formula is similar to that for Z of a series RC circuit. The same derivation that was presented in Chapter 12 is true for the series RL circuit because of the 90° phase difference between the inductor voltage and current. Since in a series circuit the same current exists in both the resistor and the inductor, *the resistor voltage is in phase with the current, and the inductor voltage is leading the current by 90°.* Thus, there is a phase difference of 90° between the resistor voltage, V_R, and the inductor voltage, V_L, as illustrated in Figure 15-1.

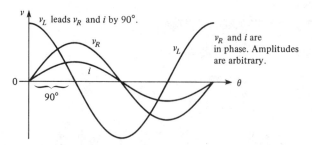

FIGURE 15-1 *Phase relation of voltages and current in a series RL circuit.*

IMPEDANCE OF A SERIES RL CIRCUIT

From Kirchhoff's voltage law, we know that the sum of the voltage drops must equal the applied voltage. However, since V_R and V_L are not in phase with each other, they cannot be added algebraically. They must be added as phasor quantities, with V_L leading V_R by 90°, as shown in Figure 15–2A. As shown in Part B of the figure, V_s is the phasor sum of V_R and V_L. Therefore, there is a *right triangle relationship* among V_s, V_R, and V_L: V_s is the hypotenuse, V_R is the adjacent side, and V_L is the opposite side, as shown in Figure 15–2C.

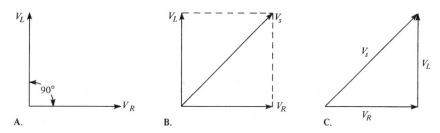

FIGURE 15–2 *Voltage phasor diagram for a series RL circuit.*

As discussed in Chapter 12, you can calculate the length of the hypotenuse of a right triangle from the other two sides by using the Pythagorean theorem. In our case, the rms values of V_R and V_L are represented by the other two sides of the right triangle. This relationship is expressed as

$$V_s = \sqrt{V_R^2 + V_L^2}$$

By Ohm's law, which also applies to impedance, $Z = V_s/I$. By substitution of $\sqrt{V_R^2 + V_L^2}$ for V_s, we get

$$Z = \sqrt{V_R^2 + V_L^2}/I = \sqrt{V_R^2 + V_L^2}/\sqrt{I^2} = \sqrt{V_R^2/I^2 + V_L^2/I^2}$$

If R^2 is now substituted for V_R^2/I^2 and X_L for V_L^2/I^2, we get

$$Z = \sqrt{R^2 + X_L^2}$$

This result proves Equation (15–1). The impedance triangle is shown in Figure 15–3, where Z is the hypotenuse, R is the adjacent side, and X_L is the opposite side.

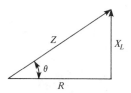

FIGURE 15–3 *The impedance triangle.*

Remember from Chapter 14 that the reactance of an inductor is *directly proportional* to the frequency of the applied signal. From Equation (15–1), you can see that Z varies directly with X_L. Thus, the impedance of a series RL circuit is directly dependent on frequency:

$$\overset{\uparrow}{Z} = \sqrt{R^2 + \overset{\uparrow}{X_L^2}} \qquad \underset{\downarrow}{Z} = \sqrt{R^2 + \underset{\downarrow}{X_L^2}}$$

The change in Z with frequency is shown in Figure 15–4, which illustrates impedance triangles for different values of frequency.

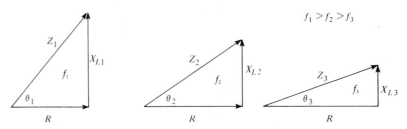

FIGURE 15–4 *Impedance varies directly with frequency.*

Example 15–1

For the series RL circuit in Figure 15–5, determine the total impedance for input frequency values of 10 kHz, 20 kHz, and 30 kHz.

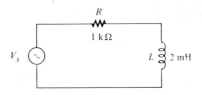

FIGURE 15–5

Solution:

For $f = 10$ kHz:

$X_L = 2\pi fL = 2\pi(10 \text{ kHz})(2 \text{ mH})$
$ = 125.66 \text{ }\Omega$
$Z = \sqrt{R^2 + X_L^2} = \sqrt{(1000 \text{ }\Omega)^2 + (125.66 \text{ }\Omega)^2}$
$ = 1007.86 \text{ }\Omega$

For $f = 20$ kHz:

$X_L = 2\pi fL = 2\pi(20 \text{ kHz})(2 \text{ mH})$
$ = 251.33 \text{ }\Omega$
$Z = \sqrt{R^2 + X_L^2} = \sqrt{(1000 \text{ }\Omega)^2 + (251.33 \text{ }\Omega)^2}$
$ = 1031.1 \text{ }\Omega$

PHASE ANGLE

For $f = 30$ kHz:
$$X_L = 2\pi fL = 2\pi(30 \text{ kHz})(2 \text{ mH})$$
$$= 376.99 \text{ }\Omega$$
$$Z = \sqrt{R^2 + X_L^2} = \sqrt{(1000 \text{ }\Omega)^2 + (376.99 \text{ }\Omega)^2}$$
$$= 1068.70 \text{ }\Omega$$

Notice that as the frequency increases, X_L and Z increase.

Review for 15–2

1. Does the applied voltage lead or lag the current in an RL circuit?
2. How many degrees of phase difference are there between the inductor voltage and the resistor voltage in a series RL circuit?
3. Sketch the impedance triangle for $R = X_L$.
4. If the frequency of the applied voltage in an RL circuit increases, what happens to the inductive reactance? To the impedance?
5. A series RL circuit has $R = 1$ kΩ and $X_L = 1$ kΩ. What is the impedance of the circuit?

15–3

PHASE ANGLE

In a series RL circuit, when X_L changes, both impedance and phase angle also change. As you know, the phase angle of any circuit is defined to be the angle between the total voltage and the total current, as illustrated in Figure 15–6A. In the corresponding impedance triangle in Part B, the phase angle θ appears between R and Z. Notice that for the series RL circuit, the phase angle is above the horizontal, indicating that the total voltage leads the current. This situation is opposite that of the series RC circuit, in which the current leads the voltage.

A.

B.

FIGURE 15–6 *Phase angle of an RL circuit.*

θ Varies with Frequency

Figure 15–7 shows how θ changes with frequency. If the frequency is decreased, X_L also decreases, resulting in a smaller phase angle. If the frequency is increased, X_L increases, causing a larger phase angle. This relationship is logical because the more reactive the circuit is, the more phase difference there is between the voltage and current. When the reactance is very much larger than the resistance, θ is close to 90°, because the circuit is almost purely inductive.

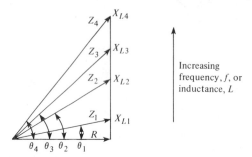

FIGURE 15–7 *Effect of frequency and inductance on phase angle.*

θ Varies with Inductance

For a fixed frequency, θ will change with the inductance value. If L is decreased, X_L decreases, resulting in a smaller phase angle. If L is increased, X_L increases, causing a larger phase angle. Figure 15–7 illustrates this relationship also.

θ Varies with Resistance

In a series RL circuit, if inductance and frequency are constant and the resistance is varied, the phase angle will also change. This relation is illustrated in Figure 15–8, in which a larger resistance produces a smaller phase angle, and a smaller resistance results in a larger phase angle. The effect of resistance on phase angle for the RL circuit is the same as that for the RC circuit, because Z varies directly with R in both cases.

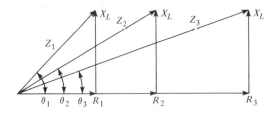

FIGURE 15–8 *Effect of resistance on phase angle.*

Calculation of Phase Angle

If the frequency, inductance, and resistance of a series RL circuit are known, the phase angle can be determined from the following equation:

PHASE ANGLE

$$\theta = \arctan\left(\frac{X_L}{R}\right) \qquad (15\text{–}2)$$

This equation is the same type that was used for the phase angle in a series RC circuit. Examples 15–2, 15–3, and 15–4 will clarify the use of this formula.

Example 15–2

In a certain series RL circuit, $R = 100\ \Omega$ and $X_L = 100\ \Omega$ at a given frequency. Determine the phase angle between the applied voltage and current. Sketch the wave forms.

Solution:

$$\theta = \arctan\left(\frac{X_L}{R}\right) = \arctan\left(\frac{100\ \Omega}{100\ \Omega}\right)$$
$$= \arctan 1 = 45°$$

The wave-form diagram is shown in Figure 15–9. The voltage leads the current by 45°.

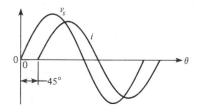

FIGURE 15–9

Example 15–3

Repeat Example 15–2 for $X_L = 50\ \Omega$.

Solution:

$$\theta = \arctan\left(\frac{50\ \Omega}{100\ \Omega}\right) = \arctan 0.5$$
$$= 26.57°$$

Notice that θ is less than 45° when X_L is less than R. A smaller phase angle corresponds to a circuit that is less reactive than resistive.

Figure 15–10 shows the wave-form diagram. The voltage leads the current by 26.57°.

Example 15–3 (continued)

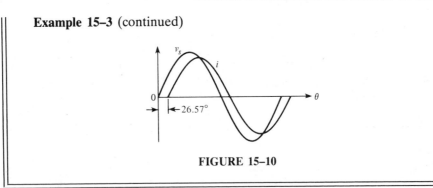

FIGURE 15–10

Example 15–4

Repeat Example 15–2 for $X_L = 150 \; \Omega$.

Solution:

$$\theta = \arctan\left(\frac{150 \; \Omega}{100 \; \Omega}\right) = \arctan 1.5$$
$$= 56.31°$$

Notice that in this case, θ is greater than 45° when X_L is greater than R. A greater phase angle corresponds to a more reactive circuit.
The wave forms are shown in Figure 15–11.

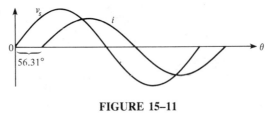

FIGURE 15–11

Review for 15–3

1. In a series RL circuit, is the current in phase or out of phase with the inductor voltage?
2. Does the current in a series RL circuit lead or lag the inductor voltage?
3. Does the current in a series RL circuit lead or lag the applied voltage?
4. If the inductive reactance in an RL circuit is increased, does the phase angle increase or decrease?
5. For a given frequency and inductance, if the resistance is increased, what happens to the phase angle?
6. When does a 45° phase angle occur?

7. Does a phase angle less than 45° mean that the reactance is greater than or less than the resistance?

15-4
VOLTAGE AND CURRENT MAGNITUDES

Voltage and current values in the series RL circuit are determined in the same way as those for the series RC circuit. The calculations are similar because both circuits act as a voltage divider, and because a certain portion of the total voltage is across the resistor and another portion across the inductor.

Remember that the phase difference between V_R and V_L requires a phasor sum rather than a direct sum of the voltages. The following are the voltage divider equations as applied to the series RL circuit:

$$V_R = \left(\frac{R}{\sqrt{R^2 + X_L^2}}\right) V_s \qquad (15\text{--}3)$$

$$V_L = \left(\frac{X_L}{\sqrt{R^2 + X_L^2}}\right) V_s \qquad (15\text{--}4)$$

The voltage drops divide in a manner directly proportional to the resistance or reactance across which they are taken. The total voltage magnitude is the *phasor sum* of V_R and V_L:

$$V_s = \sqrt{V_R^2 + V_L^2} \qquad (15\text{--}5)$$

Example 15-5

For the circuit of Figure 15–12, determine the rms values of the voltages across R and L and the current. The input is 5 V rms.

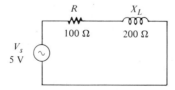

FIGURE 15–12

Solution:

First we calculate the impedance magnitude:

$$Z = \sqrt{R^2 + X_L^2} = \sqrt{(100 \ \Omega)^2 + (200 \ \Omega)^2}$$
$$= 223.6 \ \Omega$$

Example 15-5 (continued)

Now we can find the current and voltages:

$$I = \frac{V_s}{Z} = \frac{5 \text{ V}}{223.6 \text{ }\Omega}$$
$$= 22.36 \text{ mA}$$

$$V_R = \left(\frac{R}{Z}\right)V_s = \left(\frac{100 \text{ }\Omega}{223.6 \text{ }\Omega}\right)5 \text{ V}$$
$$= 2.24 \text{ V}$$

$$V_L = \left(\frac{X_L}{Z}\right)V_s = \left(\frac{200 \text{ }\Omega}{223.6 \text{ }\Omega}\right)5 \text{ V}$$
$$= 4.47 \text{ V}$$

Check: $V_s = \sqrt{V_R^2 + V_L^2} = \sqrt{(2.24 \text{ V})^2 + (4.47 \text{ V})^2} = 5 \text{ V}.$

Review for 15-4

1. Explain how the series RL circuit acts as a voltage divider.
2. If X_L is larger than R, where does the greater voltage drop occur?
3. If you are measuring the voltage across the resistor in a series RL circuit, what would you expect to see as the frequency of the source voltage is increased?
4. Calculate the rms voltages across the resistor and the inductor in Figure 15-13.

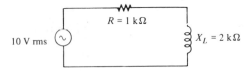

FIGURE 15-13

15-5

THE RL LAG NETWORK

Figure 15-14A shows a series RL circuit with *the output voltage taken across the resistor.* The phase angle between the current and the input voltage is also the phase angle between the resistor voltage and the input voltage, because V_R and I are in phase. Since V_L leads V_R by 90°, the phase angle between the resistor voltage and the input voltage is equal to θ, as shown in Figure 15-14B. The resistor voltage is taken as

THE RL LAG NETWORK

the output voltage, and it *lags* the input voltage by the angle ϕ. Output lags input, thus creating a *basic lag network*.

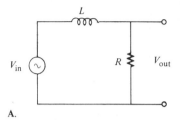
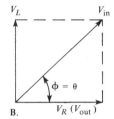

FIGURE 15–14 *RL lag network.*

If the input and output wave forms of the lag network are displayed on an oscilloscope, they will resemble the form in Figure 15–15. The amount of phase difference between the input and output depends on the relative sizes of the inductive reactance and the resistance. The magnitude of the output also depends on these relative sizes.

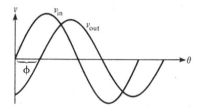

FIGURE 15–15 *Input and output voltage wave forms of an RL lag network.*

Phase Difference of Input and Output

The phase between the input voltage and the output voltage in the lag network is $\phi = \arctan(X_L/R)$, as was stated in Equation (15–2). This phase relationship is the same as the circuit phase angle between applied voltage and current because the resistor voltage (V_{out}) is in phase with the current. The phase diagam is shown in Figure 15–14B.

Example 15–6

Determine the phase angle between the input and the output voltages in the lag network of Figure 15–16.

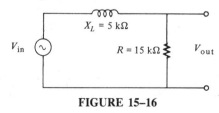

FIGURE 15–16

Example 15–6 (continued)

Solution:

$$\phi = \arctan\left(\frac{X_L}{R}\right) = \arctan\left(\frac{5 \text{ k}\Omega}{15 \text{ k}\Omega}\right)$$
$$= 18.43°$$

The output voltage (V_R) lags the input voltage by 18.43°.

Example 15–7

What is the phase angle between the input and the output for Figure 15–17?

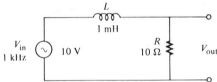

FIGURE 15–17

Solution:

$$X_L = 2\pi f L = 2\pi(1 \text{ kHz})(1 \text{ mH})$$
$$= 6.28 \text{ }\Omega$$
$$\phi = \arctan\left(\frac{X_L}{R}\right) = \arctan\left(\frac{6.28 \text{ }\Omega}{10 \text{ }\Omega}\right)$$
$$= 32.13°$$

Thus, V_{out} lags V_{in} by 32.13°.

Magnitude of Output Voltage

To determine the magnitude of the output voltage of the lag network, visualize the circuit as a voltage divider. A portion of the total input voltage is dropped across R, and a portion across X_L. The output voltage is the resistor voltage and can be calculated with Equation (15–3).

Example 15–8

For the lag network in Figure 15–17, V_{in} has an rms value of 10 V. Find the magnitude of the output voltage. Then use the value of the phase angle found in Example 15–7 to sketch the input and output wave forms in their proper relationship.

THE RL LEAD NETWORK

Solution:

$$V_{out} = \left(\frac{R}{\sqrt{R^2 + X_L^2}}\right) V_{in} = \left(\frac{10\ \Omega}{\sqrt{(10\ \Omega)^2 + (6.28\ \Omega)^2}}\right) 10\ \text{V}$$
$$= 8.47\ \text{V}$$

The wave-form relationship is shown in Figure 15–18.

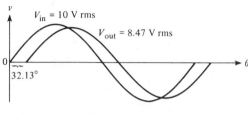

FIGURE 15–18

Review for 15–5

1. In an RL lag network, does the resistor voltage lead or lag the input?
2. In a series RL lag network, across which component is the output taken?
3. If R and X_L are equal, what is the input-to-output phase angle?
4. If the frequency of the input is increased, will the output phase angle increase or decrease?

15–6

THE RL LEAD NETWORK

If the output voltage is taken across the inductor in a series RL circuit, the circuit becomes an *RL lead network*. The output voltage leads the input in phase, as shown in Figure 15–19.

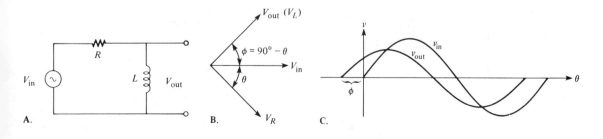

FIGURE 15–19 *RL lead network.*

Phase Difference between Input and Output

In a series RL circuit, the total current lags the applied input voltage. Also, the resistor voltage is in phase with the current, and the inductor voltage leads the current by 90°. These relationships are indicated in Figure 15–19B, and the wave-form diagrams are shown in Figure 15–19C.

The phase difference (ϕ) between the inductor voltage (output) and the input voltage is the difference between 90° and θ, as stated in the following equations:

$$\phi = 90° - \arctan\left(\frac{X_L}{R}\right) \quad (15\text{–}6)$$

or

$$\phi = \arctan\left(\frac{R}{X_L}\right) \quad (15\text{–}7)$$

The following example will illustrate calculations of the output phase angle for lead networks.

Example 15–9

Calculate the output phase angle for the RL lead network in Figure 15–20.

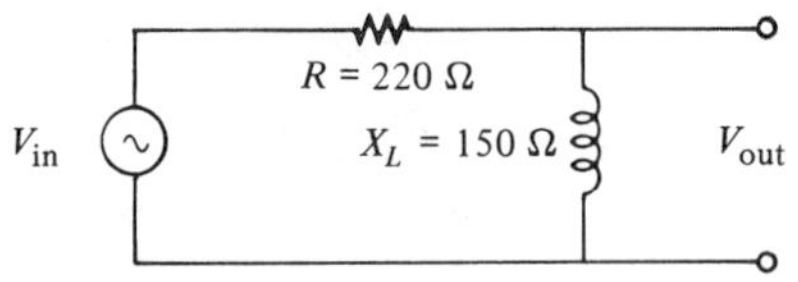

V_{in} — $R = 220\ \Omega$, $X_L = 150\ \Omega$ — V_{out}

FIGURE 15–20

Solution:

$$\phi = \arctan\left(\frac{R}{X_L}\right) = \arctan\left(\frac{220\ \Omega}{150\ \Omega}\right)$$

$$= 55.71°$$

The output leads the input by 55.71°.

Example 15–10

Find the output phase angle for the lead network in Figure 15–21.

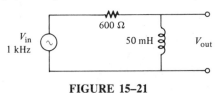

FIGURE 15–21

THE RL LEAD NETWORK

Solution:

$$X_L = 2\pi f L = 2\pi(1000 \text{ Hz})(50 \text{ mH})$$
$$= 314.16 \text{ }\Omega$$

$$\phi = \arctan\left(\frac{R}{X_L}\right) = \arctan\left(\frac{600 \text{ }\Omega}{314.16 \text{ }\Omega}\right)$$
$$= 62.36°$$

The output leads the input by 62.36°.

Magnitude of Output Voltage

The magnitude of the output voltage (V_L) can be found by use of the voltage divider formula in Equation (15–4), as Example 15–11 illustrates.

Example 15–11

The input voltage in Figure 15–21 has an rms value of 5 V. Find the rms value of the output voltage, and sketch the input and output voltage relationship in a wave-form diagram.

Solution:

$$V_{out} = \left(\frac{X_L}{\sqrt{R^2 + X_L^2}}\right) V_{in} = \left(\frac{314.16 \text{ }\Omega}{\sqrt{(600 \text{ }\Omega)^2 + (314.16 \text{ }\Omega)^2}}\right) 5 \text{ V}$$
$$= 2.32 \text{ V}$$

The peak value of the input voltage is 1.414(5 V) = 7.07 V. The peak value of the output voltage is 1.414(2.32 V) = 3.28 V.

The wave forms are shown in Figure 15–22.

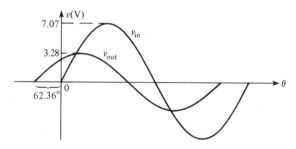

FIGURE 15–22

Review for 15-6

1. In an RL lead network, does the inductor voltage lead or lag the input?
2. In a series RL lead network, across which component is the output voltage taken?
3. If the input frequency is decreased in an RL lead network, does the output phase angle increase or decrease?
4. If the input frequency is increased in an RL lead network, does the output voltage magnitude increase or decrease?

15-7

THE SERIES RL CIRCUIT AS A FILTER

As we discussed in Chapter 12 on RC circuits, a filter is a circuit that permits certain frequencies to pass from input to output and blocks other frequencies. Now we will see how the RL circuit acts as a *simple filter*.

Low-Pass Filter

The series RL configuration that was discussed as a lag network also operates as a low-pass filter. That is, it tends to pass lower frequencies from input to output and to block higher-frequency inputs. The basic reason for this filtering action is that the inductor is in series between input and output, and its reactance increases as frequency increases. As X_L increases with frequency, more and more of the input voltage is dropped across the coil and less across the resistor. The resistor voltage is the output voltage; so it decreases with an increase in frequency.

Figure 15-23 is an example of the filtering action of an RL lag network. In Part A, the input is *zero* frequency (dc). Neglecting R_w, the coil looks essentially like a short to dc; so *all* of the input voltage appears on the output (the coil drops no voltage and the resistor drops all of it).

In Figure 15-23B, the frequency of the input has been increased to 1 kHz, causing X_L to increase to 62.83 Ω. For this case, an input voltage of 10 V rms causes an output voltage of approximately 8.47 V, which we calculate using the voltage divider formula.

In Figure 15-23C, the input frequency is increased again to 10 kHz, causing X_L to increase to 628.3 Ω. For a constant input of 10 V rms, the output is now 1.57 V.

As the input frequency is increased further, the output voltage approaches zero, because X_L becomes extremely large compared to R, as Figure 15-23D shows.

We have seen that as the frequency of the input increases, less of the input voltage is passed through to the output. This reduction occurs because R drops less voltage as X_L increases. Since the lower frequencies pass through the circuit better than the higher ones, the RL lag network is a basic form of *low-pass filter*.

A general response curve for a low-pass filter is shown in Figure 15-24. The effect is the same in either an RC or an RL low-pass circuit.

THE SERIES RL CIRCUIT AS A FILTER

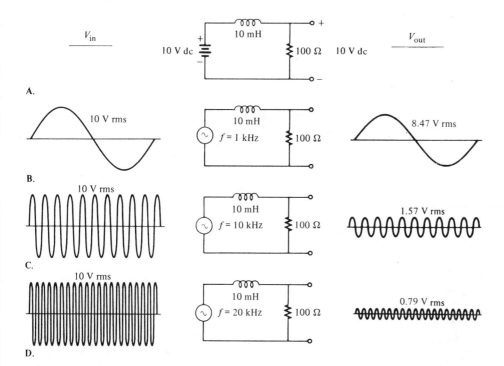

FIGURE 15–23 *Low-pass filter action of an RL circuit.*

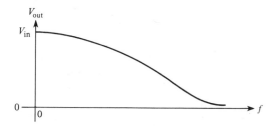

FIGURE 15–24 *Low-pass filter response curve.*

High-Pass Filter

When the output of a series RL circuit is taken across the coil, as in the lead network, the circuit acts as a high-pass filter. At low frequencies, the X_L of the coil is low, and therefore most of the input voltage is dropped across the resistor. As the input frequency increases, X_L increases, and more output voltage appears across the coil.

In Figure 15–25A, the input is dc and the output is zero volts because the coil essentially shorts the output ($X_L = 0$). Again, we neglect R_w.

In Figure 15–25B, the frequency of the input voltage has been increased to 100 Hz at 10 V. The output voltage is 0.63 V, indicating that only a small percentage of the input is getting through to the output at this frequency.

In Figure 15–25C, the input frequency is increased further to 1 kHz, causing X_L to increase and drop more voltage. The output at this frequency is 5.32 V.

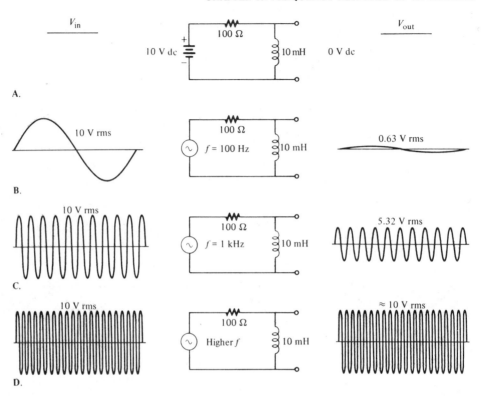

FIGURE 15-25 *High-pass filter action of an RL circuit.*

As you can see, the output voltage increases as the frequency increases. Eventually, a frequency will be reached at which X_L is extremely large compared to R. Then most of the input voltage will appear across the output, as shown in Figure 15-25D.

This circuit tends to prevent lower frequencies from passing through to the output but allows the higher frequencies to pass. The RL lead network is therefore a basic form of *high-pass filter*.

Figure 15-26 shows a plot of output amplitude versus frequency for the high-pass filter. Again, the effect is the same for either an RC or an RL low-pass circuit.

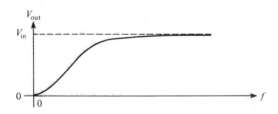

FIGURE 15-26 *High-pass filter response curve.*

Radio Frequency (rf) Chokes

A common application of the RL low-pass filter is the prevention of radio frequencies (rf) from getting on the dc power line in certain circuits. In this case the coil looks almost like a short to dc voltage and an open to rf, as illustrated in Figure 15–27.

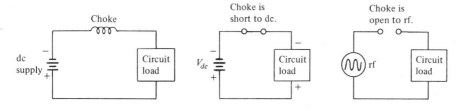

FIGURE 15–27

Review for 15–7

1. When an RL circuit is used as a low-pass filter, across which component is the output taken?
2. In a high-pass filter, when the frequency is decreased, does the output magnitude increase or decrease?
3. How would you classify the filter when an inductor is used as an rf choke?

15–8

POWER IN AN RL CIRCUIT

As you have already learned, a purely resistive circuit dissipates all of the energy that the source delivers. In a purely inductive circuit, all of the energy delivered by the ac source is stored by the magnetic field of the coil and then returned to the source. *There is no energy loss in pure inductance.* This situation is similar to the case of pure capacitance.

When *both* resistance and inductance are in a circuit, *some of the energy is stored and returned to the source by the inductor, and some is dissipated by the resistor.* Just as in the RC circuit, the amount of power in an RL circuit is determined by the relative sizes of the resistance and the reactance. If the resistance is greater than the inductive reactance, more of the total energy from the source is dissipated as heat than is stored by the inductor. Likewise, if the reactance is greater than the resistance, more of the total energy is stored and returned than is lost.

The power in a resistor, often called the *average power* (P_{avg}) and the power in an inductor, often called the *reactive power* (P_r), are given in the following equations:

$$P_{avg} = I^2 R \qquad (15\text{--}8)$$

$$P_r = I^2 X_L \qquad (15\text{--}9)$$

The Power Triangle

The power triangle for the RL circuit is developed in the same way as that for the RC circuit, except the phase angle θ lies above the horizontal because the inductor voltage leads the resistor voltage by 90°. The development of the power triangle for an RL circuit is shown in Figure 15–28. Recall that the apparent power P_a is the total power that *appears* to be delivered by the source to the load. It is not the actual power, because part of it is returned from the inductor. The unit of reactive power is the volt-ampere reactive (VAR), and the unit of apparent power is the volt-ampere (VA).

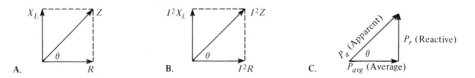

FIGURE 15–28 *Power triangle for an RL circuit.*

From the power triangle, the following equations can be stated:

$$P_{avg} = P_a \cos \theta$$
$$P_r = P_a \sin \theta$$
$$P_a = \sqrt{P_{avg}^2 + P_r^2}$$

Notice that these equations are the same as those for the RC circuit. Also, in this case the power factor is cos θ, just as in the RC circuit.

Example 15–12

Determine the average power in Figure 15–29.

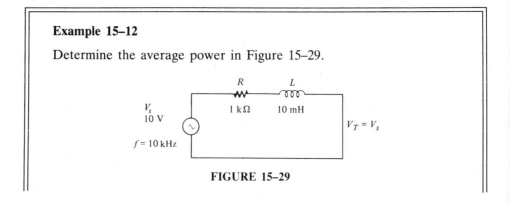

FIGURE 15–29

THE PARALLEL RL CIRCUIT

Solution:

$$X_L = 2\pi f L = 2\pi(10 \text{ kHz})(10 \text{ mH})$$
$$= 628 \text{ }\Omega$$

$$Z = \sqrt{R^2 + X_L^2} = \sqrt{(1000 \text{ }\Omega)^2 + (628 \text{ }\Omega)^2}$$
$$= 1180.8 \text{ }\Omega$$

$$I = \frac{V_s}{Z} = \frac{10 \text{ V}}{1180.8 \text{ }\Omega}$$
$$= 0.0085 \text{ A}$$

$$\theta = \arctan\left(\frac{X_L}{R}\right) = \arctan\left(\frac{628 \text{ }\Omega}{1000 \text{ }\Omega}\right)$$
$$= 32.13°$$

$$P_{avg} = V_s I \cos\theta = (10 \text{ V})(0.0085 \text{ A})\cos 32.13°$$
$$= 0.072 \text{ W}$$

Note that I^2R gives the same result.

Review for 15-8

1. What is the difference between average power and reactive power?
2. If R is less than X_L, is the average power greater than or less than the reactive power?
3. Define *apparent power*.
4. In a series RL circuit, what happens to the average power as the frequency decreases?

15-9

THE PARALLEL RL CIRCUIT

A basic parallel RL circuit is shown in Figure 15-30A. Because R and L are in parallel, the voltage across each component is the same as the source voltage in both phase and magnitude.

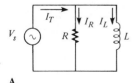

FIGURE 15-30 *Parallel RL circuit.*

Total Current

In Figure 15–30, the current through the resistor is in phase with the voltage. The current through the inductor lags the voltage by 90°. The total current is the *phasor* sum of the two branch currents, as shown in Figure 15–30B. The magnitude of the total current is

$$I_T = \sqrt{I_R^2 + I_L^2} \qquad (15\text{–}10)$$

Phase Angle

The total current lags the source voltage by the angle θ, as indicated in Figure 15–30B. The formula for this phase angle is

$$\theta = \arctan\left(\frac{R}{X_L}\right) \qquad (15\text{–}11)$$

Impedance

The impedance of a parallel RL circuit equals the total voltage divided by the total current. As for the RC circuit in Chapter 12, it can be shown that the magnitude of the impedance can also be expressed as follows:

$$Z = \frac{RX_L}{\sqrt{R^2 + X_L^2}} \qquad (15\text{–}12)$$

Example 15–13

Determine the total current in the parallel RL circuit of Figure 15–31.

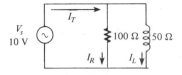

FIGURE 15–31

Solution:

$$I_R = \frac{V_s}{R} = \frac{10 \text{ V}}{100 \text{ }\Omega}$$
$$= 0.1 \text{ A}$$

$$I_L = \frac{V_s}{X_L} = \frac{10 \text{ V}}{50 \text{ }\Omega}$$
$$= 0.2 \text{ A}$$

$$I_T = \sqrt{I_R^2 + I_L^2} = \sqrt{(0.1 \text{ A})^2 + (0.2 \text{ A})^2}$$
$$= 0.22 \text{ A}$$

FORMULAS

Example 15–14

Determine the total impedance magnitude and the phase angle in Figure 15–32.

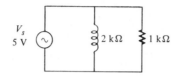

FIGURE 15–32

Solution:

$$Z = \frac{RX_L}{\sqrt{R^2 + X_L^2}} = \frac{(1\ \text{k}\Omega)(2\ \text{k}\Omega)}{\sqrt{(1\ \text{k}\Omega)^2 + (2\ \text{k}\Omega)^2}}$$
$$= 894.4\ \Omega$$

$$\theta = \arctan\left(\frac{R}{X_L}\right) = \arctan\left(\frac{1\ \text{k}\Omega}{2\ \text{k}\Omega}\right)$$
$$= 26.57°$$

I_T lags V_s by 26.57°.

Review for 15–9

1. In a parallel RL circuit, if $R = 2.2\ \text{k}\Omega$ and $X_L = 5\ \text{k}\Omega$, which component has the larger current?
2. Is the total current in a parallel RL circuit equal to $I_R + I_L$?
3. What happens to the phase angle in a parallel RL circuit if the frequency of the applied voltage is increased?

Formulas

For series RL circuits:

$$Z = \sqrt{R^2 + X_L^2} \qquad (15\text{–}1)$$

$$\theta = \arctan\left(\frac{X_L}{R}\right) \qquad (15\text{–}2)$$

$$I = \frac{V_s}{\sqrt{R^2 + X_L^2}}$$

$$V_R = \left(\frac{R}{\sqrt{R^2 + X_L^2}}\right)V_s \qquad (15\text{–}3)$$

$$V_L = \left(\frac{X_L}{\sqrt{R^2 + X_L^2}}\right)V_s \qquad (15\text{–}4)$$

$$V_s = \sqrt{V_R^2 + V_L^2} \qquad (15\text{–}5)$$

$$\phi = 90° - \arctan\left(\frac{X_L}{R}\right) \qquad (15\text{–}6)$$

$$\phi = \arctan\left(\frac{R}{X_L}\right) \qquad (15\text{–}7)$$

$$P_{avg} = I^2 R \qquad (15\text{–}8)$$

$$P_r = I^2 X_L \qquad (15\text{–}9)$$

For parallel RL circuits:

$$I_T = \sqrt{I_R^2 + I_L^2} \qquad (15\text{–}10)$$

$$\theta = \arctan\left(\frac{R}{X_L}\right) \qquad (15\text{–}11)$$

$$Z = \left(\frac{RX_L}{\sqrt{R^2 + X_L^2}}\right) \qquad (15\text{–}12)$$

Summary

1. A sine wave voltage applied to an RL circuit produces a sine wave current.
2. Current lags the voltage in an RL circuit.
3. Resistor voltage is in phase with the current.
4. Inductor voltage leads the current by 90°.
5. The magnitude of impedance in an RL circuit varies directly with frequency.
6. The phase angle is dependent on the relative values of R and X_L.
7. If X_L is negligible compared to R, the phase angle is close to zero.
8. For a given value of R, the phase angle increases as X_L increases.
9. When R and X_L are equal, the phase angle is 45°.
10. In an RL lag network, the output voltage is taken across the resistor.
11. In an RL lag network, the output voltage lags the input.
12. In an RL lead network, the output voltage is taken across the inductor.
13. In an RC lead network, the output voltage leads the input.
14. An RL lag network acts as a low-pass filter.
15. An RL lead network acts as a high-pass filter.

SELF-TEST

16. An inductor dissipates no energy except in its winding resistance.
17. When a sine wave source drives an RL circuit, part of the total power delivered by the source is resistive (average) power, and part of it is reactive power.
18. Reactive power is the power in the inductor.
19. The unit of reactive power is the volt-ampere reactive (VAR).
20. The total power being delivered by the source is the combination of average power and reactive power. Its unit is the volt-ampere (VA).
21. The power factor is the cosine of the phase angle.

Self-Test

1. If you apply a voltage sine wave to an RL circuit, what is the shape of the current wave form?
2. In a given series RL circuit, $R = 3.3\,k\Omega$ and $L = 20\,mH$. What is the magnitude of the impedance at a frequency of 10 kHz?
3. In Figure 15–33, at what frequency does a 45° phase angle between the applied voltage and the current occur?

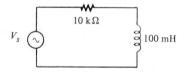

FIGURE 15–33

4. In a given series RL circuit, the inductance is 200 μH. At 5 kHz, what value of resistance will produce a 60° phase angle?
5. In a certain circuit, R is 500 Ω and X_L is 100 Ω. What is the phase angle?
6. A circuit has an inductive reactance that is twice the resistance at a certain frequency. By how many degrees does the voltage lead the current?
7. Determine the magnitudes of the inductor voltage and the resistor voltage in Figure 15–34.

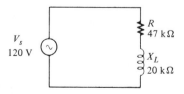

FIGURE 15–34

8. How much current is there in Figure 15–34?
9. By how much does the output voltage lag the input voltage in Figure 15–35?

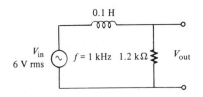

FIGURE 15–35   **FIGURE 15–36**

10. What is the magnitude of the output voltage in Figure 15–35?
11. For the lead network in Figure 15–36, determine the angle by which the output voltage leads the input voltage.
12. What is the magnitude of the output voltage in Figure 15–36?
13. For Figure 15–35, is the output voltage greater at 2 kHz or at 1 kHz?
14. Determine the following for Figure 15–36 given that the frequency is 1 kHz:
    (a) P_{avg} (b) P_r (c) P_a (d) PF
15. For the parallel circuit in Figure 15–37, find each branch current and the total current. What is the phase angle between the applied voltage and the total current?

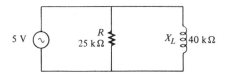

FIGURE 15–37

Problems

15–1. Calculate the impedance of the circuits in Figure 15–38.

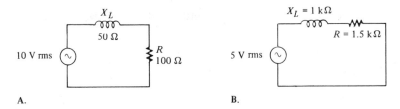

A.   B.

FIGURE 15–38

15–2. Determine the impedance of each circuit in Figure 15–39.

PROBLEMS

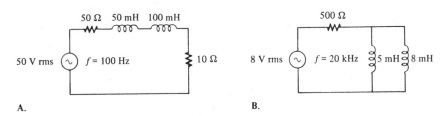

FIGURE 15-39

15-3. Find the current in each circuit in Figure 15-38.

15-4. Find the current in each circuit in Figure 15-39.

15-5. In Figure 15-40, determine the impedance for each of the following frequencies:
(a) 100 Hz (b) 500 Hz (c) 1 kHz (d) 2 kHz

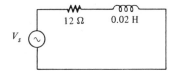

FIGURE 15-40

15-6. Calculate the phase angle for the circuits in Figure 15-38.

15-7. Calculate the phase angle for each circuit in Figure 15-39.

15-8. Find the phase angle for each of the frequencies in Problem 15-5.

15-9. Calculate θ for the circuit in Figure 15-41.

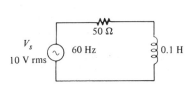   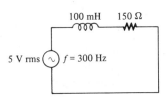

FIGURE 15-41   **FIGURE 15-42**

15-10. If the inductance in Figure 15-41 is doubled, by how many degrees does θ change? Does it increase or decrease?

15-11. On a single graph, sketch the wave forms for V_s, V_R, and V_L in Figure 15-41.

15-12. Find the magnitudes of the voltages across R and L in Figure 15-42.

15-13. For the circuit in Figure 15-42, find the rms values for V_R and V_L for each of the following frequencies:
(a) 60 Hz (b) 200 Hz (c) 500 Hz (d) 1 kHz

15–14. For the lag network in Figure 15–43, determine the phase lag of the output voltage with respect to the input for the following frequencies:
(a) 1 Hz (b) 100 Hz (c) 1 kHz (d) 10 kHz

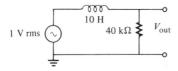

FIGURE 15–43

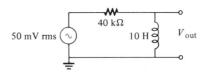

FIGURE 15–44

15–15. Draw the response curve for the circuit in Figure 15–43. Show the output voltage versus frequency in 1-kHz increments from 0 Hz to 5 kHz.

15–16. Repeat Problem 15–14 for the lead network in Figure 15–44.

15–17. Using the same procedure as in Problem 15–15, draw the response curve for Figure 15–44.

15–18. Sketch the voltage phasor diagram for each circuit in Figures 15–43 and 15–44 for a frequency of 8 kHz.

15–19. Find the phase angle between the input and output for the network in Figure 15–45 given that the frequency is 2 kHz. Does the output lead or lag the input?

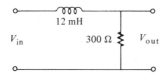

FIGURE 15–45

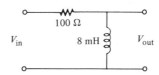

FIGURE 15–46

15–20. Calculate the output phase angle for Figure 15–46 given that $f = 5$ kHz. Is it leading or lagging?

15–21. Calculate the average power, reactive power, and apparent power in Figure 15–45 given that $V_{in} = 10$ V rms and $f = 2$ kHz.

15–22. Calculate the average power, reactive power, and apparent power in Figure 15–46 given that $V_{in} = 30$ V rms and $f = 5$ kHz.

15–23. (a) What is the power factor in Figure 15–45?
(b) What is the power factor in Figure 15–46?

15–24. (a) Sketch the power triangle for Figure 15–45.
(b) Sketch the power triangle for Figure 15–46.

15–25. Find the total current in Figure 15–47.

ANSWERS TO SECTION REVIEWS **517**

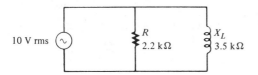

FIGURE 15–47

15–26. In Figure 15–47, what is the phase angle between the total current and the source voltage?

15–27. Find the following values for the circuit in Figure 15–48:
(a) Z_T (b) I_R (c) I_L (d) I_T (e) θ

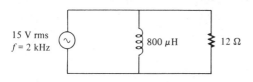

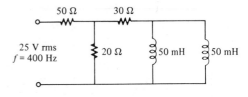

FIGURE 15–48

FIGURE 15–49

15–28. Determine the following values for the circuit in Figure 15–49:
(a) I_T (b) θ (c) P_{avg} (d) PF

Answers to Section Reviews

Section 15–1:
1. 1 kHz.

Section 15–2:
1. Lead. **2.** 90°. **3.** See Figure 15–50. **4.** It increases; it increases. **5.** 1414.2 Ω.

FIGURE 15–50

Section 15–3:
1. Out of phase. **2.** Lag. **3.** Lag. **4.** Increase. **5.** It decreases. **6.** When $X_L = R$.
7. Less.

Section 15–4:
1. The total voltage is the phasor sum of the resistor voltage and the inductor voltage. 2. Across X_L. 3. A decrease in V_R. 4. $V_R = 4.47$ V; $V_L = 8.94$ V.

Section 15–5:
1. Lag. 2. Resistor. 3. 45°. 4. Increase.

Section 15–6:
1. Lead. 2. Inductor. 3. Increase. 4. Increase.

Section 15–7:
1. Resistor. 2. Decrease. 3. Low-pass.

Section 15–8:
1. Average power is the rate of energy loss in resistance. Reactive power is the rate at which energy is stored by the reactive component and returned to the source. 2. Less. 3. The phasor sum of average and reactive power. 4. It increases.

Section 15–9:
1. R. 2. No, it is the phasor sum. 3. It decreases.

The characteristics of inductive reactance are basically opposite to those of capacitive reactance. Therefore, the responses to pulses are opposite. The approach to the topics in this chapter is similar to that used in Chapter 13, with common areas omitted.

16–1 Time Constant
16–2 Induced Voltage in a Series RL Circuit
16–3 The RL Integrator
16–4 The RL Differentiator
16–5 Relationship of Time Response to Frequency Response

16
PULSE RESPONSE
OF RL CIRCUITS

16-1

TIME CONSTANT

Because the inductor's basic action is to *oppose a change in its current,* it follows that *current cannot change instantaneously in an inductor.* That is, a certain time interval is required for the current to make a change from one value to another. The rate at which the current changes is fixed by the *time constant.* The time constant τ for a series RL circuit is the inductance L divided by the resistance R:

$$\tau = \frac{L}{R} \qquad (16\text{-}1)$$

where τ is in seconds.

Example 16-1

A series RL circuit has a resistance of 1 kΩ and an inductance of 1 mH. What is the time constant?

Solution:

$$\tau = \frac{L}{R} = \frac{1 \text{ mH}}{1 \text{ k}\Omega} = \frac{1 \times 10^{-3} \text{ H}}{1 \times 10^{3} \text{ }\Omega}$$

$$= 1 \times 10^{-6} \text{ s} = 1 \text{ }\mu\text{s}$$

Current Build-Up in an Inductor

During a time interval equal to *one* time constant (1τ), the current will change approximately 63%. Therefore, in a series RL circuit as shown in Figure 16–1, the current will increase to 63% of its full value in one time constant after the switch is closed. This build-up in current is comparable to the build-up in capacitor voltage in a series RC circuit: They both follow an exponential curve and reach the approximate percentages of full value indicated in Table 16–1 after each time-constant interval.

TABLE 16-1 *L/R time-constant current build-up.*

τ	% Full Value
1	63
2	86
3	95
4	98
5	99

TIME CONSTANT

When the current reaches full value, after about 5τ, it ceases to change. At this time, the inductor looks like a short (except for its winding resistance) to the constant current. Thus, the current reaches a full value of V_s/R (applied voltage divided by total series resistance) after 5τ. The change in current over five time-constant intervals is illustrated in Figure 16–1. In this case, the current reaches a final value of 10 mA (10 V/1 kΩ).

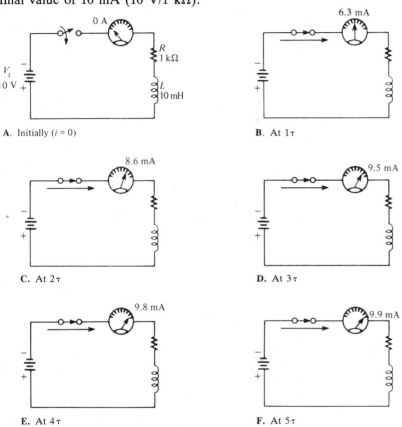

FIGURE 16–1 *Current build-up in a series RL circuit.*

Example 16–2

Calculate the time constant for Figure 16–2. Then determine the current and the time after each time-constant interval, measured from the instant the switch is closed.

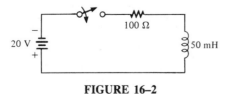

FIGURE 16–2

Example 16–2 (continued)

Solution:

$$I_{max} = \frac{V_s}{R} = \frac{20 \text{ V}}{100 \text{ }\Omega}$$
$$= 0.2 \text{ A (final value)}$$
$$\tau = \frac{L}{R} = \frac{50 \text{ mH}}{100 \text{ }\Omega}$$
$$= 0.5 \text{ ms}$$

At $1\tau = 0.5$ ms:
$$i = 0.63(0.2 \text{ A})$$
$$= 0.126 \text{ A}$$

At $2\tau = 1$ ms:
$$i = 0.86(0.2 \text{ A})$$
$$= 0.172 \text{ A}$$

At $3\tau = 1.5$ ms:
$$i = 0.95(0.2 \text{ A})$$
$$= 0.19 \text{ A}$$

At $4\tau = 2$ ms:
$$i = 0.98(0.2 \text{ A})$$
$$= 0.196 \text{ A}$$

At $5\tau = 2.5$ ms:
$$i = 0.99(0.2 \text{ A}) = 0.198 \text{ A}$$
$$\cong 0.2 \text{ A}$$

Current Decay in an Inductor

Current in an inductor decreases along an exponential curve according to the approximate percentage values for each time-constant interval in Table 16–2.

TABLE 16–2 *L/R time-constant current decay.*

τ	% Full Value
1	37
2	14
3	5
4	2
5	1

TIME CONSTANT

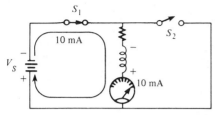

A. Initially

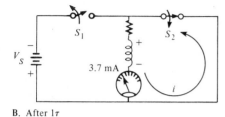

B. After 1τ

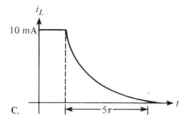

C.

FIGURE 16–3 *Decay of the inductive current.*

Figure 16-3A shows a constant current of 10 mA through the coil. If switch 2 is closed at the *same instant* that switch 1 is opened, as shown in Part B of the figure, the current will decrease to zero in five time constants ($5L/R$), as shown in Part C.

Example 16–3

In Figure 16–4, S_1 is opened at the instant that S_2 is closed.

(a) What is the time constant?
(b) What is the initial value of current?
(c) What is the current after 1τ?

Assume steady state current through the coil prior to switch change.

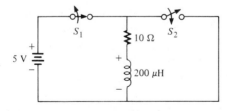

FIGURE 16–4

Solution:

(a) $\tau = L/R = 200\ \mu H/10\ \Omega = 20\ \mu s$.
(b) Current cannot change instantaneously in an inductor. Therefore, the current at the instant of the switch change is the same as the steady state coil current:

> **Example 16-3** (continued)
>
> $$i = 5\text{ V}/10\text{ }\Omega = 0.5\text{ A}$$
>
> (c) After 1τ, the current decays to 37% of its initial value:
>
> $$i = 0.37(0.5\text{ A}) = 0.185\text{ A}$$

Review for 16-1

1. In a series RL circuit, R is 2.2 kΩ and L is 1000 mH. What is τ?
2. If a 10-V battery is suddenly connected across the circuit mentioned in Problem 1, how long will it take the current to reach full value? What is the maximum current?

16-2

INDUCED VOLTAGE IN A SERIES RL CIRCUIT

When current changes in a coil, a voltage is induced. In this section, we will see what happens to the voltages across the resistor and the coil in a series circuit when a change in current occurs.

Look at the circuit in Figure 16-5A. When the switch is open, there is no current, and the resistor voltage and the coil voltage are both zero. At the instant the switch is closed, as indicated in Part B, v_R is zero and v_L is 10 V. The reason for this change is that the induced voltage across the coil is equal and opposite to the applied voltage to prevent the current from changing instantaneously. Therefore, at the instant of switch closure, L effectively acts as an *open* with all of the applied voltage across it.

During the first five time constants, the current is building up exponentially, and the induced coil voltage is decreasing. The resistor voltage increases with the current, as Figure 16-5C illustrates.

After five time constants have elapsed, the current has reached its final value, V_S/R. At this time all of the applied voltage is dropped across the resistor and none across the coil. Thus, L effectively acts as a *short* to nonchanging current, as Figure 16-5D illustrates.

Keep in mind that the inductor always reacts to a change in current by creating an induced voltage in order to counteract that change.

Now let us examine the case illustrated in Figure 16-6, where the steady state current is switched out and the inductor discharges through another path. Part A shows the steady state condition, and Part B illustrates the instant at which the source is removed by opening S_1 and a discharge path is connected with the closure of S_2. There was 1 A through L prior to this. Notice that 10 V are induced in L in the direction to try to keep 1 A in the same direction. Then, as shown in Part C, the current decays exponentially, and so do V_R and V_L. After 5τ, as shown in Part D,

INDUCED VOLTAGE IN A SERIES RL CIRCUIT

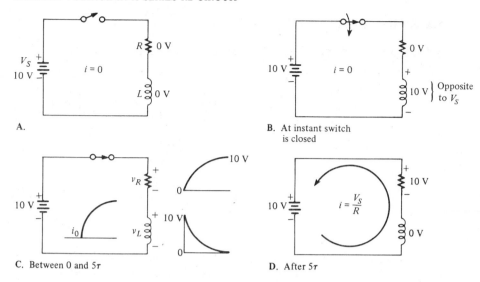

FIGURE 16-5 *Voltage in an RL circuit with the inductor charging.*

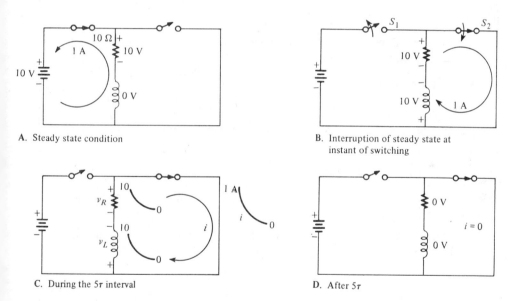

FIGURE 16-6 *Voltage in an RL circuit with the inductor discharging.*

all of the energy stored in the magnetic field of L is dissipated, and all values are zero.

Example 16-4

(a) In Figure 16-7A, what is v_L at the instant S_1 is closed? What is v_L after 5τ?

Example 16-4 (continued)

(b) In Figure 16-7B, what is v_L at the instant S_1 opens and S_2 closes? What is v_L after 5τ?

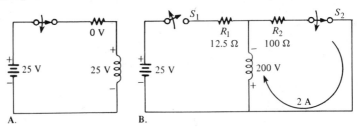

FIGURE 16-7

Solution:

(a) At the instant the switch is closed, all of the voltage is across L. Thus, $v_L = 25$ V, with the polarity as shown. After 5τ, L acts as a short; so $v_L = 0$.

(b) With S_1 closed and S_2 open, the steady state current is

$$25 \text{ V}/12.5 \text{ }\Omega = 2 \text{ A}$$

When the switches are thrown, an induced voltage is created across L sufficient to keep this current for an instant. In this case it takes $v_L = IR_2 = (2\text{ A})(100\text{ }\Omega) = 200$ V. After 5τ, the inductor voltage is zero.

These results are indicated in the circuit diagram.

Review for 16-2

1. A 20-V dc source is connected to a series RL circuit with a switch. At the instant of switch closure, what are v_R and v_L?
2. In the same circuit, after a time interval of 5τ from switch closure, what are v_R and v_L? (V_S is 20 V.)

16-3

THE RL INTEGRATOR

Figure 16-8 shows a series RL circuit with *the output taken across the resistor*. This circuit is also known as an *integrator*. Its *output wave form* under equivalent conditions is the same as that for the RC integrator discussed in Chapter 13. Recall that in the RC case, the output was across the capacitor.

THE RL INTEGRATOR

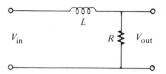

FIGURE 16–8 *The RL integrator.*

Pulse Input to the Integrator

When a pulse generator is connected to the input of the integrator and the voltage pulse goes from its low level to its high level, the inductor prevents a sudden change in current. As a result, the inductor acts as an open, and all of the input voltage is across it at the instant of the rising pulse edge. This situation is indicated in Figure 16–9A.

After the rising edge, the current builds up, and the output voltage follows the current as it increases exponentially, as shown in Figure 16–9B. The current can reach a maximum of V_{in}/R if the transient time is shorter than the pulse width (V_{in} is the pulse amplitude).

When the pulse goes from its high level to its low level, an induced voltage is created across the coil in an effort to keep a current equal to V_{in}/R. The output voltage begins to decrease exponentially, as shown in Figure 16–9C.

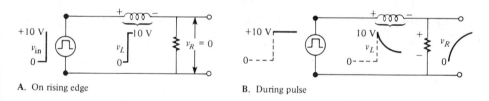

A. On rising edge   B. During pulse

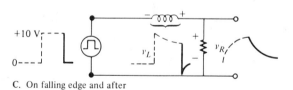

C. On falling edge and after

FIGURE 16–9 *Pulse response of RL integrator.*

The exact shape of the output depends on the L/R time constant as summarized in Figure 16–10 for various relationships between the time constant and the pulse width. You should note that the response of this RL circuit *in terms of the output* is identical to that of the RC integrator. The L/R time constant in relation to the input pulse width has the same effect as the RC time constant that we discussed in Chapter 13.

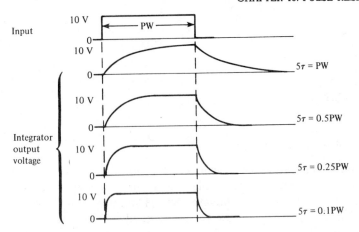

FIGURE 16–10 *Variation in output pulse shape with time constant.*

Example 16–5

A 10-V pulse with a width of 1 ms is applied to the integrator in Figure 16–11. Determine the voltage level that the output will reach during the pulse. If the source has an internal resistance of 50 Ω, how long will it take the output to decay to zero? Sketch the output voltage wave form.

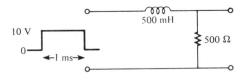

FIGURE 16–11

Solution:

During the pulse while L is charging,

$$\tau = \frac{L}{R} = \frac{500 \text{ mH}}{0.5 \text{ k}\Omega} = 1 \text{ ms}$$

Notice that the pulse width is exactly equal to τ. Thus, the output V_R will reach 63% of the full input amplitude in 1τ. Therefore, the output voltage gets to 6.3 V at the end of the pulse.

After the pulse is gone, the inductor discharges back through the 50-Ω source. The total R during discharge is 500 Ω + 50 Ω = 550 Ω; and τ = 500 mH/550 Ω = 909 μs. The source takes 5τ to completely discharge: 5τ = 5(909 μs) = 4545 μs.

The output voltage is shown in Figure 16–12.

THE RL INTEGRATOR

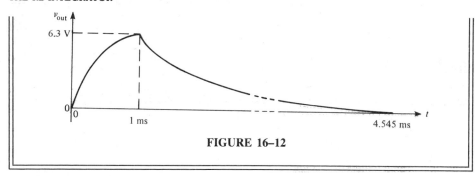

FIGURE 16–12

Example 16–6

Determine the maximum output voltage for the integrator in Figure 16–13 when a single pulse is applied as shown.

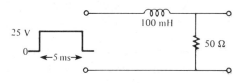

FIGURE 16–13

Solution:

Calculate the time constant:

$$\tau = \frac{L}{R} = \frac{100 \text{ mH}}{50 \text{ }\Omega} = 2 \text{ ms}$$

Because the pulse is 5 ms, the inductor charges for 2.5τ. The standard percentage time-constant table is useless in this case. We must use the exponential formula to calculate the voltage as follows:

$$v_{out} = V_F(1 - e^{-t/\tau}) = 25(1 - e^{-5/2})$$
$$= 25(1 - e^{-2.5}) = 25(1 - 0.082)$$
$$= 25(0.918) = 22.95 \text{ V}$$

Review for 16–3

1. In an RL integrator, across which component is the output voltage taken?
2. When a pulse is applied to an RL integrator, what condition must exist in order for the output voltage to reach the amplitude of the input?
3. Under what condition will the output voltage have the approximate shape of the input pulse?

16-4

THE RL DIFFERENTIATOR

Figure 16–14 shows a series RL circuit with *the output across the inductor*. Compare this circuit to the RC differentiator in which the output is the resistor voltage.

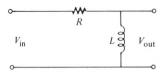

FIGURE 16–14 *The RL differentiator.*

Pulse Input to the Differentiator

A pulse generator is connected to the input of the differentiator. Initially, before the pulse, there is no current in the circuit. When the input pulse goes from its low level to its high level, the inductor prevents a sudden change in current. It does so, as you know, with an induced voltage equal and opposite to the input. As a result, L looks like an open, and *all* of the input voltage appears across it at the instant of the rising edge, as shown in Figure 16–15A with a 10-V pulse.

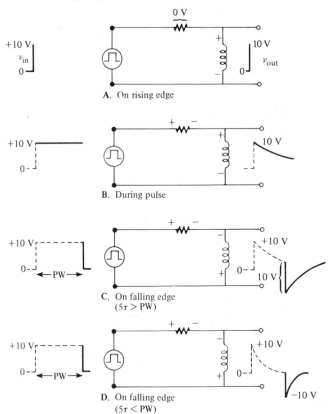

FIGURE 16–15 *Pulse response of RL differentiator.*

THE RL DIFFERENTIATOR

During the pulse, the current exponentially builds up. As a result, the inductor voltage decreases, as shown in Figure 16–15B. The rate of decrease, as you know, depends on the L/R time constant. When the falling edge of the input appears, the inductor reacts to keep the current as is, by creating an induced voltage in a direction as indicated in Part C. This reaction is seen as a sudden 10-V negative-going transition of the inductor voltage, as indicated in Parts C and D.

Two conditions are possible, as indicated in Parts C and D. In Part C, 5τ is greater than the input pulse width, and the output voltage does not have time to decay to zero. In Part D, 5τ is less than the pulse width, and so the output decays to zero before the end of the pulse. In this case a full, negative, 10-V transition occurs at the trailing edge.

Keep in mind that as far as the input and output are concerned, the RL integrator and differentiator perform the same as their RC counterparts.

A summary of the RL differentiator response for relationships of various time constants and pulse widths is shown in Figure 16–16.

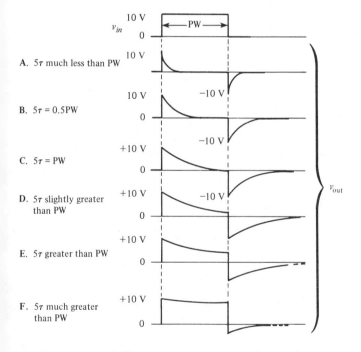

FIGURE 16–16 *Differentiator response as τ is varied.*

Example 16–7

Sketch the output voltage for the circuit in Figure 16–17.

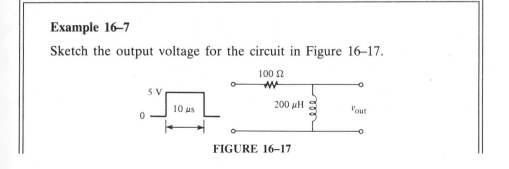

FIGURE 16–17

Example 16-7 (continued)

Solution:

First calculate the time constant:

$$\tau = \frac{L}{R} = \frac{200 \text{ μH}}{100 \text{ Ω}} = 2 \text{ μs}$$

In this case, 5τ = PW; so the output will decay to zero at the end of the pulse.

On the rising edge, the inductor voltage jumps to +5 V and then decays exponentially to zero. It reaches approximately zero at the instant of the falling edge. On the falling edge of the input, the inductor voltage jumps to −5 V and then goes back to zero. The output wave form is shown in Figure 16-18.

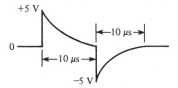

FIGURE 16-18

Example 16-8

Determine the output voltage wave form for the differentiator in Figure 16-19.

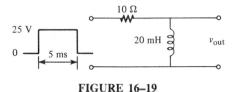

FIGURE 16-19

Solution:

First calculate the time constant:

$$\tau = \frac{L}{R} = \frac{20 \text{ mH}}{10 \text{ Ω}} = 2 \text{ ms}$$

On the rising edge, the inductor voltage immediately jumps to +25 V. Because the pulse width is 5 ms, the inductor charges for only 2.5τ; so we must use the formula for a decreasing exponential:

$$v_L = V_i e^{-t/\tau} = 25e^{-5/2}$$
$$= 25e^{-2.5} = 25(0.082)$$
$$= 2.05 \text{ V}$$

This result is the inductor voltage at the end of the 5-ms input pulse.

On the falling edge, the output immediately jumps from +2.05 V down to −22.95 V (a 25-V negative-going transition). The complete output wave form is sketched in Figure 16–20.

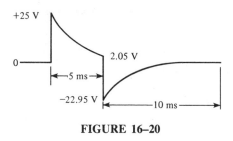

FIGURE 16–20

Review for 16–4

1. In an RL differentiator, across which component is the output taken?
2. Under what condition does the output pulse shape most closely resemble the input pulse?
3. If the inductor voltage in an RL differentiator is down to +2 V at the end of a +10-V input pulse, to what negative voltage will the output voltage go in response to the falling edge of the input?

16–5

RELATIONSHIP OF TIME RESPONSE TO FREQUENCY RESPONSE

As we discussed in Chapter 13, a pulse contains high-frequency components in its fast rising and falling edges and lower-frequency components in its flat portion. Its average value is the dc component.

Integrator as a Low-Pass Filter

Like the RC integrator, the RL integrator acts as a basic low-pass filter, because L is in series between input and output. X_L is small for low frequencies and offers little opposition. It increases with frequency; so at higher frequencies most of the total voltage is dropped across L and very little across R, the output. If the input is dc, L is like a short ($X_L = 0$). At high frequencies, L becomes like an open, as illustrated in Figure 16–21.

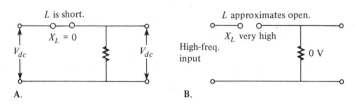

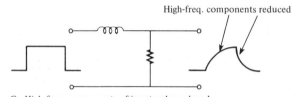

C. High-freq. components of input pulse reduced

FIGURE 16–21 *Low-pass filter action.*

Differentiator as a High-Pass Filter

Again like the RC differentiator, the RL differentiator acts as a basic high-pass filter. Because L is connected across the output, less voltage is developed across it at lower frequencies than at higher ones. There are zero volts across the output for dc. For high frequencies, most of the input voltage is dropped across the output coil ($X_L = 0$ for dc; $X_L \cong$ open for high frequencies). Figure 16–22 shows high-pass filter action.

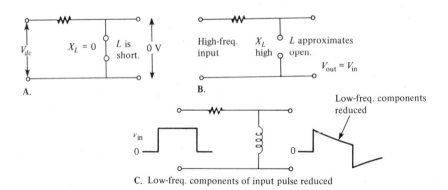

FIGURE 16–22 *High-pass filter action.*

Formula for Highest-Frequency Component in a Pulse

The formula for the highest-frequency component is the same as that given in Chapter 13:

$$f_h = \frac{0.35}{t_r} \qquad (16\text{–}2)$$

$$f_h = \frac{0.35}{t_f} \quad (16\text{–}3)$$

Review for 16–5

1. Why is an integrator a basic low-pass filter?
2. Why is a differentiator a basic high-pass filter?

Formulas

$$\tau = \frac{L}{R} \quad (16\text{–}1)$$

$$f_h = \frac{0.35}{t_r} \quad (16\text{–}2)$$

$$f_h = \frac{0.35}{t_f} \quad (16\text{–}3)$$

Summary

1. The time constant for a series RL circuit is the inductance divided by the resistance.
2. The voltage and current make a 63% change during each time-constant interval for a pulse response.
3. Energy is stored by an inductor in its magnetic field.
4. While the magnetic field is building up (current increasing), the inductor is charging (storing energy).
5. While the magnetic field is collapsing (current decaying), the inductor is discharging.
6. Five time constants (5τ) are required to fully charge or discharge an inductor in an RL circuit.
7. Currents and voltages in an RL circuit change exponentially when a pulse is applied.
8. An integrator is a series RL circuit in which the output voltage is across the resistor.
9. A differentiator is a series RL circuit in which the output voltage is across the inductor.
10. An inductor acts like an *open* to an instantaneous change. Stated another way, it acts like an open to high frequencies when X_L is extremely large.
11. An inductor acts like a short to dc because X_L is zero.

Self-Test

1. A circuit has $R = 2$ kΩ in series with $L = 10$ mH. What is the time constant?
2. A 10-V pulse is applied to a series RL circuit. The pulse width equals 1τ. To what value does the resistor voltage increase during the pulse? What are the maximum and minimum inductor voltages?
3. Repeat Problem 2 for the following values of PW:
   (a) 2τ (b) 3τ (c) 4τ (d) 5τ
4. Sketch the approximate shape of an integrator output for a single input pulse when $5\tau = 0.5$PW. Repeat for $5\tau = 5$PW.
5. Repeat Problem 4 for a differentiator.
6. Determine the output voltage for the circuit in Figure 16–23. A single input pulse is applied as shown.

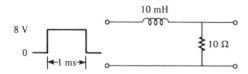

FIGURE 16–23

7. Convert the circuit in Figure 16–23 to a differentiator, and repeat Problem 6.
8. What is the output voltage in Figure 16–24A if the winding resistance is 1 Ω. Repeat for Part B of the figure.

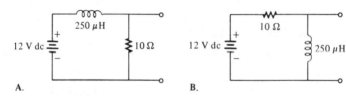

FIGURE 16–24

Problems

16–1. Determine the time constant for each of the following series RL combinations:
(a) $R = 100$ Ω, $L = 100$ μH   (b) $R = 4.7$ kΩ, $L = 10$ mH
(c) $R = 1.5$ MΩ, $L = 3$ H

16–2. In a series RL circuit, determine how long it takes the current to build up to its full value for each of the following:
(a) $R = 50$ Ω, $L = 50$ μH   (b) $R = 3300$ Ω, $L = 15$ mH
(c) $R = 22$ kΩ, $L = 100$ mH

PROBLEMS

16-3. In the circuit of Figure 16-25, there is initially no current. Determine the inductor voltage after the following time intervals when the switch is closed:

(a) 10 μs (b) 20 μs (c) 30 μs (d) 40 μs (e) 50 μs

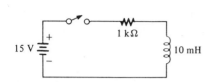

FIGURE 16-25

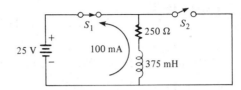

FIGURE 16-26

16-4. In Figure 16-26, there are 100 mA through the coil. When S_1 is opened and S_2 simultaneously closed, find the inductor voltage at the following times:
(a) Initially (b) After 1.5 ms (c) After 4.5 ms (d) After 6 ms

16-5. Repeat Problem 16-3 for the following time intervals:
(a) 2 μs (b) 5 μs (c) 15 μs

16-6. Repeat Problem 16-4 for the following times:
(a) 0.5 ms (b) 1 ms (c) 2 ms

16-7. In Figure 16-25, at what time is the inductor voltage 5 V?

16-8. Sketch the integrator output in Figure 16-27, showing maximum voltages.

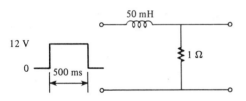

FIGURE 16-27

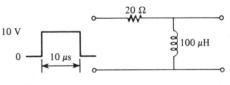

FIGURE 16-28

16-9. (a) What is τ in Figure 16-28?
(b) Sketch the output voltage.

16-10. Determine the time constant in Figure 16-29.

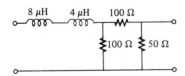

FIGURE 16-29

Answers to Section Reviews

Section 16–1:
1. 454.55 μs. **2.** 2.273 ms; 4.55 mA.

Section 16–2:
1. $v_R = 0$ V; $v_L = 20$ V. **2.** $v_R = 20$ V; $v_L = 0$ V.

Section 16–3:
1. Resistor. **2.** $5\tau \leq$ PW. **3.** 5τ much less than PW.

Section 16–4:
1. Inductor. **2.** 5τ much greater than PW. **3.** -8 V.

Section 16–5:
1. It discriminates against higher frequencies, as indicated by the rounding off of pulse edges. **2.** It discriminates against lower frequencies, as indicated by the absence of the flat portion of the pulse on the output.

In Chapter 14, we studied self-inductance. In this chapter we will study another inductive effect called *mutual inductance*, which occurs when two or more coils are in close proximity. Mutual inductance is the basis for transformer action.

In this chapter you will learn how transformers are used for increasing or decreasing voltage and current and for impedance matching. Various types of transformers will be discussed.

17-1 Mutual Inductance
17-2 The Basic Transformer
17-3 Step-Up Transformer
17-4 Step-Down Transformer
17-5 Loading the Secondary
17-6 Reflected Impedance
17-7 Impedance Matching
17-8 The Transformer as an Isolation Device
17-9 Tapped Transformers
17-10 Multiple-Winding Transformers
17-11 Autotransformers
17-12 Transformer Construction

17

TRANSFORMERS

17-1

MUTUAL INDUCTANCE

When two coils are placed close to each other, as depicted in Figure 17–1, a changing magnetic field produced by current in one coil will cause an induced voltage in the second coil. This induced voltage occurs because the magnetic lines of force (flux) of coil 1 "cut" the winding of coil 2. The two coils are thereby *magnetically linked* or *coupled*. It is important to notice that there is no electrical connection between the coils; therefore they are *electrically isolated*.

The mutual inductance L_M is a measure of how much voltage is induced in coil 2 as a result of a change in current in coil 1. Mutual inductance is measured in henries (H), just as self-inductance is. A greater L_M means that there is a greater induced voltage in coil 2 for a given change in current in coil 1.

There are several factors that determine L_M: the coefficient of coupling (k), the inductance of coil 1 (L_1), and the inductance of coil 2 (L_2).

Coefficient of Coupling

The coefficient of coupling k between two coils is the ratio of the lines of force (flux) produced by coil 1 linking coil 2, to the total flux produced by coil 1:

$$k = \frac{\text{flux linking coil 1 and coil 2}}{\text{total flux produced by coil 1}} \tag{17-1}$$

For example, if half of the total flux produced by coil 1 links coil 2, then $k = 0.5$. A greater value of k means that more voltage is induced in coil 2 for a certain rate of change of current in coil 1. Note that k has no units. *Recall that the unit of magnetic lines of force (flux) is the weber, abbreviated Wb.*

The coefficient k depends on the physical closeness of the coils and the type of core material on which they are wound. Also, the construction and shape of the cores are factors.

Formula for Mutual Inductance

The three factors influencing L_M (k, L_1, and L_2) are shown in Figure 17-2. The formula for L_M is

$$L_M = k\sqrt{L_1 L_2} \tag{17-2}$$

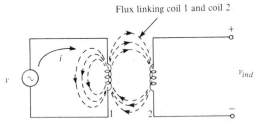

FIGURE 17–1 *Two magnetically coupled coils.*

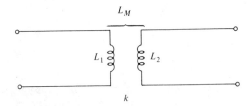

FIGURE 17–2

THE BASIC TRANSFORMER

> **Example 17–1**
>
> One coil produces a total magnetic flux of 50 μWb, and 20 μWb link coil 2. What is k?
>
> *Solution:*
>
> $$k = \frac{20\ \mu\text{Wb}}{50\ \mu\text{Wb}} = 0.4$$

> **Example 17–2**
>
> Two coils are wound on a single core, and the coefficient of coupling is 0.3. The inductance of coil 1 is 10 μH, and the inductance of coil 2 is 15 μH. What is L_M?
>
> *Solution:*
>
> $$L_M = k\sqrt{L_1 L_2} = 0.3\sqrt{(10\ \mu\text{H})(15\ \mu\text{H})}$$
> $$= 3.67\ \mu\text{H}$$

Review for 17–1

1. Define *mutual inductance*.
2. Two 50-mH coils have $k = 0.9$. What is L_M?
3. If k is increased, what happens to the voltage induced in one coil as a result of a current change in the other coil?

17–2

THE BASIC TRANSFORMER

A *basic transformer* is an electrical device constructed of two coils with mutual inductance. A schematic diagram of a transformer is shown in Figure 17–3A. As shown, one coil is called the *primary winding,* and the other is called the *secondary winding.* The source voltage is applied to the primary, and the load is connected to the secondary, as shown in Figure 17–3B.

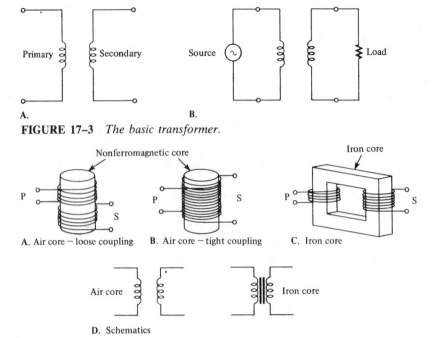

FIGURE 17-3 *The basic transformer.*

FIGURE 17-4 *Basic types of transformers and schematic symbols.*

Typical transformers are wound on a common core in several ways. The core can be either an air core or an iron core. Figure 17-4A and B shows the primary and secondary coils on a nonferromagnetic cylindrical form as examples of air core transformers. In Part A, the windings are separated; in Part B, they overlap for *tighter coupling* (higher k). Figure 17-4C illustrates an iron core transformer. The iron core increases the coefficient of coupling. Standard schematic symbols are shown for both types in Part D.

Turns Ratio

An important characteristic of a transformer is its *turns ratio*. *The turns ratio is the number of turns in the secondary winding, N_s, divided by the number of turns in the primary winding, N_p:*

$$\text{Turns ratio} = \frac{N_s}{N_p} \qquad (17\text{-}3)$$

Direction of Windings

The directions in which the primary and secondary windings are wound on the core determine the polarity of the induced secondary voltage with respect to the primary voltage. The ac voltage across the secondary can be either in phase with the primary voltage or 180° out of phase with it, depending on the winding directions. Figure 17-5 shows how winding direction affects polarity.

THE BASIC TRANSFORMER

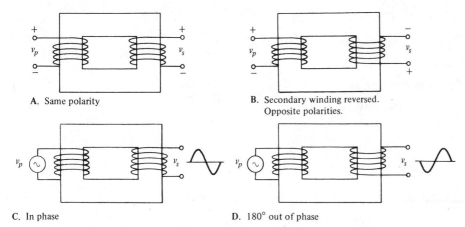

FIGURE 17–5 *How winding directions determine polarity.*

Dot Convention

In the schematic for a transformer, dots are often used to indicate polarities, as illustrated in Figure 17–6. The rules for using this convention are simple: A positive at the primary dot causes a positive at the secondary dot (Part A), and a negative at the primary dot causes a negative at the secondary dot (Part B).

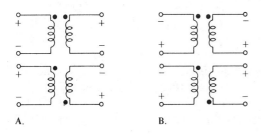

FIGURE 17–6 *Illustration of dot convention for polarity determination.*

Example 17–3

A transformer primary has 100 turns, and the secondary has 400 turns. What is the turns ratio?

Solution:

$$N_s = 400 \quad N_p = 100$$
$$\text{Turns ratio} = \frac{N_s}{N_p} = \frac{400}{100}$$
$$= 4$$

Review for 17–2

1. What is a transformer?
2. Define *turns ratio*.
3. How does the type of core affect a transformer?
4. Why are the directions of the winding important?

17–3

STEP-UP TRANSFORMER

A transformer in which the secondary voltage is greater than the primary voltage is called a step-up transformer. The amount that the voltage is stepped up depends on the turns ratio.

The ratio of secondary voltage to primary voltage is equal to the ratio of the number of secondary turns to the number of primary turns:

$$\frac{V_s}{V_p} = \frac{N_s}{N_p} \qquad (17\text{–}4)$$

where V_s is the secondary voltage and V_p is the primary voltage. From Equation (17–4), we get

$$V_s = \left(\frac{N_s}{N_p}\right) V_p \qquad (17\text{–}5)$$

This equation states that the secondary voltage is equal to the turns ratio times the primary voltage. This condition assumes that the coefficient of coupling is 1. A good iron core transformer approaches this value.

The turns ratio for a step-up transformer is always greater than 1.

Example 17–4

The transformer in Figure 17–7 has a 200-turn primary winding and a 600-turn secondary winding. What is the voltage across the secondary?

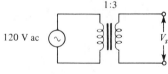

FIGURE 17–7

Solution:

$$N_p = 200 \qquad N_s = 600$$

$$\text{Turns ratio} = \frac{N_s}{N_p} = \frac{600}{200} = 3$$

STEP-DOWN TRANSFORMER

$$V_s = 3\,V_p = 3(120\text{ V})$$
$$= 360\text{ V}$$

Note that the turns ratio of 3 is indicated on the schematic as 1 : 3.

Review for 17-3

1. What does a step-up transformer do?
2. If the turns ratio is 5, how much greater is the secondary voltage than the primary voltage?

17-4

STEP-DOWN TRANSFORMER

A transformer in which the secondary voltage is less than the primary voltage is called a step-down transformer. The amount by which the voltage is stepped down depends on the turns ratio. Equation (17–5) applies to a step-down transformer, also. The turns ratio, however, is always *less than* 1.

Example 17–5

The transformer in Figure 17–8 has 50 turns in the primary and 10 turns in the secondary. What is the secondary voltage?

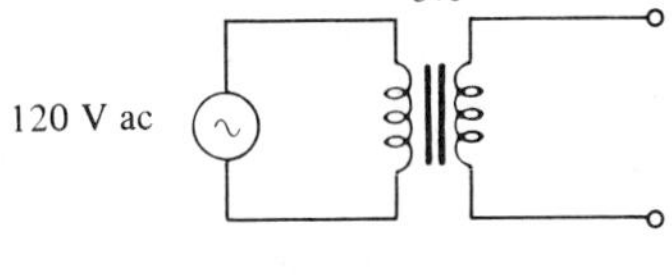

FIGURE 17–8

Solution:

$$N_p = 50 \quad N_s = 10$$
$$\text{Turns ratio} = \frac{N_s}{N_p} = \frac{10}{50}$$
$$= 0.2$$
$$V_s = 0.2 V_p = 0.2(120\text{ V})$$
$$= 24\text{ V}$$

Review for 17-4

1. What does a step-down transformer do?
2. One hundred twenty volts rms are applied to the primary of a transformer with a turns ratio of 0.5. What is the secondary rms voltage?

17-5

LOADING THE SECONDARY

When a load is connected to the secondary, as shown in Figure 17-9, secondary current will flow because of the voltage induced in the secondary coil. It can be shown that the ratio of the primary current I_p to the secondary current I_s is equal to the turns ratio, as expressed in the following equation:

$$\frac{I_p}{I_s} = \frac{N_s}{N_p} \qquad (17\text{-}6)$$

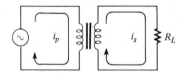

FIGURE 17-9

A manipulation of this equation gives Equation (17-7), which shows that I_s is equal to I_p times the *reciprocal* of the turns ratio:

$$I_s = \left(\frac{N_p}{N_s}\right) I_p \qquad (17\text{-}7)$$

Thus, for a step-up transformer, in which N_s/N_p is greater than 1, the secondary current is less than the primary current. For a step-down transformer, N_s/N_p is less than 1, and I_s is greater than I_p.

Example 17-6

The transformers in Figure 17-10A and B have loaded secondaries. If the primary current is 100 mA in each case, how much current is through the load?

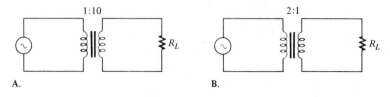

FIGURE 17-10

Solution:

Part A:

$$I_s = \left(\frac{N_p}{N_s}\right)I_p = 0.1(100 \text{ mA})$$
$$= 10 \text{ mA}$$

Part B:

$$I_s = \left(\frac{N_p}{N_s}\right)I_p = 2(100 \text{ mA})$$
$$= 200 \text{ mA}$$

Primary Power Equals Secondary Power

In a transformer the secondary power is the same as the primary power regardless of the turns ratio, as shown in the following steps:

$$P_p = V_p I_p \quad \text{and} \quad P_s = V_s I_s$$

$$I_s = \left(\frac{N_p}{N_s}\right)I_p \quad \text{and} \quad V_s = \left(\frac{N_s}{N_p}\right)V_p$$

By substitution, we obtain

$$P_s = \left(\frac{\cancel{N_p}}{\cancel{N_s}}\right)\left(\frac{\cancel{N_s}}{\cancel{N_p}}\right)V_p I_p$$

Canceling yields

$$P_s = V_p I_p = P_p$$

Review for 17–5

1. If the turns ratio of a transformer is 2, is the secondary current greater than or less than the primary current? By how much?
2. A transformer has 100 primary turns and 25 secondary turns, and I_p is 0.5 A. What is the value of I_s?

17–6

REFLECTED IMPEDANCE

When an impedance is connected across the secondary of a transformer, to a source connected to the primary, it "appears" to have a value that is dependent on the turns ratio. That is, the source "sees" a certain impedance *reflected* from the secondary into the primary circuit. This reflected impedance is determined as follows:

The impedance in the primary circuit of Figure 17–11 is $Z_p = V_p/I_p$. The impedance in the secondary circuit is $Z_L = V_s/I_s$. From the previous sections, we know that $V_s/V_p = N_s/N_p$ and $I_p/I_s = N_s/N_p$. Using these relationships, we find a formula for Z_p in terms of Z_L as follows:

$$\frac{Z_p}{Z_L} = \frac{V_p/I_p}{V_s/I_s} = \left(\frac{V_p}{V_s}\right)\left(\frac{I_s}{I_p}\right) = \left(\frac{N_p}{N_s}\right)\left(\frac{N_p}{N_s}\right)$$

$$= \frac{N_p^2}{N_s^2} = \left(\frac{N_p}{N_s}\right)^2$$

Solving for Z_p, we get

$$Z_p = \left(\frac{N_p}{N_s}\right)^2 Z_L \qquad (17\text{--}8)$$

Equation (17–8) tells us that *the primary impedance is the square of the reciprocal of the turns ratio times the load impedance.*

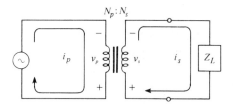

FIGURE 17–11 *Circuit for derivation of reflected impedance.*

Example 17–7

Figure 17–12 shows a source that is transformer-coupled to a load resistor of 100 Ω. The transformer has a turns ratio of 4. What is the reflected resistance seen by the source?

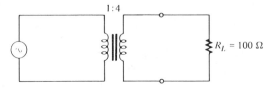

FIGURE 17–12

Solution:

In this case, the load impedance is a resistor. Therefore, the reflected resistance is determined by Equation (17–8):

$$R_p = (1/4)^2 R_L = (1/16)(100 \text{ Ω})$$
$$= 6.25 \text{ Ω}$$

The source sees a resistance of 6.25 Ω just as if it were connected directly, as shown in the equivalent circuit of Figure 17–13.

IMPEDANCE MATCHING

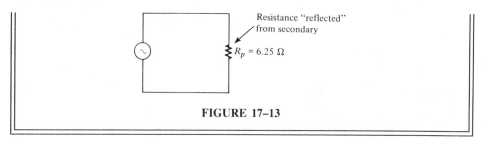

FIGURE 17-13

Example 17-8

If a transformer is used in Figure 17-12 having 40 primary turns and 10 secondary turns, what is the reflected resistance?

Solution:

Turns ratio = 0.25

$$R_p = (1/0.25)^2(100 \ \Omega) = (4)^2(100 \ \Omega)$$
$$= 1600 \ \Omega$$

This result illustrates the difference that the turns ratio makes.

Review for 17-6

1. Define *reflected impedance*.
2. What transformer characteristic determines the reflected impedance?
3. A given transformer has a turns ratio of 10, and the load is 50 Ω. How much impedance is reflected into the primary?

17-7

IMPEDANCE MATCHING

Transformers are often used to match a load to a source impedance for *maximum power transfer*. An example is illustrated in Figure 17-14 where an audio amplifier is *transformer-coupled* to a speaker. It is desirable to have the maximum possible power available from the amplifier delivered to the speaker. The amplifier is the source in this case, and it is connected to the primary. The speaker is the load, and it is connected to the secondary.

To get maximum power transfer, the reflected load impedance "seen" by the amplifier must equal the output or source resistance of the amplifier. In other words, the impedance of the load must be *matched* to that of the source. We achieve this matching by choosing the proper turns ratio for the transformer.

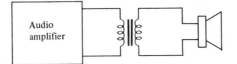

FIGURE 17–14 *Impedance matching an amplifier to a speaker.*

Example 17–9

An amplifier has a 100-Ω output resistance. In order to provide maximum power to a 16-Ω speaker, what turns ratio must we use in the coupling transformer?

Solution:

$$Z_p = \left(\frac{1}{\text{turns ratio}}\right)^2 Z_L$$

Solve for the turns ratio as follows:

$$\left(\frac{1}{\text{turns ratio}}\right)^2 = \frac{Z_p}{Z_L}$$

$$\frac{1}{\text{turns ratio}} = \sqrt{\frac{Z_p}{Z_L}}$$

$$\text{Turns ratio} = \sqrt{\frac{Z_L}{Z_p}} = \sqrt{\frac{16\ \Omega}{100\ \Omega}}$$

$$= \frac{\sqrt{16}}{\sqrt{100}} = \frac{4}{10} = 0.4$$

The diagram and its equivalent reflected circuit are shown in Figure 17–15.

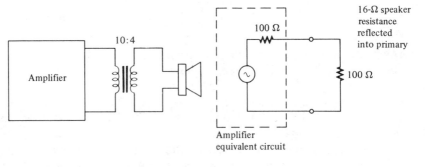

FIGURE 17–15

Review for 17-7

1. What does *impedance matching* mean?
2. What is the advantage of matching the load impedance to the impedance of a source?
3. A transformer has 100 primary turns and 50 secondary turns. What is the reflected impedance with 100 Ω across the secondary?

17-8

THE TRANSFORMER AS AN ISOLATION DEVICE

dc Isolation

As illustrated in Figure 17–16, if there is a direct current in a transformer primary, nothing happens in the secondary, because a *changing* current in the primary is necessary to induce a changing voltage in the secondary. Therefore, the transformer serves to *isolate* the secondary from any dc in the primary.

In a typical application, a transformer can be used to keep the dc voltage on the output of an amplifier stage from affecting the dc bias of the next amplifier. Only the ac signal is coupled through the transformer from one stage to the next, as Figure 17–17 illustrates.

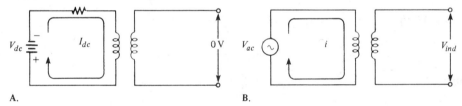

A.   B.

FIGURE 17–16 *dc isolation and ac coupling.*

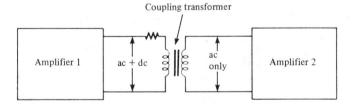

FIGURE 17–17 *Amplifier stages with transformer coupling for dc isolation.*

Power Line Isolation

Transformers are often used to isolate the 60-Hz, 120-V ac power line from a piece of electronic equipment, such as a TV set or any test instrument that operates from the 60-Hz ac power.

The reason for using a transformer to couple the 60-Hz ac to the equipment is to prevent a possible shock hazard if the "hot" side (120 V ac) of the power line is connected to the equipment chassis. This condition is possible if the line cord socket can be plugged into the outlet either way. Figure 17–18 illustrates this situation.

A transformer can prevent this hazardous condition as illustrated in Figure 17–19. With such isolation, there is no way of directly connecting the 120-V ac line to the instrument ground, no matter how the power cord is plugged into the outlet.

Many TV sets, for example, do not have isolation transformers for reasons of economy. When working on the chassis, you should exercise care by using an external isolation transformer or by plugging into the outlet so that chassis ground is not connected to the 120-V side of the outlet.

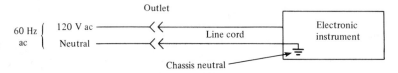

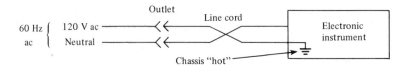

FIGURE 17–18 *Instrument powered without transformer isolation.*

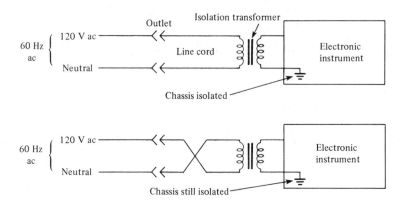

FIGURE 17–19 *Power line isolation.*

Review for 17–8

1. Name two applications of a transformer as an isolation device.
2. Will a transformer operate with a dc input?

17-9 TAPPED TRANSFORMERS

A schematic diagram of a transformer with a *center-tapped* secondary winding is shown in Figure 17–20A. The center tap (CT) is equivalent to two secondary windings with half the total voltage across each.

The voltages between either end of the secondary and the center tap are, at any instant, equal in magnitude but opposite in polarity, as illustrated in Figure 17–20B. Here, for example, at some instant on the sine wave voltage, the polarity across the entire secondary is as shown (top end +, bottom −). At the center tap, the voltage is less positive than the top end but more positive than the bottom end of the secondary. Therefore, measured *with respect to the center tap,* the top end of the secondary is positive, and the bottom end is negative. This center-tapped feature is used in power supply rectifiers in which the ac voltage is converted to dc.

Some transformers have taps on the secondary other than at the center. Also, when a transformer is to be used on a power line, several taps are sometimes brought out near one end of the primary. When connection is made to one of these taps instead of to the end of the winding, the turns ratio is adjusted to overcome line voltages that are slightly too high or too low. Example diagrams are shown in Figure 17–21.

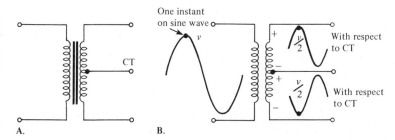

FIGURE 17–20 *A transformer with a center-tapped secondary.*

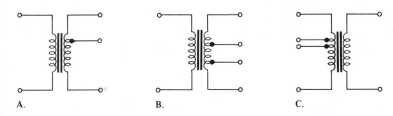

FIGURE 17–21 *Tapped transformers.*

Review for 17-9

1. What is a center tap?
2. How are the voltages at each end of the secondary related to the CT?

17-10

MULTIPLE-WINDING TRANSFORMERS

The basic transformer with one primary and one secondary is the simplest type of multiple-winding transformer; it has two windings. Other transformers may have more than one primary or more than one secondary.

Multiple Primaries

Some transformers are designed to operate from either 120-V ac or 240-V ac lines. These transformers usually have two primary windings, each of which is designed for 120 V ac. When the two are connected in series, the transformer can be used for 240-V ac operation, as illustrated in Figure 17–22.

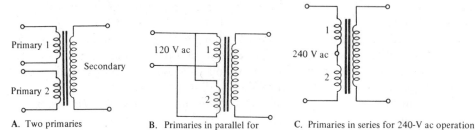

A. Two primaries
B. Primaries in parallel for 120-V ac operation
C. Primaries in series for 240-V ac operation

FIGURE 17–22 *Multiple-primary transformer.*

Multiple Secondaries

More than one secondary can be wound on a common core. Transformers with several secondaries are often used to achieve several voltages by either stepping up or stepping down the primary voltage. These types are commonly used in power supply applications in which several voltage levels are required for the operation of an electronic instrument.

A typical schematic of a multiple-secondary transformer is shown in Figure 17–23; this transformer has three secondaries. Sometimes you will find combinations of multiple-primary, multiple-secondary, and tapped transformers all in one unit.

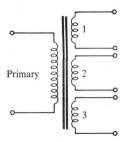

FIGURE 17–23 *Multiple-secondary transformer.*

AUTOTRANSFORMERS

Example 17-10

The transformer shown in Figure 17-24 has the numbers of turns indicated. One of the secondaries is also center-tapped. If 120 V ac are connected to the primary, determine each secondary voltage and the voltages with respect to CT on secondary 2.

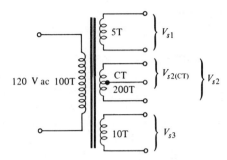

FIGURE 17-24

Solution:

$$V_{s1} = \left(\frac{5}{100}\right)120 \text{ V} = 6 \text{ V}$$

$$V_{s2} = \left(\frac{200}{100}\right)120 \text{ V} = 240 \text{ V}$$

$$V_{s2}(\text{CT}) = \frac{240 \text{ V}}{2} = 120 \text{ V}$$

$$V_{s3} = \left(\frac{10}{100}\right)120 \text{ V} = 12 \text{ V}$$

Review for 17-10

1. A certain transformer has two secondaries. The turns ratio from primary to secondary 1 is 10. The turns ratio from the primary to the other secondary is 0.2. If 240 V ac are applied to the primary, what are the secondary voltages?

17-11

AUTOTRANSFORMERS

In an autotransformer, *one winding* serves as both the primary and the secondary. The winding is tapped at the proper points to achieve the desired turns ratio for stepping up or stepping down the voltage.

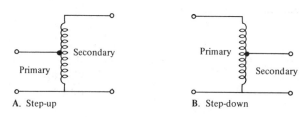

A. Step-up   B. Step-down

FIGURE 17-25 *Autotransformers.*

The autotransformer is normally smaller and lighter than a typical multiple-winding type. However, since the primary and the secondary are the same winding, there is no electrical isolation between the windings. Autotransformers are used quite often in TV receivers. Figure 17-25A shows a diagram for a step-up autotransformer, and Part B shows a step-down autotransformer diagram.

A *variac* is an adjustable autotransformer in which the output (secondary) voltage can be varied by means of an adjustable contact tap.

Review for 17-11

1. What is the difference between an autotransformer and a multiple-winding transformer?
2. What is a variac?

17-12

TRANSFORMER CONSTRUCTION

Two methods of construction for iron core transformers are shown in Figure 17-26. The core is normally made from *laminated* sheets of ferromagnetic material separated by thin insulating sheets to reduce losses in the core.

Construction of a slug core transformer is shown in Figure 17-27. An air core transformer often has this type of construction.

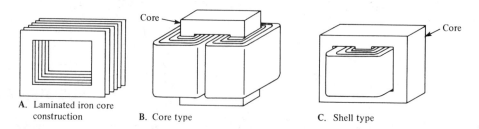

A. Laminated iron core construction   B. Core type   C. Shell type

FIGURE 17-26 *Construction of iron core transformer.*

TRANSFORMER CONSTRUCTION

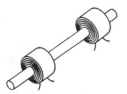

FIGURE 17–27 *Slug core transformer.*

Transformer Ratings

Typically, a transformer rating is specified as, for example, 2 kVA, 500/50, 60 Hz. The 500 and the 50 can be either secondary or primary voltages. If 500 is the primary voltage, then 50 is the secondary voltage, and vice versa. The 2-kVA value is the *apparent power rating*.

Let us assume, for example, that in this transformer, 50 V is the secondary voltage. In this case the load current is $I_L = P_a/V_s = 2$ kVA/50 V = 40 A. If, on the other hand, 500 V is the secondary voltage, then $I_L = P_a/V_s = 2$ kVA/500 V = 4 A. These are the maximum currents that the secondary can handle in either case.

The reason that the power rating is in volt-amperes (VA) rather than average power (watts) is as follows: If the transformer load is purely capacitive, for example, the average power delivered to the load is zero. However, the current for $V_s = 500$ V and $X_C = 100$ Ω at 60 Hz is $I_L = 500$ V/100 Ω = 5 A. This current exceeds the maximum that the secondary can handle, and it can cause damage to the transformer. Thus, you can see that it is meaningless to specify average power.

Figure 17–28 shows some typical types of **transformers.**

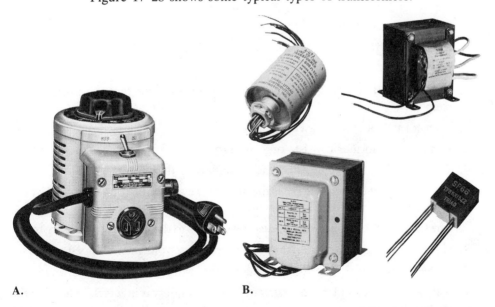

A.   **B.**

FIGURE 17–28 *Typical transformers.* **A.** *Variable. (Courtesy of The Superior Electric Company)* **B.** *Fixed. (Courtesy of Litton Triad-Utrad)*

Review for 17-12

1. What is laminated core construction?
2. What is the difference between construction of a shell type of transformer and that of a core type?

Formulas

$$k = \frac{\text{flux linking coil 1 and coil 2}}{\text{total flux produced by coil 1}} \qquad (17\text{--}1)$$

$$L_M = k\sqrt{L_1 L_2} \qquad (17\text{--}2)$$

$$\text{Turns ratio} = \frac{N_s}{N_p} \qquad (17\text{--}3)$$

$$\frac{V_s}{V_p} = \frac{N_s}{N_p} \qquad (17\text{--}4)$$

$$V_s = \left(\frac{N_s}{N_p}\right) V_p \qquad (17\text{--}5)$$

$$\frac{I_p}{I_s} = \frac{N_s}{N_p} \qquad (17\text{--}6)$$

$$I_s = \left(\frac{N_p}{N_s}\right) I_p \qquad (17\text{--}7)$$

$$Z_p = \left(\frac{N_p}{N_s}\right)^2 Z_L \qquad (17\text{--}8)$$

Summary

1. Mutual inductance is the inductance between two coils that are magnetically coupled.
2. Transformer action is based on mutual induction.
3. A basic transformer consists of a primary winding and a secondary winding.
4. The primary is the input, and the secondary is the output.
5. The coefficient of coupling (k) is the ratio of the amount of flux produced by the primary that links the secondary to the total flux produced by the primary.
6. The turns ratio is the number of secondary turns to the number of primary turns.
7. If the turns ratio is greater than 1, the primary voltage is stepped up.
8. If the turns ratio is less than 1, the primary voltage is stepped down.
9. With a loaded secondary, the current in the secondary is inversely related to the turns ratio.
10. The primary "sees" a certain impedance reflected from the secondary. The value of this impedance depends on the square of the reciprocal of the turns ratio.

11. For maximum power transfer to a load, the reflected load impedance must equal the source impedance, a condition called *impedance matching*.
12. A transformer can be used as an isolation device to keep dc from getting from the primary to the secondary or to isolate power line ground from equipment ground.
13. An autotransformer has one tapped winding used for both primary and secondary.

Self-Test

1. In a given transformer, 90% of the flux produced by the primary coil links the secondary. What is the value of the coefficient of coupling, k?
2. For the transformer in Problem 1, the inductance of the primary is 10 µH, and the inductance of the secondary is 5 µH. Determine L_M.
3. What is the turns ratio of a transformer having 12 primary turns and 36 secondary turns? Is it a step-up or a step-down transformer?
4. Determine the polarity of the secondary voltage for each transformer in Figure 17–29.

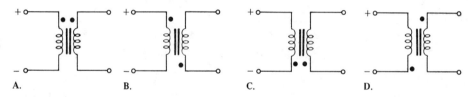

FIGURE 17–29

5. One hundred twenty volts ac are connected to the primary of a transformer with 10 primary turns and 15 secondary turns. What is the secondary voltage?
6. With 50 V ac across the primary, what is the secondary voltage if the turns ratio is 0.5?
7. Two hundred forty volts ac are applied to a transformer primary. The secondary voltage is 60 V. What is the turns ratio?
8. If a load is connected to the transformer in Problem 7, and if I_p is 0.25 A, what is I_s?
9. What is the reflected impedance in Figure 17–30?
10. If the source resistance in Figure 17–30 is 10 Ω, what must R_L be for maximum power?

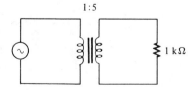

FIGURE 17–30

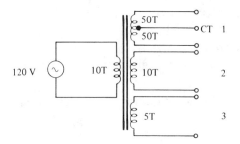

FIGURE 17-31

11. Determine the voltage for each secondary in Figure 17-31.
12. In Figure 17-31, what is the voltage from each end of the upper secondary to the center tap? Are the polarities the same?

Problems

17-1. If $k = 0.75$, $L_p = 1\ \mu\text{H}$, and $L_s = 4\ \mu\text{H}$, what is the mutual inductance?

17-2. What is k if $L_M = 1\ \mu\text{H}$, $L_p = 8\ \mu\text{H}$, and $L_s = 2\ \mu\text{H}$?

17-3. There are 25 turns in the primary of a transformer. In order to double the voltage, how many turns must be in the secondary?

17-4. To step 120 V ac down to 30 V, what must the turns ratio be?

17-5. To step 240 V ac up to 720 V, what must the turns ratio be?

17-6. For each transformer in Figure 17-32, sketch the output sine wave, showing its polarity at point A with respect to B. Also indicate its amplitude.

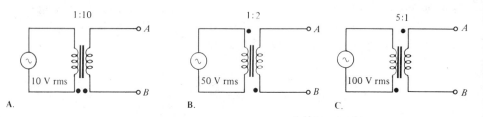

FIGURE 17-32

17-7. Determine I_s in Figure 17-33. What is R_L?

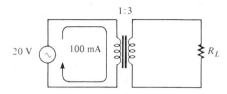

FIGURE 17-33

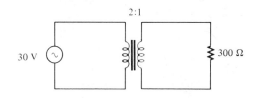

FIGURE 17-34

ANSWERS TO SECTION REVIEWS

17–8. Determine the following in Figure 17–34:
  (a) The primary current   (b) The secondary current
  (c) The secondary voltage

17–9. What is the load power in Figure 17–34?

17–10. For the circuit in Figure 17–35, find the turns ratio required to deliver maximum power to the 4-Ω speaker.

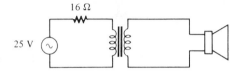

FIGURE 17–35

17–11. In Figure 17–35, what is the maximum power delivered to the speaker?

17–12. Find the secondary voltage for each autotransformer in Figure 17–36.

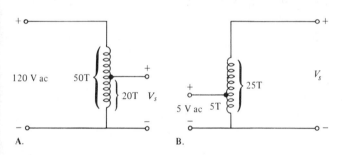 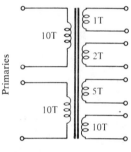

FIGURE 17–36   **FIGURE 17–37**

17–13. In Figure 17–37, each primary can accommodate 120 V ac. How should the primaries be connected for 240-V ac operation? Determine each secondary voltage.

17–14. A transformer is rated at 1 kVA. It operates on 60 Hz, 120 V ac. The secondary voltage is 600 V.
  (a) What is the maximum load current?
  (b) What is the smallest R_L that you could drive?
  (c) What is the largest capacitor that could be connected as a load?

17–15. What is the kVA rating of a transformer that can handle a maximum load current of 10 A with a secondary voltage of 2.5 kV?

Answers to Section Reviews

Section 17–1:
1. The inductance between two coils. **2.** 45 mH. **3.** It increases.

Section 17–2:
1. Two or more coils mutually coupled by electromagnetic induction. 2. Number of turns in secondary divided by number of turns in primary. 3. It influences k. 4. They determine polarities.

Section 17–3:
1. Increases primary voltage. 2. Five times greater.

Section 17–4:
1. Decreases primary voltage. 2. 60 V.

Section 17–5:
1. Less; half. 2. 2 A.

Section 17–6:
1. The impedance in the secondary reflected into the primary. 2. The turns ratio. 3. 0.5 Ω.

Section 17–7:
1. Making the load impedance equal the source impedance. 2. Maximum power is delivered to the load. 3. 400 Ω.

Section 17–8:
1. dc isolation and power line isolation. 2. No.

Section 17–9:
1. A contact at the center of a winding. 2. Each is opposite to the other in polarity.

Section 17–10:
1. 2400 V, 48 V.

Section 17–11:
1. An autotransformer has only one winding. 2. A variable autotransformer.

Section 17–12:
1. Thin sheets of a ferromagnetic material separated by sheets of insulating material.
2. The shell type has the windings on one leg of the core.

In Chapters 12 and 15, we demonstrated how to determine impedance, phase angle, voltage, and current in basic RC and RL circuits. We found that there are phase differences between various currents and voltages that make the use of phasor quantities convenient.

In this chapter we will expand on those two chapters so that you can do more complicated circuit analysis, such as analysis of circuits containing combinations of R, L, and C elements. A system of mathematics that aids the analysis of reactive circuits is introduced and used in this chapter. It is the *system of complex numbers*.

18–1 Complex Numbers
18–2 Rectangular and Polar Forms
18–3 Arithmetic of Complex Numbers
18–4 Resistance and Reactance in Complex Form
18–5 Total Impedance in Complex Form
18–6 Impedance in RLC Circuits
18–7 Complex Analysis of Reactive Circuits
18–8 Series Equivalent Circuits

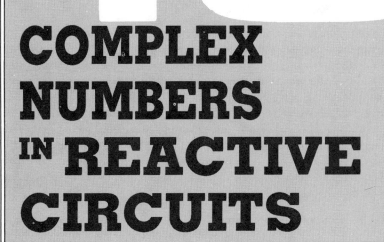

18
COMPLEX NUMBERS IN REACTIVE CIRCUITS

18-1

COMPLEX NUMBERS

The complex number system is used to determine *magnitude* and *phase angle* of electrical quantities. As you know, both of these factors are used in reactive circuits and have been represented by phasors. Essentially, complex numbers allow us to add, subtract, multiply, and divide phasor quantities in a systematic way.

Positive and Negative Numbers

Both positive and negative values can be represented on the horizontal axis of a graph, as illustrated in Figure 18–1A, where +5 and −5 are the points along the axis as shown. Also, both positive and negative values can be represented along the vertical axis, as illustrated in Figure 18–1B, where +3 and −3 are points along the axis.

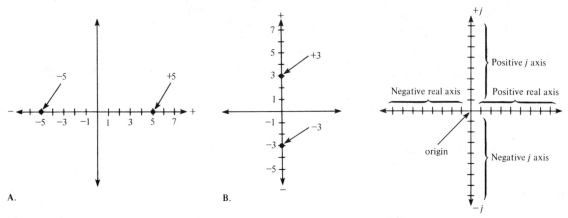

FIGURE 18–1 *Graphical representation of positive and negative numbers.*

FIGURE 18–2 *The complex plane.*

Complex Plane

To distinguish between values on the horizontal axis and values on the vertical axis, we can use what is called the *complex plane* (don't let the word *complex* frighten you). In mathematical terms, the horizontal axis is called the *real axis,* and the vertical axis is called the *imaginary axis* (there is really nothing imaginary about it, as you will see).

In electrical work, a $\pm j$ prefix is used for values that lie on the imaginary axis to distinguish them from values on the horizontal (real) axis. This prefix is known as the *j operator*. In pure mathematics, an *i* is often used instead of a *j*, but in electric circuits, the *i* can be confused with instantaneous current.

Figure 18–2 shows a complex plane. The origin is the zero value point where the axes cross.

Angles on the Complex Plane

Angular positions can be represented on the complex plane as illustrated in Figure 18–3. The positive real axis represents 0°. As we rotate

COMPLEX NUMBERS

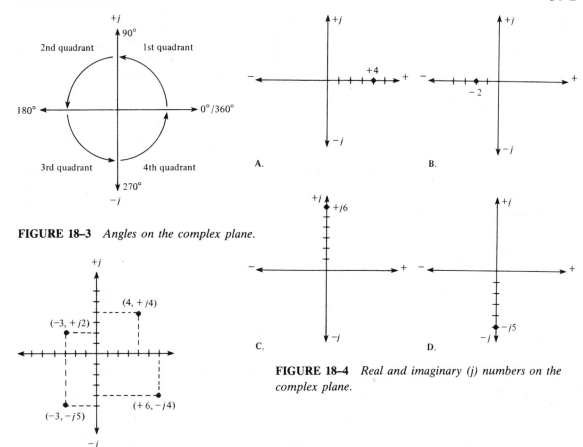

FIGURE 18-3 Angles on the complex plane.

FIGURE 18-4 Real and imaginary (j) numbers on the complex plane.

FIGURE 18-5 Points on the complex plane.

counterclockwise, the $+j$ axis represents 90°. Continuing on to the negative real axis, we have rotated through 180°. The $-j$ axis is the 270° point; then, as we return to the positive real axis, we have rotated through a complete 360°. Notice that the plane is sectioned into four *quadrants*.

Representing a Point on the Complex Plane

A point on the complex plane can be classified as real, or imaginary ($\pm j$), or a combination of the two. For example, a point located 4 units from the origin on the positive real axis is the *positive real number* $+4$, as shown in Figure 18-4A. A point 2 units from the origin on the negative real axis is the *negative real number* -2, as shown in Part B. If a point is on the $+j$ axis 6 units from the origin, as in Part C, it is designated $+j6$. Finally, if a point is 5 units along the $-j$ axis, it is designated $-j5$, as in Part D.

If a point lies not on any axis but somewhere in one of the four quadrants, it is a complex number and can be defined by its *coordinates*. For example, in Figure 18-5, the point in the first quadrant has a real value of $+4$ and a j value of $+j4$. The point in the second quadrant has coordinates -3 and $+j2$. The point in the third quadrant has coordinates -3 and $-j5$. The point in the fourth quadrant has coordinates $+6$ and $-j4$.

Value of j

The j operator has a value of $\sqrt{-1}$. For example, if we multiply the positive real value $+2$ by j, the result is $+j2$. Thus, we effectively move the $+2$ through 90° to the $+j$ axis. If we then multiply $+j2$ by j, we get

$$j^2 2 = \sqrt{-1} \times \sqrt{-1} \times 2 = -1 \times 2 = -2$$

This calculation effectively places the value on the negative real axis.

Therefore, multiplying by j^2 actually converts a positive real number to a negative real number, which, in effect, is a rotation of 180° on the complex plane. Multiplying the $+2$ by $-j$ rotates it $-90°$ to the $-j$ axis, as illustrated in Figure 18–6.

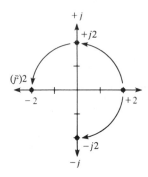

FIGURE 18–6 *Effect of the j operator on location of a number on the complex plane.*

Review for 18–1

1. Name the two axes on the complex plane.
2. What is the j operator?
3. What is the angular difference between the following numbers: $+4$, $+j4$, and -4?

18–2

RECTANGULAR AND POLAR FORMS

Complex numbers can be used to express phasor values in two ways: *rectangular* form and *polar* form. Each has certain advantages in circuit calculations which you will learn later.

A phasor quantity contains both *magnitude* and *phase*. We use V, I, X_C, X_L, R, and Z to represent the magnitudes of these quantities. In this chapter, boldface letters **V**, **I**, $\mathbf{X}_C$, $\mathbf{X}_L$, **R**, and **Z** will be used to represent the complete phasor quantities.

Rectangular Form

A phasor quantity in *rectangular form* is the sum of the *real* coordinate and the j coordinate. An "arrow" drawn from the origin to the coordinate point in the

RECTANGULAR AND POLAR FORMS

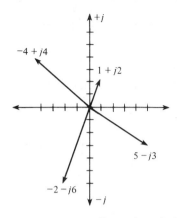

FIGURE 18–7 *Examples of phasors specified by rectangular coordinates.*

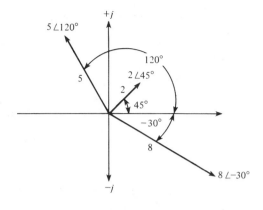

FIGURE 18–8 *Examples of phasors specified by polar values.*

complex plane is the phasor. Examples are $1 + j2, 5 - j3, -4 + j4,$ and $-2 - j6$, which are plotted on the complex plane in Figure 18–7. As you can see, the rectangular coordinates identify both the length and the direction of the phasor.

Polar Form

Phasor quantities can also be expressed in *polar form*. This form consists of the phasor magnitude and the angle relative to the positive real axis. Examples are $2 \angle 45°$, $5 \angle 120°$, and $8 \angle -30°$. The first number is the magnitude or length of the phasor, and the symbol $\angle$ precedes the angle. The angle is stated with respect to 0°. Figure 18–8 shows these phasors on the complex plane. Remember that for every phasor expressed in polar form, there is also an equivalent rectangular form expression.

Rectangular-to-Polar Conversion

Many hand-held scientific calculators have provisions for converting from rectangular form to polar form and vice versa. However, we will discuss the conversion method so that you will understand the mathematical procedure.

First, the phasor can be thought of as forming a *right triangle* in the complex plane, as indicated in Figure 18–9. The horizontal side of the triangle is the real value, A, and the vertical side is the j value, B. The hypotenuse of the triangle is the length or magnitude of the phasor, which is equal to

$$\sqrt{A^2 + B^2}$$

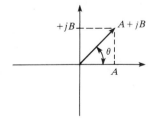

FIGURE 18–9 *Right triangle relationship in the complex plane.*

by the Pythagorean theorem. This relationship is given in the following equation:

$$C = \sqrt{A^2 + B^2} \tag{18-1}$$

where C is a general term for the phasor magnitude.

Next, the angle θ indicated in the right triangle of Figure 18–9 is

$$\theta = \arctan\left(\frac{B}{A}\right) \tag{18-2}$$

The general formula for converting from rectangular to polar form is stated as follows:

$$A \pm jB = \sqrt{A^2 + B^2} \angle \pm \arctan\left(\frac{B}{A}\right) = C \angle \pm \theta° \tag{18-3}$$

The following example will illustrate this conversion.

Example 18–1

Convert the following complex numbers from rectangular form to polar form:

(a) $8 + j6$ (b) $10 + j5$ (c) $12 - j18$ (d) $25 - j15$

Solution:

(a) The magnitude of the phasor represented by $8 + j6$ is

$$C = \sqrt{8^2 + 6^2} = \sqrt{100} = 10$$

The angle is

$$\theta = \arctan\left(\frac{6}{8}\right) = 36.87°$$

The complete polar expression for this phasor is

$$C = 10 \angle 36.87°$$

(b) The magnitude of the phasor represented by $10 + j5$ is

$$C = \sqrt{10^2 + 5^2} = \sqrt{125} = 11.18$$

The angle is

$$\theta = \arctan\left(\frac{5}{10}\right) = 26.57°$$

The complete polar expression for this phasor is

$$C = 11.18 \angle 26.57°$$

(c) The magnitude of the phasor represented by $12 - j18$ is

$$C = \sqrt{12^2 + (-18)^2} = \sqrt{468} = 21.63$$

RECTANGULAR AND POLAR FORMS

The angle is

$$\theta = \arctan\left(\frac{-18}{12}\right) = -56.31°$$

The complete polar expression for this phasor is

$$\mathbf{C} = 21.63 \angle -56.31°$$

(d) The magnitude of the phasor represented by $25 - j15$ is

$$C = \sqrt{25^2 + (-15)^2} = \sqrt{850} = 29.15$$

The angle is

$$\theta = \arctan\left(\frac{-15}{25}\right) = -30.96°$$

The complete polar expression for this phasor is

$$\mathbf{C} = 29.15 \angle -30.96°$$

Polar-to-Rectangular Conversion

To convert from polar to rectangular form, we require a reverse process. The polar form gives us the magnitude and angle, as indicated in Figure 18–10. To get the rectangular form, we need to know the A and B sides of the triangle. Using trigonometry, we can find them with the following equations:

$$A = C \cos \theta \qquad (18\text{–}4)$$

$$B = C \sin \theta \qquad (18\text{–}5)$$

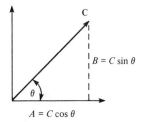

FIGURE 18–10

The general conversion formula is stated as follows:

$$C \angle \theta° = C \cos \theta \pm jC \sin \theta = A + jB \qquad (18\text{–}6)$$

The following example will demonstrate this conversion.

Example 18–2

Convert the following polar quantities to rectangular form:
(a) $10 \angle 30°$ (b) $50 \angle 60°$ (c) $200 \angle -45°$ (d) $4 \angle 135°$

Example 18–2 (continued)

Solution:

(a) The real part of the phasor represented by 10 ∠30° is
$$A = 10 \cos 30° = 10(0.866) = 8.66$$
The j part of this phasor is
$$B = 10 \sin 30° = 10(0.5) = 5$$
The complete rectangular expression is
$$8.66 + j5$$

(b) The real part of the phasor represented by 50 ∠60° is
$$A = 50 \cos 60° = 50(0.5) = 25$$
The j part is
$$B = 50 \sin 60° = 50(0.866) = 43.3$$
The complete rectangular expression is
$$25 + j43.3$$

(c) The real part of the phasor represented by 200 ∠−45° is
$$A = 200 \cos(-45°) = 141.4$$
The j part is
$$B = 200 \sin(-45°) = -141.4$$
The complete rectangular expression is
$$141.4 - j141.4$$

(d) The real part of the phasor represented by 4 ∠135° is
$$A = 4 \cos 135° = 4(-0.707) = -2.828$$
The j part is
$$B = 4 \sin 135° = 4(0.707) = 2.828$$
The complete rectangular expression is
$$-2.828 + j2.828$$

Review for 18–2

1. Name the two parts of a complex number that is in rectangular form.
2. Name the two parts of a complex number that is in polar form.
3. Convert $2 + j2$ to polar form.
4. Convert 5 ∠45° to rectangular form.

18-3 ARITHMETIC OF COMPLEX NUMBERS

Addition

Complex numbers must be in rectangular form in order for us to add them. The rule is as follows: *Add the real parts separately, and then add the j parts.*

Example 18-3

Add $8 + j5$ and $2 + j1$.

Solution:

$$(8 + j5) + (2 + j1) = (8 + 2) + j(5 + 1)$$
$$= 10 + j6$$

Example 18-4

Add $20 - j10$ and $12 + j6$.

Solution:

$$(20 - j10) + (12 + j6) = (20 + 12) + j(-10 + 6) = 32 + j(-4)$$
$$= 32 - j4$$

Subtraction

As in addition, the numbers must be in rectangular form to be subtracted. The rule is as follows: *Subtract the real parts separately, and then subtract the j parts.*

Example 18-5

Subtract $1 + j2$ from $3 + j4$.

Solution:

$$(3 + j4) - (1 + j2) = (3 - 1) + j(4 - 2)$$
$$= 2 + j2$$

Example 18–6

Subtract $10 - j8$ from $15 + j15$.

Solution:

$$(15 + j15) - (10 - j8) = (15 - 10) + j(15 - (-8))$$
$$= 5 + j23$$

Multiplication

Multiplication is done most easily in polar form. The rule is as follows: *Multiply the magnitudes, and add the angles algebraically.*

Example 18–7

Multiply $10 \angle 45°$ times $5 \angle 20°$.

Solution:

$$(10 \angle 45°)(5 \angle 20°) = (10)(5) \angle (45° + 20°)$$
$$= 50 \angle 65°$$

Example 18–8

Multiply $2 \angle 60°$ times $4 \angle -30°$.

Solution:

$$(2 \angle 60°)(4 \angle -30°) = (2)(4) \angle (60° + (-30°))$$
$$= 8 \angle 30°$$

Division

Like multiplication, division is done most easily in polar form. The rule is as follows: *Divide the magnitudes, and subtract the denominator angle from the numerator angle.*

RESISTANCE AND REACTANCE IN COMPLEX FORM

Example 18–9

Divide 100 ∠50° by 25 ∠20°.

Solution:

$$\frac{100 \angle 50°}{25 \angle 20°} = \frac{100}{25} \angle (50° - 20°)$$
$$= 4 \angle 30°$$

Example 18–10

Divide 15 ∠10° by 3 ∠−30°.

Solution:

$$\frac{15 \angle 10°}{3 \angle -30°} = \frac{15}{3} \angle (10° - (-30°))$$
$$= 5 \angle 40°$$

Review for 18–3

1. Add $1 + j2$ and $3 - j1$.
2. Subtract $12 + j18$ from $15 + j25$.
3. Multiply 8 ∠45° times 2 ∠65°.
4. Divide 30 ∠75° by 6 ∠60°.

18–4

RESISTANCE AND REACTANCE IN COMPLEX FORM

For an RC circuit, X_C is represented by a phasor −90° from **R**, as shown in Figure 18–11A. For an RL circuit, X_L is represented by a phasor +90° from **R**, as illustrated in Figure 18–11B.

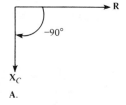
A.
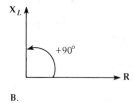
B.

FIGURE 18–11

Resistance

Since resistance causes no phase angle between voltage and current, we can assign it an angle of 0°. In polar form the resistance is expressed as $R \angle 0°$. For a pure resistance there is no j term; so in rectangular form, it is $R + j0$ or just R. Notice that in the phasor diagram of Figure 18–11, **R** lies along the real axis corresponding to 0°.

Reactance

Reactance causes a 90° phase angle between current and voltage. $\mathbf{X}_C$ causes the voltage to *lag* the current by 90°; so in polar form it is assigned a $-90°$ angle and is expressed as $X_C \angle -90°$. In rectangular form, it is expressed as $-jX_C$. Notice that $\mathbf{X}_C$ lies along the $-j$ axis in the phasor diagram.

$\mathbf{X}_L$ causes the voltage to *lead* the current by 90°; so it is assigned a $+90°$ angle in polar form. It is expressed as $X_L \angle 90°$. In rectangular form, it is expressed as jX_L. Notice that $\mathbf{X}_L$ lies along the $+j$ axis in the phasor diagram.

In the case of both $\mathbf{X}_C$ and $\mathbf{X}_L$, the j indicates the 90° phase shift.

Example 18–11

Express the resistance or reactance for each circuit in Figure 18–12 in both polar form and rectangular form.

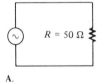
A.

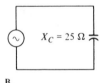
B.

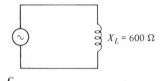
C.

FIGURE 18–12

Solution:

Part A: The resistance is expressed as $50 \angle 0°$ Ω in polar form and as 50 Ω in rectangular form.

Part B: The capacitive reactance is expressed as $25 \angle -90°$ Ω in polar form and as $-j25$ Ω in rectangular form.

Part C: The inductive reactance is expressed as $600 \angle 90°$ Ω in polar form and as $+j600$ Ω in rectangular form.

Review for 18–4

1. Write the following in polar form:
   (a) $R = 10$ Ω   (b) $X_C = 1$ kΩ   (c) $X_L = 30$ Ω

2. Write the following in rectangular form:
   (a) $R = 2 \text{ k}\Omega$ (b) $X_C = 47 \text{ }\Omega$ (c) $X_L = 5 \text{ }\Omega$

18–5

TOTAL IMPEDANCE IN COMPLEX FORM

Series RC Circuit

The impedance of the RC circuit in Figure 18–13 can be expressed in rectangular and polar forms. The rectangular form is most easily determined directly from the circuit and is stated in Equation (18–7). The polar form of the impedance is stated in Equation (18–8).

$$\mathbf{Z} = R - jX_C \tag{18–7}$$

$$\mathbf{Z} = \sqrt{R^2 + X_C^2} \angle -\arctan\left(\frac{X_C}{R}\right) \tag{18–8}$$

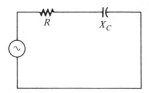

FIGURE 18–13 *Series RC circuit and its impedance triangle.*

Series RL Circuit

The impedance of the RL circuit in Figure 18–14 is expressed in rectangular form by Equation (18–9) and in polar form by Equation (18–10).

$$\mathbf{Z} = R + jX_L \tag{18–9}$$

$$\mathbf{Z} = \sqrt{R^2 + X_L^2} \angle \arctan\left(\frac{X_L}{R}\right) \tag{18–10}$$

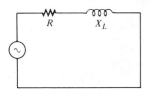

FIGURE 18–14 *Series RL circuit and its impedance triangle.*

Parallel RC Circuit

For the parallel RC circuit in Figure 18–15, the impedance is developed as follows using the "product over the sum" formula, which we used earlier in parallel resistive circuits:

$$Z = \frac{(R \angle 0°)(X_C \angle -90°)}{R - jX_C} = \frac{RX_C \angle (0° - 90°)}{\sqrt{R^2 + X_C^2} \angle -\arctan \frac{X_C}{R}}$$

$$Z = \frac{RX_C}{\sqrt{R^2 + X_C^2}} \angle \left(-90° + \arctan \frac{X_C}{R}\right) \tag{18-11}$$

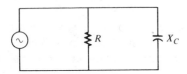

FIGURE 18–15 *Parallel RC circuit.*

Parallel RL Circuit

For the parallel RL circuit in Figure 18–16, the impedance is developed as follows, in the same way as for the parallel RC circuit:

$$Z = \frac{(R \angle 0°)(X_L \angle 90°)}{R + jX_L} = \frac{RX_L \angle (0° + 90°)}{\sqrt{R^2 + X_L^2} \angle \arctan \frac{X_L}{R}}$$

$$Z = \frac{RX_L}{\sqrt{R^2 + X_L^2}} \angle \left(90° - \arctan \frac{X_L}{R}\right) \tag{18-12}$$

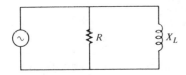

FIGURE 18–16 *Parallel RL circuit.*

Example 18–12

For each of the circuits in Figure 18–17, express the impedance in both rectangular form and polar form.

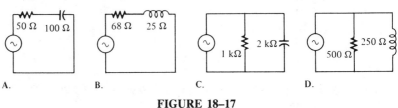

FIGURE 18–17

TOTAL IMPEDANCE IN COMPLEX FORM

Solution:

Circuit A:

$$\mathbf{Z} = R - jX_C$$
$$= 50\ \Omega - j100\ \Omega \quad \text{in rectangular form}$$
$$\mathbf{Z} = \sqrt{R^2 + X_C^2}\ \angle -\arctan\left(\frac{X_C}{R}\right)$$
$$= \sqrt{(50\ \Omega)^2 + (100\ \Omega)^2}\ \angle -\arctan\left(\frac{100}{50}\right)$$
$$= 111.8\ \angle -63.4°\ \Omega \quad \text{in polar form}$$

Circuit B:

$$\mathbf{Z} = R + jX_L$$
$$= 68\ \Omega + j25\ \Omega \quad \text{in rectangular form}$$
$$\mathbf{Z} = \sqrt{R^2 + X_L^2}\ \angle \arctan\left(\frac{X_L}{R}\right)$$
$$= \sqrt{(68\ \Omega)^2 + (25\ \Omega)^2}\ \angle \arctan\left(\frac{25}{68}\right)$$
$$= 72.45\ \angle 20.19°\ \Omega \quad \text{in polar form}$$

Circuit C:

$$\mathbf{Z} = \frac{RX_C}{\sqrt{R^2 + X_C^2}}\ \angle\left(-90° + \arctan\frac{X_C}{R}\right)$$
$$= \frac{(1\ \text{k}\Omega)(2\ \text{k}\Omega)}{\sqrt{(1\ \text{k}\Omega)^2 + (2\ \text{k}\Omega)^2}}\ \angle\left(-90° + \arctan\frac{2}{1}\right)$$
$$= 894.4\ \angle -26.57°\ \Omega \quad \text{in polar form}$$
$$\mathbf{Z} = 894.4\ \cos(-26.57°) + j894.4\ \sin(-26.57°)$$
$$= 799.9\ \Omega - j400\ \Omega \quad \text{in rectangular form}$$

Circuit D:

$$\mathbf{Z} = \frac{RX_L}{\sqrt{R^2 + X_L^2}}\ \angle\left(90° - \arctan\frac{X_L}{R}\right)$$
$$= \frac{(500\ \Omega)(250\ \Omega)}{\sqrt{(500\ \Omega)^2 + (250\ \Omega)^2}}\ \angle\left(90° - \arctan\frac{250}{500}\right)$$
$$= 223.6\ \angle 63.43°\ \Omega \quad \text{in polar form}$$
$$\mathbf{Z} = 223.6\ \cos(63.43°) + j223.6\ \sin(63.43°)$$
$$= 100\ \Omega + j199.99\ \Omega \quad \text{in rectangular form}$$

Review for 18-5

1. For Figure 18–18, express **Z** in rectangular and polar forms.

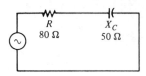

FIGURE 18–18

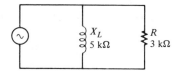

FIGURE 18–19

2. Express **Z** in polar form for Figure 18–19.

18-6

IMPEDANCE IN RLC CIRCUITS

Series RLC Circuit

A series RLC circuit is shown in Figure 18–20. It contains resistance, inductance, and capacitance. As you know, X_L causes the total current to lag the applied voltage. X_C has the opposite effect: It causes current to lead the voltage. Thus, X_L and X_C tend to offset each other. If they are equal, they cancel, and the total reactance is zero. In any case, the magnitude of the total reactance is $X_L - X_C$.

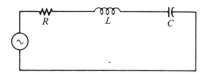

FIGURE 18–20 *Series RLC circuit.*

The total impedance for the series RLC circuit is given in rectangular form in Equation (18–13) and in polar form in Equation (18–14).

$$\mathbf{Z} = R + jX_L - jX_C \tag{18-13}$$

$$\mathbf{Z} = \sqrt{R^2 + (X_L - X_C)^2} \; \angle \arctan\left(\frac{X_L - X_C}{R}\right) \tag{18-14}$$

Example 18–13

Determine the impedance in Figure 18–21. Express it in polar and rectangular forms.

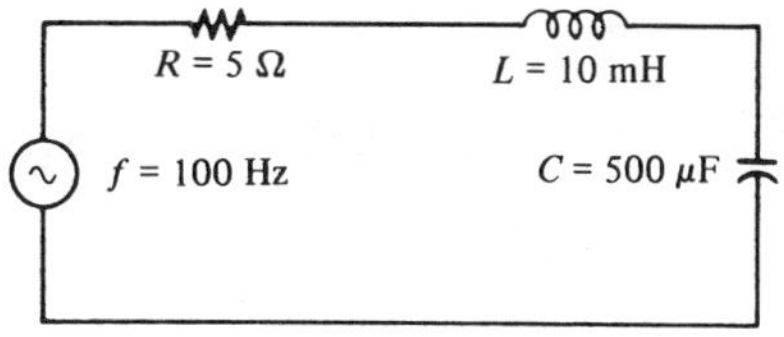

IMPEDANCE IN RLC CIRCUITS

FIGURE 18-21

Solution:

First find X_L and X_C:

$$X_L = 2\pi fL = 2\pi(100 \text{ Hz})(10 \text{ mH})$$
$$= 6.28 \text{ }\Omega$$

$$X_C = \frac{1}{2\pi fC} = \frac{1}{2\pi(100 \text{ Hz})(500 \text{ }\mu\text{F})}$$
$$= 3.18 \text{ }\Omega$$

In this case, X_L is greater, and thus the circuit is more inductive than capacitive. The net reactance magnitude is

$$X_L - X_C = 6.28 \text{ }\Omega - 3.18 \text{ }\Omega = 3.1 \text{ }\Omega$$

The rectangular form impedance is

$$\mathbf{Z} = R + jX_L - jX_C = 5 \text{ }\Omega + j6.28 \text{ }\Omega - j3.18 \text{ }\Omega$$
$$= 5 \text{ }\Omega + j3.1 \text{ }\Omega$$

The polar form impedance is

$$\mathbf{Z} = \sqrt{R^2 + X^2} \; \angle\arctan\left(\frac{X}{R}\right) = \sqrt{(5 \text{ }\Omega)^2 + (3.1 \text{ }\Omega)^2} \; \angle\arctan\left(\frac{3.1}{5}\right)$$
$$= 5.88 \; \angle 31.8° \text{ }\Omega$$

Parallel RLC Circuit

A parallel RLC circuit is shown in Figure 18-22. We can calculate its total impedance by using the *sum of the reciprocals* method, just as we did for all resistances in parallel. For a three-branch circuit, the formula is

$$\frac{1}{\mathbf{Z}} = \frac{1}{R \angle 0°} + \frac{1}{X_L \angle 90°} + \frac{1}{X_C \angle -90°} \qquad (18\text{-}15)$$

The formula can, of course, be extended to any number of parallel branches.

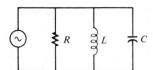

FIGURE 18–22 *Parallel RLC circuit.*

Example 18–14

For Figure 18–23, find **Z** in polar form.

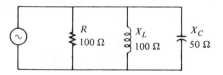

FIGURE 18–23

Solution:

$$\frac{1}{Z} = \frac{1}{R\angle 0°} + \frac{1}{X_L \angle 90°} + \frac{1}{X_C \angle -90°}$$

$$= \frac{1}{100 \angle 0°} + \frac{1}{100 \angle 90°} + \frac{1}{50 \angle -90°}$$

Applying the rule for division of polar numbers, we obtain

$$\frac{1}{Z} = 0.01 \angle 0° \text{ S} + 0.01 \angle -90° \text{ S} + 0.02 \angle 90° \text{ S}$$

When the denominator angle is brought up to the numerator, the sign changes, because it is effectively subtracted from 0°. That is, $0° - 90° = -90°$, and $0° - (-90°) = +90°$.

Converting each term to its rectangular equivalent yields

$$\frac{1}{Z} = 0.01 \text{ S} - j0.01 \text{ S} + j0.02 \text{ S}$$
$$= 0.01 \text{ S} + j0.01 \text{ S}$$

$$Z = \frac{1}{0.01 \text{ S} + j0.01 \text{ S}} = \frac{1}{\sqrt{(0.01 \text{ S})^2 + (0.01 \text{ S})^2} \angle \arctan(0.01/0.01)}$$

$$= \frac{1}{0.01414 \angle 45° \text{ S}} = 70.7 \angle -45° \text{ }\Omega$$

The negative angle shows that the circuit is more capacitive than inductive. Does this result surprise you? If so, think about it. The reason is that X_C is smaller than X_L. In a parallel circuit, the smaller quantity has the largest influence on the total current. Just as in the case of all resistances in parallel, the smaller draws the most current and has the most effect on the total R.

IMPEDANCE IN RLC CIRCUITS

In this circuit, the total current leads the total voltage by a phase angle of 45°.

Conductance, Susceptance, and Admittance

As you already know, the reciprocal of resistance is *conductance (G)*. In complex form, it is expressed as

$$\mathbf{G} = \frac{1}{R \angle 0°} = G \angle 0° = G \qquad (18\text{--}16)$$

The reciprocal of reactance is called *susceptance*, designated as B_C or B_L. The complex forms of capacitive susceptance and inductive susceptance are stated in the following equations:

$$\mathbf{B}_C = \frac{1}{X_C \angle -90°} = B_C \angle 90° = jB_C \qquad (18\text{--}17)$$

$$\mathbf{B}_L = \frac{1}{X_L \angle 90°} = B_L \angle -90° = -jB_L \qquad (18\text{--}18)$$

The reciprocal of impedance is *admittance*. The complex forms are

$$\mathbf{Y} = \frac{1}{Z \angle \pm\theta°} = Y \angle \mp\theta° = G + jB_C - jB_L \qquad (18\text{--}19)$$

The unit of each of these variables is siemens (S).

Example 18–15

Determine the conductance, capacitive susceptance, inductive susceptance, and total admittance in Figure 18–24.

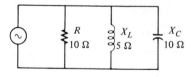

FIGURE 18–24

Solution:

$$G = \frac{1}{R \angle 0°} = \frac{1}{10 \angle 0°}$$
$$= 0.1 \angle 0° \text{ S}$$

$$\mathbf{B}_C = \frac{1}{X_C \angle -90°} = \frac{1}{10 \angle -90°}$$
$$= 0.1 \angle 90° \text{ S}$$

Example 18–15 (continued)

$$B_L = \frac{1}{X_L \angle 90°} = \frac{1}{5 \angle 90°}$$
$$= 0.2 \angle -90° \text{ S}$$
$$\mathbf{Y} = G + jB_C - jB_L = 0.1 \text{ S} + j0.1 \text{ S} - j0.2 \text{ S}$$
$$= 0.1 \text{ S} - j0.1 \text{ S} = 0.1414 \angle -45° \text{ S}$$

From **Y** we can get **Z**:

$$\mathbf{Z} = \frac{1}{\mathbf{Y}} = \frac{1}{0.1414 \angle -45°}$$
$$= 7.07 \angle 45° \; \Omega$$

Review for 18–6

1. In a given series RLC circuit, X_L is 150 Ω and X_C is 80 Ω. What is the net reactance in ohms? Is it inductive or capacitive?
2. Is a parallel RLC circuit having an X_L of 20 Ω and an X_C of 10 Ω capacitive or inductive?

18–7

COMPLEX ANALYSIS OF REACTIVE CIRCUITS

The examples in this section will show you how to apply the material in the previous sections to analyzing reactive circuits. In addition, phasor diagrams will be used in the examples to graphically illustrate the amplitude and phase relationships of currents and voltages.

Example 18–16

Find the total current and the voltage drops across each element in Figure 18–25. Express each quantity in polar form, and draw a phasor diagram of all voltages.

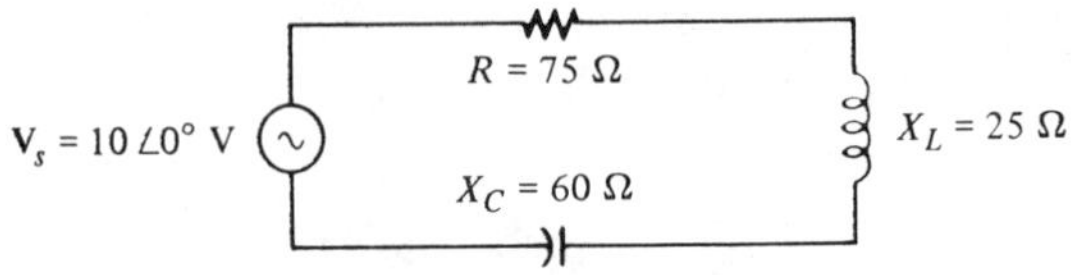

$V_s = 10 \angle 0°$ V, $R = 75 \; \Omega$, $X_C = 60 \; \Omega$, $X_L = 25 \; \Omega$

FIGURE 18–25

COMPLEX ANALYSIS OF REACTIVE CIRCUITS

Solution:

First find the total impedance:

$$\mathbf{Z} = R + jX_L - jX_C = 75\ \Omega + j25\ \Omega - j60\ \Omega$$
$$= 75\ \Omega - j35\ \Omega$$

Convert to polar form for convenience in applying Ohm's law:

$$\mathbf{Z} = \sqrt{(75\ \Omega)^2 + (35\ \Omega)^2}\ \angle\arctan\left(\frac{-35}{75}\right)$$
$$= 82.76\ \angle -25°\ \Omega$$

Apply Ohm's law to get the current:

$$\mathbf{I} = \frac{\mathbf{V}_s}{\mathbf{Z}} = \frac{10\ \angle 0°\ \text{V}}{82.76\ \angle -25°\ \Omega}$$
$$= 0.121\ \angle 25°\ \text{A}$$

Apply Ohm's law to get the voltage drops:

$$\mathbf{V}_R = \mathbf{IR} = (0.121\ \angle 25°\ \text{A})(75\ \angle 0°\ \Omega)$$
$$= 9.075\ \angle 25°\ \text{V}$$
$$\mathbf{V}_L = \mathbf{IX}_L = (0.121\ \angle 25°\ \text{A})(25\ \angle 90°\ \Omega)$$
$$= 3.025\ \angle 115°\ \text{V}$$
$$\mathbf{V}_C = \mathbf{IX}_C = (0.121\ \angle 25°\ \text{A})(60\ \angle -90°)$$
$$= 7.26\ \angle -65°\ \text{V}$$

The phasor diagram is shown in Figure 18–26. Notice that $\mathbf{V}_L$ is leading $\mathbf{V}_R$ by 90°, and $\mathbf{V}_C$ is lagging $\mathbf{V}_R$ by 90°. Also, there is a 180° phase difference between $\mathbf{V}_L$ and $\mathbf{V}_C$. If the current phasor were shown, it would be at the same angle as $\mathbf{V}_R$. The current is leading $\mathbf{V}_s$, the source voltage, by 25°, indicating a capacitive circuit ($X_C > X_L$).

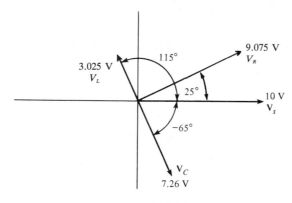

FIGURE 18–26

Example 18–17

Find each branch current and the total current in Figure 18–27.

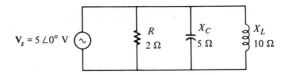

FIGURE 18–27

Solution:

We find each branch current by using Ohm's law:

$$I_R = \frac{V_s}{R} = \frac{5 \angle 0° \text{ V}}{2 \angle 0° \text{ Ω}}$$
$$= 2.5 \angle 0° \text{ A}$$

$$I_C = \frac{V_s}{X_C} = \frac{5 \angle 0° \text{ V}}{5 \angle -90° \text{ Ω}}$$
$$= 1 \angle 90° \text{ A}$$

$$I_L = \frac{V_s}{X_L} = \frac{5 \angle 0° \text{ V}}{10 \angle 90° \text{ Ω}}$$
$$= 0.5 \angle -90° \text{ A}$$

The total current is the phasor sum of the branch currents, by Kirchhoff's current law:

$$I_T = I_R + I_C + I_L = 2.5 \angle 0° \text{ A} + 1 \angle 90° \text{ A} + 0.5 \angle -90° \text{ A}$$
$$= 2.5 \text{ A} + j1 \text{ A} - j0.5 \text{ A} = 2.5 \text{ A} + j0.5 \text{ A}$$
$$= \sqrt{(2.5 \text{ A})^2 + (0.5 \text{ A})^2} \angle \arctan\left(\frac{0.5}{2.5}\right)$$
$$= 2.55 \angle 11.31° \text{ A}$$

The total current is 2.55 A, leading V_s by 11.31°.

Figure 18–28 is the phasor diagram of the currents.

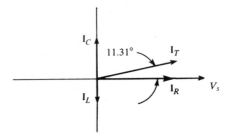

FIGURE 18–28

COMPLEX ANALYSIS OF REACTIVE CIRCUITS

Example 18–18

In Figure 18–29, find the voltage across the capacitor in polar form. Is this circuit predominately inductive or capacitive?

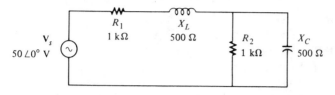

FIGURE 18–29

Solution:

Let us use the voltage divider formula in this problem. The impedance of the series R_1 and X_L will be called Z_1.

$$Z_1 = R_1 + jX_L$$
$$= 1000 \, \Omega + j500 \, \Omega \quad \text{in rectangular form}$$

$$Z_1 = \sqrt{(1000 \, \Omega)^2 + (500 \, \Omega)^2} \; \angle \arctan\left(\frac{500}{1000}\right)$$
$$= 1118 \, \angle 26.57° \, \Omega \quad \text{in polar form}$$

The impedance of R_2 and X_C in parallel will be called Z_2.

$$Z_2 = \frac{R_2 X_C}{\sqrt{R_2^2 + X_C^2}} \; \angle\left(-90° + \arctan\frac{X_C}{R}\right)$$

$$= \frac{(1000 \, \Omega)(500 \, \Omega)}{\sqrt{(1000 \, \Omega)^2 + (500 \, \Omega)^2}} \; \angle\left(-90° + \arctan\frac{500}{1000}\right)$$

$$= 447.2 \, \angle -63.4° \, \Omega \quad \text{in polar form}$$

or

$$Z_2 = 447.2 \cos(-63.4°) + j447.2 \sin(-63.4°)$$
$$= 200.24 \, \Omega - j399.87 \, \Omega \quad \text{in rectangular form}$$

The total impedance Z_T is

$$Z_T = Z_1 + Z_2 = (1000 \, \Omega + j500 \, \Omega) + (200.24 \, \Omega - j399.87 \, \Omega)$$
$$= 1200.24 \, \Omega + j100.13 \, \Omega \quad \text{in rectangular form}$$

or

$$Z_T = \sqrt{(1200.24 \, \Omega)^2 + (100.13 \, \Omega)^2} \; \angle \arctan\left(\frac{100.13}{1200.24}\right)$$

$$= 1204.41 \, \angle 4.77° \, \Omega \quad \text{in polar form}$$

Applying the voltage divider formula to get V_C yields

Example 18–18 (continued)

$$V_C = \left(\frac{Z_2}{Z_T}\right)V_s = \left(\frac{447.2 \angle -63.4°\ \Omega}{1204.41 \angle 4.77°\ \Omega}\right)50 \angle 0°\ V$$

$$= 18.57 \angle -68.17°\ V$$

Therefore, V_C is 18.57 V and lags V_s by 68.17°.

The $+j$ term in Z_T or the positive angle in its polar form tells us that the circuit is more inductive than capacitive. However, it is just slightly more because the angle is small. This result is surprising because $X_C = X_L = 500\ \Omega$. However, the capacitor is in *parallel* with a resistor; so the capacitor actually has less effect on the total impedance than does the inductor. Figure 18–30 shows the phasor relationship of V_C and V_s.

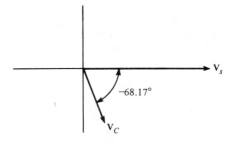

FIGURE 18–30

Example 18–19

For the reactive circuit in Figure 18–31, find the voltage at point B with respect to ground.

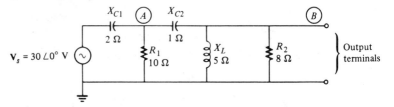

FIGURE 18–31

Solution:

We must find the voltage across the open output terminals. Let us use the voltage divider approach. To do so, we must know the voltage at point A first; so we need to find the impedance from point A to ground as a starting point.

COMPLEX ANALYSIS OF REACTIVE CIRCUITS

The parallel combination of X_L and R_2 is in series with X_{C2}. This combination is in parallel with R_1. We will call this impedance from point A to ground, $\mathbf{Z}_A$. To find $\mathbf{Z}_A$, we take the following steps:

R_2 and X_L in parallel ($\mathbf{Z}_1$):

$$\mathbf{Z}_1 = \frac{R_2 X_L}{\sqrt{R_2^2 + X_L^2}} \angle \left(90° - \arctan \frac{X_L}{R}\right)$$

$$= \frac{(8\ \Omega)(5\ \Omega)}{\sqrt{(8\ \Omega)^2 + (5\ \Omega)^2}} \angle \left(90° - \arctan \frac{5}{8}\right)$$

$$= \frac{40}{9.43} \angle (90° - 32°) = 4.24 \angle 58°\ \Omega$$

$\mathbf{Z}_1$ in series with $\mathbf{X}_{C2}$ ($\mathbf{Z}_2$):

$$\mathbf{Z}_2 = \mathbf{X}_{C2} + \mathbf{Z}_1 = 1 \angle -90°\ \Omega + 4.24 \angle 58°\ \Omega$$
$$= -j1\ \Omega + 2.25\ \Omega + j3.6\ \Omega$$
$$= 2.25\ \Omega + j2.6\ \Omega = \sqrt{(2.25\ \Omega)^2 + (2.6\ \Omega)^2} \angle \arctan\left(\frac{2.6}{2.25}\right)$$
$$= 3.44 \angle 49.13°\ \Omega$$

R_1 in parallel with $\mathbf{Z}_2$ ($\mathbf{Z}_A$):

$$\mathbf{Z}_A = \frac{(10 \angle 0°\ \Omega)(3.44 \angle 49.13°\ \Omega)}{10\ \Omega + 2.25\ \Omega + j2.6\ \Omega} = \frac{34.4 \angle 49.13°}{12.25 + j2.6}$$

$$= \frac{34.4 \angle 49.13°}{12.52 \angle 11.98°} = 2.75 \angle 37.15°\ \Omega$$

The simplified circuit is shown in Figure 18–32.

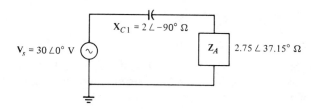

FIGURE 18–32

We can now apply the voltage divider principle to find $\mathbf{V}_A$. The total impedance is

$$\mathbf{Z}_T = \mathbf{X}_{C1} + \mathbf{Z}_A = 2 \angle -90°\ \Omega + 2.75 \angle 37.15°\ \Omega$$
$$= -j2\ \Omega + 2.19\ \Omega + j1.66\ \Omega$$
$$= 2.19\ \Omega - j0.34\ \Omega = \sqrt{(2.19\ \Omega)^2 + (0.34\ \Omega)^2} \angle -\arctan\left(\frac{0.34}{2.19}\right)$$
$$= 2.22 \angle -8.82°\ \Omega$$

Example 18–19 (continued)

$$V_A = \left(\frac{Z_A}{Z_T}\right)V_s = \left(\frac{2.75\ \angle 37.15°\ \Omega}{4.22\ \angle -8.82°\ \Omega}\right)30\ \angle 0°\ V$$

$$= 37.16\ \angle 45.97°\ V$$

Next, we find V_B by dividing V_A down, as indicated in Figure 18–33. V_B is the open terminal output voltage.

$$V_B = \left(\frac{Z_1}{Z_2}\right)V_A = \left(\frac{4.24\ \angle 58°\ \Omega}{3.44\ \angle 49.13°\ \Omega}\right)37.16\ \angle 45.97°\ V$$

$$= 45.8\ \angle 54.84°\ V$$

Surprisingly, V_A is greater than V_s, and V_B is greater than V_A! This result is possible because of the interchange of energy between the reactive components. Remember that X_C and X_L tend to cancel each other.

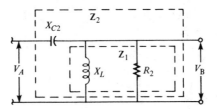

FIGURE 18–33

Review for 18–7

1. Calculate I given that $V = 10\ \angle 0°$ V and $Z = 50\ \angle 45°\ \Omega$.
2. The impedance of a circuit is 300 $\angle 28°\ \Omega$, and the applied voltage is 120 $\angle 30°$ V. What is the phase angle between the current and the voltage? What is the magnitude of the current?
3. The two voltage drops in a series circuit are 24 $\angle 0°$ V and 12 $\angle 30°$ V. Determine the total voltage.

18–8

SERIES EQUIVALENT CIRCUITS

Any reactive network can be reduced to an equivalent series circuit with an impedance of the form $Z = R \pm jX$. Some examples will demonstrate this reduction.

SERIES EQUIVALENT CIRCUITS

Example 18–20

Convert the circuit in Figure 18–34 into an equivalent series form.

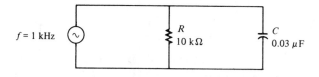

FIGURE 18–34

Solution:

First find X_C:

$$X_C = \frac{1}{2\pi f C} = \frac{1}{2\pi(1000 \text{ Hz})(0.03 \text{ }\mu\text{F})}$$
$$= 5305 \text{ }\Omega$$

Next find $\mathbf{Z}_T$:

$$\mathbf{Z}_T = \frac{RX_C}{\sqrt{R^2 + X_C^2}} \angle\left(-90° + \arctan\frac{X_C}{R}\right)$$

$$= \frac{(10{,}000 \text{ }\Omega)(5305 \text{ }\Omega)}{\sqrt{(10{,}000 \text{ }\Omega)^2 + (5305 \text{ }\Omega)^2}} \angle\left(-90° + \arctan\frac{5305}{10{,}000}\right)$$

$$= 4686 \angle -62° \text{ }\Omega$$

Now convert $\mathbf{Z}_T$ into rectangular form:

$$\mathbf{Z}_T = 4686 \cos(-62°) + j4686 \sin(-62°)$$
$$= 2199.9 \text{ }\Omega - j4137.5 \text{ }\Omega$$
$$= R_{eq} - jX_{C(eq)}$$

The equivalent series values are taken from the rectangular form of $\mathbf{Z}_T$:

$$R_{eq} = 2199.9 \text{ }\Omega$$
$$X_{C(eq)} = 4137.5 \text{ }\Omega$$
$$C_{eq} = \frac{1}{2\pi f X_{C(eq)}} = \frac{1}{2\pi(1000 \text{ Hz})(4137.5 \text{ }\Omega)}$$
$$= 0.0385 \text{ }\mu\text{F}$$

The series equivalent circuit is shown in Figure 18–35 on page 596. It is equivalent to the original parallel circuit in terms of the total impedance seen by the source.

Example 18–20 (continued)

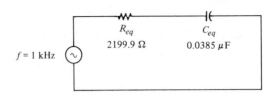

FIGURE 18–35

Example 18–21

Convert the circuit in Figure 18–36 to its series equivalent.

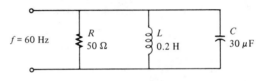

FIGURE 18–36

Solution:

$$X_L = 2\pi fL = 2\pi(60 \text{ Hz})(0.2 \text{ H})$$
$$= 75.4 \text{ }\Omega$$

$$X_C = \frac{1}{2\pi fC} = \frac{1}{2\pi(60 \text{ Hz})(30 \text{ }\mu\text{F})}$$
$$= 88.4 \text{ }\Omega$$

$$\frac{1}{\mathbf{Z}_T} = \frac{1}{R \angle 0°} + \frac{1}{X_L \angle 90°} + \frac{1}{X_C \angle -90°}$$

$$= \frac{1}{50 \angle 0° \text{ }\Omega} + \frac{1}{75.4 \angle 90° \text{ }\Omega} + \frac{1}{88.4 \angle -90° \text{ }\Omega}$$

$$= 0.02 \angle 0° \text{ S} + 0.013 \angle -90° \text{ S} + 0.011 \angle 90° \text{ S}$$

$$= 0.02 \angle 0° \text{ S} - j0.013 \text{ S} + j0.011 \text{ S} = 0.02 \text{ S} - j0.002 \text{ S}$$

$$= \sqrt{(0.02 \text{ S})^2 + (0.002 \text{ S})^2} \angle -\arctan\left(\frac{0.002}{0.02}\right)$$

$$= 0.02 \angle -5.7° \text{ S}$$

$$\mathbf{Z}_T = \frac{1}{0.02 \angle -5.7° \text{ S}} = 50 \angle 5.7° \text{ }\Omega = 49.75 \text{ }\Omega + j4.97 \text{ }\Omega$$

$$= R_{eq} + jX_{L(eq)}$$

The series equivalent values are $R_{eq} = 49.75$ Ω and $X_{L(eq)} = 4.97$ Ω. The series equivalent circuit is shown in Figure 18–37.

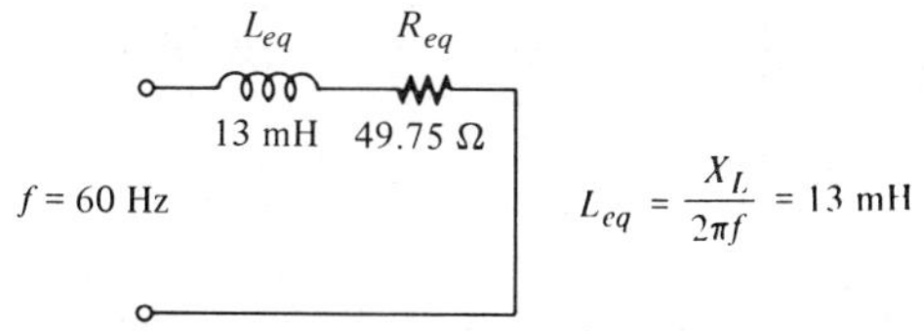

FORMULAS

L_{eq} R_{eq}
13 mH 49.75 Ω

$f = 60$ Hz

$L_{eq} = \dfrac{X_L}{2\pi f} = 13$ mH

FIGURE 18–37

Review for 18–8

1. A reactive circuit of any configuration can be converted to an equivalent series form in terms of its impedance (T or F).
2. The impedance of a certain parallel circuit is found to be 39 $\angle -50°$ Ω. Determine the equivalent series resistance and reactance.

Formulas

$$C = \sqrt{A^2 + B^2} \qquad (18\text{--}1)$$

$$\theta = \arctan\left(\frac{B}{A}\right) \qquad (18\text{--}2)$$

$$A \pm jB = \sqrt{A^2 + B^2} \angle \pm \arctan\left(\frac{B}{A}\right) = C \angle \pm \theta° \qquad (18\text{--}3)$$

$$A = C \cos \theta \qquad (18\text{--}4)$$

$$B = C \sin \theta \qquad (18\text{--}5)$$

$$C \angle \theta° = C \cos \theta \pm jC \sin \theta = A \pm jB \qquad (18\text{--}6)$$

$$\mathbf{Z} = R - jX_C \qquad (18\text{--}7)$$

$$\mathbf{Z} = \sqrt{R^2 + X_C^2} \angle -\arctan\left(\frac{X_C}{R}\right) \qquad (18\text{--}8)$$

$$\mathbf{Z} = R + jX_L \qquad (18\text{--}9)$$

$$\mathbf{Z} = \sqrt{R^2 + X_L^2} \angle \arctan\left(\frac{X_L}{R}\right) \qquad (18\text{--}10)$$

$$\mathbf{Z} = \frac{RX_C}{\sqrt{R^2 + X_C^2}} \angle \left(-90° + \arctan\frac{X_C}{R}\right) \qquad (18\text{--}11)$$

$$\mathbf{Z} = \frac{RX_L}{\sqrt{R^2 + X_L^2}} \angle \left(90° - \arctan\frac{X_L}{R}\right) \qquad (18\text{--}12)$$

$$\mathbf{Z} = R + jX_L - jX_C \qquad (18\text{--}13)$$

$$\mathbf{Z} = \sqrt{R^2 + (X_L - X_C)^2} \angle \arctan\left(\frac{X_L - X_C}{R}\right) \qquad (18\text{--}14)$$

$$\frac{1}{Z_T} = \frac{1}{R \angle 0°} + \frac{1}{X_L \angle 90°} + \frac{1}{X_C \angle -90°} \qquad (18\text{–}15)$$

$$G = \frac{1}{R \angle 0°} = G \angle 0° = G \qquad (18\text{–}16)$$

$$B_C = \frac{1}{X_C \angle -90°} = B_C \angle 90° = jB_C \qquad (18\text{–}17)$$

$$B_L = \frac{1}{X_L \angle 90°} = B_L \angle -90° = -jB_L \qquad (18\text{–}18)$$

$$Y = \frac{1}{Z \angle \pm \theta°} = Y \angle \mp \theta° = G + jB_C - jB_L \qquad (18\text{–}19)$$

Summary

1. A complex number specifies magnitude and phase angle of electrical quantities.
2. The horizontal axis on the complex plane is called the *real axis*.
3. The vertical axis on the complex plane is called the *imaginary* or *j axis*.
4. The rectangular form of a complex number consists of a real part and a *j* part.
5. The polar form of a complex number consists of a magnitude and an angle.
6. Either form (rectangular or polar) specifies the position of a phasor in the complex plane.
7. Complex numbers can be added, subtracted, multiplied, and divided.
8. Resistance is written as R in rectangular notation and as $R \angle 0°$ in polar notation.
9. Inductive reactance is written as jX_L in rectangular notation and as $X_L \angle 90°$ in polar notation.
10. Capacitive reactance is written as $-jX_C$ in rectangular notation and as $X_C \angle -90°$ in polar notation.
11. X_L and X_C have opposing effects in a circuit.
12. In a series circuit, the larger reactance value determines if the circuit is inductive or capacitive.
13. In a parallel circuit, the smaller reactance determines if the circuit is inductive or capacitive.

Self-Test

1. Indicate the following points in the complex plane:
   (a) $1 + j2$ (b) $3 + j4$ (c) $5 - j3$ (d) $-2 + j3$ (e) $-1 - j2$
2. Sketch the phasors specified by the following polar coordinates:
   (a) $2 \angle 45°$ (b) $4 \angle 0°$ (c) $5 \angle 30°$ (d) $1 \angle 90°$ (e) $3 \angle -90°$
3. Convert the following rectangular numbers into polar form:
   (a) $5 + j5$ (b) $12 + j9$ (c) $8 - j10$ (d) $100 - j50$

SELF-TEST

4. Convert the following polar numbers into rectangular form:
   (a) 1 ∠45°   (b) 12 ∠60°   (c) 100 ∠−80°   (d) 40 ∠125°
5. Find the impedance in rectangular form of each circuit in Figure 18–38.

A.

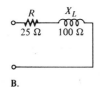
B.

FIGURE 18–38

6. How many ohms of impedance are there in each circuit in Figure 18–38?
7. Determine the current in polar form for each circuit in Figure 18–38 if a voltage of 10 V at a reference angle of 0° is applied.
8. Determine the impedance in polar form of each circuit in Figure 18–39.

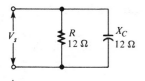
A.

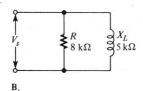
B.

FIGURE 18–39

9. Determine the total current and each branch current in Figure 18–39 if $V_s = 5 \angle 0°$ V.
10. Does the current in the RLC circuit in Figure 18–40 lead or lag the source voltage? Why?

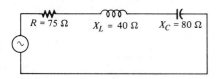

FIGURE 18–40

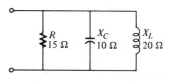

FIGURE 18–41

11. What is the impedance in ohms in Figure 18–40?
12. Determine the impedance in polar form for Figure 18–41.
13. Determine the series equivalent impedance for the circuit in Figure 18–42.

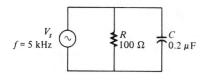

FIGURE 18–42

14. What is the phase difference between the applied voltage and the total current in Figure 18–42?

15. Convert the circuit in Figure 18–43 to its series equivalent.

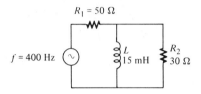

FIGURE 18–43

Problems

18–1. What is the value of the hypotenuse of a right triangle whose sides are 10 and 15?

18–2. A right triangle has a hypotenuse with a value of 50 and an adjacent angle of 55°. Determine the values of the two sides.

18–3. Two phasors of equal magnitude are 30° apart. Phasor A is at 10°, and phasor B leads. Sketch the phasor diagram.

18–4. Convert the following from polar form to rectangular form:
    (a) $1000 \angle -50°$ (b) $15 \angle 160°$ (c) $25 \angle -135°$ (d) $3 \angle 180°$

18–5. Convert the following from rectangular to polar form:
    (a) $40 - j40$ (b) $50 - j200$ (c) $35 - j20$ (d) $98 + j45$

18–6. For the circuit in Figure 18–44, determine the following in polar form:
    (a) $\mathbf{Z}_T$ (b) $\mathbf{I}_T$ (c) $\mathbf{V}_R$ (d) $\mathbf{V}_C$

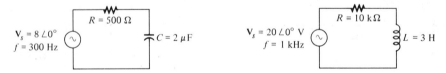

FIGURE 18–44   **FIGURE 18–45**

18–7. Sketch a phasor diagram showing the current and voltages in Problem 18–6.

18–8. For the circuit in Figure 18–45, determine the following in polar form:
    (a) $\mathbf{Z}_T$ (b) $\mathbf{I}_T$ (c) $\mathbf{V}_R$ (d) $\mathbf{V}_L$

18–9. For the circuit in Figure 18–46, find all the currents and voltages in polar form.

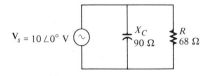

FIGURE 18–46

PROBLEMS

18–10. Repeat Problem 18–9 for Figure 18–47.

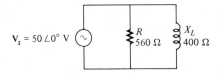

FIGURE 18–47

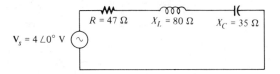

FIGURE 18–48

18–11. Sketch the current phasor diagrams for the circuits in Figures 18–46 and 18–47.

18–12. For the circuit in Figure 18–48, find I_T, V_R, V_L, and V_C in polar form. Is the circuit capacitive or inductive? What is the power in R?

18–13. Find the total impedance for each circuit in Figure 18–49.

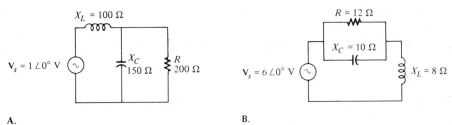

A. B.

FIGURE 18–49

18–14. For each circuit in Figure 18–49, determine the phase angle between the source voltage and the total current.

18–15. Determine the voltages in polar form across each element in Figure 18–50.

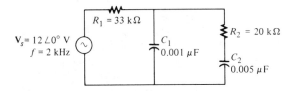

FIGURE 18–50

18–16. Is the circuit in Figure 18–50 more resistive than capacitive?

18–17. What is the current through R_2 in Figure 18–51?

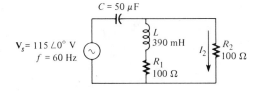

FIGURE 18–51

18–18. In Problem 18–17, what is the phase angle between I_2 and the source voltage?

18–19. Determine the equivalent series impedance in Figure 18–51.

18–20. Determine the total resistance and the total reactance in Figure 18–52.

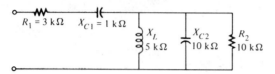

FIGURE 18–52

Answers to Section Reviews

Section 18–1:
1. Real axis, j (imaginary) axis. **2.** It indicates a $+90°$ or a $-90°$ phase and has a value of $\sqrt{-1}$.
3. 90° between $+4$ and $+j4$, 90° between -4 and $+j4$, and 180° between $+4$ and -4.

Section 18–2:
1. Real and j parts. **2.** Magnitude and angle. **3.** 2.828 $\angle 45°$. **4.** $3.536 + j3.536$.

Section 18–3:
1. $4 + j1$. **2.** $3 + j7$. **3.** 16 $\angle 110°$. **4.** 5 $\angle 15°$.

Section 18–4:
1. (a) 10 $\angle 0°$ Ω (b) 1k$\angle -90°$ Ω (c) 30 $\angle 90°$ Ω. **2.** (a) 2 kΩ (b) $-j47$ Ω (c) $j5$Ω.

Section 18–5:
1. 80 Ω $- j50$ Ω, 94.34 $\angle -32°$ Ω. **2.** 2572.48 $\angle 30.96°$ Ω.

Section 18–6:
1. 70 Ω; inductive. **2.** Capacitive.

Section 18–7:
1. 0.2 $\angle -45°$ A. **2.** 28°; 0.4 A. **3.** 34.9 $\angle 9.9°$ V.

Section 18–8:
1. T. **2.** $R = 25$ Ω; $X_C = 29.9$ Ω.

Resonance is a very important aspect of electronic applications. It is the basis for frequency selectivity in communications. For example, the ability of a radio or television receiver to select a certain frequency transmitted by a particular station and eliminate frequencies from other stations is based on resonance.

Resonance occurs under certain specific conditions in circuits having combinations of inductance and capacitance. In this chapter we will discuss the conditions that produce resonance, and we will study the characteristics of resonant circuits.

19–1 Series Resonance
19–2 Parallel Resonance
19–3 Bandwidth of a Resonant Circuit
19–4 Quality Factor (Q) of a Resonant Circuit
19–5 Applications of Resonant Circuits

RESONANCE

19-1

SERIES RESONANCE

In a series RLC circuit, series resonance occurs when $X_L = X_C$. The frequency at which resonance occurs is called the *resonant frequency*, f_r. Figure 19-1 illustrates the series resonant condition.

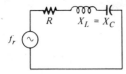

FIGURE 19-1 *Series resonant circuit.*

In a series resonant circuit, the total impedance is

$$Z_T = R + jX_L - jX_C$$

Since $X_L = X_C$, the j terms cancel, and *the impedance is purely resistive* ($Z = R$). These resonant conditions are stated in the following equations:

$$X_L = X_C \tag{19-1}$$

$$Z_T = R \tag{19-2}$$

Example 19-1

For the series RLC circuit in Figure 19-2, determine X_C and Z_T at resonance.

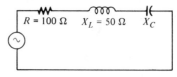

FIGURE 19-2

Solution:

$$X_C = X_L \text{ at the resonant frequency}$$

Thus,

$$X_C = 50 \text{ }\Omega$$
$$Z_T = R + jX_L - jX_C = 100 + j50 - j50$$
$$= 100 \angle 0° \text{ }\Omega$$

The impedance is equal to the resistance because $\mathbf{X}_L$ and $\mathbf{X}_C$ cancel.

Series Resonant Frequency

For a given series RLC circuit, resonance happens at only one specific frequency. A formula for this resonant frequency is developed as follows: At resonance, $X_L = X_C$. Substituting the reactance formulas, we have

$$2\pi f_r L = \frac{1}{2\pi f_r C}$$

Solving for f_r^2 yields

$$(2\pi f_r L)(2\pi f_r C) = 1$$
$$4\pi^2 f_r^2 LC = 1$$
$$f_r^2 = \frac{1}{4\pi^2 LC}$$

Taking the square root of both sides, we obtain

$$f_r = \frac{1}{2\pi\sqrt{LC}} \tag{19-3}$$

Example 19-2

Find the series resonant frequency for Figure 19-3.

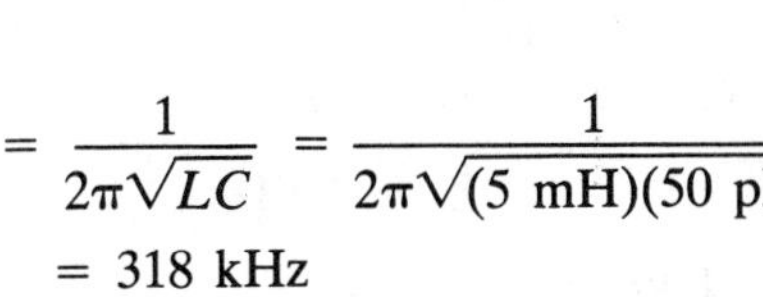

(Circuit: V_s source in series with $R = 10\,\Omega$, $C = 50$ pF, $L = 5$ mH)

FIGURE 19-3

Solution:

$$f_r = \frac{1}{2\pi\sqrt{LC}} = \frac{1}{2\pi\sqrt{(5\text{ mH})(50\text{ pF})}}$$
$$= 318 \text{ kHz}$$

Series RLC Impedance

At frequencies below f_r, $X_C > X_L$, and thus the circuit is capacitive. At the resonant frequency, $X_C = X_L$; so the circuit is purely resistive. At frequencies above f_r, $X_L > X_C$; so the circuit is inductive.

The impedance magnitude is minimum at resonance ($Z_T = R$) and increases in value above and below the resonant point. The graph in Figure 19-4 illustrates how impedance changes with frequency.

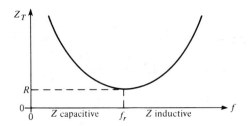

FIGURE 19–4 *Series RLC impedance as a function of frequency.*

Example 19–3

For the circuit in Figure 19–5, determine the impedance magnitude at resonance, at 1000 Hz below resonance, and at 1000 Hz above resonance.

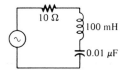

FIGURE 19–5

Solution:

$$f_r = \frac{1}{2\pi\sqrt{LC}} = \frac{1}{2\pi\sqrt{(100 \text{ mH})(0.01 \text{ }\mu\text{F})}}$$
$$= 5.03 \text{ kHz}$$

At 1000 Hz below f_r:

$$f_r - 1 \text{ kHz} = 4.03 \text{ kHz}$$

At 1000 Hz above f_r:

$$f_r + 1 \text{ kHz} = 6.03 \text{ kHz}$$

Impedance at resonance is equal to R:

$$Z_T = 10 \text{ }\Omega$$

Impedance at $f_r - 1$ kHz:

$$Z_T = \sqrt{R^2 + (X_L - X_C)^2} = \sqrt{(10 \text{ }\Omega)^2 + (2.53 \text{ k}\Omega - 3.95 \text{ k}\Omega)^2}$$
$$= 1.42 \text{ k}\Omega$$

Notice that X_C is greater than X_L; so Z_T is more capacitive.

Impedance at $f_r + 1$ kHz:

$$Z_T = \sqrt{R^2 + (X_L - X_C)^2} = \sqrt{(10 \text{ }\Omega)^2 + (3.79 \text{ k}\Omega - 2.64 \text{ k}\Omega)^2}$$
$$= 1.15 \text{ k}\Omega$$

X_L is greater than X_C in this case; so Z_T is more inductive.

SERIES RESONANCE

Current and Voltages

At the series resonant frequency, *the circuit current is maximum* ($I_{max} = V_s/R$). Above and below resonance, the current decreases because the impedance increases. A *response curve* showing the plot of current versus frequency is shown in Figure 19–6A.

If a series RLC circuit is connected as in Figure 19–7, with the resistor voltage as the output, V_R follows the current and is maximum at resonance, as in Figure 19–6B. The general shapes of the V_L and V_C curves are indicated in Figure 19–6C and D. Notice that V_C is equal to V_s when $f = 0$ and also that V_L approaches V_s as f becomes very high.

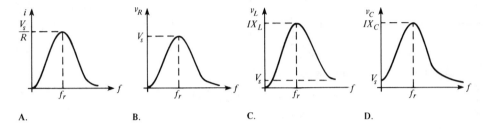

FIGURE 19–6 *Current and voltage magnitudes as a function of frequency in a series RLC circuit.*

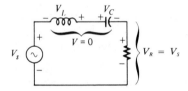

FIGURE 19–7 *Series RLC circuit at resonance.*

The voltages are maximum at resonance but drop off above and below f_r. Thus, the circuit responds best to signals having the resonant frequency and tends to reject signals with frequencies above and below f_r. Because of this type of response, the RLC circuit is a basic type of *band-pass filter*.

The voltages across L and C at resonance are exactly *equal in magnitude but 180° out of phase; so they cancel*. Thus, the voltage across both L and C together is *zero*, and V_R is equal to V_s at resonance, as indicated in Figure 19–7. Individually, V_L and V_C can be much greater than the source voltage, as we will see later.

Example 19–4

Find I, V_R, V_L, and V_C at resonance in Figure 19–8.

Example 19-4 (continued)

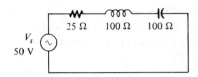

FIGURE 19-8

Solution:

At resonance, I is maximum and equal to V_s/R:

$$I = \frac{V_s}{R} = \frac{50 \text{ V}}{25 \text{ }\Omega}$$
$$= 2 \text{ A}$$

Applying Ohm's law, we obtain the following:

$$V_R = IR = (2 \text{ A})(25 \text{ }\Omega)$$
$$= 50 \text{ V}$$
$$V_L = IX_L = (2 \text{ A})(100 \text{ }\Omega)$$
$$= 200 \text{ V}$$
$$V_C = IX_C = (2 \text{ A})(100 \text{ }\Omega)$$
$$= 200 \text{ V}$$

Notice that all of the source voltage is dropped across the resistor. Also, of course, V_L and V_C are equal in magnitude. Their out-of-phase relationship causes them to cancel; so the *total* reactive voltage drop is zero.

Phase Characteristic

At resonance, *the current and the source voltage are in phase,* because the impedance is purely resistive. For frequencies below resonance, the circuit is capacitive ($X_C > X_L$), and the current *leads* the voltage. For frequencies above resonance, the circuit is inductive ($X_C < X_L$), and the current *lags* the voltage.

A phase plot is shown in Figure 19-9. Notice in the phase plot that at f_r, the phase angle θ is $0°$. As frequency approaches zero, θ approaches $-90°$. As frequency becomes very high, θ approaches $+90°$.

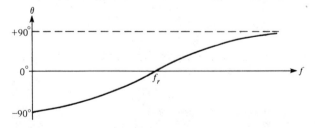

FIGURE 19-9 *Phase angle between current and voltage versus frequency in a series RLC circuit.*

PARALLEL RESONANCE

Review for 19-1

1. List the conditions for series resonance.
2. For what frequencies is a series RLC circuit inductive?
3. For what frequencies is a series RLC circuit capacitive?
4. At resonance, all of the source voltage is across R (T or F).
5. Why is the current at a maximum at the resonant frequency?
6. State the formula for resonant frequency.

19-2 PARALLEL RESONANCE

A basic parallel resonant circuit is shown in Figure 19-10. The resistance normally represents the coil's winding resistance. The parallel resonant circuit is often called a *tank circuit* because energy is stored by the inductance and capacitance.

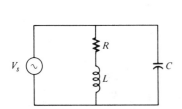

FIGURE 19-10 Parallel resonant (tank) circuit.

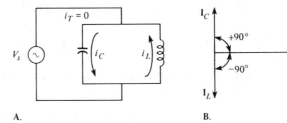

FIGURE 19-11 Ideal parallel resonant circuit.

Condition for Parallel Resonance

Parallel resonance occurs when $X_L = X_C$. Notice that this condition is the same as that for series resonance. It is based on the assumption that the resistance is much less than X_L at the resonant frequency so that there is very little energy loss. This assumption is a practical one to make for many applications.

Ideally, the current in the inductor and the current in the capacitor are equal in magnitude at resonance, but they flow in opposite directions. Let us assume an ideal tank circuit with $R = 0$, as shown in Figure 19-11A. Since $X_L = X_C$ and the source voltage is across both branches, then $I_L = I_C$. I_L lags V_s by 90°, and I_C leads V_s by 90°. Thus, the currents are equal in magnitude but opposite in phase, as indicated in Figure 19-11B. Therefore, they cancel, and the total current is ideally *zero* as indicated in Part A. Thus, in the ideal tank circuit, there is a current that circulates back and forth between L and C, but no current is drawn from the source.

Actually, because there is some small resistance in series with L, some energy is lost. Thus, the currents are not exactly equal, and the total current has a small value. There must be a small amount of current from the source to replace the energy lost in the resistance.

Resonant Frequency

The formula for the resonant frequency of the parallel LC circuit is the same as that for the series circuit. It is the same, however, only if R is much smaller than X_L at resonance so that there is negligible energy loss; but for most cases where a high-Q coil is used, the formula is valid. It is restated as follows:

$$f_r = \frac{1}{2\pi\sqrt{LC}}$$

Impedance

At the parallel resonant frequency, *the total impedance is maximum and appears to be purely resistive.* The total current is therefore minimum (ideally zero). A formula for the impedance of a parallel resonant circuit is developed as follows where R is the winding resistance:

Using the reciprocal formula for parallel circuits, we have

$$\frac{1}{\mathbf{Z}_T} = \frac{1}{-jX_C} + \frac{1}{R + jX_L}$$

$$= j\frac{1}{X_C} + \frac{R - jX_L}{(R + jX_L)(R - jX_L)}$$

$$= j\frac{1}{X_C} + \frac{R - jX_L}{R^2 + X_L^2}$$

Splitting the numerator of the second term yields

$$\frac{1}{\mathbf{Z}_T} = j\left(\frac{1}{X_C}\right) - j\left(\frac{X_L}{R^2 + X_L^2}\right) + \left(\frac{R}{R^2 + X_L^2}\right)$$

At resonance, $\mathbf{Z}_T$ is purely resistive; so it has no j part (the j terms in the last expression cancel). Thus, only the real part is left, as stated in the following equation for $\mathbf{Z}_T$ at resonance:

$$\mathbf{Z}_T = \frac{R^2 + X_L^2}{R} \tag{19-4}$$

Since the j terms in the formula preceding Equation (19–4) are equal, we can substitute $X_L X_C = R^2 + X_L^2$ into Equation (19–4) and get an expression for the magnitude of the total impedance in terms of R, L, and C:

$$\mathbf{Z}_T = \frac{R^2 + X_L^2}{R} = \frac{X_L X_C}{R}$$

$$= \frac{(2\pi f_r L)(1/2\pi f_r C)}{R}$$

Canceling yields

$$\mathbf{Z}_T = \frac{L}{RC} \tag{19-5}$$

PARALLEL RESONANCE

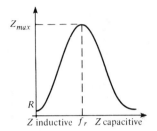

FIGURE 19–12 *Impedance of parallel resonant circuit versus frequency.*

Figure 19–12 shows that the impedance of a parallel resonant circuit is maximum at f_r and decreases above or below resonance. At frequencies below resonance, $X_L < X_C$; thus, the parallel circuit is inductive (the smaller value in parallel is predominant). At frequencies above resonance, $X_C < X_L$; thus, the circuit is capacitive. Notice that this characteristic is opposite to that of series resonant impedance.

Example 19–5

What is Z_T at resonance for the circuit in Figure 19–13?

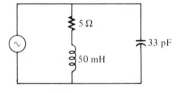

FIGURE 19–13

Solution:

$$Z_T = \frac{L}{RC} = \frac{50 \text{ mH}}{(5 \text{ }\Omega)(33 \text{ pF})}$$

$$= 303 \text{ M}\Omega$$

In an *ideal* tank circuit, R is zero, and Z_T is therefore infinite.

Example 19–6

For the tank circuit in Figure 19–13, determine the total impedance at 1000 Hz below the resonant frequency.

Solution:

$$f_r = \frac{1}{2\pi\sqrt{LC}} = 123.9 \text{ kHz}$$

Example 19-6 (continued)

At $f = f_r - 1$ kHz $= 122.9$ kHz:

$$X_L = 2\pi fL = 2\pi(122.9 \text{ kHz})(50 \text{ mH})$$
$$= 38.6 \text{ k}\Omega$$

$$X_C = \frac{1}{2\pi fC} = \frac{1}{2\pi(122.9 \text{ kHz})(33 \text{ pF})}$$
$$= 39.2 \text{ k}\Omega$$

Neglecting R because it is so small compared to X_L, we determine **Z** as follows:

$$\mathbf{Z} = \frac{(jX_L)(-jX_C)}{jX_L - jX_C} = \frac{(j38.6 \text{ k}\Omega)(-j39.2 \text{ k}\Omega)}{j38.6 \text{ k}\Omega - j39.2 \text{ k}\Omega}$$
$$= j2.52 \times 10^6 \text{ }\Omega$$
$$= 2.52 \angle 90° \text{ M}\Omega$$

This result shows that the magnitude of the impedance is considerably less than its resonant value of 303 MΩ, and the circuit is inductive.

Review for 19-2

1. At parallel resonance, $X_L = 1500$ Ω. What is X_C?
2. A parallel tank circuit has the following values: $R = 4$ Ω, $L = 50$ mH, and $C = 10$ pF. Calculate f_r and Z_T at resonance.
3. In a parallel resonant circuit, impedance is maximum, and the total current from the source is minumum at the resonant frequency (T or F).
4. At parallel resonance, the inductive current and the capacitive current are in phase (T or F).

19-3

BANDWIDTH OF A RESONANT CIRCUIT

Series Resonant Circuit

As you learned, the current in a series RLC circuit is maximum at the resonant frequency and drops off on either side of this frequency. *The bandwidth is defined to be the range of frequencies for which the current is equal to or greater than 70.7% of its resonant value.* Bandwidth, sometimes abbreviated BW, is an important characteristic of a resonant circuit.

BANDWIDTH OF A RESONANT CIRCUIT

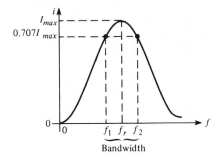

FIGURE 19–14 *Bandwidth on series resonant response curve for I.*

Figure 19–14 illustrates bandwidth on the response curve of a series RLC circuit. Notice that the frequency f_1 below f_r is the point at which the current is $0.707I_{max}$ and is commonly called the *lower cutoff frequency*. The frequency f_2 above f_r where the current is again $0.707I_{max}$ is the *upper cutoff frequency*. Other names for f_1 and f_2 are *band frequencies* and *half-power frequencies*. We will discuss the significance of this latter term later in this chapter.

Example 19–7

A given series resonant circuit has a maximum current of 100 mA at the resonant frequency. What is the value of the current at the cutoff frequencies?

Solution:

Current at the cutoff frequencies is 70.7% of maximum:

$$I_{f1} = I_{f2} = 0.707I_{max}$$
$$= 0.707(100 \text{ mA}) = 70.7 \text{ mA}$$

Parallel Resonant Circuit

For a parallel resonant circuit, *the impedance is maximum at the resonant frequency;* so the total current is minimum. The bandwidth can be defined in relation to the impedance curve in the same manner that the current curve was used in the series circuit. Of course, f_r is the frequency at which Z_T is maximum; f_1 is the lower cutoff frequency at which $Z_T = 0.707Z_{max}$; and f_2 is the upper cutoff frequency at which again $Z_T = 0.707Z_{max}$. The bandwidth (BW) is the range of frequencies between f_1 and f_2, as shown in Figure 19–15.

Formula for Bandwidth

The bandwidth for either series or parallel resonant circuits is the range of frequencies between the cutoff frequencies for which the response curve (*I* or

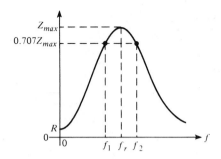

FIGURE 19–15 *Bandwidth on parallel resonant response curve for Z_T.*

Z_T) is 0.707 of the maximum value. Thus, the bandwidth is actually the *difference* between f_2 and f_1:

$$\text{BW} = f_2 - f_1 \qquad (19\text{–}6)$$

Ideally, f_r is the *center frequency* midway between f_1 and f_2 and can be found with either of the following equations:

$$f_r = f_1 + \frac{\text{BW}}{2} \qquad (19\text{–}7)$$

$$f_r = f_2 - \frac{\text{BW}}{2} \qquad (19\text{–}8)$$

Example 19–8

A resonant circuit has a lower cutoff frequency of 8 kHz and an upper cutoff frequency of 12 kHz. What is its bandwidth?

Solution:

$$\text{BW} = f_2 - f_1 = 12 \text{ kHz} - 8 \text{ kHz}$$
$$= 4 \text{ kHz}$$

Example 19–9

What is the resonant frequency of a circuit having an upper cutoff frequency of 100 kHz and a bandwidth of 2 kHz?

Solution:

$$f_r = f_2 - \frac{\text{BW}}{2} = 100 \text{ kHz} - \frac{2 \text{ kHz}}{2} = 100 \text{ kHz} - 1 \text{ kHz}$$
$$= 99 \text{ kHz}$$

BANDWIDTH OF A RESONANT CIRCUIT

Half-Power Frequencies

As mentioned previously, the upper and lower cutoff frequencies are sometimes called the *half-power frequencies*. This term is derived from the fact that the power from the source at these frequencies is one-half the power delivered at the resonant frequency. The following steps show that this is true for a series circuit. The same result applies to a parallel circuit.

At resonance,

$$P_{max} = I_{max}^2 R$$

The power at f_1 or f_2 is

$$P_{f1} = I_{f1}^2 R = (0.707 I_{max})^2 R = (0.707)^2 I_{max}^2 R$$
$$= 0.5 I_{max}^2 R = 0.5 P_{max}$$

Selectivity

The response curves in Figures 19–14 and 19–15 are also called *selectivity curves*. *Selectivity* defines how well a resonant circuit responds to a certain frequency and discriminates against all others. *The smaller the bandwidth is, the greater the selectivity is.*

We normally assume that a resonant circuit accepts frequencies within its bandwidth and completely eliminates frequencies outside the bandwidth. Such is not actually the case, however, because signals with frequencies outside the bandwidth are not completely eliminated. However, their magnitudes are greatly reduced. The further the frequencies are from the cutoff frequencies, the greater is the reduction, as illustrated in Figure 19–16A. An ideal selectivity curve is shown in Figure 19–16B.

As you can see in Figure 19–16, another factor that influences selectivity is the *sharpness* of the slopes of the curve. The faster the curve drops off at the cutoff frequencies, the more selective the circuit is, because it responds only to the frequencies within the bandwidth. Figure 19–17 compares three general response curves with varying degrees of selectivity.

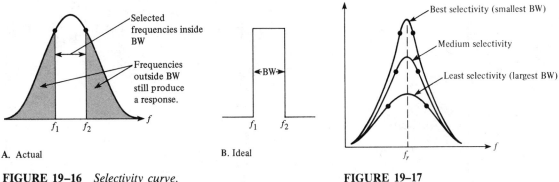

FIGURE 19–16 *Selectivity curve.*

FIGURE 19–17

Review for 19-3

1. Define *bandwidth*.

2. What is the center frequency?
3. Define *cutoff frequencies*.
4. If $f_1 = 3$ kHz and $f_2 = 4$ kHz, what is BW?

19-4

QUALITY FACTOR (Q) OF A RESONANT CIRCUIT

The *quality factor Q* is the ratio of the reactive power in the inductor (or capacitor) to the resistive power. It is a ratio of the power in L to the power in R. The quality factor is an important factor in both series and parallel resonant circuits. A formula for Q is developed as follows:

$$Q = \frac{\text{reactive power}}{\text{resistive power}} = \frac{\text{energy stored}}{\text{energy lost}}$$

$$= \frac{I^2 X_L}{I^2 R}$$

Since I is the same in L and R, the I^2 terms cancel, giving

$$Q = \frac{X_L}{R} \qquad (19\text{-}9)$$

When the resistance is just the winding resistance of the coil, the circuit Q and the coil Q are the same. Since Q varies with frequency because X_L varies, we are interested mainly in Q at resonance.

Example 19-10

Determine Q for each circuit in Figure 19-18 at resonance.

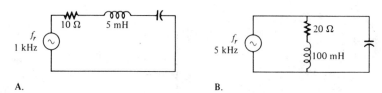

A.

B.

FIGURE 19-18

Solution:

Circuit A:

$$X_L = 2\pi f_r L$$
$$= 31.4 \ \Omega$$

$$Q = \frac{X_L}{R} = \frac{31.4 \ \Omega}{10 \ \Omega}$$
$$= 3.14$$

QUALITY FACTOR (Q) OF A RESONANT CIRCUIT

Circuit B:
$$X_L = 3.14 \text{ k}\Omega$$
$$Q = \frac{X_L}{R} = \frac{3.14 \text{ k}\Omega}{20 \text{ }\Omega}$$
$$= 157$$

Q "Magnifies" Applied Voltage in Series RLC Circuit

In a series RLC circuit, the voltages across L and across C depend on the resonant Q. At resonance, the total source voltage is dropped across R because the combined voltages across X_L and X_C are zero. By Ohm's law, $V_L = IX_L$ and $V_C = IX_C$. Figure 19–19 illustrates these voltages. The following steps give us expressions for V_L and V_C at resonance in terms of Q:

$$V_L = IX_L$$
$$Q = \frac{X_L}{R}$$

or
$$X_L = QR$$

Substituting QR for X_L gives
$$V_L = IQR = IRQ$$

Since $IR = V_s$ at resonance,
$$V_L = QV_s \qquad (19\text{–}10)$$

The same holds true for V_C:
$$V_C = QV_s \qquad (19\text{–}11)$$

These equations tell us that the circuit Q times the source voltage is the voltage across L or C at resonance. If Q is high, then V_L and V_C can be much larger than V_s! Keep in mind, however, that V_L and V_C are out of phase and cancel, leaving a net reactive voltage of zero.

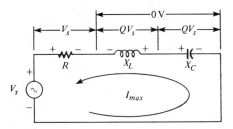

FIGURE 19–19 *Voltages in a series resonant circuit. Instantaneous polarities are shown.*

Example 19–11

Find V_L and V_C in the circuit of Figure 19–20 at the resonant frequency.

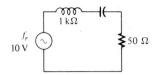

FIGURE 19–20

Solution:

$$Q = \frac{X_L}{R} = \frac{1000\ \Omega}{50\ \Omega}$$
$$= 20$$
$$V_L = V_C = QV_S = (20)10\ \text{V}$$
$$= 200\ \text{V}$$

Q Affects Bandwidth

A higher value of circuit Q results in a smaller bandwidth. A lower value of Q causes a larger bandwidth. Therefore, a higher Q means better selectivity. A formula for the bandwidth of a resonant circuit in terms of Q is stated in the following equation:

$$\text{BW} = \frac{f_r}{Q} \tag{19–12}$$

Example 19–12

What is the bandwidth of each circuit in Figure 19–21?

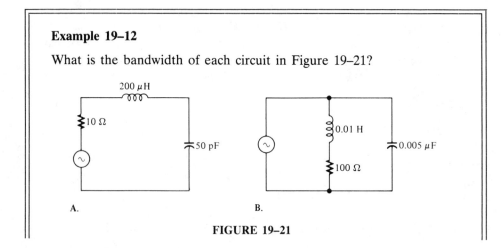

FIGURE 19–21

QUALITY FACTOR (Q) OF A RESONANT CIRCUIT

Solution:

Circuit A:

$$f_r = \frac{1}{2\pi\sqrt{LC}}$$
$$= 1.59 \text{ MHz}$$

$$Q = \frac{X_L}{R} = \frac{2 \text{ k}\Omega}{10 \text{ }\Omega}$$
$$= 200$$

$$BW = \frac{f_r}{Q} = \frac{1.59 \text{ MHz}}{200}$$
$$= 7.95 \text{ kHz}$$

Circuit B:

$$f_r = \frac{1}{2\pi\sqrt{LC}}$$
$$= 22.5 \text{ kHz}$$

$$Q = \frac{X_L}{R} = \frac{1.41 \text{ k}\Omega}{100 \text{ }\Omega}$$
$$= 14.1$$

$$BW = \frac{f_r}{Q} = \frac{22.5 \text{ kHz}}{14.1}$$
$$= 1.6 \text{ kHz}$$

Review for 19–4

1. Define *quality factor* of a circuit.

2. If $X_L = 200 \text{ }\Omega$ and $R = 10 \text{ }\Omega$, what is Q?

3. A 5-V source is applied to a series RLC circuit with a Q of 10 at the resonant frequency. What are V_L and V_C?

4. Does a larger Q mean a smaller or a larger bandwidth?

5. A large value of Q indicates good selectivity (T or F).

19-5

APPLICATIONS OF RESONANT CIRCUITS

Resonant circuits are widely used in filter applications; we will cover this topic in the next chapter. Now we will briefly examine the use of resonant circuits in tuning applications.

Tuned Circuits

A common example of the application of a parallel resonant circuit is in a radio or TV receiver. When you "tune" the receiver, you are actually varying the L or C in a tank circuit in order to change the resonant frequency. This tuning is typically accomplished with a variable capacitance as shown in Figure 19–22A. As the capacitance is varied, the resonant frequency changes, but the bandwidth remains constant, as illustrated in Part B of the figure. In this manner, the receiver can be made to select a narrow band of frequencies transmitted by the desired station and eliminate the other transmitted frequencies from unwanted stations.

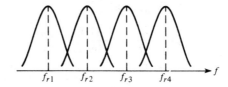

A.   B.

FIGURE 19–22 *Tuning a circuit for frequency selection.*

Actually, a receiver circuit is much more complex than indicated in Figure 19–22. The resonant circuit is normally combined with *amplifier circuits* in some manner.

Review for 19–5

1. What does *tuning* mean?
2. How does a resonant circuit select a particular band of frequencies?

Formulas

$$X_L = X_C \qquad (19\text{--}1)$$

$$\mathbf{Z}_T = R \qquad (19\text{--}2)$$

$$f_r = \frac{1}{2\pi\sqrt{LC}} \qquad (19\text{--}3)$$

$$\mathbf{Z}_T = \frac{R^2 + X_L^2}{R} \qquad (19\text{--}4)$$

$$Z_T = \frac{L}{RC} \qquad (19\text{--}5)$$

$$BW = f_2 - f_1 \qquad (19\text{--}6)$$

$$f_r = f_1 + \frac{BW}{2} \qquad (19\text{--}7)$$

$$f_r = f_2 - \frac{BW}{2} \qquad (19\text{--}8)$$

$$Q = \frac{X_L}{R} \qquad (19\text{--}9)$$

$$V_L = QV_s \qquad (19\text{--}10)$$

$$V_C = QV_s \qquad (19\text{--}11)$$

$$BW = \frac{f_r}{Q} \qquad (19\text{--}12)$$

Summary

1. At resonance, the inductive and capacitive reactances are equal.
2. The impedance of a series RLC circuit is purely resistive at the resonant frequency.
3. In a series RLC circuit, maximum current occurs at resonance.
4. The reactive voltages V_L and V_C cancel at resonance in a series circuit because they are equal in magnitude and 180° out of phase.
5. In a parallel resonant circuit, the impedance is maximum at the resonant frequency.
6. In a parallel resonant circuit, the current from the source is minimum at the resonant frequency.
7. A parallel resonant circuit is commonly called a *tank circuit*.
8. The impedance of a parallel resonant circuit is purely resistive at resonance.
9. The bandwidth of a series resonant circuit is the range of frequencies for which the current is $0.707 I_{max}$ or greater.
10. The bandwidth of a parallel resonant circuit is the range of frequencies for which the impedance is $0.707 Z_{max}$ or greater.
11. The cutoff frequencies are the frequencies above and below resonance where the circuit response is 70.7% of the maximum response.
12. A smaller bandwidth means better selectivity.
13. The quality factor (Q) of a circuit is a ratio of stored power to lost power.
14. The inductive voltage and the capacitive voltage are equal to Q times the applied voltage at series resonance.
15. A higher Q means a smaller bandwidth.
16. *Tuned circuit* is a common term for resonant circuit.

Self-Test

1. At resonance, X_L is 500 Ω. What is X_C?
2. A given series RLC circuit has $R = 15$ Ω, $L = 20$ mH, and $C = 12$ pF. Determine the resonant frequency. What is the total impedance at resonance?
3. What are the impedance and the phase angle 10 kHz above resonance in Problem 2?
4. Find the following in Figure 19–23 at the resonant frequency:
   (a) I_{max} (b) f_r (c) V_R (d) V_L (e) V_C (f) θ

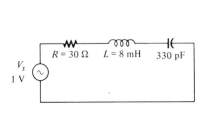

FIGURE 19–23

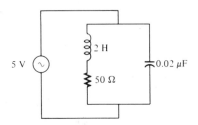

FIGURE 19–24

5. Determine Z_{max} for the parallel circuit in Figure 19–24.
6. In Figure 19–24, how much current is drawn from the source at resonance?
7. Determine Q and BW of the circuit in Figure 19–25.

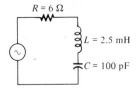

FIGURE 19–25

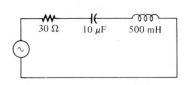

FIGURE 19–26

8. Find the cutoff frequencies for the circuit in Figure 19–25.
9. What is the impedance at the cutoff frequencies in Figure 19–26?
10. A tank circuit has a lower cutoff frequency of 8 kHz and a bandwidth of 2 kHz. What is its resonant frequency?

Problems

19–1. Find X_L, X_C, Z_T, and I at the resonant frequency in Figure 19–27.

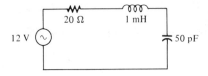

FIGURE 19–27

PROBLEMS

19–2. In Figure 19–27, what is the value of the circuit Q at resonance?

19–3. A given series resonant circuit has a maximum current of 50 mA and a V_L of 100 V. The applied voltage is 10 V. What is Z_T? What are X_L and X_C?

19–4. The bandwidth of a given series resonant circuit is 200 Hz, and the resonant frequency is 5 kHz. What are the cutoff frequencies?

19–5. For the series RLC circuit in Figure 19–28, determine the resonant frequency and the cutoff frequencies.

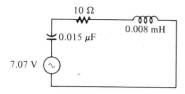

FIGURE 19–28

19–6. What is the value of the current at the half-power points in Figure 19–28?

19–7. Determine the phase angle between the applied voltage and the current at the cutoff frequencies in Figure 19–28. What is the phase angle at resonance?

19–8. What is the theoretical impedance of a parallel resonant circuit that has no resistance in either branch?

19–9. Find Z_T at resonance and f_r for the tank circuit in Figure 19–29.

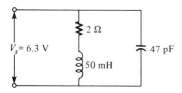

FIGURE 19–29

19–10. How much current is drawn from the source in Figure 19–29 at resonance? What are the inductive current and the capacitive current at the resonant frequency?

19–11. At resonance, $X_L = 2$ kΩ and $R_w = 25$ Ω. The resonant frequency is 5 kHz. Determine BW.

19–12. The Q of a series resonant circuit is 25, and the source voltage is 15 V. Determine V_L and V_C at resonance. What is V_R?

19–13. What values of L and C should be used in a tank circuit to obtain a resonant frequency of 8 kHz? The BW must be 800 Hz. Assume the resistance of the coil to be 10 Ω.

19–14. A parallel resonant circuit has a Q of 50 and a BW of 400 Hz. If Q is doubled, what is the bandwidth for the same f_r?

19–15. If the lower cutoff frequency is 2400 Hz and the upper cutoff frequency is 2800 Hz, what is the bandwidth? What is the resonant frequency? What is Q?

Answers to Section Reviews

Section 19–1:
1. $X_L = X_C$, $Z_T = R$, $\theta = 0°$. 2. Above f_r. 3. Below f_r. 4. T. 5. Because the impedance is minimum. 6. $f_r = 1/(2\pi\sqrt{LC})$.

Section 19–2:
1. 1500 Ω. 2. $f_r = 225$ kHz, $Z_T = 1250$ MΩ. 3. T. 4. F.

Section 19–3:
1. The range of frequencies for which the response is 70.7% of its maximum or greater. 2. The resonant frequency. 3. The frequencies at which the response curve is at 70.7% of maximum. 4. ·1 kHz.

Section 19–4:
1. X_L/R; ratio of reactive power of inductor (or capacitor) to resistive power. 2. 20. 3. 50 V. 4. Smaller. 5. T.

Section 19–5:
1. Changing the resonant frequency of a circuit. 2. It has a maximum response at only one frequency. The response drops off above and below this frequency.

Filters are widely used in electronics for selecting certain frequencies and rejecting others. In this chapter we will discuss *passive filters*. Passive filters use various combinations of resistors, capacitors, and inductors. They can be placed in four general categories according to their function: *low-pass, high-pass, band-pass,* and *band-stop*. Within each functional category, there are several common types which are discussed in this chapter.

20–1 Low-Pass Filters
20–2 Other Types of Low-Pass Filters
20–3 High-Pass Filters
20–4 Other Types of High-Pass Filters
20–5 Band-Pass Filters
20–6 Band-Stop Filters
20–7 Filter Response Characteristics

20 FILTERS

20-1

LOW-PASS FILTERS

As you already know, a low-pass filter allows signals with lower frequencies to pass from input to output while rejecting higher frequencies. A block diagram and a general response curve for a low-pass filter appear in Figure 20-1.

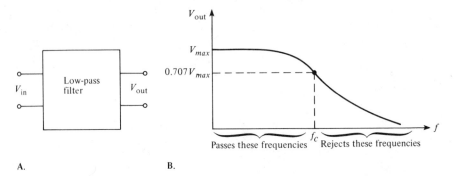

FIGURE 20-1

Pass Band

The range of low frequencies passed by a low-pass filter within a specified limit is called the *pass band* of the filter. The point considered to be the *upper end* of the pass band is at the *cutoff frequency*, f_c, as illustrated in Figure 20-1B. The cutoff frequency is the frequency at which the filter's output voltage is 70.7% of the maximum. The filter cutoff frequency is sometimes called the *break frequency*. The output is often said to be *down 3 dB* at this frequency. This term is a commonly used one which you should understand; so we will discuss decibel (dB) measurement in Section 20-7.

Capacitor to Ground

A capacitor connected from some point in a circuit to ground is a simple but commonly used low-pass filter. The basic concept is illustrated in Figure 20-2.

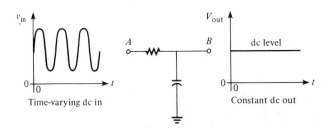

FIGURE 20-2 *Capacitor to ground as an ideal low-pass filter.*

LOW-PASS FILTERS

In many circuits you will find that a *time-varying* voltage is superimposed on a *dc* voltage to form a pulsating or varying dc voltage. This case is graphically illustrated at point *A* in Figure 20–2.

In many circuits the *variation* of the dc voltage is undesirable and must be significantly reduced. The capacitor presents a *low-impedance* path (low X_C) to ground to the time-varying voltage because of its frequency. However, the capacitor looks like an *open* to the constant dc voltage. As a result, the time-varying voltage is effectively shorted to ground, and the constant dc remains on the output. This type of arrangement is often used in electronic power supplies to significantly reduce the variations on the dc voltage, which are a natural result of *rectification*. This process is used for converting ac to dc.

Example 20–1

In Figure 20–3, the output voltage of the power supply rectifier is a pulsating dc with a frequency of 120 Hz. Choose a capacitor value that presents a maximum of 10 Ω to the 120-Hz voltage, and show the capacitor connected in the proper place.

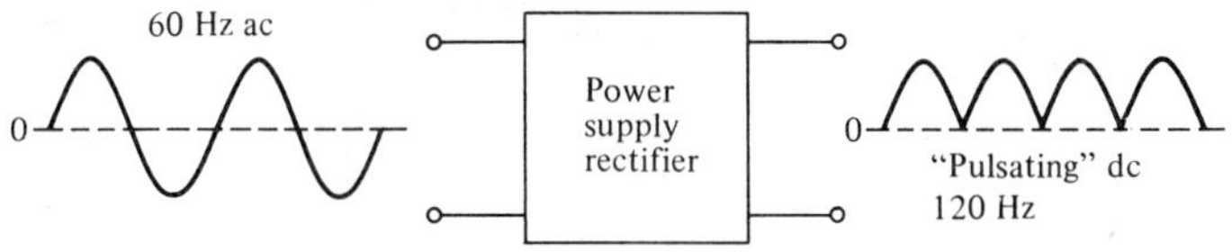

FIGURE 20–3

Solution:

$$X_C = 10 \text{ } \Omega \text{ by requirement}$$

$$X_C = \frac{1}{2\pi f C}$$

$$C = \frac{1}{2\pi f X_C} = \frac{1}{2\pi (120 \text{ Hz})(10 \text{ } \Omega)}$$

$$= 133 \text{ } \mu\text{F}$$

In practice, you should use the next larger standard value and connect it as shown in Figure 20–4.

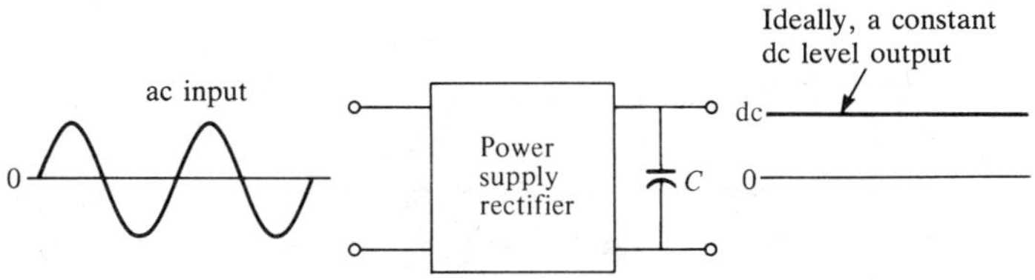

FIGURE 20–4

Example 20–2

In a transistor amplifier, it is required that the signal voltage at a certain point (the emitter) be eliminated and the constant dc voltage be retained for proper operation of the circuit. Determine the value of a filter capacitor to be connected from point A to ground in Figure 20–5 so that its reactance is no greater than $0.1R_E$ at a frequency of 10 kHz. In this application the capacitor is called a *bypass capacitor*.

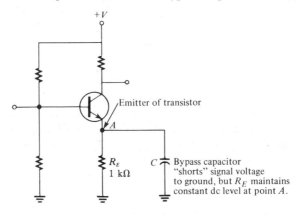

FIGURE 20–5

Solution:

$$X_C = 0.1R_E = 0.1(1000 \ \Omega)$$
$$= 100 \ \Omega$$

$$C = \frac{1}{2\pi f X_C} = \frac{1}{2\pi (10 \ \text{kHz})(100 \ \Omega)}$$
$$= 0.159 \ \mu\text{F}$$

X_C provides a parallel path to ground for the signal voltage, but it presents an open to dc. Therefore, only the dc voltage remains at point A because most of the signal is shorted to ground through the capacitor. That is, very little signal voltage is developed across the capacitor because of its low X_C.

Choke in Series

An inductor is commonly called a *choke*, especially in filter applications. A choke in series between two points in a circuit can be used as a low-pass filter, as illustrated in Figure 20–6. At point A in the figure, there is a varying dc voltage. When X_L is high compared to R at the frequency of the time-varying signal, only the constant dc will get through to the output at point B.

LOW-PASS FILTERS

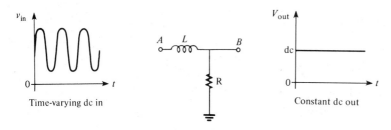

FIGURE 20-6 *Series choke as an ideal low-pass filter.*

That is, practically all of the varying voltage is dropped across the X_L and none across the resistor. Practically all of the constant dc is dropped across R and none across X_L.

The *reduction* of an unwanted frequency or frequencies by a filter is called *attenuation*. It is a commonly used term that applies to all types of filters.

Example 20-3

Determine the output voltage for the low-pass filter in Figure 20-7. The input is a 2-V peak-to-peak sine wave riding on a 5-V constant dc level. $X_L = 1000\ \Omega$ at the frequency of the sine wave.

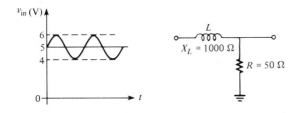

FIGURE 20-7

Solution:

The sine wave will be attenuated by the filter, but the constant dc level of 5 V will pass unattenuated.

We can find the sine wave output voltage by applying the voltage divider formula:

$$V_{out} = \left(\frac{R}{\sqrt{R^2 + X_L^2}}\right)V_{in} = \left(\frac{50\ \Omega}{\sqrt{(50\ \Omega)^2 + (1000\ \Omega)^2}}\right)2\ \text{V}$$

$$= 0.1\ \text{V pp}$$

The total output voltage is a 0.1-V peak-to-peak sine wave riding on the 5-V dc level. As you can see, the varying part of the input has been almost eliminated (filtered out).

Review for 20-1

1. What does a low-pass filter do?
2. Define *pass band* of a low-pass filter.
3. Define *cutoff frequency*.
4. What does the term *down 3 dB* mean?

20-2

OTHER TYPES OF LOW-PASS FILTERS

Several configurations of low-pass filters other than the capacitor to ground and the series choke are common and will be discussed in this section.

Inverted-L Type

Figure 20–8 shows a capacitor and an inductor connected in what is often called an *inverted-L* configuration. The name, of course, comes from the shape of the inverted letter L formed by the two components when they are drawn in a schematic.

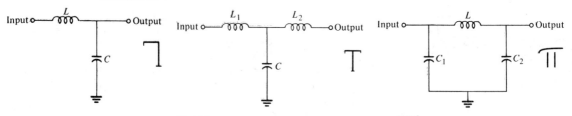

FIGURE 20–8 *Inverted-L low-pass filter.*

FIGURE 20–9 *T-type low-pass filter.*

FIGURE 20–10 *π-type low-pass filter.*

The inductive reactance of the choke acts to block higher frequencies. The capacitive reactance acts to short higher frequencies to ground. Thus, only frequencies below the cutoff frequency are passed without significant attenuation. This type of filter is more effective than the RC or RL types because the high-frequency response decreases at a faster rate beyond f_c. We will discuss frequency response in more detail in a later section.

T-Type

The T-type low-pass filter uses two chokes and a capacitor, as shown in Figure 20–9. This filter has certain advantages over the inverted-L type because the filtering action is improved by the additional choke.

π-Type

The π-type low-pass filter is shown in Figure 20–10. The additional capacitor again improves filtering over that in the inverted-L type. Notice that in all

HIGH-PASS FILTERS

of these configurations, the choke is in series from input to output and the capacitors are connected to ground.

Review for 20–2

1. Name three configurations of low-pass filters.
2. Sketch each of the above types.

20–3
HIGH-PASS FILTERS

A high-pass filter allows signals with higher frequencies to pass from input to output while rejecting lower frequencies. A block diagram and a general response curve for a high-pass filter are shown in Figure 20–11.

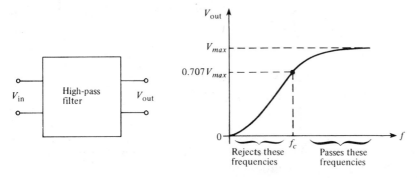

FIGURE 20–11

The frequency considered to be the *lower* end of the band is called the *cutoff frequency*. Just as in the low-pass filter, it is the frequency at which the output is 70.7% of the maximum, as indicated in the figure.

Choke to Ground

A choke connected from some point in a circuit to ground acts as a simple high-pass filter. The basic concept is illustrated in Figure 20–12.

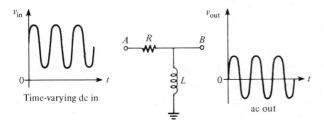

FIGURE 20–12 *Choke to ground as an ideal high-pass filter.*

We will use the same example as we used for the low-pass filter. A time-varying voltage in Figure 20–12 is superimposed on a constant dc voltage at point A. The choke presents a high impedance to the sine wave between point B and ground. However, the choke looks like a short to the constant dc voltage; so it is shorted to ground. Thus, only the sine wave is left at point B, the output. Notice that the output voltage now varies about a zero volt level at B. This circuit has filtered out the zero frequency (dc) and has allowed the higher-frequency signal to pass from input to output.

Example 20–4

Determine the output voltage for the high-pass filter in Figure 20–13. The input is a 2-V peak-to-peak sine wave riding on a 4-V constant dc level. $X_L = 1000 \, \Omega$ at the frequency of the sine wave. Neglect the coil's winding resistance.

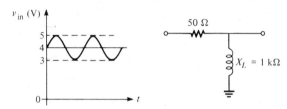

FIGURE 20–13

Solution:

The constant dc level will be attenuated by the filter, but the sine wave will pass through with very little attenuation.

The sine wave output is calculated as follows:

$$V_{out} = \left(\frac{X_L}{\sqrt{R^2 + X_L^2}}\right)V_{in} = \left(\frac{1000 \, \Omega}{\sqrt{(50 \, \Omega)^2 + (1000 \, \Omega)^2}}\right)2 \text{ V}$$
$$= 1.998 \text{ V}$$

The output is a 1.998-V peak-to-peak sine wave (almost the same as the input), varying about zero volts. The constant 4-V dc has been filtered out.

Series Capacitor

A capacitor connected between two points in a circuit acts as a high-pass filter, as illustrated in Figure 20–14. To illustrate, we again use a sine wave superimposed on a constant dc level. When X_C is low compared to the resistance, the sine wave passes through to point B, the output. However, the capacitor blocks the constant dc. Thus, we have a high-pass filter.

HIGH-PASS FILTERS

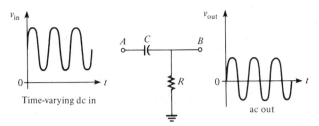

FIGURE 20–14 *Series capacitor as a high-pass filter.*

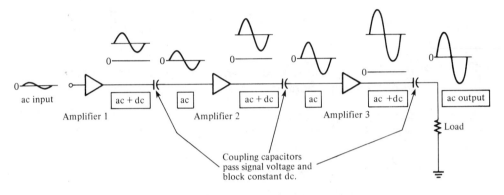

FIGURE 20–15 *Example of a coupling capacitor application.*

A very common application of the series capacitor high-pass filter is as a *coupling capacitor* between audio amplifier stages. It is used to pass the amplified audio signal and block the constant dc level (bias voltage) from one stage to the next, as illustrated in Figure 20–15.

Example 20–5

(a) In Figure 20–16, find the value of C so that X_C is ten times less than R at an input frequency of 10 kHz.
(b) If a 5-V sine wave with a dc level of 10 V is applied, what is the output voltage?

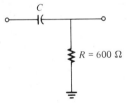

FIGURE 20–16

Solution:
(a) The value of C is determined as follows:

Example 20–5 (continued)

$$X_C = 0.1R$$
$$= 60 \text{ }\Omega$$
$$C = \frac{1}{2\pi f X_C} = \frac{1}{2\pi(10 \text{ kHz})(60 \text{ }\Omega)}$$
$$= 0.265 \text{ }\mu\text{F}$$

(b) The sine wave output is determined as follows:

$$V_{out} = \left(\frac{R}{\sqrt{R^2 + X_C^2}}\right)V_{in} = \left(\frac{600 \text{ }\Omega}{\sqrt{(600 \text{ }\Omega)^2 + (60 \text{ }\Omega)^2}}\right)5 \text{ V}$$
$$= 4.98 \text{ V}$$

The sine wave output is almost equal to the input. The constant dc level of 10 V (the average value of the input) is missing.

Review for 20–3

1. What is the purpose of a high-pass filter?
2. A capacitor between two points in a circuit passes constant dc voltage and blocks any changing signal (T or F).

20–4

OTHER TYPES OF HIGH-PASS FILTERS

The same basic forms of LC filters are used for high-pass as well as low-pass filters, except the component positions are opposite.

Inverted-L Type

Figure 20–17 shows an inverted-L high-pass filter. At lower frequencies the capacitive reactance is large and the inductive reactance is small. Thus, most of the input voltage is dropped across the capacitor and very little across the choke at lower frequencies. Therefore, the output voltage is much less than the input. As the frequency is increased, X_C becomes less and X_L becomes greater, causing the output voltage to increase. Thus, high frequencies are passed while low frequencies are attenuated.

T-Type

A T-type high-pass filter uses two capacitors and a choke, as shown in Figure 20–18.

BAND-PASS FILTERS

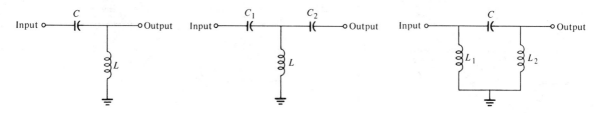

FIGURE 20–17 *Inverted-L high-pass filter.*

FIGURE 20–18 *T-type high-pass filter.*

FIGURE 20–19 *π-type high-pass filter.*

π-Type

The π-type high-pass filter is shown in Figure 20–19. Notice that in all of the high-pass configurations, the capacitors are in series between input and output, and the chokes go to ground. The capacitors can be viewed as shorts to high frequencies, thus passing them from input to output, but as open to low frequencies, thus blocking them. The chokes can be thought of as open to high frequencies and as shorts to ground for low frequencies.

Review for 20–4

1. Sketch each type of high-pass filter.
2. Describe the basis for high-pass filter action.

20–5

BAND-PASS FILTERS

A band-pass filter allows a certain *band* of frequencies to pass and attenuates or rejects all frequencies below and above the pass band. A typical band-pass response curve is shown in Figure 20–20.

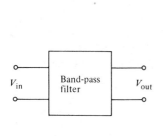
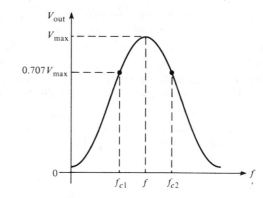

FIGURE 20–20 *Typical band-pass response curve.*

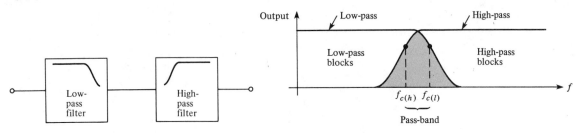

FIGURE 20–21 *Low-pass and high-pass filters used to form a band-pass filter.*

FIGURE 20–22 *Overlapping response curves of high-pass/low-pass filter.*

Low-Pass/High-Pass Filter

A combination of a low-pass and a high-pass filter can be used to form a band-pass filter, as illustrated in Figure 20–21.

If the cutoff frequency of the low-pass ($f_{c(l)}$) is higher than the cutoff of the high-pass ($f_{c(h)}$), the responses overlap. Thus, all frequencies except those between $f_{c(h)}$ and $f_{c(l)}$ are eliminated, as shown in Figure 20–22.

Example 20–6

A high-pass filter with $f_c = 2$ kHz and a low-pass filter with $f_c = 2.5$ kHz are used to construct a band-pass filter. What is the bandwidth of the pass band?

Solution:

$$\text{BW} = f_{c(l)} - f_{c(h)} = 2.5 \text{ kHz} - 2 \text{ kHz}$$
$$= 500 \text{ Hz}$$

Series Resonant Filter

A type of series resonant band-pass filter is shown in Figure 20–23. As you learned in the last chapter, a series resonant circuit has *minimum impedance* and *maximum current* at the resonant frequency, f_r. Thus, most of the input voltage is dropped across the resistor at the resonant frequency. Therefore, the output across R has a band-pass characteristic with a maximum output at the frequency of resonance. The bandwidth is determined by the circuit Q.

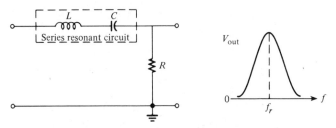

FIGURE 20–23 *Series resonant band-pass filter.*

BAND-PASS FILTERS

Example 20-7

Determine the output voltage at the resonant frequency and the bandwidth for the filter in Figure 20-24.

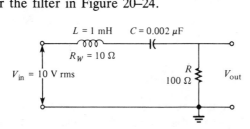

FIGURE 20-24

Solution:

At f_r, the impedance of the resonant circuit is equal to the winding resistance, R_W. By the voltage divider formula,

$$V_{out} = \left(\frac{R}{R + R_W}\right) V_{in} = \left(\frac{100 \ \Omega}{110 \ \Omega}\right) 10 \text{ V}$$

$$= 9.09 \text{ V}$$

The resonant frequency is

$$f_r = \frac{1}{2\pi\sqrt{LC}}$$

$$= 112.5 \text{ kHz}$$

The circuit Q is

$$Q = \frac{X_L}{R_T} = \frac{707 \ \Omega}{110 \ \Omega}$$

$$= 6.4$$

The bandwidth is

$$BW = \frac{f_r}{Q} = \frac{112.5 \text{ kHz}}{6.4}$$

$$= 17.6 \text{ kHz}$$

Parallel Resonant Filter

A type of band-pass filter using a parallel resonant circuit is shown in Figure 20-25. Recall that a parallel resonant circuit has *maximum impedance* at resonance. The circuit in Figure 20-25 acts as a voltage divider. At resonance, the impedance of the tank is much greater than R. Thus, most of the input voltage is across the tank, producing a maximum output voltage at the resonant frequency.

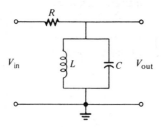

FIGURE 20-25 *Parallel resonant band-pass filter.*

For frequencies above or below resonance, the tank impedance drops off, and more of the input voltage is across *R*. As a result, the output voltage across the tank drops off, creating a band-pass characteristic.

Example 20-8

What is the center frequency of the filter in Figure 20-26?

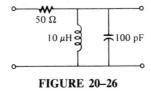

FIGURE 20-26

Solution:

The center frequency of the filter is its resonant frequency:

$$f_r \cong \frac{1}{2\pi\sqrt{LC}} = \frac{1}{2\pi\sqrt{(10\ \mu H)(100\ pF)}}$$

$$= 5.03\ \text{MHz}$$

Review for 20-5

1. What is the purpose of the band-pass filter?
2. Name three basic ways to construct a band-pass filter.

20-6

BAND-STOP FILTERS

A band-stop filter is essentially the opposite of a band-pass in terms of the responses. It allows *all* frequencies to pass except those lying within a certain *stop band*. A general band-stop response curve is shown in Figure 20-27.

BAND-STOP FILTERS

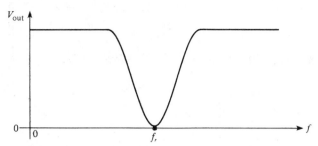

FIGURE 20–27 *General band-stop response curve.*

Low-Pass/High-Pass Filter

A band-stop filter can be formed from a low-pass and a high-pass filter, as shown in Figure 20–28.

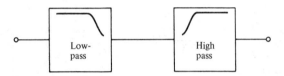

FIGURE 20–28 *Low-pass and high-pass filters used to form a band-stop filter.*

If the low-pass cutoff frequency $f_{c(l)}$ is set lower than the high-pass cutoff frequency $f_{c(h)}$, a band-stop characteristic is formed as illustrated in Figure 20–29.

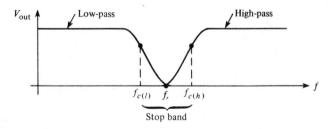

FIGURE 20–29

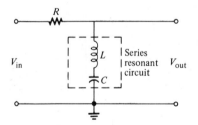

FIGURE 20–30 *Series resonant band-stop filter.*

Series Resonant Filter

A series resonant circuit used in a band-stop configuration is shown in Figure 20–30. Basically, it works as follows: At the resonant frequency, the impedance is minimum, and therefore the output voltage is minimum. Most of the input voltage is dropped across R. At frequencies above and below resonance, the impedance increases, causing more voltage across the output.

Example 20-9

Find the output voltage at f_r and the bandwidth in Figure 20-31.

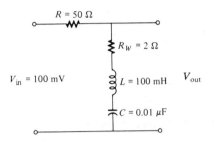

FIGURE 20-31

Solution:

Since $X_L = X_C$ at resonance, then

$$V_{out} = \left(\frac{R_W}{R + R_W}\right)V_{in} = \left(\frac{2\,\Omega}{52\,\Omega}\right)100\text{ mV}$$

$$= 3.85\text{ mV}$$

$$f_r = \frac{1}{2\pi\sqrt{LC}} = \frac{1}{2\pi\sqrt{(100\text{ mH})(0.01\text{ }\mu\text{F})}}$$

$$= 5.03\text{ kHz}$$

$$Q = \frac{X_L}{R} = \frac{3160\,\Omega}{52\,\Omega}$$

$$= 60.8$$

$$BW = \frac{f_r}{Q} = \frac{5.03\text{ kHz}}{60.8}$$

$$= 82.7\text{ Hz}$$

Parallel Resonant Filter

A parallel resonant circuit used in a band-stop configuration is shown in Figure 20-32. At the resonant frequency, the tank impedance is maximum, and so most of the input voltage is dropped across it. Very little voltage is across R at resonance. As the tank impedance decreases above and below resonance, the output voltage increases.

BAND-STOP FILTERS

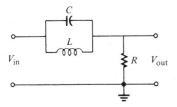

FIGURE 20-32 *Parallel resonant band-stop filter.*

Example 20-10

Find the center frequency of the filter in Figure 20-33. Sketch the output response curve showing the minimum and maximum voltages.

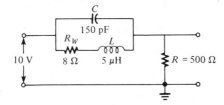

FIGURE 20-33

Solution:

$$f_r = \frac{1}{2\pi\sqrt{LC}} = \frac{1}{2\pi\sqrt{(5\ \mu H)(150\ pF)}}$$

$$= 5.8\ \text{MHz}$$

From Equation (19-5), at the resonant frequency,

$$Z_{tank} = \frac{L}{R_W C}$$

$$= \frac{5\ \mu H}{(8\ \Omega)(150\ pF)}$$

$$= 4.167\ \text{k}\Omega \quad \text{purely resistive}$$

We use the voltage divider formula to find the *minimum* output voltage:

$$V_{out(min)} = \left(\frac{R}{R + Z_{tank}}\right) V_{in} = \left(\frac{500\ \Omega}{4.667\ \text{k}\Omega}\right) 10\ \text{V}$$

$$= 1.07\ \text{V}$$

Example 20–10 (continued)

At zero frequency, the impedance of the tank is R_w because $X_C = \infty$ and $X_L = 0\ \Omega$. Therefore, the maximum output voltage is

$$V_{out(max)} = \left(\frac{R}{R + R_w}\right)V_{in} = \left(\frac{500\ \Omega}{508\ \Omega}\right)10\ \text{V}$$
$$= 9.84\ \text{V}$$

As the frequency increases much higher than f_r, X_C approaches $0\ \Omega$, and V_{out} approaches V_{in}.

Figure 20–34 is the response curve.

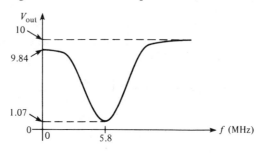

FIGURE 20–34

Review for 20–6

1. How does a band-stop filter differ from a band-pass filter?
2. Name three basic ways to construct a band-stop filter.

20–7

FILTER RESPONSE CHARACTERISTICS

In this section we will examine the filter response curve in more detail and will identify some important characteristics associated with basic RC and RL filters.

Decibel (dB) Measurement

The decibel (dB) is a logarithmic measurement of voltage or power ratios which can be used to express the input-to-output relationship of a filter. The following equation expresses a *voltage* ratio in decibels:

$$\text{dB} = 20 \log\left(\frac{V_{out}}{V_{in}}\right) \tag{20–1}$$

The following equation is the decibel formula for a *power* ratio:

FILTER RESPONSE CHARACTERISTICS

$$dB = 10 \log\left(\frac{P_{out}}{P_{in}}\right) \quad (20\text{-}2)$$

Example 20–11

The output voltage of a filter is 5 V and the input is 10 V. Express the voltage ratio in decibels.

Solution:

$$20 \log\left(\frac{V_{out}}{V_{in}}\right) = 20 \log\left(\frac{5 \text{ V}}{10 \text{ V}}\right) = 20 \log(0.5)$$
$$= -6.02 \text{ dB}$$

The −3 dB Frequencies

The output of a filter is said to be *down 3 dB* at the cutoff frequencies. Actually, this frequency is the point at which the output voltage is 70.7% of the maximum voltage, as shown in Figure 20–35A for a low-pass filter. The same applies to high-pass and band-pass filters, as shown in Parts B and C of the figure.

We can show that the 70.7% point is the same as 3 dB below maximum (or −3 dB) as follows. The maximum voltage is the zero dB reference:

$$20 \log\left(\frac{0.707 V_{max}}{V_{max}}\right) = 20 \log(0.707)$$
$$= -3 \text{ dB}$$

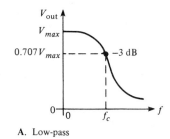
A. Low-pass

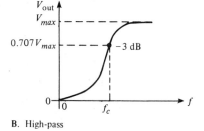
B. High-pass

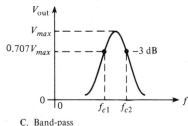
C. Band-pass

FIGURE 20–35 *Response curves showing −3 dB points.*

Formula for Cutoff Frequencies

For a basic RC filter, $X_C = R$ at the cutoff frequency f_c. This condition can be written as $1/(2\pi f_c C) = R$. Solving for f_c, we get

$$f_c = \frac{1}{2\pi RC} \quad (20\text{-}3)$$

Also, for a basic RL filter, $X_L = R$ at the cutoff frequency. This condition can be stated as $2\pi f_c L = R$. Solving for f_c, we get

$$f_c = \frac{1}{2\pi(L/R)} \qquad (20\text{--}4)$$

The condition that the reactance equals the resistance at the cutoff frequency can be shown as follows: When $X_C = R$, the output voltage can be expressed as $X_C V_{in}/\sqrt{R^2 + X_C^2}$ by the voltage divider formula. This formula is for a low-pass RC filter.

If $X_C = R$, then

$$\frac{X_C V_{in}}{\sqrt{R^2 + X_C^2}} = \frac{RV_{in}}{\sqrt{R^2 + R^2}} = \frac{RV_{in}}{\sqrt{2R^2}}$$

$$= \frac{RV_{in}}{R\sqrt{2}} = \frac{V_{in}}{\sqrt{2}} = 0.707 V_{in}$$

These calculations show that the output is 70.7% of the input when $X_C = R$. The frequency at which this occurs is, by definition, the cutoff frequency, as we have previously discussed. Similar expressions can be made for a high-pass filter and also for RL filters.

Keep in mind that Equations (20–3) and (20–4) are for RC and RL filters only. Analysis of the more complex filter configurations is beyond the scope of this book.

Example 20–12

Determine the cutoff frequency for the low-pass RC filter in Figure 20–36.

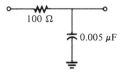

FIGURE 20–36

Solution:

$$f_c = \frac{1}{2\pi RC} = \frac{1}{2\pi(100\ \Omega)(0.005\ \mu F)}$$

$$= 318.3\ \text{kHz}$$

This result means that the output voltage is 3 dB below V_{in} at this frequency (V_{out} has a maximum value of V_{in}). Also, the bandwidth of this filter goes from zero frequency to 318.3 kHz, or BW = 318.3 kHz.

FILTER RESPONSE CHARACTERISTICS

"Roll-Off" of the Response Curve

The dashed lines in Figure 20–37 show an actual response curve for a low-pass filter. The maximum output is defined to be 0 dB as a reference. Zero decibels corresponds to $V_{out} = V_{in}$, because $20 \log(V_{out}/V_{in}) = 20 \log 1 = 0$ dB. The output drops from 0 dB to -3 dB at the cutoff frequency and then continues to decrease at a *fixed* rate. This pattern of decrease is called the *roll-off* of the frequency response. The solid line shows an ideal output response that is considered to be "flat" out to f_c. The output then decreases at the fixed rate.

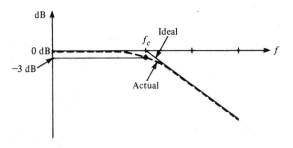

FIGURE 20–37 *Actual and ideal response curves for a low-pass filter.*

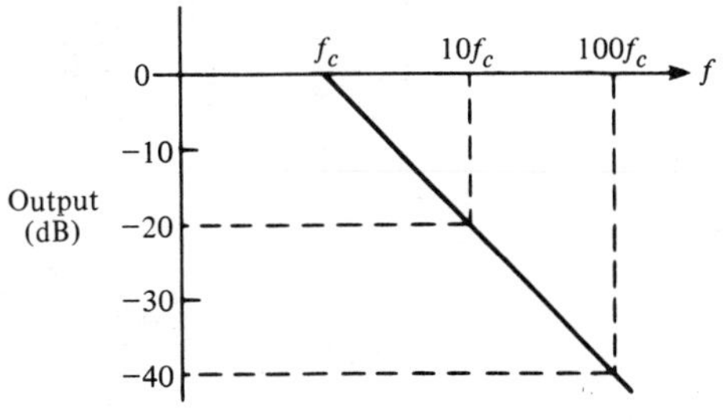

FIGURE 20–38 *Frequency roll-off for a low-pass RC filter (Bode plot).*

It can be shown that the response decreases at a constant rate of -20 dB *per decade of frequency*. A one-decade increase in frequency is a *tenfold* increase. This roll-off is a constant for a basic RC or RL filter. Figure 20–38 shows a frequency response plot on a semilog scale where each interval on the horizontal axis represents a tenfold increase in frequency. This frequency response curve is called a *Bode plot*. It can also be done for a high-pass filter, as shown in Example 20–13.

Example 20–13

Make a Bode plot for the filter in Figure 20–39 for three decades of frequency. Use semilog graph paper.

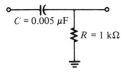

FIGURE 20–39

Solution:

The cutoff frequency for this high-pass filter is

$$f_c = \frac{1}{2\pi RC} = \frac{1}{2\pi (1 \text{ k}\Omega)(0.005 \text{ }\mu\text{F})}$$
$$= 31.8 \text{ kHz}$$

Example 20-13 (continued)

The idealized Bode plot is shown with the solid line on the semilog graph in Figure 20-40. The approximate actual response curve is shown with the dashed lines. Notice first that the horizontal scale is logarithmic and the vertical scale is linear. The frequency is on the logarithmic scale, and the filter output in decibels is on the vertical.

The output is flat beyond f_c (31.8 kHz). As the frequency is reduced below f_c, the output drops at a -20 dB/decade rate. Thus, for the ideal curve, every time the frequency is reduced by ten, the output is reduced by 20 dB. A slight variation from this occurs in actual practice. The output is actually at -3 dB rather than 0 dB at the cutoff frequency.

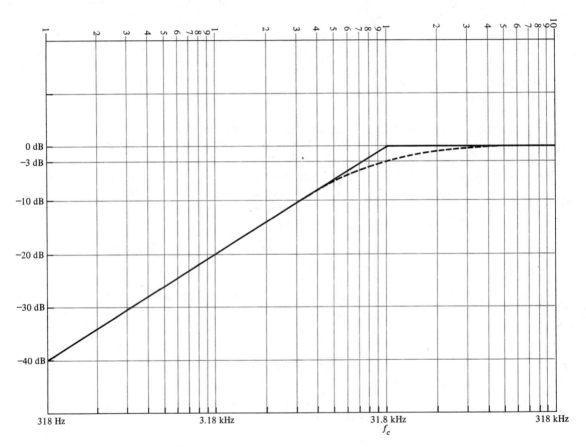

FIGURE 20-40 *Bode plot for Example 20-13.*

Review for 20-7

1. What does "dB" stand for?

SELF-TEST

2. At f_c, how many decibels less is the output of a filter than at its mid-range or maximum value?
3. What does the term *roll-off* mean?
4. At what rate does the output of a basic low-pass RC or RL filter change beyond f_c?
5. What is a decade change in frequency?

Formulas

$$dB = 20 \log\left(\frac{V_{out}}{V_{in}}\right) \qquad (20\text{--}1)$$

$$dB = 10 \log\left(\frac{P_{out}}{P_{in}}\right) \qquad (20\text{--}2)$$

$$f_c = \frac{1}{2\pi RC} \qquad (20\text{--}3)$$

$$f_c = \frac{1}{2\pi(L/R)} \qquad (20\text{--}4)$$

Summary

1. A capacitor to ground forms a basic type of low-pass filter.
2. A choke between input and output forms a basic type of low-pass filter.
3. A choke to ground forms a basic type of high-pass filter.
4. A capacitor between input and output forms a basic type of high-pass filter.
5. A band-pass filter passes frequencies between the lower and upper cutoff frequencies and rejects all others.
6. A band-stop filter rejects frequencies between its lower and upper cutoff frequencies and passes all others.
7. The bandwidth of a resonant filter is determined by the quality factor (Q) of the circuit and the resonant frequency.
8. Cutoff frequencies are also called -3 dB frequencies.
9. The output voltage is 70.7% of its maximum at the cutoff frequencies.
10. The roll-off rate of a basic RC or RL filter is -20 dB per decade.

Self-Test

1. The maximum output voltage of a given low-pass filter is 10 V. What is the output voltage at the cutoff frequency?
2. A sine wave with a peak-to-peak value of 15 V is applied to an RC low-pass filter. The average value of the sine wave is 10 V. If the reactance at the sine wave frequency is assumed to be zero, what is the output voltage?

3. The same signal in Problem 2 is applied to an RC high-pass filter. If the reactance is zero at the signal frequency, what is the output voltage of the filter?
4. Calculate the value of C used as a *bypass* if its reactance must be at least ten times smaller than the resistance of 2.5 kΩ at 1 kHz.
5. A low-pass RL filter has $X_L = 100\ \Omega$ and $R = 5\ \Omega$ at the input frequency. What is the exact output voltage if the input is a 5-V peak-to-peak sine wave riding on an 8-V constant dc level? Assume a winding resistance of 0 Ω.
6. Sketch an inverted-L type of low-pass filter and high-pass filter.
7. Sketch a π-type low-pass filter and high-pass filter.
8. Sketch a series resonant band-pass filter and a series resonant band-stop filter.
9. Sketch a parallel resonant band-pass filter and a parallel resonant band-stop filter.
10. The resonant frequency of a band-pass filter is 5 kHz. The circuit Q is 20. What is the bandwidth?
11. The output voltage of a filter is 4 V, and the input is 12 V. Express this relationship in dB.
12. The output power of a circuit is 5 W, and the input power is 10 W. What is the dB power ratio?
13. What is the cutoff frequency of an RC filter with $R = 1\ k\Omega$ and $C = 0.02\ \mu F$?
14. What is the cutoff frequency of an RL filter with $R = 5\ k\Omega$ and $L = 0.01\ mH$?

Problems

20–1. In a low-pass filter, X_C is 500 Ω and R is 2 kΩ. What is the output voltage if the input is 10 V rms?

20–2. For the same values in Problem 20–1, determine the output voltage of a high-pass filter.

20–3. A certain low-pass filter has a cutoff frequency of 3 kHz. Determine which of the following frequencies are passed and which are rejected:
 (a) 100 Hz   (b) 1 kHz   (c) 2 kHz   (d) 3 kHz   (e) 5 kHz

20–4. A high-pass filter has a cutoff frequency of 50 Hz. Determine which of the following frequencies are passed and which are rejected:
 (a) 1 Hz   (b) 20 Hz   (c) 50 Hz   (d) 60 Hz   (e) 30 kHz

20–5. Determine the output voltage of each filter in Figure 20–41 at the specified frequency if the input to each is 10 V.

20–6. What is f_c for each filter in Figure 20–41? What are the output voltages at this frequency?

20–7. Classify each filter in Figure 20–41 as a high-pass or a low-pass.

PROBLEMS   **653**

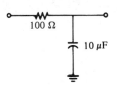

A. $f = 60$ Hz

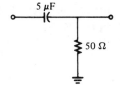

B. $f = 400$ Hz

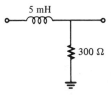

C. $f = 1$ kHz

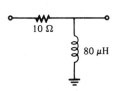

D. $f = 2$ kHz

FIGURE 20–41

20–8. Determine the center frequency for each filter in Figure 20–42.

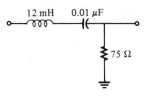

A.

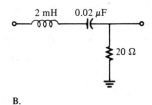

B.

FIGURE 20–42

20–9. Assuming that the coils in Figure 20–42 have a winding resistance of 10 Ω, find the BW for each filter.

20–10. What are the lower and upper cutoff frequencies for each filter in Figure 20–42?

20–11. For each filter in Figure 20–43, find the center frequency of the pass band.

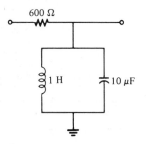

A.

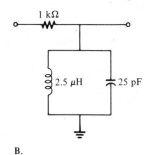

B.

FIGURE 20–43

20–12. If the coils in Figure 20–43 have a winding resistance of 4 Ω, what is the output voltage at resonance if $V_{in} = 120$ V?

20–13. For the filter in Figure 20–44, calculate the value of C required for each of the following cutoff frequencies:
(a) 60 Hz (b) 500 Hz (c) 1 kHz (d) 5 kHz

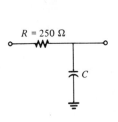

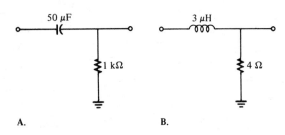

FIGURE 20–44

FIGURE 20–45

20–14. Find the cutoff frequency for each filter in Figure 20–45.

20–15. For each following case, express the voltage ratio in dB:
(a) $V_{in} = 1$ V, $V_{out} = 1$ V (b) $V_{in} = 5$ V, $V_{out} = 3$ V
(c) $V_{in} = 10$ V, $V_{out} = 7.07$ V (d) $V_{in} = 25$ V, $V_{out} = 5$ V

20–16. The input voltage to a low-pass RC filter is 8 V rms. Find the output voltage at the following dB levels:
(a) -1 dB (b) -3 dB (c) -6 dB (d) -20 dB

20–17. For a basic RC low-pass filter, find the output voltage in dB relative to a 0-dB input for the following frequencies (f_c is 1 kHz):
(a) 10 kHz (b) 100 kHz (c) 1 MHz

Answers to Section Reviews

Section 20–1:
1. It passes lower frequencies and rejects higher frequencies. **2.** Zero to f_c. **3.** The frequency at which the output is 70.7% of its maximum. **4.** The point at which the output voltage is 3 dB below its maximum; equivalent to the 70.7% value.

Section 20–2:
1. Inverted-L, T, and π. **2.** See Figure 20–46.

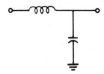 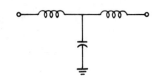 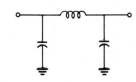

FIGURE 20–46

Section 20–3:
1. To pass higher frequencies and reject lower frequencies. **2.** F.

ANSWERS TO SECTION REVIEWS

Section 20–4:
1. See Figure 20–47. **2.** A capacitor in series presents low impedance to higher frequencies and high impedance to lower frequencies. A choke to ground presents high impedance to higher frequencies and and low impedance to lower frequencies.

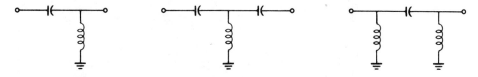

FIGURE 20–47

Section 20–5:
1. To pass a specified range of frequencies and reject all frequencies above and below the band. **2.** High-pass/low-pass combination, series resonance, and parallel resonance.

Section 20–6:
1. It rejects a certain band of frequencies rather than passing them. **2.** High-pass/low-pass combination, series resonance, and parallel resonance.

Section 20–7:
1. Decibel, a logarithmic measurement. **2.** 3 dB. **3.** The rate of decrease of the frequency response. **4.** -20 dB per decade of frequency. **5.** A tenfold change.

Test equipment is indispensable to the electronic technician and engineer. New circuit and systems designs are checked and problems are traced down with the help of various types of instruments in the laboratory and on the service bench.

To learn the practical use of test instruments, you must have actual "hands-on" experience in the laboratory. The intent of this chapter is to give you the basic principles of three important categories of test equipment: (1) meters, (2) signal generators, and (3) oscilloscopes. There are many other types of instruments, but these are the ones that you will probably use most often.

21-1 Meter Movements
21-2 The Ammeter
21-3 The Voltmeter
21-4 ac Meters
21-5 The Ohmmeter
21-6 Multimeters
21-7 Signal Generators
21-8 The Oscilloscope
21-9 Oscilloscope Controls

21

TEST AND MEASUREMENT INSTRUMENTS

21-1

METER MOVEMENTS

Moving-Coil Movement

In this type of meter movement, known as the *d'Arsonval movement,* the pointer is deflected in proportion to the amount of current through a coil. Figure 21–1A shows a basic d'Arsonval meter movement. It consists of a coil wound on a bearing-mounted assembly placed between the poles of a permanent magnet. A pointer is attached to the moving assembly. With no current through the coil, a spring mechanism keeps the pointer at its left-most position. When there is current through the coil, forces act on the coil, causing a rotation to the right. The amount of rotation depends on the amount of current. Figure 21–1B shows a construction view of the parts of a typical movement.

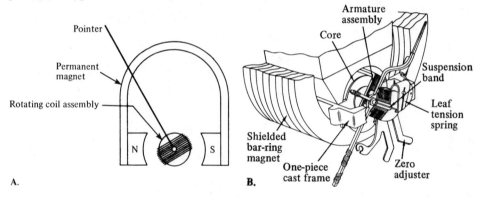

FIGURE 21–1 *Basic d'Arsonval movement.* (**B,** *courtesy of the Triplett Corporation*)

Figure 21–2 illustrates how the interaction of magnetic fields produces rotation of the coil assembly. The current is in at the "cross" and out at the "dot" in the single winding shown. The current outward produces a clockwise field that reinforces the permanent field at the top of the right side of the coil and weakens the field at the bottom. The result is a downward force on the right conductor as shown. An upward force is developed on the left side of the coil where the current is inward. These forces produce a clockwise rotation of the coil assembly.

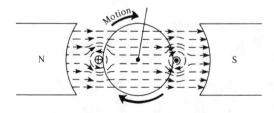

FIGURE 21–2 *Magnetic forces on the coil.*

METER MOVEMENTS

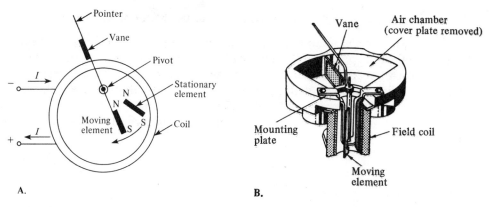

FIGURE 21-3 *Iron-vane movement.* (**B,** *courtesy of the Triplett Corporation*)

Iron-Vane Movement

This type of movement consists basically of two iron bars placed within a coil. The magnetic field produced by the current in the coil induces a north pole and a south pole in the iron bars. The like poles repel each other, causing the moving element to move away from the stationary element. The attached pointer is deflected by the movement of the element in proportion to the current through the coil. A "vane" is attached to the movement and is housed in an air chamber for damping purposes. Figure 21-3A illustrates the basic mechanism, and Part B shows the construction of a basic movement.

Electrodynamometer Movement

Figure 21-4A shows the basic electrodynamometer movement. It differs from the d'Arsonval movement in that it uses an electromagnetic field rather than a permanent magnetic field. The electromagnetic field is produced by current in the stationary coil. The pointer is attached to the moving coil. This type of movement is commonly used in wattmeters. Figure 21-4B shows the construction of a typical movement.

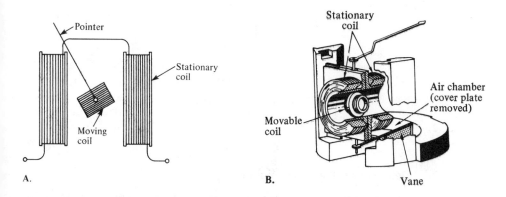

FIGURE 21-4 *Electrodynamometer movement.* (**B,** *courtesy of the Triplett Corporation*)

Current Sensitivity and Resistance of the Meter Movement

The *current sensitivity* of a meter movement is the amount of current required to deflect the pointer full scale (all the way to its right-most position). For example, a 1-mA sensitivity means that when there is 1 mA through the meter coil, the needle is at its maximum deflection. If there is 0.5 mA through the coil, the needle is at the halfway point of its full deflection.

The *movement resistance* is simply the resistance of the coil of wire used.

Review for 21-1

1. List three types of meter movements.
2. Define *current sensitivity* of a meter movement.

21-2

THE AMMETER

A typical d'Arsonval movement might have a current sensitivity of 1 mA and a resistance of 50 Ω. In order to measure more than 1 mA, additional circuitry must be used with the basic meter movement. Figure 21–5A shows a simple ammeter with a *shunt* (parallel) resistor across the movement. The purpose of the shunt resistor is to bypass current in excess of 1 mA around the movement. For example, let us assume that this meter must measure currents up to 10 mA. Thus, for *full-scale* deflection, the movement must carry 1 mA, and the shunt resistor must carry 9 mA, as indicated in Figure 21–5B.

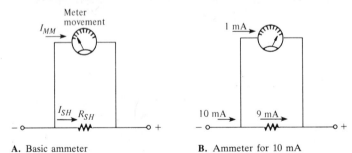

A. Basic ammeter

B. Ammeter for 10 mA

FIGURE 21–5

Determining the Shunt Value

In our example, a proper value of shunt resistance must be used. The following calculations illustrate how this resistance is determined.

THE AMMETER

Since the shunt resistor R_{SH} and the 50-Ω meter movement are in parallel, the voltage drops across them are the same; that is,

$$V_{SH} = V_{MM}$$

By Ohm's law,

$$I_{SH}R_{SH} = I_{MM}R_{MM}$$

Solving for R_{SH}, we get

$$R_{SH} = \frac{I_{MM}R_{MM}}{I_{SH}} = \frac{(1 \text{ mA})(50 \text{ Ω})}{9 \text{ mA}}$$

$$= 5.56 \text{ Ω}$$

Multiple-Range Ammeter

The example meter just discussed has only *one* range. It can measure currents from 0 to 10 mA and no higher. However, most practical ammeters have several ranges. Each range must have a different shunt resistance which is selected with a switch. For example, Figure 21–6 shows a two-range ammeter. A 100-mA range is incorporated with the 10-mA range previously described. When the switch is in the 10-mA position, the meter indicates 10 mA at full-scale deflection of the pointer. When the switch is in the 100-mA position, 100 mA is indicated at full scale.

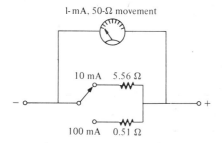

FIGURE 21–6 *Ammeter with two ranges.*

The value of the 100-mA shunt resistor is determined in the same manner used for the 10-mA shunt. At full scale the voltage across the movement is (1 mA)(50 Ω) = 50 mV. Therefore, since the shunt must carry 99 mA at full scale, R_{SH} = 50 mV/99 mA = 0.51 Ω. We can obtain other current ranges by switching in appropriate values of shunt resistance.

Example 21–1

Show a three-range basic ammeter that has a 1-A range in addition to the 10-mA and 100-mA ranges of the example meter just discussed.

Example 21-1 (continued)

Solution:

First we must find the shunt resistance for the 1-A range. Again, $V_{SH} = (1\text{ mA})(50\text{ }\Omega) = 50\text{ mV}$. The shunt resistor must carry all of the 1 A except the 1 mA to operate the movement at full scale. Thus,

$$I_{SH} = 1\text{ A} - 1\text{ mA} = 0.999\text{ A}$$

$$R_{SH} = \frac{50\text{ mV}}{0.999\text{ A}} = 0.05\text{ }\Omega$$

The three-range meter is shown in Figure 21-7.

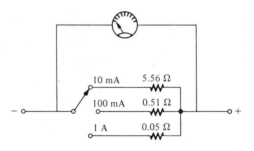

FIGURE 21-7

Effect of the Ammeter on the Circuit

As you know, we connect an ammeter *in series* to measure the current in a circuit. Ideally, the meter should not alter the current it is intended to measure. In practice, however, the meter unavoidably has some effect on the circuit, because its *internal resistance* is connected in series with the circuit resistance. However, in most cases, the meter's internal resistance is small enough compared to the circuit resistance that it can be neglected.

The internal resistance of the ammeter is the shunt resistance in parallel with the coil resistance of the movement. In our example meter, it is approximately 0.05 Ω on the 1-A range. Figure 21-8B shows the meter on the 1-A range connected to measure the current in the circuit of Figure 21-8A. The 0.05-Ω internal resistance (R_{INT}) of the meter is negligible compared to the 100-Ω circuit resistance. Therefore, the meter does not significantly alter the actual circuit current. This precision is necessary, of course, because we do not want the measuring instrument to change the quantity that it is supposed to measure accurately.

Ammeter Scales

A typical ammeter or milliammeter has more than one scale, each corresponding to different range switch positions. Figure 21-9 shows a two-scale meter as an example. This particular meter has four ranges, as indicated on the range switch in the diagram.

THE VOLTMETER

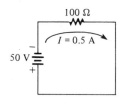 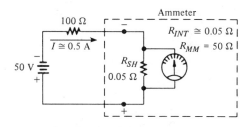 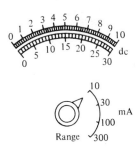

A. Circuit to be measured for current.

B. Ammeter connected in circuit. R_{INT} has negligible effect on I.

FIGURE 21–8 *Measurement of current with an ammeter.*

FIGURE 21–9 *Ammeter scale and range switch.*

The scales are read in conjunction with the range switch as follows: If the range switch is set at 10 mA or 100 mA, the top scale is read. If the range switch is set at 30 mA or 300 mA, the bottom scale is read. The range switch setting always corresponds to the *full-scale* deflection current. For example, if the range switch is set on 100 mA, the "10" mark on the top scale represents 100 mA.

Review for 21–2

1. What is the purpose of the shunt resistor in an ammeter?
2. A multiple-range ammeter has one shunt for each range (T or F).

21–3

THE VOLTMETER

The voltmeter utilizes the same type of movement as the ammeter. Different external circuitry is added so that the movement will function as a voltmeter.

As you have seen, the voltage drop across the meter coil is dependent on the current and the coil resistance. For example, a 50-μA, 1000-Ω movement has a voltage drop of $(50\,\mu A)(1000\,\Omega) = 50\,mV$ at full scale. To use the meter to indicate voltages greater than 50 mV, we must add a *series resistance* to drop any additional voltage. This resistance is called the *multiplier resistance*, R_M.

A basic voltmeter is shown in Figure 21–10 with a single multiplier resistor for one range. To make this meter measure 1 V full scale, we must determine the value of the multiplier resistance as follows: The movement drops 50 mV at a full-scale current of 50 μA. Therefore, the multiplier resistor R_M must drop the remaining voltage of $1\,V - 50\,mV = 950\,mV$. Since R_M is in series with the movement, it also carries 50 μA at full scale. Thus,

$$R_M = \frac{950\,mV}{50\,\mu A} = 19\,k\Omega$$

FIGURE 21–10 *Basic voltmeter with one range.*

Therefore, for 1-V full-scale deflection, the *total resistance* of the voltmeter is 20 kΩ (the multiplier resistance plus the coil resistance).

Voltmeter Sensitivity

Voltmeter sensitivity is defined in terms of resistance per volt (Ω/V). The example meter just discussed has a sensitivity of 20 kΩ/V because it has a total resistance of 20 kΩ and a full-scale deflection of 1 V. This is a common sensitivity figure for many commercial meters.

Multiple-Range Voltmeter

The meter in Figure 21–10 has only one voltage range (1-V); that is, it can measure voltages from 0 V to 1 V. In order to measure higher voltages with the same movement, we must use additional multiplier resistors. One multiplier resistor is required for each additional range.

For the 50-μA movement, the *total resistance* required is 20 kΩ for *each* volt of the full-scale reading. In other words, the sensitivity for the 50-μA movement is always 20 kΩ/V *regardless of the range selected*. Thus, the full-scale meter current is 50 μA in *any* range. For any range, we find the *total* meter resistance by multiplying the *sensitivity* and the *full-scale voltage* for that range. For example, for a 10-V range, $R_T = (20 \text{ k}\Omega/\text{V})(10 \text{ V}) = 200 \text{ k}\Omega$.

The total resistance for the 1-V range is 20 kΩ; so R_M for the 10-V range must be 200 kΩ − 20 kΩ = 180 kΩ. This *two-range* voltmeter is shown in Figure 21–11. Additional ranges require the appropriate values of multiplier resistance added in series.

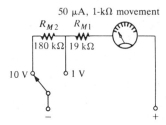

FIGURE 21–11 *Two-range voltmeter.*

Example 21–2

Show the circuit for a basic voltmeter having 1-V, 10-V, and 100-V ranges.

THE VOLTMETER

Solution:

We have already determined R_M for the 1-V and the 10-V ranges. We need only to calculate the additional R_M required for the 100-V range. This calculation is done as follows:

$$R_T = (20 \text{ k}\Omega/\text{V})(100 \text{ V}) = 2 \text{ M}\Omega$$

Now we subtract the meter resistance of the existing two-range meter from 2 MΩ to get the R_M required for the 100-V range:

$$R_{M3} = R_T - R_{M2} - R_{M1} - R_{MM}$$
$$= 2 \text{ M}\Omega - 180 \text{ k}\Omega - 19 \text{ k}\Omega - 1 \text{ k}\Omega = 2 \text{ M}\Omega - 200 \text{ k}\Omega$$
$$= 1.8 \text{ M}\Omega$$

The schematic for this three-range voltmeter is shown in Figure 21–12.

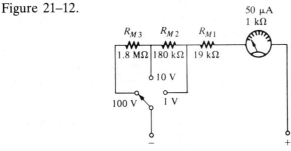

FIGURE 21–12

Loading Effect of a Voltmeter

As you know, a voltmeter is always connected in parallel with a circuit component across which the voltage is to be measured. Thus, it is much easier to use a voltmeter than an ammeter, because we must break a circuit to insert an ammeter in series. We simply connect a voltmeter across the circuit without disrupting the circuit or breaking a connection.

Since some current is required through the voltmeter to operate the movement, the voltmeter has some effect on the circuit to which it is connected. This effect is called *loading*. However, as long as the meter resistance is *much greater* than the resistance across which it is connected, the loading effect is negligible. This precision is necessary because we do not want the measuring instrument to change the voltage that it measures.

Figure 21–13A shows a simple resistive circuit. Part B of the figure shows the same circuit but with much higher resistor values. Assume that we wish to measure the voltage across R_2 in both circuits with a 20 kΩ/V voltmeter using the 10-V range. When the voltmeter is connected across R_2 in circuit A, the meter's internal resistance of 200 kΩ appears in parallel with 1 kΩ. The voltage across R_2 is still approximately 6 V because the meter does not significantly affect the circuit resistance (200 kΩ in parallel with 1 kΩ is still approximately 1 kΩ).

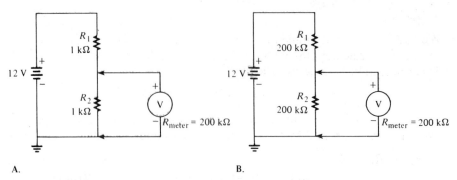

A.   B.

FIGURE 21-13

When the meter is connected across R_2 in circuit B, the meter's internal resistance of 200 kΩ appears in parallel with the 200 kΩ in the circuit. In this case the voltage across R_2, which is normally 6 V, is *reduced* because of the loading effect of the voltmeter. An inaccurate measurement results.

This example demonstrates the importance of using a voltmeter with an internal resistance that is very much higher than the resistance of the circuit across which it is connected. The following example will illustrate this idea further.

Example 21-3

Determine the exact voltage that would be measured with a 20 kΩ/V voltmeter across R_2 in the circuits of Figure 21-13.

Solution:

In circuit A, the meter is in parallel with the 1-kΩ R_2. The combined resistance of the meter and R_2 is

$$\frac{(200 \text{ k}\Omega)(1 \text{ k}\Omega)}{200 \text{ k}\Omega + 1 \text{ k}\Omega} = 0.995 \text{ k}\Omega$$

Using the voltage divider rule, we determine V_2 as follows:

$$V_2 = \left(\frac{0.995 \text{ k}\Omega}{1.995 \text{ k}\Omega}\right) 12 \text{ V} = 5.985 \text{ V}$$

Without the meter's loading effect, V_2 is 6 V. The meter loading produces an error of 15 mV, which in most cases is not enough to worry about. Figure 21-14 illustrates this situation.

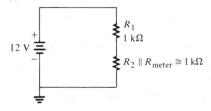

FIGURE 21-14

THE VOLTMETER

In circuit B the meter is in parallel with the 200-kΩ R_2. The combined resistance of the meter and R_2 is 200 kΩ/2 = 100 kΩ. Using the voltage divider rule, we find V_2 as follows:

$$V_2 = \left(\frac{100 \text{ k}\Omega}{300 \text{ k}\Omega}\right) 12 \text{ V} = 4 \text{ V}$$

In this situation, the error is 2 V, which is unacceptable. The voltmeter loading is significant, and this voltmeter could not be used in this case. We would have to use a higher-impedance voltmeter. Figure 21–15 illustrates this second case.

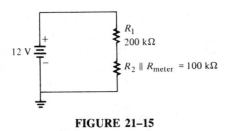

FIGURE 21–15

Voltmeter Scales

Like the ammeter, a typical voltmeter has more than one scale. For example, the voltmeter in Figure 21–16 has four ranges, as indicated on the range switch and two scales. If the range switch is set at 10 V or 100 V, the top scale is read. If it is set at 30 V or 300 V, the bottom scale is read. The range switch setting always corresponds to the *full-scale* voltage. For example, if the range switch is set on 300 V, the "30" mark on the bottom scale represents 300 V.

FIGURE 21–16 *Voltmeter scale and range switch.*

Review for 21–3

1. What is the purpose of the multiplier resistors in a voltmeter?
2. The impedance of a voltmeter should be high compared to that of the circuit across which it is connected (T or F).

21-4

ac METERS

The meters discussed in the previous sections can handle only *dc measurements*, because current must be in only one direction through the meter coil in order to deflect the pointer from left to right. For *ac measurement* of current or voltage, additional circuitry is required for the basic ammeter and voltmeter. This additional circuitry consists of a *rectifier*. Basically, rectifiers convert ac to dc. There are two types of rectifiers: half-wave and full-wave.

A half-wave rectifier is shown in block form in Figure 21-17A with its input and output. The ac input is converted to *pulsating dc* on every positive half-cycle. The full-wave rectifier converts the ac to pulsating dc on both the positive and the negtive half-cycles, as indicated in Figure 21-17B. Half-wave and full-wave rectifiers are discussed in detail in the next chapter.

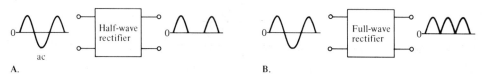

FIGURE 21-17 *Converting ac to pulsating dc.*

Basically, in an ac meter the rectifier precedes the meter movement. The movement responds to the *average value* of the pulsating dc. The scale can be calibrated to show rms, average, or peak values because these relationships are fixed mathematically, as you have learned. Figure 21-18 shows a basic meter with a full-wave rectifier for converting ac to dc.

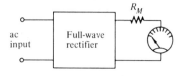

FIGURE 21-18 *Basic ac voltmeter.*

Review for 21-4

1. What does a rectifier do?
2. Basically, what is the difference between a dc meter and an ac meter?

21-5

THE OHMMETER

The meter movement used for the ammeter and the voltmeter functions can also be used for the ohmmeter. The ohmmeter is used to measure resistance values.

A basic one-range ohmmeter is shown in Figure 21–19A. It contains a battery and a variable resistor in series with the movement. To measure resistance, we connect the leads across the external resistor to be measured, as shown in Part B. This connection completes the circuit, allowing the internal battery to produce current through the coil movement and causing a deflection of the pointer proportional to the value of the external resistance.

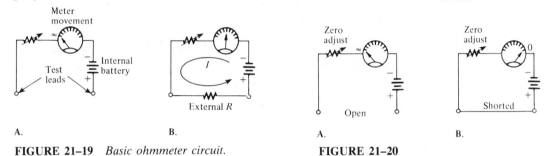

FIGURE 21–19 *Basic ohmmeter circuit.*   **FIGURE 21–20**

Zero Adjustment

When the ommeter leads are *open*, as in Figure 21–20A, the pointer is at full left scale, indicating *infinite* (∞) resistance (open circuit). When the leads are *shorted*, as in Figure 21–20B, the pointer is at full right scale, indicating *zero* resistance.

The purpose of the variable resistor is to adjust the current so that the pointer is at exactly zero when the leads are shorted. It is used to compensate for changes in the internal battery voltage due to aging.

Ohmmeter Scale

Figure 21–21 shows one type of ohmmeter scale. Between zero and infinity (∞), the scale is marked to indicate various resistor values. Because the values increase from right to left, this scale is called a *back-off scale*.

Let us assume that a certain ohmmeter uses a 50-μA, 1000-Ω movement and an internal 1.5-V battery. A current of 50 μA produces a full-scale deflection when the test leads are shorted. To have 50 μA, the *total* ohmmeter resistance is 1.5 V/50 μA = 30 kΩ. Therefore, since the coil resistance is 1 kΩ, the variable zero adjustment resistor must be set at 30 kΩ − 1 kΩ = 29 kΩ.

Suppose that a 120-kΩ resistor is connected to the ohmmeter leads. Combined with the 30-kΩ internal meter resistance, the total R is 150 kΩ. The current is 1.5 V/150 kΩ = 10 μA, which is 20% of the full-scale current and which appears on the scale as shown in Figure 21–21. Now, for example, a 45-kΩ resistor connected to the ohmmeter leads results in a current of 1.5 V/75 kΩ = 20 μA, which is 40% of the full-scale current and which is marked on the scale as shown. Additional

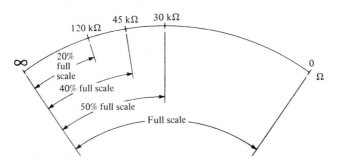

FIGURE 21–21 *Simplified ohmmeter scale illustrating nonlinearity with a few example values.*

similar calculations will show that the scale is *nonlinear*. It is more crowded or compressed toward the left side than the right.

The *center scale* point corresponds to an external resistance equal to the internal meter resistance (30 kΩ in this case). The reason is as follows: With 30 kΩ connected to the leads, the current is 1.5 V/60 kΩ = 25 μA, which is half of the full-scale current of 50 μA.

Multiple-Range Ohmmeter

An ohmmeter usually has several ranges. These typically are labeled RX1, RX10, RX100, RX1k, RX10k, RX100k, and RX1M, although some ohmmeters may not have all of the ranges mentioned. These range settings are interpreted differently from those of the ammeter or voltmeter. *The reading on the ohmmeter scale is multiplied by the factor indicated by the range setting.* For example, if the pointer is at 20 on the scale *and* the range switch is set at RX100, the actual resistance measurement is 20 × 100, or 2 kΩ. This example is illustrated in Figure 21–22 for a typical scale.

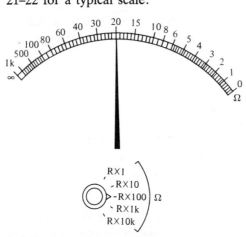

FIGURE 21–22 *Example of reading an ohmmeter scale in conjunction with the range setting (R = 2 kΩ).*

To measure small resistance values, we must use a higher ohmmeter current than we need for measuring large resistance values. Shunt resistors are used to provide multiple ranges on the ohmmeter to accommodate both small and large

MULTIMETERS

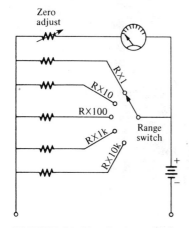

FIGURE 21-23 *Basic multiple-range ohmmeter circuit.*

resistance measurements. For each range, a different value of shunt resistance is switched in. The shunt resistance increases for higher ohm ranges and is always equal to the center scale reading on any range. In some meters, a higher battery voltage is used for the highest ohm range. A typical circuit is shown in Figure 21-23.

Review for 21-5

1. Why is an internal battery required in an ohmmeter?
2. What does the term *back-off scale* mean?
3. How do you zero adjust an ohmmeter?
4. If the pointer indicates "8" on the ohmmeter scale and the range switch is set at RX1k, what resistance is being read?

21-6

MULTIMETERS

Generally, the ammeter, voltmeter, and ohmmeter functions are combined into a single instrument for economy and convenience. This instrument is called a *multimeter* or sometimes a volt-ohm-milliammeter, abbreviated VOM.

Two typical multimeters are shown in Figure 21-24. Notice that both meters have a range switch and a function switch for selecting the desired range of current, ac or dc voltage, and resistance.

Digital Multimeter

Digital voltmeters (DVM) and digital multimeters (DMM) offer the advantages of easier reading and greater accuracy over conventional analog (pointer) meters. Many digital multimeters have a four- or five-digit readout with

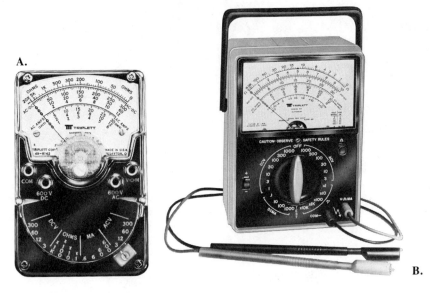

FIGURE 21-24 *Typical portable multimeters. (Courtesy of the Triplett Corporation)*

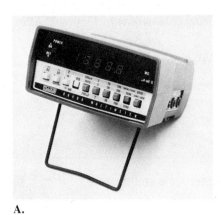

FIGURE 21-25 A. *Typical digital multimeters. (Courtesy of John Fluke Mfg. Co., Inc.)* B. *Typical digital volt-ohm-milliammeter. (Courtesy of the Triplett Corporation)*

accuracies of ±0.01% or better. Sometimes you will see a reference to a 3½-digit meter. This term means that the fourth (left-most) digit is displayed only as a "1" to handle overflow. Typical digital multimeters and a digital VOM are shown in Figure 21–25.

The internal principles of digital meters are considerably different from those of the analog (pointer) meters previously discussed. An understanding of DVMs, and DMMs requires some knowledge of digital circuits, which is beyond the scope of this text.

Review for 21–6

1. Name two advantages of a digital multimeter over a pointer multimeter.
2. How many readout positions does a 3½-digit meter have?

21–7

SIGNAL GENERATORS

The signal generator is an instrument that produces electrical signals for use in testing or controlling electronic circuits and systems. There are a variety of signal generators, ranging from special-purpose instruments that produce only one type of wave form in a limited frequency range, to programmable instruments that produce a wide range of frequencies and a variety of wave forms. In this section we will discuss several important types.

Audio Frequency Generators

The audible range of frequencies is typically from about 20 Hz to about 20 kHz. As a minimum requirement, an audio frequency (af) generator must provide sine wave voltages within this range. Many af generators or *audio oscillators,* however, have frequency ranges much greater than the actual audible range. A typical laboratory af generator produces frequencies from less than 1 Hz to greater than 1 MHz.

Figure 21–26 shows one type of audio generator. Notice that there is a range switch for selecting the desired frequency range. The frequency control sets the exact frequency within the selected range. The amplitude control adjusts the output voltage. Once it is set, it should remain essentially constant over the frequency range of the instrument. The maximum specified output voltage occurs when there is no load connected across the output terminals of the generator.

Many af generators have an *output impedance* of 600 Ω. Thus, if a load resistance of 600 Ω is connected to the output terminals, the voltage will be half of its open terminal value. Smaller load values further reduce the maximum achievable output amplitude.

FIGURE 21-26 *Audio oscillator. (Courtesy of Tektronix, Inc.)*

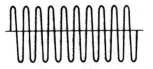

A. Continuous wave (CW)

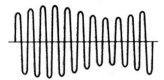

B. Amplitude modulation (AM)

FIGURE 21-27 *CW and amplitude modulated rf signals.*

Radio Frequency Generators

Many rf generators cover frequencies ranging from 30 kHz to 3000 MHz. The lower end of the range, of course, overlaps with the range of many af generators. Most rf generators produce at least two types of output signal: *continuous-wave* and *modulated.*

A continuous-wave or CW signal is a single-frequency sine wave with a steady amplitude. A modulated signal is a sine wave with an amplitude that varies sinusoidally at a much lower frequency, called *amplitude modulation* (AM). The two types of signals are illustrated in Figure 21-27.

There are many types of rf generators. All have provisions for adjusting the frequency over the specified range and for setting the output voltage amplitude. The percentage of modulation can also be set, as illustrated in Figure 21-28. Some rf generators also have provisions for frequency modulation (FM). Figure 21-29 shows two typical rf generators.

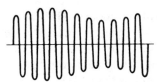

Lower percentage of modulation

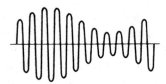

Higher percentage of modulation

FIGURE 21-28 *Variations in modulation.*

Sweep Generators

Sweep generators are a specializd type of rf generator that can produce a continuously varying frequency, called a *sweep frequency.* That is, the frequency changes continuously from a lower limit to an upper limit. This type of generator is

SIGNAL GENERATORS

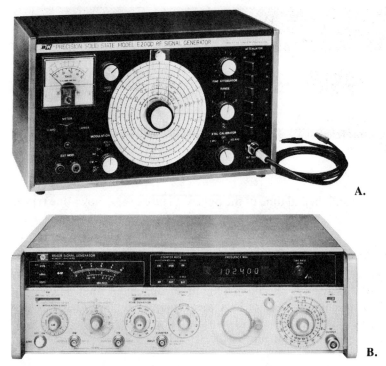

FIGURE 21–29 *Typical rf signal generators. (A, courtesy of B&K-Precision Test Instruments, Dynascan Corp. B, courtesy of Hewlett-Packard Company)*

FIGURE 21–30 *Microwave sweep oscillator. (Courtesy of Hewlett-Packard Company)*

widely used in testing of the band-pass circuits in communications receivers such as TV and radar. A typical sweep generator is shown in Figure 21–30.

Pulse Generators

Pulse generators are nonsinusoidal generators used extensively in testing digital circuits and systems. A typical pulse generator produces output pulses with variable pulse width, frequency, and amplitude. Many also have provisions for

FIGURE 21-31 *Pulse generator. (Courtesy of Hewlett-Packard Company)*

adjusting the rise time and the fall time of the pulses. Figure 21-31 shows one type of pulse generator.

Multipurpose Instruments

Figure 21-32 shows a multipurpose instrument that is actually several instruments in one. It uses a modular plug-in arrangement that allows the instruments to be interchanged for various measurement requirements. It also provides the convenience of having several instruments in one package for excellent portability. Therefore, this type of unit is ideal for field use.

FIGURE 21-32 *Multipurpose test system with plug-in instruments. (Courtesy of Tektronix, Inc.)*

Function Generators

A function generator is a versatile instrument that provides several types of output wave forms, including *square waves, triangular waves, ramps,* and *sine waves*. A function switch on the front panel selects the type of output wave form desired. Other controls select frequency, amplitude, and slope of ramps. Typical function generators are shown in Figure 21-33A and B. Part C of the figure shows the wave forms that are normally produced by many function generators.

SIGNAL GENERATORS

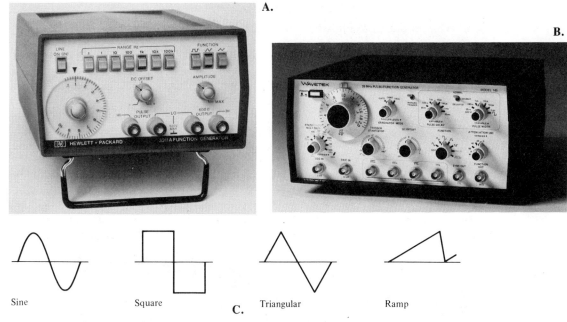

FIGURE 21-33 Function generators and typical output wave forms. (**A**, courtesy of Hewlett-Packard Company. **B**, courtesy of Wavetek)

Programmable Wave-Form Generators

With this type of generator, any wave form can be programmed in and then generated as an output. Irregular and standard wave forms can be programmed and stored. A typical programmable wave-form generator is shown in Figure 21-34.

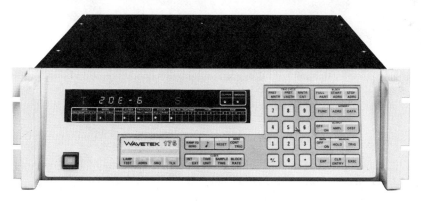

FIGURE 21-34 Programmable wave-form generator. (Courtesy of Wavetek)

Review for 21-7

1. Typically, what is the audible range of frequencies?
2. What do "af" and "rf" stand for?

3. What is a sweep frequency signal?

4. What is the difference between a pulse generator and a function generator?

21-8

THE OSCILLOSCOPE

The oscilloscope, or *scope* for short, is one of the most widely used and versatile test instruments. It displays on a screen the actual wave shape of a voltage from which amplitude, time, and frequency measurements can be made.

Figure 21-35 shows two typical oscilloscopes. The one in Part A is a simpler and less expensive model. The one in Part B has better performance characteristics and plug-in modules that provide a variety of specialized functions.

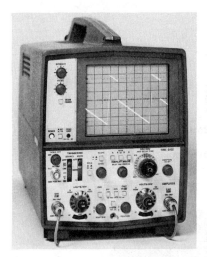 

A.   B.

FIGURE 21-35 *Oscilloscopes. (Courtesy of Tektronix, Inc.)*

Cathode Ray Tube (CRT)

The oscilloscope is built around the cathode ray tube (CRT), which is the device that displays the wave forms. The screen of the scope is the front of the CRT.

The CRT is a vacuum device containing an *electron gun* that emits a narrow, focused *beam* of electrons. A phosphorescent coating on the face of the tube forms the *screen*. The beam is electronically focused and accelerated so that it strikes the screen, causing light to be emitted at the point of impact.

Figure 21-36 shows the basic construction of a CRT. The *electron gun* assembly contains a *heater,* a *cathode,* a *control grid,* and *accelerating* and *focusing grids.* The heater carries current that indirectly heats the cathode. The heated

THE OSCILLOSCOPE

cathode emits electrons. The amount of voltage on the control grid determines the flow of electrons and thus the *intensity* of the beam. The electrons are accelerated by the accelerating grid and are focused by the focusing grid into a narrow beam that converges at the screen. The beam is further accelerated to a high speed after it leaves the electron gun by a high voltage on the *anode* surfaces of the CRT.

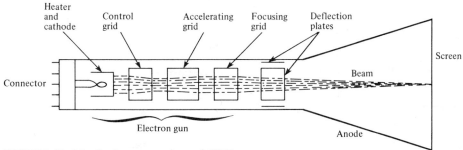

FIGURE 21–36 *Basic construction of CRT.*

Deflection of the Beam

The purpose of the *deflection plates* in the CRT is to produce a "bending" or deflection of the electron beam. This deflection allows the position of the point of impact on the screen to be varied. There are two sets of deflection plates. One set is for *vertical deflection,* and the other set is for *horizontal deflection.*

Figure 21–37 shows a front view of the CRT's deflection plates. One plate from each set is normally grounded as shown. If there is *no voltage* on the other plates, as in Figure 21–37A, the beam is not deflected and hits the center of the screen. If a *positive* voltage is on the vertical plate, the beam is attracted upward, as indicated in Part B of the figure. Remember that opposite charges attract. If a *negative* voltage is applied, the beam is deflected downward because like charges repel, as shown in Part C.

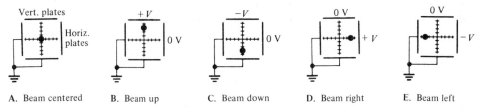

FIGURE 21–37 *Deflection of electron beam.*

Likewise, a positive or a negative voltage on the horizontal plate deflects the beam right or left, respectively, as shown in Figure 21–37D and E. The amount of deflection is proportional to the amount of voltage on the plates.

Sweeping the Beam Horizontally

In normal oscilloscope operation, the beam is horizontally deflected from left to right across the screen at a certain rate. This *sweeping action* produces a horizontal line or *trace* across the screen, as shown in Figure 21–38.

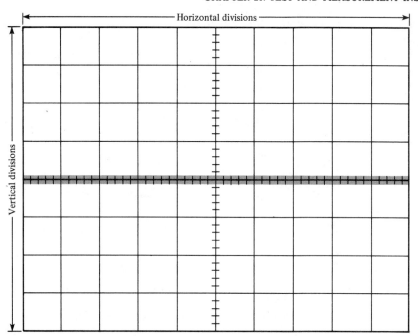

FIGURE 21–38 *Scope screen with horizontal trace (8 cm × 10 cm).*

The *rate* at which the beam is swept across the screen establishes a *time base*. The scope screen is divided into horizontal (and vertical) divisions, as shown in Figure 21–38. For a given time base, each horizontal division represents a fixed interval of time. For example, if the beam takes 1 second for a full left-to-right sweep, then each division represents 0.1 second. All scopes have provisions for selecting various sweep rates, which will be discussed later.

The actual sweeping of the beam is accomplished by application of a *sawtooth voltage* across the horizontal plates. The basic idea is illustrated in Figure 21–39. When the sawtooth is at its maximum negative peak, the beam is deflected to its left-most screen position. This deflection is due to maximum repulsion from the right deflection plate.

As the sawtooth voltage *increases,* the beam moves toward the center of the screen. When the sawtooth voltage is *zero,* the beam is at the center of the screen because there is no repulsion or attraction from the plate.

As the voltage increases *positively,* the plate attracts the beam, causing it to move toward the right side of the screen. At the positive peak of the sawtooth, the beam is at its right-most screen position.

The rate at which the sawtooth goes from negative to positive is determined by its frequency. This rate in turn establishes the *sweep rate* of the beam.

When the sawtooth makes the abrupt change from positive back to negative, the beam is rapidly returned to the left side of the screen, ready for another sweep. During this "flyback" time, the beam is *blanked* out and thus does not produce a trace on the screen.

THE OSCILLOSCOPE

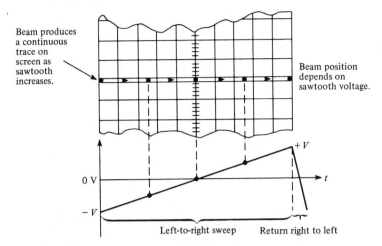

FIGURE 21–39 *Sweeping the beam across the screen.*

How a Wave-Form Pattern Is Produced

The main purpose of the scope is to display the wave form of a voltage under test. To do so, we apply the voltage under test across the *vertical plates* through a vertical amplifier circuit. As you have seen, a voltage across the vertical plates causes a vertical deflection of the beam. A negative voltage causes the beam to go below the center of the screen, and a positive voltage makes it go above center.

Assume, for example, that a sine wave voltage is applied across the vertical plates. As a result, the beam will move up and down on the screen. The amount that the beam goes above or below center depends on the peak value of the sine wave voltage.

At the same time that the beam is being deflected vertically, it is also sweeping horizontally, causing the vertical voltage wave form to be traced out across the screen as illustrated in Figure 21–40. All scopes provide for the calibrated adjustment of the vertical deflection; so each vertical division represents a known amount of voltage.

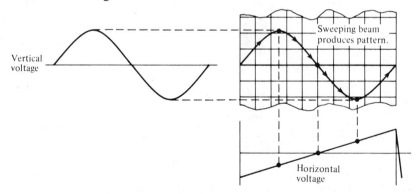

FIGURE 21–40 *Vertical and horizontal deflection combined to produce a pattern on the screen.*

Review for 21-8

1. What is the basic purpose of an oscilloscope?
2. What does "CRT" stand for?
3. Name the basic components of a CRT.

21-9

OSCILLOSCOPE CONTROLS

There are a wide variety of oscilloscopes available, ranging from relatively simple instruments having limited capabilities to much more sophisticated models that provide a variety of optional functions and precision measurements. Regardless of their complexity, however, all scopes have certain operational features in common. In this section, we will examine the most common front panel controls. Figure 21–41 shows a representative oscilloscope front panel. Each control and its basic function will be described.

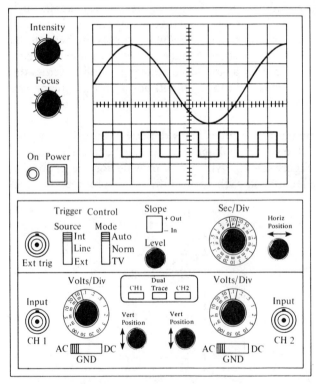

FIGURE 21–41 *Representative dual trace oscilloscope front panel.*

Screen

In the upper portion of Figure 21–41 is the CRT screen. There are eight vertical divisions and ten horizontal divisions indicated with grid lines or graticules.

OSCILLOSCOPE CONTROLS

A standard screen size is 8 cm × 10 cm. The screen is coated with phosphor that emits light when struck by the electron beam.

Power Switch and Light

This switch turns the power on and off to the scope. The light indicates when the power is on.

Intensity

This control knob varies the brightness of the trace on the screen. Caution should be used so that the intensity is not left too high for an extended period of time, especially when the beam forms a motionless dot on the screen. Damage to the screen can result from excessive intensity.

Focus

This control focuses the beam so that it converges to a tiny point at the screen. An out-of-focus condition results in a fuzzy trace.

Horizontal Position

This control knob adjusts the neutral horizontal position of the beam. It is used to horizontally reposition a wave-form display for more convenient viewing or measurement.

Seconds/Division

This selector switch sets the horizontal sweep rate. It is the *time base control*. The switch selects the time interval that is to be represented by each horizontal division in seconds, milliseconds, or microseconds. The setting in Figure 21–41 is at 10 μs. Thus, each of the ten horizontal divisions represents 10 μs; so there are 100 μs from the extreme left of the screen to the extreme right.

One cycle of the displayed sine wave covers eight horizontal divisions. Therefore, the period of the sine wave is (8 div)(10 μs/div) = 80 μs. From this, the frequency can be calculated as $f = 1/T = 1/80 \mu s = 12.5$ kHz. If the sec/div switch is moved to a different setting, the displayed sine wave will change correspondingly. If it is moved to a lower time setting, fewer cycles will be displayed. If it is moved to a higher time setting, more cycles will be displayed.

Trigger Control

These controls allow the beam to be triggered from various selected sources. The triggering of the beam causes it to begin its sweep across the screen. It can be triggered from an internally generated signal derived from an input signal, or from the line voltage, or from an externally applied trigger signal. The *modes* of triggering are auto, normal, and TV. In the auto mode, sweep occurs in the absence of an adequate trigger signal. In the normal mode, a trigger signal must be present for the sweep to occur. The TV mode provides triggering on the TV field or TV line signals. The *slope* switch allows the triggering to occur on either the positive-going slope or the negative-going slope of the trigger wave form. The *level* control selects the amplitude point on the trigger signal at which the triggering occurs.

Basically, the trigger controls provide for synchronization of the sweep wave form and the input signal wave form. As a result, the display of the input signal is stable on the screen, rather than appearing to drift across the screen.

Volts/Division

The example scope in Figure 21–41 is a *dual-trace* type which allows two wave forms to be displayed simultaneously. Many scopes have only single-trace capability. Notice that there are two identical volts/div selectors. There is a set of controls for each of the two *input channels*. We will describe only one, but the same applies to the other as well.

The volts/div selector switch sets the number of volts to be represented by each division on the *vertical* scale. For example, the displayed sine wave is applied to channel 1 and covers four vertical divisions from the positive peak to the negative peak. The volts/div switch for channel 1 is set at 50 mV, which means that each vertical division represents 50 mV. Therefore, the peak-to-peak value of the sine wave is (4 div) (50 mV/div) = 200 mV. If a lower setting were selected, the displayed wave would cover more vertical divisions. If a higher setting were selected, the displayed wave would cover fewer vertical divisions.

Notice that there is a set of three switches for selecting channel 1 (CH 1), channel 2 (CH 2), or dual trace. Either input signal can be displayed separately, or both can be displayed as illustrated.

Vertical Position

The two vertical position controls move the traces up or down for easier measurement or observation.

ac-gnd-dc Switch

This switch, located below the volts/div control, allows the input signal to be ac coupled, dc coupled, or grounded. The ac coupling eliminates any dc component on the input signal. The dc coupling permits dc values to be displayed. The ground position allows a zero volt reference to be established on the screen.

Input

The signals to be displayed are connected into the channel 1 and channel 2 input connectors. This connection is normally done via a special *probe* that minimizes the loading effect of the scope's input resistance and capacitance on the circuit being measured.

Review for 21-9

1. Name the common controls found on all oscilloscopes.
2. Which control adjusts the vertical deflection?
3. Which control adjusts the sweep rate of the horizontal trace?

Summary

1. Three types of meter movement are d'Arsonval, iron-vane, and electrodynamometer.
2. An ammeter uses shunt resistors in parallel with the movement.
3. An ammeter presents a very low resistance when connected in series in a circuit.
4. A voltmeter uses multiplier resistors in series with the movement.
5. A voltmeter has a very high internal resistance to prevent loading a circuit.
6. Voltmeter sensitivity is specified in ohms/volt. It gives the internal resistance of the meter on any given range.
7. An ac meter uses a rectifier to convert ac to dc.
8. An ohmmeter uses an internal battery to produce current through the movement when an external resistor is connected.
9. The ohms scale is nonlinear.
10. A multimeter combines an ammeter, a voltmeter, and an ohmmeter in one package.
11. Signal generators can be classified broadly as audio frequency (af), radio frequency (rf), and nonsinusoidal.
12. Oscilloscopes display a wave-form picture.
13. The CRT (cathode ray tube) is the basic element in an oscilloscope.

Self-Test

1. If 25 μA flow through a 50-μA movement, how much will the pointer be deflected?
2. If an ammeter is to measure a full-scale current of 1 mA, what shunt resistance is required for a 1-mA movement? For a 50-μA, 1-kΩ movement?
3. A voltmeter has a sensitivity of 20,000 Ω/V. What is its internal resistance on the 100-V range?
4. What multiplier resistance is used in the meter in Problem 3 on the 100-V range? Assume that the resistance of the movement is 1 kΩ.
5. The voltmeter in Figure 21–42 has a sensitivity of 20,000 Ω/V. It is set on the 1-V range to measure the voltage across R_2. What voltage will be read on the meter? What is the actual voltage across R_2 with the voltmeter disconnected? Explain the difference.

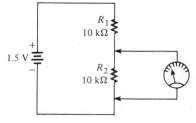

FIGURE 21–42

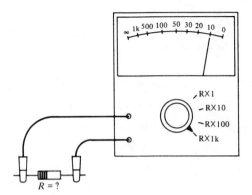

FIGURE 21-43

6. How much resistance is the ohmmeter in Figure 21-43 showing?
7. What are the amplitude and the frequency of the square wave on the scope in Figure 21-41? Assume that it is the channel 2 input.
8. What variable is the multimeter in Figure 21-44 measuring, and what is its value?

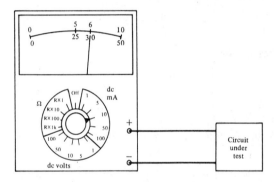

FIGURE 21-44

9. The volts/div switch on an oscilloscope is set at 0.1 V. The wave form displayed extends over 5 vertical divisions from its positive to its negative peak. What is its peak-to-peak value?
10. A particular scope has ten horizontal divisions. If the time/div switch is set at 1 μs, how much time is represented by the full ten divisions?
11. How many cycles of a 1-kHz sine wave are displayed on a scope screen if the time/div switch is set at 0.1 ms? At 1 ms? Assume ten horizontal divisions.

Problems

21-1. A 50-μA, 1000-Ω movement is used in an ammeter. Determine the full-scale shunt current (I_{SH}) on each of the following ranges:
  (a) 100 μA   (b) 1 mA   (c) 10 mA   (d) 100 mA   (e) 1 A

21-2. Repeat Problem 21-1 for a 1-mA, 50-Ω movement.

PROBLEMS

21–3. Calcuate the shunt resistor value (R_{SH}) for each range in Problem 21–1.

21–4. Calculate the multiplier resistor values for a 20,000 Ω/V voltmeter on the following ranges: 0.1 V, 1 V, 5 V, 10 V, and 50 V.

21–5. What range setting would you use to measure the voltage in Figure 21–45? How much is the error due to the loading effect of the voltmeter?

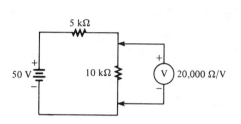

FIGURE 21–45

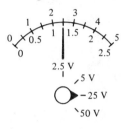

A.

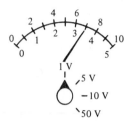

B.

FIGURE 21–46

21–6. What are the voltage readings in Figure 21–46?

21–7. An ohmmeter uses a 50-μA, 1000-Ω movement. It has a 3-V internal battery. What is the value of resistance that it is measuring when the pointer is at center scale, assuming no internal shunt?

21–8. Determine the resistance indicated by each of the following ohmmeter readings and range settings:
   (a) Pointer at 2, range setting at RX10
   (b) Pointer at 15, range setting at RX10M
   (c) Pointer at 45, range setting at RX100

21–9. A multimeter has the following ranges: 1 mA, 10 mA, 100 mA; 100 mV, 1 V, 10 V; RX1, RX10, RX100. Indicate schematically how you would connect the multimeter in Figure 21–47 to measure the following:
   (a) Current through R_1 (b) Voltage across R_1
   (c) Resistance of R_1
   In each case indicate the *function* on which you would set the meter and the *range* that you would use.

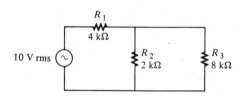

FIGURE 21–47

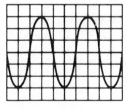

FIGURE 21–48

21–10. Determine how many cycles of a 5-kHz sine wave are displayed on the screen for each of the following time/div settings. Assume ten horizontal divisions.
   (a) 10 μs   (b) 0.1 ms   (c) 1 ms

21–11. Determine the peak and rms values and the frequency of the sine wave displayed in Figure 21–48 for the control settings of 0.5 volt/div and 10 microseconds/div.

21-12. On a typical 8 × 10 division scope screen you expect to observe a 1-V rms wave with a frequency of 2.5 kHz. What are the minimum settings of the volts/division and time/division controls so that at least one full cycle of the complete wave form is displayed?

Answers to Section Reviews

Section 21–1:
1. d'Arsonval, iron-vane, and electrodynamometer. **2.** That value of coil current producing a full-scale deflection.

Section 21–2:
1. To bypass current exceeding full-scale current. **2.** T.

Section 21–3:
1. To provide multiple-range capability. **2.** T.

Section 21–4:
1. Converts ac to pulsating dc. **2.** The ac meter uses a rectifier circuit.

Section 21–5:
1. To provide current through the movement. **2.** The scale reads from right to left. **3.** Short the test leads, and use the zero adjust knob to set the pointer to zero on the scale. **4.** 8 kΩ.

Section 21–6:
1. It is easier to read and has greater accuracy. **2.** 4.

Section 21–7:
1. 20 Hz to 20 kHz. **2.** Audio frequency and radio frequency. **3.** The frequency continuously increases between two limits. **4.** A function generator produces several types of wave forms; a pulse generator produces only rectangular pulses.

Section 21–8:
1. To display a voltage wave form. **2.** Cathode ray tube. **3.** Electron gun, deflection plates, anode, screen, and glass envelope.

Section 21–9:
1. Power swtich, intensity, focus, vertical input, position, volts/division, time/division, trigger. **2.** Volts/division. **3.** Time/division.

Appendix A	Table of Standard 10% Resistor Values
Appendix B	Batteries
Appendix C	A Computer Program for Circuit Analysis
Appendix D	rms (Effective) Value of a Sine Wave
Appendix E	Average Value of a Half-Cycle Sine Wave
Appendix F	Reactance Derivations
Appendix G	Table of Trigonometric Functions
Appendix H	Table of Exponential Functions

APPENDICES

APPENDIX A

TABLE OF STANDARD 10% RESISTOR VALUES (commercially available)

Ohms (Ω)				Kilohms (kΩ)		Megohms (MΩ)	
1.0	10	100	1000	10	100	1.0	10
1.2	12	120	1200	12	120	1.2	12
1.5	15	150	1500	15	150	1.5	15
1.8	18	180	1800	18	180	1.8	18
2.2	22	220	2200	22	220	2.2	22
2.7	27	270	2700	27	270	2.7	
3.3	33	330	3300	33	330	3.3	
3.9	39	390	3900	39	390	3.9	
4.7	47	470	4700	47	470	4.7	
5.6	56	560	5600	56	560	5.6	
6.8	68	680	6800	68	680	6.8	
8.2	82	820	8200	82	820	8.2	

APPENDIX B

BATTERIES

Batteries are an important source of dc voltage. They are available in two basic categories: the *wet cell* and the *dry cell*. A battery generally is made up of several individual cells.

A cell consists basically of two *electrodes* immersed in an *electrolyte*. A voltage is developed between the electrodes as a result of the *chemical* action between the electrodes and the electrolyte. The electrodes typically are two dissimilar metals, and the electrolyte is a chemical solution.

Simple Wet Cell

Figure B–1 shows a simple copper-zinc (Cu-Zn) chemical cell. One electrode is made of copper, the other of zinc. These electrodes are immersed in a solution of water and hydrochloric acid (HCl), which is the electrolyte.

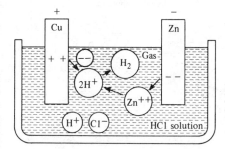

FIGURE B–1 *Simple chemical cell.*

Positive hydrogen ions (H^+) and negative chlorine ions (Cl^-) are formed when the HCl ionizes in the water. Since zinc is more active than hydrogen, zinc atoms leave the zinc electrode and form zinc ions (Zn^{++}) in the solution. When a zinc ion is formed, two excess electrons are left on the zinc electrode, and two hydrogen ions are displaced from the solution. These two hydrogen ions will migrate to the copper electrode, take two electrons from the copper, and form a molecule of hydrogen gas (H_2). As a result of this reaction, a negative charge develops on the zinc electrode and a positive charge develops on the copper electrode, creating a potential difference or voltage between the two electrodes.

In this copper-zinc cell, the hydrogen gas given off at the copper electrode tends to form a layer of bubbles around the electrodes, insulating the copper from the electrolyte. This effect, called *polarization,* results in a reduction in the voltage produced by the cell. Polarization can be remedied by the addition of an agent to the electrolyte to remove hydrogen gas or by the use of an electrolyte that does not form hydrogen gas.

Lead-Acid Cell: The positive electrode of a lead-acid cell is lead peroxide (PbO_2), and the negative electrode is spongy lead (Pb). The electrolyte is sulfuric acid (H_2SO_4) in water. Thus, the lead-acid cell is classified as a wet cell.

Two positive hydrogen ions ($2H^+$) and one negative sulfate ion (SO_4^{--}) are formed when the sulfuric acid ionizes in the water. Lead ions (Pb^{++}) from both electrodes displace the hydrogen ions in the electrolyte solution. When the lead ion from the spongy lead electrode enters the solution, it combines with a sulfate ion (SO_4^{--}) to form lead sulfate ($PbSO_4$), and it leaves two excess electrons on the electrode.

When a lead ion from the lead peroxide electrode enters the solution, it also leaves two excess electrons on the electrode and forms lead sulfate in the solution. However, because this electrode is lead peroxide, two free oxygen atoms are created when a lead atom leaves and enters the solution as a lead ion. These two oxygen atoms take four electrons from the lead peroxide electrode and become oxygen ions (O^{--}). This process creates a deficiency of two electrons on this electrode (there were initially two excess electrons).

The two oxygen ions ($2O^{--}$) combine in the solution with four hydrogen ions ($4H^+$) to produce two molecules of water ($2H_2O$). This process dilutes the electrolyte over a period of time. Also, there is a buildup of lead sulfate on the electrodes. These two factors result in a reduction in the voltage produced by the cell and necessitate periodic recharging.

As you have seen, for each departing lead ion there is an excess of two electrons on the spongy lead electrode, and there is a deficiency of two electrons on the lead peroxide electrode. Therefore, the lead peroxide electrode is positive and the spongy lead electrode is negative. This chemical reaction is pictured in Figure B–2.

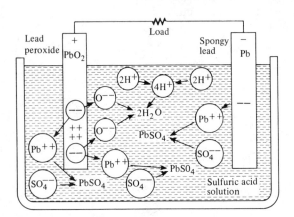

FIGURE B–2 *Chemical reaction in a discharging lead-acid cell.*

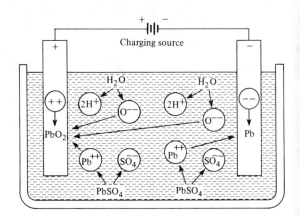

FIGURE B–3 *Recharging a lead-acid cell.*

As mentioned, the dilution of the electrolyte by the formation of water and lead sulfate requires that the lead-acid cell be recharged to *reverse* the chemical process. A chemical cell that can be recharged is called a *secondary cell*. One that cannot be recharged is called a *primary cell*.

The cell is recharged by connection of an external voltage source to the electrodes, as shown in Figure B–3. The formula for the chemical reaction in a lead-acid cell is as follows:

$$Pb + PbO_2 + 2H_2SO_4 \longrightarrow 2PbSO_4 + 2H_2O$$

Dry Cell

In the dry cell, some of the disadvantages of a liquid electrolyte are overcome. Actually, the electrolyte in a typical dry cell is not dry but rather is in the form of a moist paste. This electrolyte is a combination of granulated carbon, powdered manganese dioxide, and ammonium chloride solution.

A typical carbon-zinc dry cell is illustrated in Figure B–4. The zinc container or can is dissolved by the electrolyte. As a result of this reaction, an excess of electrons accumulates on the container, making it the negative electrode.

APPENDIX B

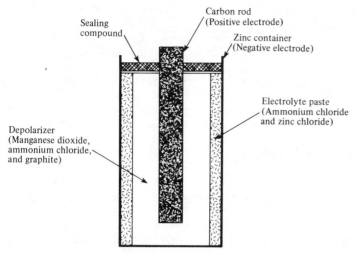

FIGURE B-4 *Simplified construction of dry cell.*

The hydrogen ions in the electrolyte take electrons from the carbon rod, making it the positive electrode. Hydrogen gas is formed near the carbon electrode, but this gas is eliminated by reaction with manganese dioxide (called a *depolarizing agent*). This depolarization prevents bursting of the container due to gas formation. Because the chemical reaction is not reversible, the carbon-zinc cell is a primary cell.

Types of Chemical Cells

Although only two common types of battery cells have been discussed, there are several types, listed in Table B-1.

TABLE B-1 *Types of battery cells.*

Type	+ electrode	− electrode	Electrolyte	Volts	Comments
Carbon-zinc	Carbon	Zinc	Ammonium and zinc chloride	1.5	Dry, primary
Lead-acid	Lead peroxide	Spongy lead	Sulfuric acid	2.0	Wet, secondary
Manganese-alkaline	Manganese dioxide	Zinc	Potassium hydroxide	1.5	Dry, primary or secondary
Mercury	Zinc	Mercuric oxide	Potassium hydroxide	1.3	Dry, primary
Nickel-cadmium	Nickel	Cadmium hydroxide	Potassium hydroxide	1.25	Dry, secondary
Nickel-iron (Edison cell)	Nickel oxide	Iron	Potassium hydroxide	1.36	Wet, secondary

APPENDIX C

A COMPUTER PROGRAM FOR CIRCUIT ANALYSIS

In Chapter 9, you learned how to use several methods of circuit analysis. The analyses were limited to two unknown currents or voltages and two equations. You also learned how to set up three equations to solve for three unknowns. It is important that you understand the principles of these analysis methods so that you can apply them to a wide variety of circuit problems.

The more unknown quantities there are in a circuit, the more difficult and tedious it becomes to calculate the solutions. In such a situation, *computer-aided analysis* is extremely helpful. If you understand the circuit principles and know how to apply the various analysis methods, then you can program a computer to do the tedious calculations and give you almost immediate answers.

Although computer programming is beyond the scope of this book, a computer program is presented for solving simultaneous equations and for doing mesh or node analysis. The main purpose of this program is to familiarize you with the use of a computer as a circuit analysis tool by an example of what can be done.

When you learn programming and understand circuit theory and analysis methods, the possibilities are almost limitless for use of the computer as an analysis and design tool. The computer does, however, have one limitation: It cannot think for you.

The Program

The program presented here can do three operations:

1. It can solve two or three simultaneous equations for unknown currents or voltages.
2. It can solve for two or three loop currents using the mesh analysis method when the resistor values and voltages are entered.
3. It can solve for two or three unknown node voltages using the node analysis method when the resistor values and voltages sources are entered.

This program is written in Level II BASIC for the TRS-80 computer. It can be run on many other BASIC computers with perhaps only minor changes. If you wish to convert to other BASIC "dialects," see *The Basic Handbook: An Encyclopedia of the Basic Computer Language,* by David A. Lien. This book is available from CompuSoft Publishing, A Division of CompuSoft, Inc., 8643 Navajo Rd., San Diego, California 92119.

How to Enter the Program in the Computer

Refer to your computer manual for instructions on entering the program into the computer. Once the program is entered and running properly, you should store it on magnetic tape or disk to eliminate the task of entering it from the keyboard each time you want to use it. Again, refer to your manual for this procedure. Comments and instructional statements are used throughout the program, making it completely self-explanatory. You will find it very useful to use this program in conjunction with the problems in Chapter 9.

The program, of course, is limited and is not a universal circuit analysis program. Its purpose is to illustrate the potential of the computer in circuit analysis.

APPENDIX C

Program for Mesh/Node Analysis

```
10   CLEAR(1000): DIMA(20,20)
20   CLS
30   PRINT @ 384, CHR$(23) "PROGRAM FOR MESH/NODE ANALYSIS"
40   FOR X=1 TO 5000: NEXT: CLS
50   GO TO 530
60   PRINT "THIS PROGRAM WILL SOLVE FOR 2 OR 3 UNKNOWN CURRENTS
     OR VOLTAGES. FIRST YOU MUST SET UP THE EQUATIONS USING THE
     MESH ANALYSIS OR THE NODE ANALYSIS METHOD."
70   PRINT "FOR 2 UNKNOWNS, THE EQUATIONS MUST BE IN THE FOLLOWING
     FORMAT."
80   PRINT TAB(20) "AX+BY=C"
90   PRINT TAB(20) "DX+EY=F"
100  PRINT "FOR 3 UNKNOWNS, THE EQUATIONS MUST BE IN THE FOLLOWING
     FORMAT."
110  PRINT TAB(20) "AX+BY+CZ=D"
120  PRINT TAB(20) "EX+FY+GZ=H": PRINT TAB(20) "IX+JY+KZ=L"
130  INPUT "WHEN YOUR EQUATIONS ARE READY, PRESS 'ENTER'"; W: CLS
140  INPUT "HOW MANY UNKNOWNS"; N
150  IF N=2 THEN A(3,3)=1: C$="AX+BY=C"
160  IF N=3 THEN C$="AX+BY+CZ=D"
170  PRINT "THE EQUATIONS MUST BE IN THE FORM"; C$
180  INPUT "ARE THE UNKNOWNS VOLTAGE OR CURRENT"; A$
190  CLS
200  INPUT "ARE THE UNITS VOLTS OR AMPS"; B$
210  FOR I=1 TO N
220  PRINT "THE COEFFICIENTS FOR EQUATION "; I
230  FOR J=1 TO N
240  PRINT "WHAT IS THE COEFFICIENT FOR "; A$; J
250  INPUT A(I,J)
260  NEXT J
270  INPUT "WHAT IS THE VALUE OF THE CONSTANT"; A(I,4)
280  CLS
290  NEXT I
300  CLS
310  PRINT "FOR VERIFICATION, THE COEFFICIENTS AND CONSTANTS OF
     THE EQUATIONS ARE"
320  FOR I=1 TO N
330  FOR J=1 TO N
340  PRINT A(I,J),
350  NEXT J
360  PRINT A(I,4)
370  NEXT I
380  PRINT "IF INCORRECT TYPE NO"
390  INPUT "IF CORRECT TYPE YES"; N$: IF N$="NO" THEN 20
400  D=A(1,1)*(A(2,2)*A(3,3)-A(3,2)*A(2,3))-A(1,2)*(A(2,1)*A(3,3)
     -A(3,1)*A(2,3))+A(1,3)*(A(2,1)*A(3,2)-A(3,1)*A(2,2))
410  N(1)=A(1,4)*(A(2,2)*A(3,3)-A(3,2)*A(2,3))-A(1,2)*(A(2,4)*
     A(3,3)-A(3,4)*A(2,3))+A(1,3)*(A(2,4)*A(3,2)-A(3,4)*A(2,2))
420  N(2)=A(1,1)*(A(2,4)*A(3,3)-A(3,4)*A(2,3))-A(1,4)*(A(2,1)*
     A(3,3)-A(3,1)*A(2,3))+A(1,3)*(A(2,1)*A(3,4)-A(3,1)*A(2,4))
430  N(3)=A(1,1)*(A(2,2)*A(3,4)-A(3,2)*A(2,4))-A(1,2)*(A(2,1)*
     A(3,4)-A(3,1)*A(2,4))+A(1,4)*(A(2,1)*A(3,2)-A(3,1)*A(2,2))
440  CLS
450  FOR I=1 TO N
460  PRINT @ 275+64*I, A$;I;"=";N(I)/D; B$
470  NEXT
```

```
480    FOR X=10 TO 100: SET(X,10): NEXT
490    FOR X=10 TO 100: SET(X,30): NEXT
500    FOR Y=10 TO 30: SET(10,Y): NEXT
510    FOR Y=10 TO 30: SET(100,Y): NEXT
520    GO TO 520
530    PRINT "****THIS PROGRAM PROVIDES 3 OPTIONS FOR CIRCUIT
       ANALYSIS ****"
540    PRINT
550    PRINT "OPTION 1: SOLUTION OF 2 OR 3 SIMULTANEOUS MESH OR NODE
       EQUATIONS. USE THIS OPTION WHEN YOU HAVE ALREADY SET UP THE
       EQUATIONS."
560    PRINT
570    PRINT "OPTION 2: MESH ANALYSIS FOR A 2 OR 3 LOOP CIRCUIT.
       THIS OPTION ALLOWS YOU TO ENTER THE CIRCUIT COMPONENT VALUES
       AFTER YOU HAVE DEFINED THE LOOPS, FOR RESISTIVE CIRCUITS ONLY."
580    PRINT
590    PRINT "OPTION 3: NODE ANALYSIS FOR CIRCUIT WITH 2 OR 3
       UNKNOWN NODES. THIS OPTION ALLOWS YOU TO ENTER THE CIRCUIT
       COMPONENT VALUES AFTER YOU HAVE DEFINED THE NODES AT WHICH
       THE VOLTAGES ARE NOT KNOWN. FOR RESISTIVE CIRCUITS ONLY."
600    PRINT
610    PRINT "IF YOU ARE READY TO SELECT YOUR OPTION PRESS 'ENTER'":
       INPUT W
620    CLS: PRINT "TO SOLVE SIMULTANEOUS EQUATIONS TYPE 1"
630    PRINT "TO USE MESH ANALYSIS TYPE 2"
640    PRINT "TO USE NODE ANALYSIS TYPE 3"
650    INPUT Q: CLS: ON Q GO TO 60, 660, 1010
660    PRINT "*************** MESH ANALYSIS INSTRUCTIONS ***************"
       : A$="I"
670    PRINT
680    PRINT "1. ASSIGN LOOPS IN YOUR CIRCUIT."
690    PRINT
700    PRINT "2. ASSIGN LOOP CURRENTS CLOCKWISE."
710    PRINT
720    PRINT "3. VOLTAGE SOURCES - TO + IN THE DIRECTION OF CURRENT
       ARE POSITIVE. OTHERWISE NEGATIVE."
730    PRINT
740    PRINT "TO CONTINUE PRESS 'ENTER'": INPUT W
750    CLS
760    PRINT "HOW MANY LOOPS?"
770    INPUT N
780    IF N=2 THEN A(3,3)=1
790    FOR I= 1 TO N
800    FOR J= I TO N
810    S=1
820    IF I=J PRINT "HOW MANY RESISTORS IN LOOP ";I ELSE PRINT "HOW
       MANY RESISTORS COMMON TO LOOP ";I;" AND LOOP ";J: S=-1
830    INPUT M
840    FOR K=1 TO M
850    INPUT "RESISTOR VALUE"; R
860    A(I,J)=A(I,J)+R
870    NEXT K
880    A(I,J)=S*A(I,J)
890    A(J,I)=A(I,J)
900    CLS
910    NEXT J
920    PRINT "HOW MANY VOLTAGE SOURCES IN LOOP "; I
930    INPUT P
940    FOR L=1 TO P: IF P=0 THEN E=0
950    IF P>0 THEN INPUT "VOLTAGE SOURCE VALUE WITH PROPER SIGN"; E
960    A(I,4)=A(I,4)+E
```

APPENDIX C

```
970   NEXT L
980   CLS
990   NEXT I
1000  GO TO 300
1010  PRINT "************** NODE ANALYSIS INSTRUCTIONS **************"
1020  PRINT
1030  PRINT "1. IDENTIFY AND NUMBER THE UNKNOWN NODES."
1040  PRINT
1050  PRINT "2. VOLTAGE SOURCES WITH + TERMINAL TOWARD NODE ARE
      POSITIVE, OTHERS ARE NEGATIVE."
1060  PRINT
1070  PRINT
1080  PRINT "TO CONTINUE PRESS 'ENTER'": INPUT W
1090  A$="V"
1100  CLS
1110  PRINT "HOW MANY NODES?"
1120  INPUT N
1130  IF N=2 THEN A(3,3)=1
1140  FOR I=1 TO N
1150  FOR J=I TO N
1160  S=1
1170  IF I=J PRINT "HOW MANY RESISTORS ARE CONNECTED TO NODE ";I
      ELSE PRINT "HOW MANY RESISTORS BETWEEN NODE ";I;"AND NODE ";
      J; S=-1
1180  INPUT M
1190  FOR K=1 TO M
1200  INPUT "RESISTOR VALUE";R
1210  A(I,J)=A(I,J)+1/R
1220  NEXT K
1230  A(I,J)=S*A(I,J)
1240  A(J,I)=A(I,J)
1250  CLS
1260  NEXT J
1270  PRINT "HOW MANY VOLTAGE SOURCES CONNECTED TO NODE ";I;"THROUGH
      A SERIES RESISTOR?"
1280  INPUT P
1290  FOR L=1 TO P: IF P=0 THEN E=0 AND R=1
1300  IF P>0 THEN INPUT "VOLTAGE SOURCE VALUE WITH PROPER SIGN"; E
1310  IF P>0 THEN INPUT "SERIES RESISTOR VALUE";R
1320  A(I,4)=A(I,4)+E/R
1330  NEXT L
1340  CLS
1350  NEXT I
1360  CLS
1370  GO TO 300
```

APPENDIX D

rms (EFFECTIVE) VALUE OF A SINE WAVE

The abbreviation "rms" stands for the *root mean square* process by which this value is derived. In the process, we first square the equation of a sine wave:

$$v^2 = V_p^2 \sin^2\theta$$

Next, we obtain the mean or average value of v^2 by dividing the area under a half-cycle of the curve by π (see Figure D–1). The area is found by integration and trigonometric identities:

$$V_{avg}^2 = \frac{\text{area}}{\pi}$$

$$= \frac{1}{\pi}\int_0^\pi V_p^2 \sin^2\theta \, d\theta$$

$$= \frac{V_p^2}{2\pi}\int_0^\pi (1 - \cos 2\theta) \, d\theta$$

$$= \frac{V_p^2}{2\pi}\int_0^\pi 1 \, d\theta - \frac{V_p^2}{2\pi}\int_0^\pi (-\cos 2\theta) \, d\theta$$

$$= \frac{V_p^2}{2\pi}(d\theta - \tfrac{1}{2}\sin 2\theta)_0^\pi$$

$$= \frac{V_p^2}{2\pi}(\pi - 0)$$

$$= \frac{V_p^2}{2}$$

Finally, the square root of V_{avg}^2 is V_{rms}:

$$V_{rms} = \sqrt{V_{avg}^2}$$

$$= \sqrt{V_p^2/2}$$

$$= \frac{V_p}{\sqrt{2}}$$

$$= 0.707 V_p$$

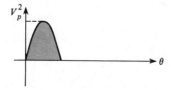

FIGURE D–1

APPENDIX E

AVERAGE VALUE OF A HALF-CYCLE SINE WAVE

The average value of a sine wave is determined for a half-cycle because the average over a full-cycle is zero.

The equation for a sine wave is

$$v = V_p \sin \theta$$

The average value of the half-cycle is the area under the curve divided by the distance of the curve along the horizontal axis (see Figure E–1):

$$V_{avg} = \frac{\text{area}}{\pi}$$

To find the area, we use integral calculus:

$$\begin{aligned}
V_{avg} &= \frac{1}{\pi} \int_0^\pi V_p \sin \theta \, d\theta \\
&= \frac{V_p}{\pi}(-\cos \theta)\Big|_0^\pi \\
&= \frac{V_p}{\pi}[-\cos \pi - (-\cos 0)] \\
&= \frac{V_p}{\pi}[-(-1) - (-1)] \\
&= \frac{V_p}{\pi}(2) \\
&= \frac{2}{\pi} V_p \\
&= 0.637 V_p
\end{aligned}$$

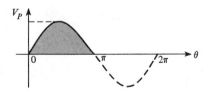

FIGURE E–1

APPENDIX F

REACTANCE DERIVATIONS

Derivation of Capacitive Reactance

$$\theta = 2\pi ft = \omega t$$

$$i = C\frac{dV}{dt} = C\frac{d(V_p \sin \theta)}{dt}$$

$$= C\frac{d(V_p \sin \omega t)}{dt}$$

$$= \omega C(V_p \cos \omega t)$$

$$I_{rms} = \omega C V_{rms}$$

$$X_C = \frac{V_{rms}}{I_{rms}} = \frac{V_{rms}}{\omega C V_{rms}} = \frac{1}{\omega C}$$

$$= \frac{1}{2\pi fC}$$

Derivation of Inductive Reactance

$$v = L\frac{di}{dt} = L\frac{d(I_p \sin \omega t)}{dt}$$

$$= \omega L(I_p \cos \omega t)$$

$$V_{rms} = \omega L I_{rms}$$

$$X_L = \frac{V_{rms}}{I_{rms}} = \frac{\omega L I_{rms}}{I_{rms}} = \omega L$$

$$= 2\pi fL$$

APPENDIX G

TABLE OF TRIGONOMETRIC FUNCTIONS

Deg.	Sin	Tan	Cot	Cos	Deg.	Deg.	Sin	Tan	Cot	Cos	Deg.
0.0	.00000	.00000	∞	1.0000	90.0	4.0	.06976	.06993	14.301	0.9976	86.0
.1	.00175	.00175	573.0	1.0000	89.9	.1	.07150	.07168	13.951	.9974	85.9
.2	.00349	.00349	286.5	1.0000	.8	.2	.07324	.07344	13.617	.9973	.8
.3	.00524	.00524	191.0	1.0000	.7	.3	.07498	.07519	13.300	.9972	.7
.4	.00698	.00698	143.24	1.0000	.6	.4	.07672	.07695	12.996	.9971	.6
.5	.00873	.00873	114.59	1.0000	.5	.5	.07846	.07870	12.706	.9969	.5
.6	.01047	.01047	95.49	0.9999	.4	.6	.08020	.08046	12.429	.9968	.4
.7	.01222	.01222	81.85	.9999	.3	.7	.08194	.08221	12.163	.9966	.3
.8	.01396	.01396	71.62	.9999	.2	.8	.08368	.08397	11.909	.9965	.2
.9	.01571	.01571	63.66	.9999	89.1	.9	.08542	.08573	11.664	.9963	85.1
1.0	.01745	.01746	57.29	0.9998	89.0	5.0	.08716	.08749	11.430	0.9962	85.0
.1	.01920	.01920	52.08	.9998	88.9	.1	.08889	.08925	11.205	.9960	84.9
.2	.02094	.02095	47.74	.9998	.8	.2	.09063	.09101	10.988	.9959	.8
.3	.02269	.02269	44.07	.9997	.7	.3	.09237	.09277	10.780	.9957	.7
.4	.02443	.02444	40.92	.9997	.6	.4	.09411	.09453	10.579	.9956	.6
.5	.02618	.02619	38.19	.9997	.5	.5	.09585	.09629	10.385	.9954	.5
.6	.02792	.02793	35.80	.9996	.4	.6	.09758	.09805	10.199	.9952	.4
.7	.02967	.02968	33.69	.9996	.3	.7	.09932	.09981	10.019	.9951	.3
.8	.03141	.03143	31.82	.9995	.2	.8	.10106	.10158	9.845	.9949	.2
.9	.03316	.03317	30.14	.9995	88.1	.9	.10279	.10334	9.677	.9947	84.1
2.0	.03490	.03492	28.64	0.9994	88.0	6.0	.10453	.10510	9.514	0.9945	84.0
.1	.03664	.03667	27.27	.9993	87.9	.1	.10626	.10687	9.357	.9943	83.9
.2	.03839	.03842	26.03	.9993	.8	.2	.10800	.10863	9.205	.9942	.8
.3	.04013	.04016	24.90	.9992	.7	.3	.10973	.11040	9.058	.9940	.7
.4	.04188	.04191	23.86	.9991	.6	.4	.11147	.11217	8.915	.9938	.6
.5	.04362	.04366	22.90	.9990	.5	.5	.11320	.11394	8.777	.9936	.5
.6	.04536	.04541	22.02	.9990	.4	.6	.11494	.11570	8.643	.9934	.4
.7	.04711	.04716	21.20	.9989	.3	.7	.11667	.11747	8.513	.9932	.3
.8	.04885	.04891	20.45	.9988	.2	.8	.11840	.11924	8.386	.9930	.2
.9	.05059	.05066	19.74	.9987	87.1	.9	.12014	.12101	8.264	.9928	83.1
3.0	.05234	.05241	19.081	0.9986	87.0	7.0	.12187	.12278	8.144	0.9925	83.0
.1	.05408	.05416	18.464	.9985	86.9	.1	.12360	.12456	8.028	.9923	82.9
.2	.05582	.05591	17.886	.9984	.8	.2	.12533	.12633	7.916	.9921	.8
.3	.05756	.05766	17.343	.9983	.7	.3	.12706	.12810	7.806	.9919	.7
.4	.05931	.05941	16.832	.9982	.6	.4	.12880	.12988	7.700	.9917	.6
.5	.06105	.06116	16.350	.9981	.5	.5	.13053	.13165	7.596	.9914	.5
.6	.06279	.06291	15.895	.9980	.4	.6	.13226	.13343	7.495	.9912	.4
.7	.06453	.06467	15.464	.9979	.3	.7	.13399	.13521	7.396	.9910	.3
.8	.06627	.06642	15.056	.9978	.2	.8	.13572	.13698	7.300	.9907	.2
.9	.06802	.06817	14.669	.9977	86.1	.9	.13744	.13876	7.207	.9905	82.1
Deg.	Cos	Cot	Tan	Sin	Deg.	Deg.	Cos	Cot	Tan	Sin	Deg.

Deg.	Sin	Tan	Cot	Cos	Deg.	Deg.	Sin	Tan	Cot	Cos	Deg.
8.0	.13917	.14054	7.115	0.9903	82.0	13.0	.2250	.2309	4.331	0.9744	77.0
.1	.14090	.14232	7.026	.9900	81.9	.1	.2267	.2327	4.297	.9740	76.9
.2	.14263	.14410	6.940	.9898	.8	.2	.2284	.2345	4.264	.9736	.8
.3	.14436	.14588	6.855	.9895	.7	.3	.2300	.2364	4.230	.9732	.7
.4	.14608	.14767	6.772	.9893	.6	.4	.2317	.2382	4.198	.9728	.6
.5	.14781	.14945	6.691	.9890	.5	.5	.2334	.2401	4.165	.9724	.5
.6	.14954	.15124	6.612	.9888	.4	.6	.2351	.2419	4.134	.9720	.4
.7	.15126	.15302	6.535	.9885	.3	.7	.2368	.2438	4.102	.9715	.3
.8	.15299	.15481	6.460	.9882	.2	.8	.2385	.2456	4.071	.9711	.2
.9	.15471	.15660	6.386	.9880	81.1	.9	.2402	.2475	4.041	.9707	76.1
9.0	.15643	.15838	6.314	0.9877	81.0	14.0	.2419	.2493	4.011	0.9703	76.0
.1	.15816	.16017	6.243	.9874	80.9	.1	.2436	.2512	3.981	.9699	75.9
.2	.15988	.16196	6.174	.9871	.8	.2	.2453	.2530	3.952	.9694	.8
.3	.16160	.16376	6.107	.9869	.7	.3	.2470	.2549	3.923	.9690	.7
.4	.16333	.16555	6.041	.9866	.6	.4	.2487	.2568	3.895	.9686	.6
.5	.16505	.16734	5.976	.9863	.5	.5	.2504	.2586	3.867	.9681	.5
.6	.16677	.16914	5.912	.9860	.4	.6	.2521	.2605	3.839	.9677	.4
.7	.16849	.17093	5.850	.9857	.3	.7	.2538	.2623	3.812	.9673	.3
.8	.17021	.17273	5.789	.9854	.2	.8	.2554	.2642	3.785	.9668	.2
.9	.17193	.17453	5.730	.9851	80.1	.9	.2571	.2661	3.758	.9664	75.1
10.0	.1736	.1763	5.671	0.9848	80.0	15.0	0.2588	0.2679	3.732	0.9659	75.0
.1	.1754	.1781	5.614	.9845	79.9	.1	.2605	.2698	3.706	.9655	74.9
.2	.1771	.1799	5.558	.9842	.8	.2	.2622	.2717	3.681	.9650	.8
.3	.1788	.1817	5.503	.9839	.7	.3	.2639	.2736	3.655	.9646	.7
.4	.1805	.1835	5.449	.9836	.6	.4	.2656	.2754	3.630	.9641	.6
.5	.1822	.1853	5.396	.9833	.5	.5	.2672	.2773	3.606	.9636	.5
.6	.1840	.1871	5.343	.9829	.4	.6	.2689	.2792	3.582	.9632	.4
.7	.1857	.1890	5.292	.9826	.3	.7	.2706	.2811	3.558	.9627	.3
.8	.1874	.1908	5.242	.9823	.2	.8	.2723	.2830	3.534	.9622	.2
.9	.1891	.1926	5.193	.9820	79.1	.9	.2740	.2849	3.511	.9617	74.1
11.0	.1908	.1944	5.145	0.9816	79.0	16.0	0.2756	0.2867	3.487	0.9613	74.0
.1	.1925	.1962	5.097	.9813	78.9	.1	.2773	.2886	3.465	.9608	73.9
.2	.1942	.1980	5.050	.9810	.8	.2	.2790	.2905	3.442	.9603	.8
.3	.1959	.1998	5.005	.9806	.7	.3	.2807	.2924	3.420	.9598	.7
.4	.1977	.2016	4.959	.9803	.6	.4	.2823	.2943	3.398	.9593	.6
.5	.1994	.2035	4.915	.9799	.5	.5	.2840	.2962	3.376	.9588	.5
.6	.2011	.2053	4.872	.9796	.4	.6	.2857	.2981	3.354	.9583	.4
.7	.2028	.2071	4.829	.9792	.3	.7	.2874	.3000	3.333	.9578	.3
.8	.2045	.2089	4.787	.9789	.2	.8	.2890	.3019	3.312	.9573	.2
.9	.2062	.2107	4.745	.9785	78.1	.9	.2907	.3038	3.291	.9568	73.1
12.0	.2079	.2126	4.705	0.9781	78.0	17.0	0.2924	0.3057	3.271	0.9563	73.0
.1	.2096	.2144	4.665	.9778	77.9	.1	.2940	.3076	3.251	.9558	72.9
.2	.2113	.2162	4.625	.9774	.8	.2	.2957	.3096	3.230	.9553	.8
.3	.2130	.2180	4.586	.9770	.7	.3	.2974	.3115	3.211	.9548	.7
.4	.2147	.2199	4.548	.9767	.6	.4	.2990	.3134	3.191	.9542	.6
.5	.2164	.2217	4.511	.9763	.5	.5	.3007	.3153	3.172	.9537	.5
.6	.2181	.2235	4.474	.9759	.4	.6	.3024	.3172	3.152	.9532	.4
.7	.2198	.2254	4.437	.9755	.3	.7	.3040	.3191	3.133	.9527	.3
.8	.2215	.2272	4.402	.9751	.2	.8	.3057	.3211	3.115	.9521	.2
.9	.2233	.2290	4.366	.9748	77.1	.9	.3074	.3230	3.096	.9516	72.1
Deg.	Cos	Cot	Tan	Sin	Deg.	Deg.	Cos	Cot	Tan	Sin	Deg.

APPENDIX G

Deg.	Sin	Tan	Cot	Cos	Deg.	Deg.	Sin	Tan	Cot	Cos	Deg.
18.0	0.3090	0.3249	3.078	0.9511	72.0	23.0	0.3907	0.4245	2.356	0.9205	67.0
.1	.3107	.3269	3.060	.9505	71.9	.1	.3923	.4265	2.344	.9198	66.9
.2	.3123	.3288	3.042	.9500	.8	.2	.3939	.4286	2.333	.9191	.8
.3	.3140	.3307	3.024	.9494	.7	.3	.3955	.4307	2.322	.9184	.7
.4	.3156	.3327	3.006	.9489	.6	.4	.3971	.4327	2.311	.9178	.6
.5	.3173	.3346	2.989	.9483	.5	.5	.3987	.4348	2.300	.9171	.5
.6	.3190	.3365	2.971	.9478	.4	.6	.4003	.4369	2.289	.9164	.4
.7	.3206	.3385	2.954	.9472	.3	.7	.4019	.4390	2.278	.9157	.3
.8	.3223	.3404	2.937	.9466	.2	.8	.4035	.4411	2.267	.9150	.2
.9	.3239	.3424	2.921	.9461	71.1	.9	.4051	.4431	2.257	.9143	66.1
19.0	0.3256	0.3443	2.904	0.9455	71.0	24.0	0.4067	0.4452	2.246	0.9135	66.0
.1	.3272	.3463	2.888	.9449	70.9	.1	.4083	.4473	2.236	.9128	65.9
.2	.3289	.3482	2.872	.9444	.8	.2	.4099	.4494	2.225	.9121	.8
.3	.3305	.3502	2.856	.9438	.7	.3	.4115	.4515	2.215	.9114	.7
.4	.3322	.3522	2.840	.9432	.6	.4	.4131	.4536	2.204	.9107	.6
.5	.3338	.3541	2.824	.9426	.5	.5	.4147	.4557	2.194	.9100	.5
.6	.3355	.3561	2.808	.9421	.4	.6	.4163	.4578	2.184	.9092	.4
.7	.3371	.3581	2.793	.9415	.3	.7	.4179	.4599	2.174	.9085	.3
.8	.3387	.3600	2.778	.9409	.2	.8	.4195	.4621	2.164	.9078	.2
.9	.3404	.3620	2.762	.9403	70.1	.9	.4210	.4642	2.154	.9070	65.1
20.0	0.3420	0.3640	2.747	0.9397	70.0	25.0	0.4226	0.4663	2.145	0.9063	65.0
.1	.3437	.3659	2.733	.9391	69.9	.1	.4242	.4684	2.135	.9056	64.9
.2	.3453	.3679	2.718	.9385	.8	.2	.4258	.4706	2.125	.9048	.8
.3	.3469	.3699	2.703	.9379	.7	.3	.4274	.4727	2.116	.9041	.7
.4	.3486	.3719	2.689	.9373	.6	.4	.4289	.4748	2.106	.9033	.6
.5	.3502	.3739	2.675	.9367	.5	.5	.4305	.4770	2.097	.9026	.5
.6	.3518	.3759	2.660	.9361	.4	.6	.4321	.4791	2.087	.9018	.4
.7	.3535	.3779	2.646	.9354	.3	.7	.4337	.4813	2.078	.9011	.3
.8	.3551	.3799	2.633	.9348	.2	.8	.4352	.4834	2.069	.9003	.2
.9	.3567	.3819	2.619	.9342	69.1	.9	.4368	.4856	2.059	.8996	64.1
21.0	0.3584	0.3839	2.605	0.9336	69.0	26.0	0.4384	0.4877	2.050	0.8988	64.0
.1	.3600	.3859	2.592	.9330	68.9	.1	.4399	.4899	2.041	.8980	63.9
.2	.3616	.3879	2.578	.9323	.8	.2	.4415	.4921	2.032	.8973	.8
.3	.3633	.3899	2.565	.9317	.7	.3	.4431	.4942	2.023	.8965	.7
.4	.3649	.3919	2.552	.9311	.6	.4	.4446	.4964	2.014	.8957	.6
.5	.3665	.3939	2.539	.9304	.5	.5	.4462	.4986	2.006	.8949	.5
.6	.3681	.3959	2.526	.9298	.4	.6	.4478	.5008	1.997	.8942	.4
.7	.3697	.3979	2.513	.9291	.3	.7	.4493	.5029	1.988	.8934	.3
.8	.3714	.4000	2.500	.9285	.2	.8	.4509	.5051	1.980	.8926	.2
.9	.3730	.4020	2.488	.9278	68.1	.9	.4524	.5073	1.971	.8918	63.1
22.0	0.3746	0.4040	2.475	0.9272	68.0	27.0	0.4540	0.5095	1.963	0.8910	63.0
.1	.3762	.4061	2.463	.9265	67.9	.1	.4555	.5117	1.954	.8902	62.9
.2	.3778	.4081	2.450	.9259	.8	.2	.4571	.5139	1.946	.8894	.8
.3	.3795	.4101	2.438	.9252	.7	.3	.4586	.5161	1.937	.8886	.7
.4	.3811	.4122	2.426	.9245	.6	.4	.4602	.5184	1.929	.8878	.6
.5	.3827	.4142	2.414	.9239	.5	.5	.4617	.5206	1.921	.8870	.5
.6	.3843	.4163	2.402	.9232	.4	.6	.4633	.5228	1.913	.8862	.4
.7	.3859	.4183	2.391	.9225	.3	.7	.4648	.5250	1.905	.8854	.3
.8	.3875	.4204	2.379	.9219	.2	.8	.4664	.5272	1.897	.8846	.2
.9	.3891	.4224	2.367	.9212	67.1	.9	.4679	.5295	1.889	.8838	62.1
Deg.	Cos	Cot	Tan	Sin	Deg.	Deg.	Cos	Cot	Tan	Sin	Deg.

Deg.	Sin	Tan	Cot	Cos	Deg.	Deg.	Sin	Tan	Cot	Cos	Deg.
28.0	0.4695	0.5317	1.881	0.8829	62.0	33.0	0.5446	0.6494	1.5399	0.8387	57.0
.1	.4710	.5340	1.873	.8821	61.9	.1	.5461	.6519	1.5340	.8377	56.9
.2	.4726	.5362	1.865	.8813	.8	.2	.5476	.6544	1.5282	.8368	.8
.3	.4741	.5384	1.857	.8805	.7	.3	.5490	.6569	1.5224	.8358	.7
.4	.4756	.5407	1.849	.8796	.6	.4	.5505	.6594	1.5166	.8348	.6
.5	.4772	.5430	1.842	.8788	.5	.5	.5519	.6619	1.5108	.8339	.5
.6	.4787	.5452	1.834	.8780	.4	.6	.5534	.6644	1.5051	.8329	.4
.7	.4802	.5475	1.827	.8771	.3	.7	.5548	.6669	1.4994	.8320	.3
.8	.4818	.5498	1.819	.8763	.2	.8	.5563	.6694	1.4938	.8310	.2
.9	.4833	.5520	1.811	.8755	61.1	.9	.5577	.6720	1.4882	.8300	56.1
29.0	0.4848	0.5543	1.804	0.8746	61.0	34.0	0.5592	0.6745	1.4826	0.8290	56.0
.1	.4863	.5566	1.797	.8738	60.9	.1	.5606	.6771	1.4770	.8281	55.9
.2	.4879	.5589	1.789	.8729	.8	.2	.5621	.6796	1.4715	.8271	.8
.3	.4894	.5612	1.782	.8721	.7	.3	.5635	.6822	1.4659	.8261	.7
.4	.4909	.5635	1.775	.8712	.6	.4	.5650	.6847	1.4605	.8251	.6
.5	.4924	.5658	1.767	.8704	.5	.5	.5664	.6873	1.4550	.8241	.5
.6	.4939	.5681	1.760	.8695	.4	.6	.5678	.6899	1.4496	.8231	.4
.7	.4955	.5704	1.753	.8686	.3	.7	.5693	.6924	1.4442	.8221	.3
.8	.4970	.5727	1.746	.8678	.2	.8	.5707	.6950	1.4388	.8211	.2
.9	.4985	.5750	1.739	.8669	60.1	.9	.5721	.6976	1.4335	.8202	55.1
30.0	0.5000	0.5774	1.7321	0.8660	60.0	35.0	0.5736	0.7002	1.4281	0.8192	55.0
.1	.5015	.5797	1.7251	.8652	59.9	.1	.5750	.7028	1.4229	.8181	54.9
.2	.5030	.5820	1.7182	.8643	.8	.2	.5764	.7054	1.4176	.8171	.8
.3	.5045	.5844	1.7113	.8634	.7	.3	.5779	.7080	1.4124	.8161	.7
.4	.5060	.5867	1.7045	.8625	.6	.4	.5793	.7107	1.4071	.8151	.6
.5	.5075	.5890	1.6977	.8616	.5	.5	.5807	.7133	1.4019	.8141	.5
.6	.5090	.5914	1.6909	.8607	.4	.6	.5821	.7159	1.3968	.8131	.4
.7	.5105	.5938	1.6842	.8599	.3	.7	.5835	.7186	1.3916	.8121	.3
.8	.5120	.5961	1.6775	.8590	.2	.8	.5850	.7212	1.3865	.8111	.2
.9	.5135	.5985	1.6709	.8581	59.1	.9	.5864	.7239	1.3814	.8100	54.1
31.0	0.5150	0.6009	1.6643	0.8572	59.0	36.0	0.5878	0.7265	1.3764	0.8090	54.0
.1	.5165	.6032	1.6577	.8563	58.9	.1	.5892	.7292	1.3713	.8080	53.9
.2	.5180	.6056	1.6512	.8554	.8	.2	.5906	.7319	1.3663	.8070	.8
.3	.5195	.6080	1.6447	.8545	.7	.3	.5920	.7346	1.3613	.8059	.7
.4	.5210	.6104	1.6383	.8536	.6	.4	.5934	.7373	1.3564	.8049	.6
.5	.5225	.6128	1.6319	.8526	.5	.5	.5948	.7400	1.3514	.8039	.5
.6	.5240	.6152	1.6255	.8517	.4	.6	.5962	.7427	1.3465	.8028	.4
.7	.5255	.6176	1.6191	.8508	.3	.7	.5976	.7454	1.3416	.8018	.3
.8	.5270	.6200	1.6128	.8499	.2	.8	.5990	.7481	1.3367	.8007	.2
.9	.5284	.6224	1.6066	.8490	58.1	.9	.6004	.7608	1.3319	.7997	53.1
32.0	0.5299	0.6249	1.6003	0.8480	58.0	37.0	0.6018	0.7536	1.3270	0.7986	53.0
.1	.5314	.6273	1.5941	.8471	57.9	.1	.6032	.7563	1.3222	.7976	52.9
.2	.5329	.6297	1.5880	.8462	.8	.2	.6046	.7590	1.3175	.7965	.8
.3	.5344	.6322	1.5818	.8453	.7	.3	.6060	.7618	1.3127	.7955	.7
.4	.5358	.6346	1.5757	.8443	.6	.4	.6074	.7646	1.3079	.7944	.6
.5	.5373	.6371	1.5697	.8434	.5	.5	.6088	.7673	1.3032	.7934	.5
.6	.5388	.6395	1.5637	.8425	.4	.6	.6101	.7701	1.2985	.7923	.4
.7	.5402	.6420	1.5577	.8415	.3	.7	.6115	.7729	1.2938	.7912	.3
.8	.5417	.6445	1.5517	.8406	.2	.8	.6129	.7757	1.2892	.7902	.2
.9	.5432	.6469	1.5458	.8396	57.1	.9	.6143	.7785	1.2846	.7891	52.1
Deg.	Cos	Cot	Tan	Sin	Deg.	Deg.	Cos	Cot	Tan	Sin	Deg.

Deg.	Sin	Tan	Cot	Cos	Deg.	Deg.	Sin	Tan	Cot	Cos	Deg.
38.0	0.6157	0.7813	1.2799	0.7880	52.0	42.0	0.6691	0.9004	1.1106	0.7431	48.0
.1	.6170	.7841	1.2753	.7869	51.9	.1	.6704	.9036	1.1067	.7420	47.9
.2	.6184	.7869	1.2708	.7859	.8	.2	.6717	.9067	1.1028	.7408	.8
.3	.6198	.7898	1.2662	.7848	.7	.3	.6730	.9099	1.0990	.7396	.7
.4	.6211	.7926	1.2617	.7837	.6	.4	.6743	.9131	1.0951	.7385	.6
.5	.6225	.7954	1.2572	.7826	.5	.5	.6756	.9163	1.0913	.7373	.5
.6	.6239	.7983	1.2527	.7815	.4	.6	.6769	.9195	1.0875	.7361	.4
.7	.6252	.8012	1.2482	.7804	.3	.7	.6782	.9228	1.0837	.7349	.3
.8	.6266	.8040	1.2437	.7793	.2	.8	.6794	.9260	1.0799	.7337	.2
.9	.6280	.8069	1.2393	.7782	51.1	.9	.6807	.9293	1.0761	.7325	47.1
39.0	0.6293	0.8098	1.2349	0.7771	51.0	43.0	0.6820	0.9325	1.0724	0.7314	47.0
.1	.6307	.8127	1.2305	.7760	50.9	.1	.6833	.9358	1.0686	.7302	46.9
.2	.6320	.8156	1.2261	.7749	.8	.2	.6845	.9391	1.0649	.7290	.8
.3	.6334	.8185	1.2218	.7738	.7	.3	.6858	.9424	1.0612	.7278	.7
.4	.6347	.8214	1.2174	.7727	.6	.4	.6871	.9457	1.0575	.7266	.6
.5	.6361	.8243	1.2131	.7716	.5	.5	.6884	.9490	1.0538	.7254	.5
.6	.6374	.8273	1.2088	.7705	.4	.6	.6896	.9523	1.0501	.7242	.4
.7	.6388	.8302	1.2045	.7694	.3	.7	.6909	.9556	1.0464	.7230	.3
.8	.6401	.8332	1.2002	.7683	.2	.8	.6921	.9590	1.0428	.7218	.2
.9	.6414	.8361	1.1960	.7672	50.1	.9	.6934	.9623	1.0392	.7206	46.1
40.0	0.6428	0.8391	1.1918	0.7660	50.0	44.0	0.6947	0.9657	1.0355	0.7193	46.0
.1	.6441	.8421	1.1875	.7649	49.9	.1	.6959	.9691	1.0319	.7181	45.9
.2	.6455	.8451	1.1833	.7638	.8	.2	.6972	.9725	1.0283	.7169	.8
.3	.6468	.8481	1.1792	.7627	.7	.3	.6984	.9759	1.0247	.7157	.7
.4	.6481	.8511	1.1750	.7615	.6	.4	.6997	.9793	1.0212	.7145	.6
						.5	.7009	.9827	1.0176	.7133	.5
40.5	0.6494	0.8541	1.1708	0.7604	49.5	.6	.7022	.9861	1.0141	.7120	.4
.6	.6508	.8571	1.1667	.7593	.4	.7	.7034	.9896	1.0105	.7108	.3
.7	.6521	.8601	1.1626	.7581	.3	.8	.7046	.9930	1.0070	.7096	.2
.8	.6534	.8632	1.1585	.7570	.2	.9	.7059	.9965	1.0035	.7083	45.1
.9	.6547	.8662	1.1544	.7559	49.1	45.0	0.7071	1.0000	1.0000	0.7071	45.0
41.0	0.6561	0.8693	1.1504	0.7547	49.0	Deg.	Cos	Cot	Tan	Sin	Deg.
.1	.6574	.8724	1.1463	.7536	48.9						
.2	.6587	.8754	1.1423	.7524	.8						
.3	.6600	.8785	1.1383	.7513	.7						
.4	.6613	.8816	1.1343	.7501	.6						
.5	.6626	.8847	1.1303	.7490	.5						
.6	.6639	.8878	1.1263	.7478	.4						
.7	.6652	.8910	1.1224	.7466	.3						
.8	.6665	.8941	1.1184	.7455	.2						
.9	.6678	.8972	1.1145	.7443	48.1						
Deg.	Cos	Cot	Tan	Sin	Deg.						

APPENDIX H

TABLE OF EXPONENTIAL FUNCTIONS

VALUES OF e^x AND e^{-x}

x	Function	0.00	0.01	0.02	0.03	0.04	0.05	0.06	0.07	0.08	0.09
0.0	e^x	1.0000	1.0101	1.0202	1.0305	1.0408	1.0513	1.0618	1.0725	1.0833	1.0942
	e^{-x}	1.0000	0.9900	0.9802	0.9704	0.9608	0.9512	0.9418	0.9324	0.9231	0.9139
0.1	e^x	1.1052	1.1163	1.1275	1.1388	1.1503	1.1618	1.1735	1.1853	1.1972	1.2093
	e^{-x}	0.9048	0.8958	0.8869	0.8781	0.8694	0.8607	0.8521	0.8437	0.8353	0.8270
0.2	e^x	1.2214	1.2337	1.2461	1.2546	1.2712	1.2840	1.2969	1.3100	1.3231	1.3364
	e^{-x}	0.8187	0.8106	0.8025	0.7945	0.7856	0.7788	0.7711	0.7634	0.7558	0.7483
0.3	e^x	1.3499	1.3634	1.3771	1.3910	1.4049	1.4191	1.4333	1.4477	1.4623	1.4770
	e^{-x}	0.7408	0.7334	0.7261	0.7189	0.7118	0.7047	0.6977	0.6907	0.6839	0.6771
0.4	e^x	1.4918	1.5068	1.5220	1.5373	1.5527	1.5683	1.5841	1.6000	1.6161	1.6323
	e^{-x}	0.6703	0.6637	0.6570	0.6505	0.6440	0.6376	0.6313	0.6250	0.6188	0.6126
0.5	e^x	1.6487	1.6653	1.6820	1.6989	1.7160	1.7333	1.7507	1.7683	1.7860	1.8040
	e^{-x}	0.6065	0.6005	0.5945	0.5886	0.5827	0.5769	0.5712	0.5655	0.5599	0.5543
0.6	e^x	1.8221	1.8404	1.8589	1.8776	1.8965	1.9155	1.9348	1.9542	1.9739	1.9939
	e^{-x}	0.5488	0.5434	0.5379	0.5326	0.5273	0.5220	0.5169	0.5117	0.5066	0.5017
0.7	e^x	2.0138	2.0340	2.0544	2.0751	2.0959	2.1170	2.1383	2.1598	2.1815	2.2034
	e^{-x}	0.4966	0.4916	0.4868	0.4819	0.4771	0.4724	0.4677	0.4630	0.4584	0.4538
0.8	e^x	2.2255	2.2479	2.2705	2.2933	2.3164	2.3396	2.3632	2.3869	2.4109	2.4351
	e^{-x}	0.4493	0.4449	0.4404	0.4360	0.4317	0.4274	0.4232	0.4190	0.4148	0.4107
0.9	e^x	2.4596	2.4843	2.5093	2.5345	2.5600	2.5857	2.6117	2.6379	2.6645	2.6912
	e^{-x}	0.4066	0.4025	0.3985	0.3946	0.3906	0.3867	0.3829	0.3791	0.3753	0.3716
1.0	e^x	2.7183	2.7456	2.7732	2.8011	2.8292	2.8577	2.8864	2.9154	2.9447	2.9743
	e^{-x}	0.3679	0.3642	0.3606	0.3570	0.3535	0.3499	0.3465	0.3430	0.3396	0.3362
1.1	e^x	3.0042	3.0344	3.0649	3.0957	3.1268	3.1582	3.1899	3.2220	3.2544	3.2871
	e^{-x}	0.3329	0.3296	0.3263	0.3230	0.3198	0.3166	0.3135	0.3104	0.3073	0.3042
1.2	e^x	3.3201	3.3535	3.3872	3.4212	3.4556	3.4903	3.5254	3.5609	3.5966	3.6328
	e^{-x}	0.3012	0.2982	0.2952	0.2923	0.2894	0.2865	0.2837	0.2808	0.2780	0.2753
1.3	e^x	3.6693	3.7062	3.7434	3.7810	3.8190	3.8574	3.8962	3.9354	3.9749	4.0149
	e^{-x}	0.2725	0.2698	0.2671	0.2645	0.2618	0.2592	0.2567	0.2541	0.2516	0.2491
1.4	e^x	4.0552	4.0960	4.1371	4.1787	4.2207	4.2631	4.3060	4.3492	4.3929	4.4371
	e^{-x}	0.2466	0.2441	0.2417	0.2393	0.2369	0.2346	0.2322	0.2299	0.2276	0.2254
1.5	e^x	4.4817	4.5267	4.5722	4.6182	4.6646	4.7115	4.7588	4.8066	4.8550	4.9037
	e^{-x}	0.2231	0.2209	0.2187	0.2165	0.2144	0.2122	0.2101	0.2080	0.2060	0.2039
1.6	e^x	4.9530	5.0028	5.0531	5.1039	5.1552	5.2070	5.2593	5.3122	5.3656	5.4195
	e^{-x}	0.2019	0.1999	0.1979	0.1959	0.1940	0.1920	0.1901	0.1882	0.1864	0.1845
1.7	e^x	5.4739	5.5290	5.5845	5.6407	5.6973	5.7546	5.8124	5.8709	5.9299	5.9895
	e^{-x}	0.1827	0.1809	0.1791	0.1773	0.1755	0.1738	0.1720	0.1703	0.1686	0.1670
1.8	e^x	6.0496	6.1104	6.1719	6.2339	6.2965	6.3598	6.4237	6.4883	6.5535	6.6194
	e^{-x}	0.1653	0.1637	0.1620	0.1604	0.1588	0.1572	0.1557	0.1541	0.1526	0.1511
1.9	e^x	6.6859	6.7531	6.8210	6.8895	6.9588	7.0287	7.0993	7.1707	7.2427	7.3155
	e^{-x}	0.1496	0.1481	0.1466	0.1451	0.1437	0.1423	0.1409	0.1395	0.1381	0.1367

APPENDIX H

VALUES OF e^x AND e^{-x}

x	Function	0.00	0.01	0.02	0.03	0.04	0.05	0.06	0.07	0.08	0.09
2.0	e^x	7.3891	7.4633	7.5383	7.6141	7.6906	7.7679	7.8460	7.9248	8.0045	8.8049
	e^{-x}	0.1353	0.1340	0.1327	0.1313	0.1300	0.1287	0.1275	0.1262	0.1249	0.1237
2.1	e^x	8.1662	8.2482	8.3311	8.4149	8.4994	8.5849	8.6711	8.7583	8.8463	8.9352
	e^{-x}	0.1225	0.1212	0.1200	0.1188	0.1177	0.1165	0.1153	0.1142	0.1130	0.1119
2.2	e^x	9.0250	9.1157	9.2073	9.2999	9.3933	9.4877	9.5831	9.6794	9.7767	9.8749
	e^{-x}	0.1108	0.1097	0.1086	0.1075	0.1065	0.1054	0.1044	0.1033	0.1023	0.1013
2.3	e^x	9.9742	10.074	10.716	10.278	10.381	10.486	10.591	10.697	10.805	10.913
	e^{-x}	0.1003	0.0993	0.0983	0.0973	0.0963	0.0954	0.0944	0.0935	0.0926	0.0916
2.4	e^x	11.023	11.134	11.246	11.359	11.473	11.588	11.705	11.822	11.941	12.061
	e^{-x}	0.0907	0.0898	0.0889	0.0880	0.0872	0.0863	0.0854	0.0846	0.0837	0.0829
2.5	e^x	12.182	12.305	12.429	12.553	12.680	12.807	12.936	13.066	13.197	13.330
	e^{-x}	0.0821	0.0813	0.0805	0.0797	0.0789	0.0781	0.0773	0.0765	0.0758	0.0750
2.6	e^x	13.464	13.599	13.736	13.874	14.013	14.154	14.296	14.440	14.585	14.732
	e^{-x}	0.0743	0.0735	0.0728	0.0721	0.0714	0.0707	0.0699	0.0693	0.0686	0.0679
2.7	e^x	14.880	15.029	15.180	15.333	15.487	15.643	15.800	15.959	16.119	16.281
	e^{-x}	0.0672	0.0665	0.0659	0.0652	0.0646	0.0639	0.0633	0.0627	0.0620	0.0614
2.8	e^x	16.445	16.610	16.777	16.945	17.116	17.288	17.462	17.637	17.814	17.993
	e^{-x}	0.0608	0.0602	0.0596	0.0590	0.0584	0.0578	0.0573	0.0567	0.0561	0.0556
2.9	e^x	18.174	18.357	18.541	18.728	18.916	19.106	19.298	19.492	19.688	19.886
	e^{-x}	0.0550	0.0545	0.0539	0.0534	0.0529	0.0523	0.0518	0.0513	0.0508	0.0503
3.0	e^x	20.086	20.287	20.491	20.697	20.905	21.115	21.328	21.542	21.758	21.977
	e^{-x}	0.0498	0.0493	0.0488	0.0483	0.0478	0.0474	0.0469	0.0464	0.0460	0.0455
3.1	e^x	22.198	22.421	22.646	22.874	23.104	23.336	23.571	23.807	24.047	24.288
	e^{-x}	0.0450	0.0446	0.0442	0.0437	0.0433	0.0429	0.0424	0.0420	0.0416	0.0412
3.2	e^x	24.533	24.779	25.028	25.280	25.534	25.790	26.050	26.311	26.576	26.843
	e^{-x}	0.0408	0.0404	0.0400	0.0396	0.0392	0.0388	0.0384	0.0380	0.0376	0.0373
3.3	e^x	27.113	26.385	27.660	27.938	28.219	28.503	28.789	29.079	29.371	29.666
	e^{-x}	0.0369	0.0365	0.0362	0.0358	0.0354	0.0351	0.0347	0.0344	0.0340	0.0337
3.4	e^x	29.964	30.265	30.569	30.877	31.187	31.500	31.817	32.137	32.460	32.786
	e^{-x}	0.0334	0.0330	0.0327	0.0324	0.0321	0.0317	0.0314	0.0311	0.0308	0.0305
3.5	e^x	33.115	33.448	33.784	34.124	34.467	34.813	35.163	35.517	35.874	36.234
	e^{-x}	0.0302	0.0299	0.0296	0.0293	0.0290	0.0287	0.0284	0.0282	0.0279	0.0276
3.6	e^x	36.598	36.966	37.338	37.713	38.092	38.475	38.861	39.252	39.646	40.045
	e^{-x}	0.0273	0.0271	0.0268	0.0265	0.0263	0.0260	0.0257	0.0255	0.0252	0.0250
3.7	e^x	40.447	40.854	41.264	41.679	42.098	42.521	42.948	43.380	43.816	44.256
	e^{-x}	0.0247	0.0245	0.0242	0.0240	0.0238	0.0235	0.0233	0.0231	0.0228	0.0226
3.8	e^x	44.701	45.150	45.604	46.063	46.525	46.993	47.465	47.942	48.424	48.911
	e^{-x}	0.0224	0.0221	0.0219	0.0217	0.0215	0.0213	0.0211	0.0209	0.0207	0.0204
3.9	e^x	49.402	49.899	50.400	50.907	51.419	51.935	52.457	52.985	53.517	54.055
	e^{-x}	0.0202	0.0200	0.0198	0.0196	0.0195	0.0193	0.0191	0.0189	0.0187	0.0185

VALUES OF e^x AND e^{-x}

x	Function	0.00	0.01	0.02	0.03	0.04	0.05	0.06	0.07	0.08	0.09
4.0	e^x	54.598	55.147	55.701	56.261	56.826	57.397	57.974	58.557	59.145	59.740
	e^{-x}	0.0183	0.0181	0.0180	0.0178	0.0176	0.0174	0.0172	0.0171	0.0169	0.0167
4.1	e^x	60.340	60.947	61.559	62.178	62.803	63.434	64.072	64.714	65.366	66.023
	e^{-x}	0.0166	0.0164	0.0162	0.0161	0.0159	0.0158	0.0156	0.0155	0.0153	0.0151
4.2	e^x	66.686	67.357	68.033	68.717	69.408	70.105	70.810	71.522	72.240	72.966
	e^{-x}	0.0150	0.0148	0.0147	0.0146	0.0144	0.0143	0.0141	0.0140	0.0138	0.0137
4.3	e^x	73.700	74.440	75.189	75.944	76.708	77.478	78.257	79.044	79.838	80.640
	e^{-x}	0.0136	0.0134	0.0133	0.0132	0.0130	0.0129	0.0128	0.0127	0.0125	0.0124
4.4	e^x	81.451	82.269	83.096	83.931	84.775	85.627	86.488	87.357	88.235	89.121
	e^{-x}	0.0123	0.0122	0.0120	0.0119	0.0118	0.0117	0.0116	0.0114	0.0113	0.0112
4.5	e^x	90.017	90.922	91.836	92.759	93.691	94.632	95.583	96.544	97.514	98.494
	e^{-x}	0.0111	0.0110	0.0109	0.0108	0.0107	0.0106	0.0105	0.0104	0.0103	0.0102
4.6	e^x	99.484	100.48	101.49	102.51	103.54	104.58	105.64	106.70	107.77	108.85
	e^{-x}	0.0101	0.0100	0.0099	0.0098	0.0097	0.0096	0.0095	0.0094	0.0093	0.0092
4.7	e^x	109.95	111.05	112.17	113.30	114.43	115.58	116.75	117.92	119.10	120.30
	e^{-x}	0.0091	0.0090	0.0089	0.0088	0.0087	0.0087	0.0086	0.0085	0.0084	0.0083
4.8	e^x	121.51	122.73	123.97	125.21	126.47	127.74	129.02	130.32	131.63	132.95
	e^{-x}	0.0082	0.0081	0.0081	0.0080	0.0079	0.0078	0.0078	0.0077	0.0076	0.0075
4.9	e^x	134.29	135.64	137.00	138.38	139.77	141.17	142.59	144.03	145.47	146.94
	e^{-x}	0.0074	0.0074	0.0073	0.0072	0.0072	0.0071	0.0070	0.0069	0.0069	0.0068
5.0	e^x	148.41	149.90	151.41	152.93	154.47	156.02	157.59	159.17	160.77	162.39
	e^{-x}	0.0067	0.0067	0.0066	0.0065	0.0065	0.0064	0.0063	0.0063	0.0062	0.0062
5.1	e^x	164.02	165.67	167.34	169.02	170.72	172.43	174.16	175.91	177.68	179.47
	e^{-x}	0.0061	0.0060	0.0060	0.0059	0.0059	0.0058	0.0057	0.0057	0.0056	0.0056
5.2	e^x	181.27	183.09	184.93	186.79	188.67	190.57	192.48	194.42	196.37	198.34
	e^{-x}	0.0055	0.0055	0.0054	0.0054	0.0053	0.0052	0.0052	0.0051	0.0051	0.0050
5.3	e^x	200.34	202.35	204.38	206.44	208.51	210.61	212.72	214.86	217.02	219.20
	e^{-x}	0.0050	0.0049	0.0049	0.0048	0.0048	0.0047	0.0047	0.0047	0.0046	0.0046
5.4	e^x	221.41	223.63	225.88	228.15	230.44	232.76	235.10	237.46	239.85	242.26
	e^{-x}	0.0045	0.0045	0.0044	0.0044	0.0043	0.0043	0.0043	0.0042	0.0042	0.0041
5.5	e^x	244.69	247.15	249.64	252.14	254.68	257.24	259.82	262.43	265.07	267.74
	e^{-x}	0.0041	0.0040	0.0040	0.0040	0.0039	0.0039	0.0038	0.0038	0.0038	0.0037
5.6	e^x	270.43	273.14	275.89	278.66	281.46	284.29	287.15	290.03	292.95	295.89
	e^{-x}	0.0037	0.0037	0.0036	0.0036	0.0036	0.0035	0.0035	0.0034	0.0034	0.0034
5.7	e^x	298.87	301.87	304.90	307.97	311.06	314.19	317.35	320.54	323.76	327.01
	e^{-x}	0.0033	0.0033	0.0033	0.0032	0.0032	0.0032	0.0032	0.0031	0.0031	0.0031
5.8	e^x	330.30	333.62	336.97	340.36	343.78	347.23	350.72	354.25	357.81	361.41
	e^{-x}	0.0030	0.0030	0.0030	0.0029	0.0029	0.0029	0.0029	0.0028	0.0028	0.0028
5.9	e^x	365.04	368.71	372.41	376.15	379.93	383.75	387.61	391.51	395.44	399.41
	e^{-x}	0.0027	0.0027	0.0027	0.0027	0.0026	0.0026	0.0026	0.0026	0.0025	0.0025

Admittance A measure of the ability of a reactive circuit to permit current. The reciprocal of impedance.

Alternating current (ac) Current that reverses direction in response to a change in voltage polarity.

Ammeter An electrical instrument used to measure current.

Ampere (A or amp) The unit of electrical current.

Ampere-hour (Ah) The measure of the capacity of a battery to supply electrical current.

Amplification The process of increasing the power, voltage, or current of an electrical signal.

Amplifier An electronic circuit having the capability of amplification and designed specifically for that purpose.

Amplitude The voltage or current value of an electrical signal which, in some cases, implies the maximum value.

Analog Characterizing a linear process in which a variable takes on a *continuous* set of values within a given range.

Apparent power The power that *appears* to be being delivered to a reactive circuit. The product of volts and amperes with units of VA (volt-amperes).

Arc tangent An inverse trigonometric function meaning "the angle whose tangent is." Also called *inverse tangent*.

Asynchronous Having no fixed time relationship, such as two wave forms that are not related to each other in terms of their time variations.

Atom The smallest particle of an element possessing the unique characteristics of that element.

Attenuation The process of reducing the power, voltage, or current value of an electrical signal. It can be thought of as negative amplification.

Audio Related to ability of the human ear to detect sound. Audio frequencies are those that can be heard by the human ear, typically from 20 Hz to 20 kHz.

Autotransformer A transformer having only one coil for both its primary and its secondary.

Average power The average rate of energy consumption. In an electrical circuit, average power occurs only in the resistance and represents a net energy loss.

AWG American Wire Gage, a standardization of wire sizes according to the diameter of the wire.

Band-pass The characteristic of a certain type of filter whereby frequencies within a certain range are passed through the circuit and all others are blocked.

Band-stop The characteristics of a certain type of filter whereby frequencies within a certain range are blocked from passing through the circuit.

Base One of the semiconductor regions in a bipolar transistor.

Baseline The normal level of a pulse wave form. The level in the absence of a pulse.

Battery An energy source that uses a chemical reaction to convert chemical energy into electrical energy.

Bias The application of a dc voltage to a diode, transistor, or other electronic device to produce a desired mode of operation.

Bode plot An idealized graph of the gain, in dB, versus frequency. Used to illustrate graphically the response of an amplifier or filter circuit.

Branch One of the current paths in a parallel circuit.

Capacitance The ability of a capacitor to store electrical charge.

Capacitor An electrical device possessing the property of capacitance.

Cascade An arrangement of circuits in which the output of one circuit becomes the input to the next.

Center tap (CT) A connection at the midpoint of the secondary of a transformer.

Charge An electrical property of matter in which an attractive force or a repulsive force is created between two particles. Charge can be positive or negative.

Chassis The metal framework or case upon which an electrical circuit or system is constructed.

Chip A tiny piece of semiconductor material upon which an integrated circuit is constructed.

Choke An inductor. The term is more commonly used in connection with inductors used to block or *choke off* high frequencies.

Circuit An interconnection of electrical components designed to produce a desired result.

Circuit breaker A resettable protective electrical device used for interrupting current in a circuit when the current has reached an excessive level.

Circular mil (CM) The unit of the cross-sectional area of a wire. A wire with a diameter of 0.001 inch has a cross-sectional area of one circular mil.

Closed circuit A circuit with a complete current path.

Coefficient of coupling A constant associated with transformers which specifies the magnetic field in the secondary as a result of that in the primary.

Coil A common term for an inductor or for the primary or secondary winding of a transformer.

Common A term sometimes used for the ground or reference point in a circuit.

Conductance The ability of a circuit to allow current. It is the reciprocal of resistance. The units are siemens.

Conductor A material that allows electrical current to exist with relative ease. An example is copper.

Computer An electronic system that can process data and perform calculations at a very fast rate. It operates under the direction of a *stored program*.

Core The material within the windings of an inductor which influences the electromagnetic characteristics of the inductor.

Coulomb (C) The unit of electrical charge.

CRT Cathode ray tube.

Current The rate of flow of electrons in a circuit.

Cycle The repetition of a pattern.

Decibel (dB) The unit of the logarithmic expression of a ratio, such as power or voltage gain.

Degree The unit of angular measure corresponding to 1/360 of a complete revolution.

Derivative The rate of change of a function, determined mathematically.

Dielectric The insulating material used between the plates of a capacitor.

Differentiator An RC or RL circuit that produces an output which approaches the mathematical derivative of the input.

Digital Characterizing a nonlinear process in which a variable takes on discrete values within a given range.

Diode An electronic device that permits current flow in only one direction. It can be a semiconductor or a tube.

Direct current (dc) Current in only one direction.

DMM Digital multimeter.

Duty cycle A characteristic of a pulse wave form which indicates the percentage of time that a pulse is present during a cycle.

DVM Digital voltmeter.

Effective value A measure of the heating effect of a sine wave. Also known as the rms (root mean square) value.

Electrical Related to the use of electrical voltage and current to achieve desired results.

Electromagnetic Related to the production of a magnetic field by an electrical current in a conductor.

Electron The basic particle of electrical charge in matter. The electron possesses negative charge.

Electronic Related to the movement and control of free electrons in semiconductors or vacuum devices.

Element One of the unique substances making up the known universe. Each element is characterized by a unique atomic structure.

Emitter One of the three regions in a bipolar transistor.

Energy The ability to do work.

Exponent The number of times a given number is multiplied by itself. The exponent is called the *power*, and the given number is called the *base*.

Falling edge The negative-going transition of a pulse.

Fall time The time interval required for a pulse to change from 90% of its amplitude to 10% of its amplitude.

Farad (F) The unit of capacitance.

Field The invisible forces existing between oppositely charged particles (electric field) or between the north and south poles of a magnet or electromagnet (magnetic field).

Filter A type of electrical circuit that passes certain frequencies and rejects all others.

Flux The lines of force in a magnetic field.

Flux density The amount of flux per unit area in a magnetic field.

Free electron A valence electron that has broken away from its parent atom and is free to move from atom to atom within the atomic structure of a material.

Frequency A measure of the rate of change of a periodic function. The electrical unit of frequency is hertz (Hz).

Function generator An electronic test instrument that is capable of producing several types of electrical wave forms, such as sine, triangular, and square waves.

Fuse A protective electrical device that burns open when there is excessive current in a circuit.

Gain The amount by which an electrical signal is increased or amplified.

Generator A general term for any of several types of devices or instruments that are sources for electrical signals.

Germanium A semiconductor material.

Giga A prefix used to designate 10^9 (one thousand million).

Ground In electrical circuits, the common or reference point. It can be chassis ground or earth ground.

Harmonics The frequencies contained in a composite wave form which are integer multiples of the repetition frequency or fundamental frequency of the wave form.

Henry (H) The unit of inductance.

Hertz (Hz) The unit of frequency. One hertz is one cycle per second.

High-pass The characteristics of a certain type of filter whereby higher frequencies are passed and lower frequencies are rejected.

Hypotenuse The longest side of a right triangle.

Impedance The total opposition to current in a reactive circuit. The unit is ohms.

Induced voltage Voltage produced as a result of a changing magnetic field.

Inductance The property of an inductor whereby a change in current causes the inductor to produce an opposing voltage.

Inductor An electrical device having the property of inductance. Also known as a *coil* or a *choke*.

Infinite Having no bounds or limits.

Input The voltage, current, or power applied to an electrical circuit to produce a desired result.

Instantaneous value The value of a variable at a given instant in time.

Insulator A material that does not allow current flow under normal conditions.

Integrated circuit (IC) A type of circuit in which all of the components are constructed on a single, tiny piece of semiconductor material.

Integrator A type of RC or RL circuit that produces an output which approaches the mathematical integral of the input.

Joule (J) The unit of energy.

Kilo A prefix used to designate 10^3 (one thousand).

Kilowatt-hour (kWh) A common unit of energy used mainly by utility companies.

Kirchhoff's laws A set of circuit laws that describe certain voltage and current relationships in a circuit.

Lag A condition of the phase or time relationship of wave forms in which one wave form is behind the other in phase or time.

Lead A wire or cable connection to an electrical or electronic device or instrument. Also, a condition of the phase or time relationship of wave forms in which one wave form is ahead of the other in phase or time.

Leading edge The first step or transition of a pulse.

Linear Characterized by a straight-line relationship.

Load The device upon which work is performed.

Logarithm The exponent to which a base number must be raised to produce a given number. For example, the logarithm of 100 is 2 ($10^2 = 100$).

Loop A closed path in a circuit.

Low-pass The characteristics of a certain type of filter whereby lower frequencies are passed and higher frequencies are rejected.

Magnetic Related to or possessing characteristics of magnetism. Having a north and a south pole with lines of force extending between the two.

Magnetomotive force (mmf) The force produced by a current in a coiled wire in establishing a magnetic field. The unit is ampere-turns.

Magnitude The value of a quantity, such as the number of amperes of current or the number of volts of voltage.

Mega A prefix designating 10^6 (one million).

Mesh An arrangement of loops in a circuit.

Micro A prefix designating 10^{-6} (one-millionth).

Milli A prefix designating 10^{-3} (one-thousandth).

Modulation The process whereby a signal containing information (such as voice) is used to modify the amplitude (AM) or the frequency (FM) of a much higher frequency sine wave (carrier).

Multimeter A measurement instrument used to measure current, voltage, and resistance.

Mutual inductance The inductance between two separate coils, such as in a transformer.

Nano A prefix designating 10^{-9} (one thousand-millionth).

Network A circuit.

Neutron An atomic particle having no electrical charge.

Node A point or junction in a circuit where two or more components connect.

Ohm The unit of resistance.

Ohmmeter An instrument for measuring resistance.

Open circuit A circuit in which there is not a complete current path.

Oscillator An electronic circuit that internally produces a time-varying signal without an external signal input.

Oscilloscope A measurement instrument that displays signal wave forms on a screen.

Output The voltage, current, or power produced by a circuit in response to an input or to a particular set of conditions.

Overshoot A short duration of excessive amplitude occurring on the positive-going transition of a pulse.

Parallel The relationship in electric circuits in which two or more current paths are connected between the same two points.

Peak value The maximum value of an electrical wave form, particularly in relation to sine waves.

Period The time interval of one cycle of a periodic wave form.

Periodic Characterized by a repetition at fixed intervals.

Permeability A measure of the ease with which a magnetic field can be established within a material.

Phase The relative displacement of a time-varying wave form in terms of its occurrence.

Pico A prefix designating 10^{-12} (one-billionth).

Potentiometer A three-terminal variable resistor.

Power The rate of energy consumption. The unit is watts.

Power factor The relationship between volt-amperes and average power or watts. Volt-amperes times the power factor equals average power.

Power supply An electronic instrument that produces voltage, current, and power from the ac power line or batteries in a form suitable for use in various applications to power electronic equipment.

p/s Pulses per second, a unit of frequency measurement for pulse wave forms.

PRF Pulse repetition frequency.

Primary The input winding of a transformer.

Proton A positively charged atomic particle.

Pulse A type of wave form consisting of two equal and opposite steps in voltage or current, separated by a time interval.

Pulse width The time interval between the opposite steps of an ideal pulse. Also, the time between the 50% points on the leading and trailing edges of a nonideal pulse.

Quality factor (Q) The ratio of reactive power to average power in a coil or a resonant circuit.

Radian A unit of angular measurement. There are 2π radians in a complete revolution. One radian equals 57.3°.

Ramp A type of wave form characterized by a linear increase or decrease in voltage or current.

Reactance The opposition of a capacitor or an inductor to sinusoidal current. The unit is ohms.

Reactive power The rate at which energy is stored and alternately returned to the source by a reactive component.

Rectifier An electronic circuit that converts ac into pulsating dc.

Reluctance The opposition to the establishment of a magnetic field in an electromagnetic circuit.

Resistance Opposition to current. The unit is ohms.

Resistivity The resistance that is characteristic of a given material.

Resistor An electrical component possessing resistance.

Resonance In an LC circuit, the condition when the impedance is minimum (series) or maximum (parallel).

Response In electronic circuits, the reaction of a circuit to a given input.

Rheostat A two-terminal, variable resistor.

Right angle A 90° angle.

Ringing An unwanted oscillation on a wave form.

Rise time The time interval required for a pulse to change from 10% of its amplitude to 90% of its amplitude.

Rising edge The positive-going transition of a pulse.

rms Root mean square. The value of a sine wave indicating its heating effect. Also known as *effective value*.

Sawtooth A type of electrical wave form composed of ramps.

Secondary The output winding of a transformer.

Semiconductor A material having a conductance value between that of a conductor and that of an insulator. Silicon and germanium are examples.

Series In an electrical circuit, a relationship of components in which the components are connected such that they provide a single current path between two points.

Short A zero resistance connection between two points.

Siemen (S) The unit of conductance.

Signal A time-varying electrical wave form.

Silicon A semiconductor material used in transistors.

Sine wave A type of alternating electrical wave form.

Slope The vertical change in a line for a given horizontal change.

Source Any device that produces energy.

Steady state An equilibrium condition in a circuit.

Step A voltage or current transition from one level to another.

Susceptance The ability of a reactive component to permit current flow. The reciprocal of reactance.

Switch An electrical or electronic device for opening and closing a current path.

Synchronous Having a fixed time relationship.

Tangent A trigonometric function that is the ratio of the opposite side of a right triangle to the adjacent side.

Tank A parallel resonant circuit.

Tapered Nonlinear, such as a tapered potentiometer.

Temperature coefficient A constant specifying the amount of change in the value of a quantity for a given change in temperature.

Tesla (T) The unit of flux density. Also, webers per square meter.

Tetrode A vacuum tube having four elements.

Thermistor A resistor whose resistance decreases with an increase in temperature.

Tilt A slope on the normally flat portion of a pulse.

Time constant A fixed time interval, set by R, C, and L values, that determines the time response of a circuit.

Tolerance The limits of variation in the value of an electrical component.

Trailing edge The last edge to occur in a pulse.

Transformer An electrical device that operates on the principle of electromagnetic induction. It is used for increasing or decreasing an ac voltage and for various other applications.

Transient A temporary or passing condition in a circuit. A sudden and temporary change in circuit conditions.

Transistor A semiconductor device used for amplification and switching applications in electronic circuits.

Triangular wave A type of electrical wave form consisting of ramps.

Trimmer A small, variable resistor or capacitor.

Trouble-shooting The process and technique of identifying and locating faults in an electrical or electronic circuit.

Turns ratio The ratio of the number of secondary turns to the number of primary turns in the transformer windings.

Undershoot The opposite of overshoot, occurring on the negative-going edge of a pulse.

Valence Related to the outer shell or orbit or an atom.

Volt The unit of voltage or electromotive force (emf).

Voltage The amount of energy available to move a certain amount of electrons from one point to another in an electrical circuit.

Watt The unit of power.

Wave form The pattern of variations of a voltage or a current.

Wavelength The length in space occupied by one cycle of an electromagnetic wave.

Weber The unit of magnetic flux.

Winding The loops of wire or coil in an inductor or transformer.

Wiper The variable contact in a potentiometer or other device.

ANSWERS TO SELECTED ODD-NUMBERED PROBLEMS

ANSWERS TO SELECTED ODD-NUMBERED PROBLEMS

Chapter 1

1–1. (a) 3×10^3 (b) 75×10^3 (c) 2×10^6

1–3. (a) $8.4 \times 10^3 = 0.84 \times 10^4 = 0.084 \times 10^5$
(b) $99 \times 10^3 = 9.9 \times 10^4 = 0.99 \times 10^5$
(c) $200 \times 10^3 = 20 \times 10^4 = 2 \times 10^5$

1–5. (a) 0.0000025 (b) 5000 (c) 0.39

1–7. (a) 126×10^6 (b) 5.00085×10^3
(c) 6060×10^{-9}

1–9. (a) 20×10^8 (b) 36×10^{14}
(c) 15.4×10^{-15}

1–11. (a) 31 mA (b) 5.5 kV (c) 200 pF

Chapter 2

2–1. 0.2 A

2–3. 0.15 C

2–5. 30 J

2–7. (a) 10 V (b) 2.5 V (c) 4 V
(d) 0.2 V

2–9. (a) 10 Ω (b) 2 Ω (c) 50 Ω
(d) 0.333 Ω

2–11. (a) $R_{max} = 34.65$ Ω, $R_{min} = 31.35$ Ω
(b) $R_{max} = 51,700$ Ω, $R_{min} = 42,300$ Ω

2–13. 2250 kV

2–15. 15.6 Ω

Chapter 3

3–1. (a) 5 A (b) 1.5 A (c) 0.5 A
(d) 2 mA (e) 50 μA

3–3. (a) 36 V (b) 250 V (c) 1500 V
(d) 28.2 V (e) 50 V

3–5. (a) 5 Ω (b) 2 Ω (c) 10 Ω
(d) 0.55 Ω (e) 300 Ω

3–7. 1.2 A

3–9. 3 kΩ

3–13. (a) 7.5 mA (b) 10 mA (c) 15 mA
(d) 20 mA (e) 50 mA

3–15. A: 15.15 mA
B: 19.23 mA
C: 21.28 mA

Chapter 4

4–1. 350 W

4–3. 20,000 W

4–5. (a) 1 MW (b) 3 MW (c) 150 MW
(d) 8.7 MW

4–7. (a) 2,000,000 μW (b) 500 μW
(c) 250 μW (d) 6.67 μW

4–9. 37.5 Ω

4–11. 345 W

4–13. 100 μW

4–15. 0.045 W

4–17. 0.68 W. Use 1-W resistor.

4–19. 8640 J

4–21. 216 kWh

4–23. 83.33 kWh

4–25. 100 W

4–27. 36 Ah

4–29. 13.54 mA

4–31. 4.25 W

Chapter 5

5–3. 5 mA

5–5. (a) 1560 Ω (b) 103 Ω (c) 13.7 kΩ
(d) 3.971 MΩ

5–7. 6 kΩ

5–9. 1144 Ω

5–11. A: 0.6875 mA
B: 4 μA

5–13. 500 Ω

5–15. 9 V

5–17. 26 V

5–19. 20 Ω

5–21. (a) 4 V (b) 6.67 V

5–23. $V_{AF} = 100$ V, $V_{BF} = 80$ V, $V_{CF} = 70$ V, $V_{DF} = 20$ V, $V_{EF} = 5$ V

5–25. 250 mW

Chapter 6

6–3. 100 V

6–5. (a) 358.97 Ω (b) 25.55 Ω
(c) 818.86 Ω (d) 996.48 Ω

6–7. 300 Ω

6–9. 2.5 Ω

6–11. 10 A

CHAPTER 7

6–13. 30 mA

6–15. 1350 mA

6–17. $R_2 = 25\ \Omega$, $R_3 = 100\ \Omega$, $R_4 = 12.5\ \Omega$

6–19. A: $I_1 = 6.67\ \mu A$, $I_2 = 3.33\ \mu A$
B: $I_1 = 5$ mA, $I_2 = 2.5$ mA,
$I_3 = 1.67$ mA, $I_4 = 0.833$ mA

6–21. 200 mW

6–23. $I_{each} = 0.682$ A, $I_T = 4.092$ A

6–25. Yes, the 500-Ω one.

Chapter 7

7–3. 2003 Ω

7–5. A: R_3 and R_4 are in parallel, and this combination is in series with R_2. This combination is in parallel with R_1.
B: All resistors are in parallel.
C: R_1 and R_2 are in series with the parallel combination of R_3 and R_4, which is in parallel with the series combination of R_5, R_8, and R_6 and R_7 in parallel.

7–7. A: 687.5 Ω
B: 400 kΩ
C: 3.2 kΩ

7–9. $V_A = 3.85$ V, $V_B = 42.31$ V, $V_C = 50$ V, $V_D = 4.55$ V

7–11. No, it should be 4.39 V.

7–13. The 2-kΩ resistor is open.

7–15. 7.5 V unloaded; 7.32 V loaded.

7–17. 50-kΩ load.

7–19. $R_1 = 1000\ \Omega$; $R_2 = R_3 = 500\ \Omega$;
$V_{5V} = 3.12$ V(loaded);
$V_{2.5V} = 1.25$ V (loaded)

7–21. $R_T = 6$ kΩ, $V_A = 3$ V, $V_B = 1.5$ V, $V_C = 0.75$ V

7–23. $V_1 = 1.67$ V, $V_2 = 6.68$ V, $V_3 = 1.67$ V, $V_4 = 3.34$ V, $V_5 = 0.4175$ V, $V_6 = 2.505$ V, $V_7 = 0.4175$ V, $V_8 = 1.67$ V, $V_9 = 1.67$ V

7–25. 1800 Ω

Chapter 8

8–1. $I_S = 6$ A, $R_S = 50\ \Omega$

8–3. $V_S = 720$ V, $R_S = 1.2$ kΩ

8–5. 0.905 mA

8–7. 32.8 mA

8–9. A: $R_{TH} = 75\ \Omega$, $V_{TH} = 8.33$ V
B: $R_{TH} = 184.2\ \Omega$, $V_{TH} = 2.21$ V
C: $R_{TH} = 33.33$ kΩ, $V_{TH} = 1.67$ V
D: $R_{TH} = 1.2$ kΩ, $V_{TH} = 80$ V

8–11. 7.93 V

8–13. 0.1 mA

8–15. $R_{eq} = 0.545\ \Omega$, $V_{eq} = -2.45$ V

8–17. A: 12 Ω
B: 8 kΩ
C: 5.67 Ω
D: 650 Ω

8–19. A: $R_1 = 167$ kΩ, $R_2 = 500$ kΩ, $R_3 = 250$ kΩ
B: $R_1 = 0.364\ \Omega$, $R_2 = 0.909\ \Omega$, $R_3 = 0.455\ \Omega$

Chapter 9

9–3. $I_1 - I_2 - I_3 = 0$

9–5. $V_1 = 5.676$ V, $V_2 = 6.365$ V, $V_3 = 0.365$ V

9–7. $I_1 = 0.738$ A, $I_2 = -0.527$ A, $I_3 = -0.469$ A

9–9. $I_1 = 0$ A, $I_2 = 2$ A

9–11. $I_{1000\Omega} = -5.18$ mA left,
$I_{800\Omega} = 3.53$ mA left,
$I_{500\Omega} = -1.65$ mA up

9–13. $V_A = 29.1$ V

9–15. $60I_1 - 10I_2 = -1.5$ V
$-10I_1 + 40I_2 - 5I_3 = -3$ V
$-5I_2 + 20I_3 = -1.5$ V

9–17. $I_1 = -6.645$ mA, $I_2 = -0.0892$ A, $I_3 = -0.1006$ A

Chapter 10

10–1. (a) 1 Hz (b) 5 Hz (c) 20 Hz
(d) 1000 Hz (e) 2 kHz (f) 100 kHz

10–3. 2 μs

10–5. (a) 8.484 V (b) 24 V (c) 7.644 V

10–7. (a) 7.07 mA (b) 6.37 mA (c) 10 mA
(d) 20 mA (e) 10 mA

10–9. 15°; A leads B.

10–11. (a) 7.32 V (b) 15.40 V (c) 21.66 V (d) 26.57 V (e) 26.57 V (f) 16.22 V (g) −26.57 V (h) −16.22 V

10–13. (a) 3.83 V (b) 7.07 V (c) 10 V (d) 7.07 V (e) 0 V (f) −10 V (g) 0 V

10–15. 30°: 0 V. 45°: 3.88 V. 90°: 12.99 V. 180°: 7.5 V. 200°: 2.60 V. 300°: −15 V

10–19. t_r = 3.5 ms, t_f = 3.5 ms, PW = 12.5 ms, A = 5 V

10–21. (a) 16.7 ms (b) 1 ms (c) 0.4 ms (d) 0.5 μs

10–23. A: −0.375 V B: 3.01 V

10–25. f_1 = 25 kHz, f_3 = 75 kHz, f_5 = 125 kHz, f_7 = 175 kHz, f_9 = 225 kHz, f_{11} = 275 kHz, f_{13} = 325 kHz

Chapter 11

11–1. 27.8 μC

11–3. C_1 = 25 μF, C_2 = 16.67 μF

11–5. (a) 0.001 μF (b) 0.0035 μF (c) 0.00025 μF

11–7. 221.25 pF

11–9. 990 pF

11–11. A: 0.02 μF B: 0.047 μF C: 0.001 μF D: 220 pF

11–13. (a) 0.67 μF (b) 68.97 pF (c) 2.7 μF

11–15. 2 μF

11–17. A: 2.5 μF B: 716.66 pF C: 1.6 μF

11–19. A: 3.386 kΩ B: 6.366 kΩ C: 5.305 kΩ

11–21. A: 33.86 kHz, 3.386 kHz B: 63.66 Hz, 6.366 Hz C: 3.183 kHz, 318.3 Hz

11–23. 3.386 V

11–25. 694 μF

Chapter 12

12–1. A: 269.26 Ω B: 1166.19 Ω

12–3. A: 37.14 mA B: 4.29 mA

12–5. (a) 797.35 kΩ (b) 166.83 kΩ (c) 93.98 kΩ (d) 59.27 kΩ

12–7. A: 21.8°, V lagging I B: 59°, V lagging I

12–9. (a) 86.4°, V lagging I (b) 72.56°, V lagging I (c) 57.86°, V lagging I (d) 32.48°, V lagging I

12–11. 27.95°, V lagging I

12–13. 27.95°, V lagging I

12–15. V_R = 2.13 V, V_C = 4.52 V

12–17. (a) 0.06° (b) 5.74° (c) 45.15° (d) 84.32°

12–19. (a) 89.96° (b) 86.4° (c) 57.86° (d) 9.04°

12–23. 46.7°, output leading

12–25. P_{avg} = 0.6272 W, P_r = 0.6655 VAR, P_a = 0.9145 VA

12–29. 9.7 mA

12–31. (a) 351.5 Ω (b) 0.024 A (c) 0.045 A (d) 0.051 A (e) 62.05°

12–33. (a) 0.045 A (b) 73.74° (c) 0.188 W (d) 0.28

Chapter 13

13–1. 4%

13–3. (a) 100 μs (b) 23.5 μs (c) 0.5 ms (d) 15 ms

13–5. (a) 9.45 V (b) 12.9 V (c) 14.25 V (d) 14.7 V (e) 14.85 V

13–7. (a) 2.72 V (b) 5.91 V (c) 11.66 V

13–9. 13.86 μs

13–11. 7.62 μs

13–15. 15 V constant dc

13–19. The output has the same shape as the input does, but its average value is 0 V.

Chapter 14

14–1. 10 μWb

14–3. 400 turns

14–5. (a) 1000 mH (b) 0.25 mH (c) 0.01 mH (d) 0.5 mH

14–7. 0.05 V

Chapter 15

14–9. 89 turns
14–11. 18 mH
14–13. 24 mH
14–15. A: 57.33 mH
B: 4 mH
C: 5.33 mH
14–17. A: 144 Ω
B: 10.1 Ω
C: 13.4 Ω
14–19. 50.26 kHz
14–21. 0.125 μJ

Chapter 15

15–1. A: 111.8 Ω
B: 1.8 kΩ
15–3. A: 89.45 mA
B: 2.78 mA
15–5. (a) 17.38 Ω (b) 63.97 Ω (c) 126.23 Ω
(d) 251.62 Ω
15–7. A: 57.52°
B: 37.72°
15–9. 37.02°
15–13. (a) V_R = 4.85 V, V_L = 1.22 V
(b) V_R = 3.83 V, V_L = 3.21 V
(c) V_R = 2.15 V, V_L = 4.51 V
(d) V_R = 1.16 V, V_L = 4.86 V
15–19. 26.69°. Output lags.
15–21. P_{avg} = 0.266 W, P_r = 0.134 VAR, P_a = 0.298 VA
15–23. (a) 0.89 (b) 0.37
15–25. 5.37 mA
15–27. (a) 7.73 Ω (b) 1.25 A (c) 1.49 A
(d) 1.95 A (e) 49.91°

Chapter 16

16–1. (a) 1 μs (b) 2.13 μs (c) 2 μs
16–3. (a) 5.55 V (b) 2.1 V (c) 0.75 V
(d) 0.3 V (e) 0.15 V
16–5. (a) 12.28 V (b) 9.1 V (c) 3.35 V
16–7. 10.99 μs
16–9. (a) 5 μs

Chapter 17

17–1. 1.5 μH
17–3. 50
17–5. 3
17–7. I_s = 33.33 mA, R_L = 1.8 kΩ
17–9. 0.75 W
17–11. 9.77 W
17–13. The secondaries from top to bottom: 12 V, 24 V, 60 V, 120 V.
17–15. 25 kVA

Chapter 18

18–1. 18.03
18–5. (a) 56.57 ∠ −45°
(b) 206.16 ∠ −75.96°
(c) 40.3 ∠ −29.74°
(d) 107.84 ∠24.66°
18–9. I_R = 0.147 ∠0° A, I_C = 0.111 ∠90° A,
I_T = 0.184 ∠37.1° A,
$V_R = V_C = V_s$ = 10 ∠0° V
18–13. A: 72.1 ∠3.18° Ω
B: 5.33 ∠23.2° Ω
18–15. V_{R1} = 8.08 ∠19.14° V,
V_{C1} = 5.11 ∠ −31.21° V,
V_{R2} = 3.99 ∠7.31° V,
V_{C2} = 3.18 ∠ −82.69° V
18–17. I_2 = 1.12 ∠42.87° A
18–19. 67.53 Ω − j29.22 Ω. A 67.53-Ω resistor in series with a capacitive reactance of 29.22 Ω at 60 Hz.

Chapter 19

19–1. $X_C = X_L$ = 4.472 kΩ,
Z_T = 20 Ω, I = 0.6 A
19–3. Z = 200 Ω, $X_C = X_L$ = 2 kΩ
19–5. f_r = 459.44 kHz, f_1 = 359.995 kHz,
f_2 = 558.885 kHz
19–7. At f_1: 48.72°, current leading
At f_2: 42.3°, current lagging
At f_r: 0°, V and I in phase
19–9. f_r = 103.82 kHz, Z_T = 531.9 MΩ
19–11. 62.5 Hz
19–13. L = 1.99 mH, C = 0.199 μF
19–15. BW = 400 Hz, f_r = 2600 Hz, Q = 6.5

Chapter 20

20–1. 2.43 V rms

20–3. (a) Passed (b) Passed
  (c) 2 kHz passed (d) 3 kHz passed
  (e) 5 kHz rejected

20–5. A: 9.36 V
  B: 5.32 V
  C: 9.95 V
  D: 0.995 V

20–7. A: Low-pass
  B: High-pass
  C: Low-pass
  D: High-pass

20–9. A: 1.13 kHz
  B: 2.4 kHz

20–11. A: 50.33 Hz
  B: 20.13 MHz

20–13. (a) 10.6 μF (b) 1.27 μF
  (c) 0.637 μF (d) 0.127 μF

20–15. (a) 0 dB (b) −4.44 dB
  (c) −3.01 dB (d) −13.98 dB

20–17. (a) −20 dB (b) −40 dB
  (c) −60 dB

Chapter 21

21–1. (a) 50 μA (b) 950 μA (c) 9.95 mA
  (d) 99.95 mA (e) 999.95 mA

21–3. (a) 1 kΩ (b) 52.63 Ω (c) 5.025 Ω
  (d) 0.5 Ω (e) 0.05 Ω

21–5. 50-V range; 0.33%

21–7. 60 kΩ

21–9. (a) In series with R_1 on 10-mA range
  (b) Across R_1 on 10-V range
  (c) Across R_1 (removed) on $R \times 100$ range

21–11. $V_p = 1.5$ V, $V_{rms} = 1.061$ V, and $f = 25$ kHz

SOLUTIONS TO SELF-TESTS

Chapter 1

1. Current: amperes. Voltage: volts. Resistance: ohms. Power: watts. Energy: joules.

2. Amperes: A. Volts: V. Ohms: Ω. Watts: W. Joules: J.

3. Current: I. Voltage: V. Resistance: R. Power: P. Energy: $\mathscr{E}$.

4. (a) $100 = 1 \times 10^2$
   (b) $12,000 = 1.2 \times 10^4$
   (c) $5,600,000 = 5.6 \times 10^6$
   (d) $78,000,000 = 7.8 \times 10^7$

5. (a) $0.03 = 3 \times 10^{-2}$
   (b) $0.0005 = 5 \times 10^{-4}$
   (c) $0.00058 = 5.8 \times 10^{-4}$
   (d) $0.0000224 = 2.24 \times 10^{-5}$

6. (a) $7 \times 10^4 = 70,000$
   (b) $45 \times 10^3 = 45,000$
   (c) $100 \times 10^{-3} = 0.1$
   (d) $4 \times 10^{-1} = 0.4$

7. Convert 12×10^5 to 1.2×10^6.
   $1.2 \times 10^6 + 25 \times 10^6 = 26.2 \times 10^6$

8. $8 \times 10^{-3} - 5 \times 10^{-3} =$
   $(8 - 5) \times 10^{-3} = 3 \times 10^{-3}$

9. $(33 \times 10^3)(20 \times 10^{-4}) =$
   $(33 \times 20) \times (10^{3+(-4)}) =$
   $660 \times 10^{-1} = 66$

10. $(4 \times 10^2)/(2 \times 10^{-3}) =$
    $(4/2) \times 10^{2-(-3)} = 2 \times 10^5$

Chapter 2

1. 3
2. 6
3. $I = Q/t = 50 \text{ C}/5 \text{ s} = 10 \text{ A}$
4. $Q = It = (2 \text{ A})(10 \text{ s}) = 20 \text{ C}$
5. $V = \mathscr{E}/Q = 500 \text{ J}/100 \text{ C} = 5 \text{ V}$
6. $G = 1/R = 1/10 \Omega = 0.1 \text{ S}$
7. Blue: 6, 1st digit
   Green: 5, 2nd digit
   Red: 2, number of zeros
   Gold: 5%, tolerance
   $R = 6500 \Omega \pm 5\%$
8. 500 Ω. Center position is one-half the resistance.

9. Breakdown strength of mica = 2000 kV/cm.
   $V_{max} = (2000 \text{ kV/cm})(10 \text{ cm}) = 20,000 \text{ kV}$

10. $d = 0.003$ in. $= 3$ mils
    $A = d^2 = 9$ CM
    $R = \rho l/A = (10.4 \text{ CM-}\Omega/\text{ft})(10 \text{ ft})/9 \text{ CM}$
    $= 11.56 \Omega$

Chapter 3

1. $V = IR$, $I = V/R$, $R = V/I$

2. To find current, use $I = V/R$.
   To find voltage, use $V = IR$.
   To find resistance, use $R = V/I$.

3. $I = V/R = 10 \text{ V}/5 \Omega = 2 \text{ A}$
4. $I = V/R = 75 \text{ V}/10 \text{ k}\Omega = 7.5 \text{ mA}$
5. $V = IR = (2.5 \text{ A})(20 \Omega) = 50 \text{ V}$
6. $V = IR = (30 \text{ mA})(2.2 \text{ k}\Omega) = 66 \text{ V}$
7. $R = V/I = 12 \text{ V}/6 \text{ A} = 2 \Omega$
8. $R = V/I = 9 \text{ V}/100 \text{ }\mu\text{A} = 0.09 \text{ M}\Omega$
   $= 90 \text{ k}\Omega$
9. $I = V/R = 50 \text{ kV}/2.5 \text{ k}\Omega = 20 \text{ A}$
10. $I = V/R = 15 \text{ mV}/10 \text{ k}\Omega = 1.5 \text{ }\mu\text{A}$
11. $V = IR = (4 \text{ }\mu\text{A})(100 \text{ k}\Omega) = 400 \text{ mV}$
12. $V = IR = (100 \text{ mA})(1 \text{ M}\Omega) = 100 \text{ kV}$
13. $R = V/I = 5 \text{ V}/2 \text{ mA} = 2.5 \text{ k}\Omega$
14. $I = V/R = 1.5 \text{ V}/5 \text{ k}\Omega = 0.3 \text{ mA}$
15. A: $I = V/R = 25 \text{ V}/10 \text{ k}\Omega = 2.5 \text{ mA}$
    B: $I = V/R = 5 \text{ V}/2 \text{ M}\Omega = 2.5 \text{ }\mu\text{A}$
    C: $I = V/R = 1.5 \text{ kV}/15 \text{ k}\Omega = 0.1 \text{ A}$
16. A: $V = IR = (3 \text{ mA})(27 \text{ k}\Omega) = 81 \text{ V}$
    B: $V = IR = (5 \text{ }\mu\text{A})(100 \text{ M}\Omega) = 500 \text{ V}$
    C: $V = IR = (2.5 \text{ A})(50 \Omega) = 125 \text{ V}$
17. A: $R = V/I = 8 \text{ V}/2 \text{ A} = 4 \Omega$
    B: $R = V/I = 12 \text{ V}/4 \text{ mA} = 3 \text{ k}\Omega$
    C: $R = V/I = 30 \text{ V}/150 \text{ }\mu\text{A} = 0.2 \text{ M}\Omega$
18. $V_2/30 \text{ mA} = 10 \text{ V}/50 \text{ mA}$
    $V_2 = (10 \text{ V})(30 \text{ mA})/50 \text{ mA}$
    $= 6 \text{ V}$ (new value)
    It decreased 4 V, from 10 V.
19. The current increase is 50%; so the voltage increase must be the same, that is,
    $(0.5)(20 \text{ V}) = 10 \text{ V}$.
    $V_2 = 20 \text{ V} + (0.5)(20 \text{ V})$
    $= 30 \text{ V}$ (new value)

CHAPTER 4

20. $R = 100\text{ V}/750\text{ mA} = 0.133\text{ k}\Omega = 133\text{ }\Omega$
$R = 100\text{ V}/1\text{ A} = 100\text{ }\Omega$

Chapter 4

1. $P = \mathscr{E}/t = 200\text{ J}/10\text{ s} = 20\text{ W}$
2. $P = \mathscr{E}/t = 10{,}000\text{ J}/300\text{ ms} = 33.33\text{ kW}$
3. $50\text{ kW} = 50 \times 10^3\text{ W} = 50{,}000\text{ W}$
4. $0.045\text{ W} = 45 \times 10^{-3}\text{ W} = 45\text{ mW}$
5. $P_R = VI = (10\text{ V})(350\text{ mA})$
   $= 3500\text{ mW} = 3.5\text{ W}$
   $P_{batt} = 3.5\text{ W}$
6. $P = V^2/R = (50\text{ V})^2/1000\text{ }\Omega = 2.5\text{ W}$
7. $P = I^2R = (0.5\text{ A})^2(5000\text{ }\Omega) = 1250\text{ W}$
8. $P = VI = (115\text{ V})(2\text{ A}) = 230\text{ W}$
9. $\mathscr{E} = Pt = (15\text{ W})(60\text{ s}) = 900\text{ J}$
10. $P = 500\text{ W} = 0.5\text{ kW}$
    $t = 24\text{ h}$
    $\mathscr{E} = Pt = (0.5\text{ kW})(24\text{ h}) = 12\text{ kWh}$
11. $(75\text{ W})(10\text{ h}) = 750\text{ Wh}$
12. There are 30 days in April.
    $(30\text{ days})(24\text{ h/day}) = 720\text{ h}$
    $750\text{ W} = 0.75\text{ kW}$
    $(0.75\text{ kW})(720\text{ h}) = 540\text{ kWh}$
    Meter reading $= 1000\text{ kWh} + 540\text{ kWh}$
    $= 1540\text{ kWh}$
13. $\dfrac{1{,}500{,}000\text{ Ws}}{(3600\text{ s/h})(1000\text{ W/kW})} = 0.4167\text{ kWh}$
14. See Figure S-1.

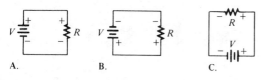

A.   B.   C.

FIGURE S-1

15. Power supply 2 has a greater load because it must supply more current than power supply 1. Number 2 has a smaller resistive load because it supplies more current for the same voltage.
16. 1: $P = VI = (25\text{ V})(0.1\text{ A}) = 2.5\text{ W}$
    2: $P = VI = (25\text{ V})(0.5\text{ A}) = 12.5\text{ W}$
17. $I_L = 12\text{ V}/600\text{ }\Omega = 0.02\text{ A}$
    $50\text{ Ah}/0.02\text{ A} = 2500\text{ h}$
18. $(8\text{ A})(2.5\text{ h}) = 20\text{ Ah}$
19. $P_{out} = P_{in} - P_{lost} = 2\text{ W} - 0.25\text{ W}$
    $= 1.75\text{ W}$
20. Efficiency $= (P_{out}/P_{in}) \times 100$
    $= (0.5\text{ W}/0.6\text{ W}) \times 100$
    $= 83.33\%$

Chapter 5

1. See Figure S-2.

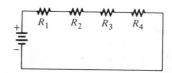

FIGURE S-2

2. Since current is the same at *all* points in a series circuit, 5 A flow out of the sixth and tenth resistors as well as all of the others.
3. See Figure S-3.

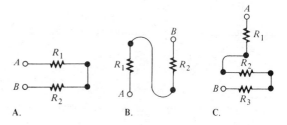

A.   B.   C.

FIGURE S-3

4. $I_{AB} = I_{BC} = I_{CD} = I_{DE} = I_{EF} = I_{FG} = I_{GH}$
   $= I_{HA} = 3\text{ A}$
5. $R_T = 75\text{ }\Omega + 470\text{ }\Omega = 545\text{ }\Omega$
6. $R_T = (8)(56\text{ }\Omega) = 448\text{ }\Omega$
7. $R_{unk} =$
   $20\text{ k}\Omega - (1\text{ k}\Omega + 1\text{ k}\Omega + 5\text{ k}\Omega + 3\text{ k}\Omega)$
   $= 10\text{ k}\Omega$
8. $R_T = (3)(1\text{ k}\Omega) = 3\text{ k}\Omega$
   $I = 5\text{ V}/3\text{ k}\Omega = 1.67\text{ mA}$
9. A: $R_T = 100\text{ }\Omega + 47\text{ }\Omega + 68\text{ }\Omega = 215\text{ }\Omega$
   $I = 5\text{ V}/215\text{ }\Omega = 0.023\text{ A}$
   B: $R_T = 50\text{ }\Omega + 175\text{ }\Omega + 33\text{ }\Omega = 258\text{ }\Omega$
   $I = 8\text{ V}/258\text{ }\Omega = 0.031\text{ A}$
   Circuit B has more current.

10. $R_T = V/I = 12\text{ V}/2\text{ mA} = 6\text{ k}\Omega$
    $R_{each} = 6\text{ k}\Omega/6 = 1\text{ k}\Omega$
11. $V_T = 8\text{ V} + 5\text{ V} + 1.5\text{ V} = 14.5\text{ V}$
12. $V_T = 9\text{ V} - 9\text{ V} = 0\text{ V}$
13. A: $V_T = 50\text{ V} - 30\text{ V} - 10\text{ V} = 10\text{ V}$
    See Figure S–4.

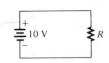

FIGURE S–4

B: $V_T =$
    $1.5\text{ V} + 3\text{ V} + 4.5\text{ V} - 9\text{ V} - 1.5\text{ V}$
    $= -1.5\text{ V}$
    See Figure S–5.

FIGURE S–5

14. Since the resistors are all of equal value, they each drop 2 V.
    $V_T = (5)(2\text{ V}) = 10\text{ V}$
15. $V_S = V_1 + V_2 + V_3$
    $V_3 = V_S - V_1 - V_2$
    $= 50\text{ V} - 10\text{ V} - 15\text{ V} = 25\text{ V}$
16. $R_T = V/I = 30\text{ V}/10\text{ mA} = 3\text{ k}\Omega$
    $R_T = R_1 + R_2 + R_3 + R_4 + R_5$
    $R_3 = R_T - (R_1 + R_2 + R_4 + R_5)$
    $= 3\text{ k}\Omega - 2\text{ k}\Omega = 1\text{ k}\Omega$
17. $V_x = (R_x/R_T)(V_T)$
    $V_{50\Omega} = (50\text{ }\Omega/200\text{ }\Omega)(20\text{ V}) = 5\text{ V}$
18. $V_x = (R_x/R_T)(V_T)$
    $V_1 = (30\text{ }\Omega/100\text{ }\Omega)(90\text{ V}) = 27\text{ V}$
    $V_2 = (20\text{ }\Omega/100\text{ }\Omega)(90\text{ V}) = 18\text{ V}$
    $V_3 = (50\text{ }\Omega/100\text{ }\Omega)(90\text{ V}) = 45\text{ V}$
19. $V_x = (R_x/R_T)(V_T)$
    $3\text{ V} = (R_x/100\text{ k}\Omega)(9\text{ V})$
    $R_{x1} = (3\text{ V})(100\text{ k}\Omega)/(9\text{ V}) = 33.33\text{ k}\Omega$
    $R_{x2} = R_T - R_{x1} = 100\text{ k}\Omega - 33.33\text{ k}\Omega$
    $= 66.67\text{ k}\Omega$
20. $P_T = P_1 + P_2 + P_3$
    $= 2.5\text{ W} + 5\text{ W} + 1.2\text{ W} = 8.7\text{ W}$

Chapter 6

1. See Figure S–6.

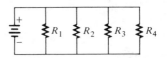

FIGURE S–6

2. $I_{br} = I_T/2 = 5\text{ A}/2 = 2.5\text{ A}$
3. See Figure S–7.

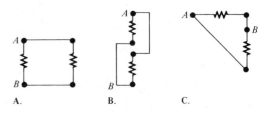

FIGURE S–7

4. $V_{AB} = V_{AC} = V_{EC} = V_{DC} = -5\text{ V}$
   (because all of the resistors are in parallel)
5. $R_T = \dfrac{(80\text{ }\Omega)(150\text{ }\Omega)}{80\text{ }\Omega + 150\text{ }\Omega} = 52.17\text{ }\Omega$
6. $1/R_T = (1/1000\text{ }\Omega) + (1/800\text{ }\Omega) + (1/500\text{ }\Omega)$
   $+ (1/200\text{ }\Omega) + (1/100\text{ }\Omega) = 0.01925\text{ S}$
   $R_T = 1/0.01925\text{ S} = 51.95\text{ }\Omega$
7. $R_T = R/n = 56\text{ }\Omega/8 = 7\text{ }\Omega$
8. See Figure S–8.

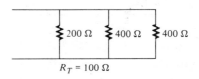

FIGURE S–8

9. $R_T = R/n = 600\text{ }\Omega/3 = 200\text{ }\Omega$
   $I_T = V/R_T = 5\text{ V}/200\text{ }\Omega = 0.025\text{ A}$
   $= 25\text{ mA}$
10. A: $R_T = (50\text{ }\Omega)(25\text{ }\Omega)/75\text{ }\Omega = 16.67\text{ }\Omega$
       $I_T = 10\text{ V}/16.67\text{ }\Omega = 0.6\text{ A} = 600\text{ mA}$
    B: $R_T = (15\text{ }\Omega)(100\text{ }\Omega)/115\text{ }\Omega = 13.04\text{ }\Omega$
       $I_T = 8\text{ V}/13.04\text{ }\Omega = 0.613\text{ A}$
       $= 613\text{ mA}$
    Circuit B has the greater total current.

CHAPTER 7

11. $R_T = V/I_T = 12$ V/3 mA = 4 kΩ
    $R_T = R/n$
    4 kΩ = $R_{each}/6$
    $R_{each} = (6)(4$ k$\Omega) = 24$ kΩ
12. $I_T = 1$ A $- 2$ A $+ 4$ A = 3 A
13. $I_T = 1$ A $- 1$ A = 0 A
14. $I_T = (5)(25$ mA$) = 125$ mA
15. $I_T = I_1 + I_2 + I_3$
    $I_3 = I_T - (I_1 + I_2)$
    $= 0.5$ A $- (0.1$ A $+ 0.2$ A$)$
    $= 0.5$ A $- 0.3$ A $= 0.2$ A
16. $V_S = V_1 = V_2 = V_3$
    $V_S = V_1 = I_1 R_1 = (1$ A$)(10$ $\Omega) = 10$ V
    $R_2 = V_S/I_2 = 10$ V/0.5 A $= 20$ Ω
    $I_3 = I_T - (I_1 + I_2) = 5$ A $- 1.5$ A
    $= 3.5$ A
    $R_3 = V_S/I_3 = 10$ V/3.5 A $= 2.86$ Ω
17. $1/R_T = (1/2000$ $\Omega) + (1/6000$ $\Omega)$
    $+ (1/3000$ $\Omega) + (1/1000$ $\Omega) = 0.002$ S
    $R_T = 1/0.002$ S $= 500$ Ω
    $I_x = (R_T/R_x)(I_T)$
    $I_{2k\Omega} = (500$ $\Omega/2000$ $\Omega)(1$ A$) = 0.25$ A
    $I_{6k\Omega} = (500$ $\Omega/6000$ $\Omega)(1$ A$) = 0.083$ A
    $I_{3k\Omega} = (500$ $\Omega/3000$ $\Omega)(1$ A$) = 0.167$ A
    $I_{1k\Omega} = (500$ $\Omega/1000$ $\Omega)(1$ A$) = 0.5$ A
18. $I_1 = \left(\dfrac{R_2}{R_1 + R_2}\right)I_T = \left(\dfrac{20\text{ k}\Omega}{25\text{ k}\Omega}\right)750$ mA
    $= 600$ mA
    $I_2 = \left(\dfrac{R_1}{R_1 + R_2}\right)I_T = \left(\dfrac{5\text{ k}\Omega}{25\text{ k}\Omega}\right)750$ mA
    $= 150$ mA
19. $P_1 = I_1^2 R_1 = (0.6$ A$)^2(5$ k$\Omega) = 1800$ W
    $P_2 = I_2^2 R_2 = (0.15$ A$)^2(20$ k$\Omega) = 450$ W
    $P_T = P_1 + P_2 = 1800$ W $+ 450$ W
    $= 2250$ W
20. $I_{200\Omega} = 10$ V/200 $\Omega = 0.05$ A $= 50$ mA
    $I_{500\Omega} = 10$ V/500 $\Omega = 0.02$ A $= 20$ mA
    The total current should be 70 mA, but it is equal to the current in the 500-Ω branch only. This finding indicates that the 200-Ω branch is *open*.
21. $R_x = \dfrac{R_A R_T}{R_A - R_T} = \dfrac{(100\text{ }\Omega)(40\text{ }\Omega)}{100\text{ }\Omega - 40\text{ }\Omega}$
    $= 66.67$ Ω

Chapter 7

1. R_2, R_3, and R_4 are in parallel, and this parallel combination is in series with both R_1 and R_5.
2. (a) $R_T = R_1 + R_5 + (R_2/3)$;
    $R_2 = R_3 = R_4 = 30$ Ω.
    $R_T = 10$ $\Omega + 10$ $\Omega + 10$ $\Omega = 30$ Ω
   (b) $I_T = V_S/R_T = 30$ V/30 $\Omega = 1$ A
   (c) $I_3 = I_T/3 = 1$ A/3 $= 0.333$ A
   (d) $V_4 = I_4 R_4$; $I_4 = I_3 = 0.333$ A
    $V_4 = (0.333$ A$)(30$ $\Omega) = 10$ V
3. (a) $R_1 \| R_2 = 2$ k$\Omega/2 = 1$ kΩ
    $R_3 \| R_4 = (300$ $\Omega)(600$ $\Omega)/900$ $\Omega = 200$ Ω
    $R_5 \| R_6 = 1.6$ k$\Omega/2 = 800$ Ω
    $R_1 \| R_2 + R_3 \| R_4 + R_5 \| R_6 =$
    1000 $\Omega + 200$ $\Omega + 800$ $\Omega = 2$ kΩ
   (b) $I_T = 10$ V/2 k$\Omega = 5$ mA
   (c) $I_1 = I_T/2 = 5$ mA/2 $= 2.5$ mA
    (because I_T splits equally between R_1 and R_2)
   (d) $I_6 = I_1 = 2.5$ mA
    $V_6 = (2.5$ mA$)(1.6$ k$\Omega) = 4$ V
4. (a) $R_4 \| R_5 = 5$ k$\Omega/2 = 2.5$ kΩ
    $R_4 \| R_5 + R_3 = 2.5$ k$\Omega + 3.5$ kΩ
    $= 6$ kΩ
    6 k$\Omega \| R_2 = (6$ k$\Omega)(3$ k$\Omega)/9$ k$\Omega = 2$ kΩ
    $R_T = 2$ k$\Omega \| R_1 = (2$ k$\Omega)(10$ k$\Omega)/12$ kΩ
    $= 1.67$ kΩ
   (b) $I_T = V_S/R_T = 6$ V/1.67 k$\Omega = 3.59$ mA
   (c) The resistance to the right of AB is 2 kΩ. The current through this part of the circuit is $I = 6$ V/2 k$\Omega = 3$ mA.
    $I_3 = \left(\dfrac{R_2}{R_2 + 6\text{ k}\Omega}\right)3$ mA $= \left(\dfrac{3\text{ k}\Omega}{9\text{ k}\Omega}\right)3$ mA
    $= 1$ mA
    $I_5 = I_3/2 = 1$ mA/2 $= 0.5$ mA
   (d) $V_2 = V_S = 6$ V
5. $V_A = 25$ V
    R from point B to gnd $= 80$ Ω
    $V_B = (80$ $\Omega/130$ $\Omega)(25$ V$) = 15.38$ V
    $V_C = (60$ $\Omega/160$ $\Omega)(15.38$ V$) = 5.77$ V
6. $V_A = 15$ V
    $V_B = 0$ V
    $V_C = 0$ V
7. $V_{3k\Omega} = (1.5$ k$\Omega/2.5$ k$\Omega)(10$ V$) = 6$ V
    The 7.5-V reading is incorrect.
    $V_{2k\Omega} = (2$ k$\Omega/3$ k$\Omega)(6$ V$) = 4$ V
    The 5-V reading is incorrect. The 3-kΩ

resistor must be open to produce these incorrect voltages.

8. $V_{out} = (20 \text{ k}\Omega/30 \text{ k}\Omega)(30 \text{ V})$
   $= 20 \text{ V}$ unloaded
   $20 \text{ k}\Omega \| 200 \text{ k}\Omega = 18.18 \text{ k}\Omega$
   $V_{out} = (18.18 \text{ k}\Omega/28.18 \text{ k}\Omega)(30 \text{ V})$
   $= 19.35 \text{ V}$ loaded with 200 kΩ

9. $V_A = (6 \text{ k}\Omega/9 \text{ k}\Omega)(10 \text{ V})$
   $= 6.67 \text{ V}$ switch open
   $6 \text{ k}\Omega \| 10 \text{ k}\Omega = 3.75 \text{ k}\Omega$
   $V_A = (3.75 \text{ k}\Omega/6.75 \text{ k}\Omega)(10 \text{ V})$
   $= 5.56 \text{ V}$ switch closed

10. (a) $R_3 + R_4 = 75 \text{ }\Omega + 25 \text{ }\Omega = 100 \text{ }\Omega$
    $(R_3 + R_4) \| R_2 = 100 \text{ }\Omega \| 100 \text{ }\Omega$
    $= 100 \text{ }\Omega/2 = 50 \text{ }\Omega$
    $R_T = (R_3 + R_4) \| R_2 + R_1$
    $= 50 \text{ }\Omega + 50 \text{ }\Omega = 100 \text{ }\Omega$
    (b) $I_T = 50 \text{ V}/100 \text{ }\Omega = 0.5 \text{ A}$
    (c) $I_3 = \left(\dfrac{R_2}{R_2 + R_3 + R_4}\right)I_T = \left(\dfrac{100 \text{ }\Omega}{200 \text{ }\Omega}\right)0.5 \text{ A}$
    $= 0.25 \text{ A}$
    (d) $I_4 = I_3 = 0.25 \text{ A}$
    (e) $V_A = I_3(R_3 + R_4) = (0.25 \text{ A})(100 \text{ }\Omega)$
    $= 25 \text{ V}$
    (f) $V_B = I_4 R_4 = (0.25 \text{ A})(25 \text{ }\Omega) = 6.25 \text{ V}$

11. $R_{C \text{ to gnd}} = 8 \text{ k}\Omega$
    $R_{B \text{ to gnd}} = 16 \text{ k}\Omega/2 = 8 \text{ k}\Omega$
    $R_{A \text{ to gnd}} = 16 \text{ k}\Omega/2 = 8 \text{ k}\Omega$
    $V_A = (8 \text{ k}\Omega/16 \text{ k}\Omega)(10 \text{ V}) = 5 \text{ V}$
    $V_B = (8 \text{ k}\Omega/16 \text{ k}\Omega)(5 \text{ V}) = 2.5 \text{ V}$
    $V_C = (8 \text{ k}\Omega/16 \text{ k}\Omega)(2.5 \text{ V}) = 1.25 \text{ V}$

12. $R_{unk} = R_V(R_2/R_4) = (10 \text{ k}\Omega)(1 \text{ k}\Omega/8 \text{ k}\Omega)$
    $= 1.25 \text{ k}\Omega$

Chapter 8

1. $I_S = V_S/R_S = 25 \text{ V}/5 \text{ }\Omega = 5 \text{ A}$
   $R_S = 5 \text{ }\Omega$

2. $I_S = 30 \text{ V}/10 \text{ }\Omega = 3 \text{ A}; R_S = 10 \text{ }\Omega$
   See Figure S-9.

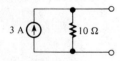

FIGURE S-9

3. $V_S = (5 \text{ mA})(25 \text{ k}\Omega) = 125 \text{ V}; R_S = 25 \text{ k}\Omega$
   See Figure S-10.

FIGURE S-10

4. For 1-V source:
   $R_T = 116.67 \text{ }\Omega$
   $I_T = 8.57 \text{ mA}$
   $I_3 = \left(\dfrac{R_2}{R_2 + R_3}\right)I_T = \left(\dfrac{50 \text{ }\Omega}{75 \text{ }\Omega}\right)8.57 \text{ mA}$
   $= 5.71 \text{ mA}$ down

   For 1.5-V source:
   $R_T = 70 \text{ }\Omega$
   $I_T = 21.4 \text{ mA}$
   $I_3 = \left(\dfrac{R_1}{R_1 + R_3}\right)I_T = \left(\dfrac{100 \text{ }\Omega}{125 \text{ }\Omega}\right)21.4 \text{ mA}$
   $= 17.12 \text{ mA}$ down

   I_3(total) $= 5.71 \text{ mA} + 17.12 \text{ mA}$
   $= 22.83 \text{ mA}$

5. For 1-V source:
   $I_2 = \left(\dfrac{R_3}{R_2 + R_3}\right)I_T = \left(\dfrac{25 \text{ }\Omega}{75 \text{ }\Omega}\right)8.57 \text{ mA}$
   $= 2.86 \text{ mA}$ down

   For 1.5-V source:
   $I_2 = I_T = 21.4 \text{ mA}$ up
   I_2(total) $= 21.4 \text{ mA} - 2.86 \text{ mA}$
   $= 18.54 \text{ mA}$

6. $R_{TH} = 1 \text{ k}\Omega/2 = 500 \text{ }\Omega$ (5-kΩ and 10-kΩ resistors are "shorted" by source)
   $V_{TH} = (1 \text{ k}\Omega/2 \text{ k}\Omega)(12 \text{ V}) = 6 \text{ V}$
   See Figure S-11.

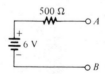

FIGURE S-11

7. Looking in at *open* terminals AB:
   $R_{TH} = (20 \text{ k}\Omega)(100 \text{ k}\Omega)/(120 \text{ k}\Omega)$
   $= 16.67 \text{ k}\Omega$
   $V_{TH} = (20 \text{ k}\Omega/120 \text{ k}\Omega)(15 \text{ V}) = 2.5 \text{ V}$
   With R_L connected:
   $I_L = V_{TH}/(R_{TH} + R_L) = 2.5 \text{ V}/516.67 \text{ k}\Omega$
   $= 4.84 \text{ }\mu\text{A}$

8. $I_N = V_{TH}/R_{TH} = 2.5\text{ V}/16.67\text{ k}\Omega = 0.15\text{ mA}$
$R_N = R_{TH} = 16.67\text{ k}\Omega$
See Figure S–12.

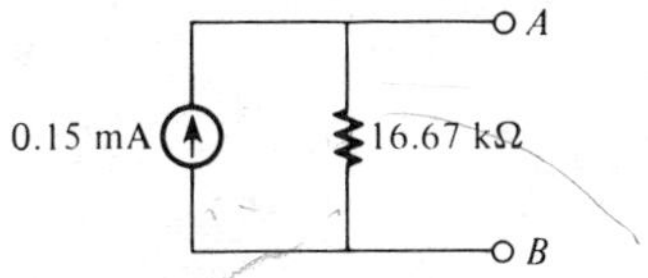

FIGURE S-12

9. $R_{eq} =$

$$\frac{1}{(1/20\ \Omega) + (1/10\ \Omega) + (1/50\ \Omega) + (1/10\ \Omega)}$$
$= 3.7\ \Omega$

$V_{eq} =$

$$\frac{(12\text{ V}/20\ \Omega) + (6\text{ V}/10\ \Omega) - (10\text{ V}/50\ \Omega) - (5\text{ V}/10\ \Omega)}{0.27}$$
$= 0.5/0.27 = 1.85\text{ V}$

See Figure S–13.

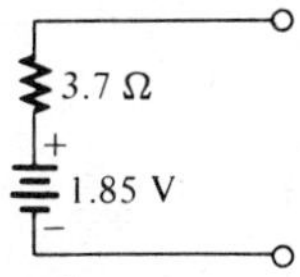

FIGURE S–13

10. $R_L = R_{eq} = 3.7\ \Omega$

11. (a) $R_1 = (R_A R_C)/(R_A + R_B + R_C)$
$= (1\text{ k}\Omega)(500\ \Omega)/2.5\text{ k}\Omega = 200\ \Omega$
$R_2 = (R_B R_C)/(R_A + R_B + R_C)$
$= (1\text{ k}\Omega)(500\ \Omega)/2.5\text{ k}\Omega = 200\ \Omega$
$R_3 = (R_A R_B)/(R_A + R_B + R_C)$
$= (1\text{ k}\Omega)(1\text{ k}\Omega)/2.5\text{ k}\Omega = 400\ \Omega$
See Figure S–14.

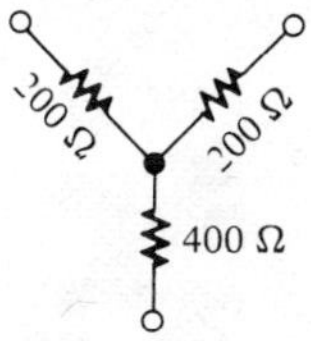

FIGURE S–14

(b) $R_A = \dfrac{R_1R_2 + R_1R_3 + R_2R_3}{R_2}$

$$= \frac{(50\ \Omega)(50\ \Omega) + (50\ \Omega)(100\ \Omega) + (50\ \Omega)(100\ \Omega)}{50\ \Omega}$$
$= 250\ \Omega$

$R_B = 12{,}500/R_1 = 12{,}500/50 = 250\ \Omega$
$R_C = 12{,}500/R_3 = 12{,}500/100 = 125\ \Omega$
See Figure S–15.

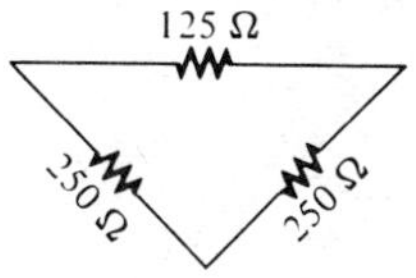

FIGURE S–15

Chapter 9

1. 2 loops, 5 nodes
2. The currents are shown in Figure S–16. The calculations are as follows.

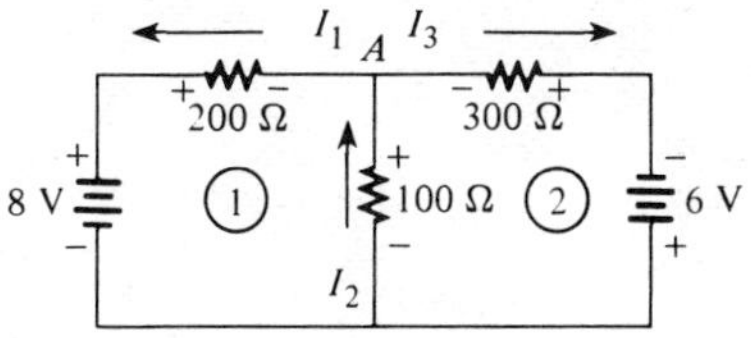

FIGURE S–16

Loop 1:
$200I_1 + 100I_2 - 8 = 0$

Loop 2:
$100I_2 + 300I_3 + 6 = 0$

Node A:
$I_1 - I_2 + I_3 = 0$

Solving, we obtain
$I_1 = I_2 - I_3$
$200\,(I_2 - I_3) + 100I_2 = 8$
$300I_2 - 200I_3 = 8$
$I_2 = (-6 - 300I_3)/100$

$$300\left(\frac{-6 - 300I_3}{100}\right) - 200I_3 = 8$$
$-18 - 900I_3 - 200I_3 = 8$
$1100I_3 = -26$
$I_3 = -0.024\text{ A}$
$100I_2 + 300\,(-0.024) = -6$
$I_2 = (-6 + 7.2)/100 = 0.012\text{ A}$
$I_1 = I_2 - I_3 = 0.012\text{ A} - (-0.024\text{ A})$
$= 0.036\text{ A}$

3. $I_1 = (15 - 5I_2)/2$
$6(15 - 5I_2)/2 - 8I_2 = 10$

$45 - 15I_2 - 8I_2 = 10$
$23I_2 = 35$
$I_2 = 35/23 = 1.52$ A
Substituting, we have
$2I_1 + 5(1.52) = 15$
$I_1 = (15 - 7.6)/2 = 3.7$ A

4. $I_1 = \dfrac{\begin{vmatrix} 25 & -12 \\ 18 & 3 \end{vmatrix}}{\begin{vmatrix} 15 & -12 \\ 9 & 3 \end{vmatrix}} = \dfrac{75 - (-216)}{45 - (-108)} = \dfrac{291}{153}$
$= 1.9$ A

5. The currents are shown in Figure S–17. The calculations are as follows.

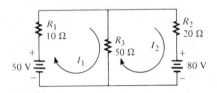

FIGURE S–17

$60I_1 - 50I_2 = -50$
$-50I_1 + 70I_2 = 80$

$I_1 = \dfrac{\begin{vmatrix} -50 & -50 \\ 80 & 70 \end{vmatrix}}{\begin{vmatrix} 60 & -50 \\ -50 & 70 \end{vmatrix}} = \dfrac{-3500 + 4000}{4200 - 2500}$

$= \dfrac{-500}{1700} = 0.294$ A

$I_2 = \dfrac{\begin{vmatrix} 60 & -50 \\ -50 & 80 \end{vmatrix}}{1700} = \dfrac{4800 - (2500)}{1700}$

$= \dfrac{2300}{1700} = 1.35$ A

$I_{R3} = I_2 - I_1 = 1.35$ A $- 0.294$ A
$= 1.056$ A up

6. The calculations are as follows. Also see Figure S–18.

$I_1 + I_2 = I_3$

$\dfrac{-15 - V_A}{100} + \dfrac{-15 - V_A}{300} - \dfrac{V_A}{600} = 0$

$\dfrac{-V_A}{100} - \dfrac{V_A}{300} - \dfrac{V_A}{600} = \dfrac{15}{100} + \dfrac{15}{300}$

$\dfrac{-6V_A - 2V_A - V_A}{600} = \dfrac{45 + 15}{300}$

$\dfrac{-9V_A}{600} = \dfrac{60}{300}$

$V_A = -\dfrac{(600)(60)}{(9)(300)} = -13.33$ V

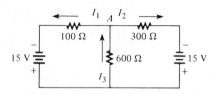

FIGURE S–18

Chapter 10

1. $f = 1/T = 1/200$ ms $= 5$ Hz
2. $T = 1/f = 1/25$ kHz $= 40$ μs
3. $T = 5$ μs
   $f = 1/T = 1/5$ μs $= 0.2$ MHz
4. $\Delta V/\Delta t = (1.5$ V $- 1$ V$)/1$ μs $= 0.5$ V/1 μs
   $= 0.5$ V/μs
5. Derivative of v with respect to t, the instantaneous rate of change of the voltage.
6. $V_p = 25$ V
   $V_{pp} = 50$ V
   $V_{rms} = 0.707V_p = 17.68$ V
   $V_{avg} = 0.637V_p = 15.93$ V
7. $V_p = 1.414V_{rms} = 1.414(115$ V$) = 162.6$ V
8. $I_{rms} = V_{rms}/R = 10$ V$/100$ Ω $= 0.1$ A
9. See Figure S–19.

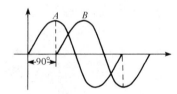

FIGURE S–19

10. $v = V_p \sin \theta$
    $= 20 \sin 10° = 3.47$ V
    $= 20 \sin 25° = 8.45$ V
    $= 20 \sin 30° = 10$ V
    $= 20 \sin 90° = 20$ V
    $= 20 \sin 180° = 0$ V
    $= 20 \sin 210° = -10$ V
    $= 20 \sin 300° = -17.32$ V

CHAPTER 11

11. $V_p = 5$ V
   $v = 5 \sin 30° = 2.5$ V
   $= 5 \sin 60° = 4.33$ V
   $= 5 \sin 100° = 4.92$ V
   $= 5 \sin 225° = -3.54$ V

12. A: $10 \sin 30° = 5$
    B: $10 \sin 90° = 10$
    C: $10 \sin 200° = -3.42$

13. $20° \equiv -340°$
    $60° \equiv -300°$
    $135° \equiv -225°$
    $200° \equiv -160°$
    $315° \equiv -45°$
    $330° \equiv -30°$

14. $\omega = 2\pi f$
    $f = \omega/(2\pi) = (1000 \text{ rad/sec})/(2\pi \text{ rad/cycle})$
    $= 159$ Hz

15. $t_r \cong 1$ μs $- 0.2$ μs $= 0.8$ μs
    $t_f \cong 5.9$ μs $- 5.1$ μs $= 0.8$ μs
    PW $\cong 5.6$ μs $- 0.6$ μs $= 5$ μs
    $A = 2$ V

16. PRF $= 1/T = 1/25$ μs $= 40$ kp/s

17. % duty cycle $= (PW/T) \times 100$
    $T = 1/PRF = 1/(2 \text{kp/s}) = 0.5$ ms $= 500$ μs
    % duty cycle $= (1 \text{ μs}/500 \text{ μs}) \times 100$
    $= 0.2\%$

18. V_{avg} = Baseline + (duty cycle)(amplitude)
    Duty cycle = PW/T = 1 μs/6 μs = 0.167
    $V_{avg} = 5$ V $+ (0.167)(5$ V$) = 5.835$ V

19. $f = $ PRF $= 1/T = 1/6$ μs $= 166.67$ kHz

20. $f_{2nd} = 2 \times 25$ kHz $= 50$ kHz

Chapter 11

1. (a) True (b) True (c) False
   (d) False

2. (a) False (b) True (c) False
   (d) True

3. $Q = CV = (0.006 \text{ μF})(10 \text{ V}) = 0.06$ μC

4. $C = Q/V = (50 \times 10^{-6} \text{ C})/5$ V
   $= 10 \times 10^{-6}$ F $= 10$ μF

5. $V = Q/C = (5 \times 10^{-8} \text{ C})/0.01$ μF
   $= (5 \times 10^{-8} \text{ C})/(0.01 \times 10^{-6}$ F$)$
   $= 500 \times 10^{-2}$ V $= 5$ V

6. $(0.005 \text{ μF})(10^6 \text{ pF/μF}) = 5000$ pF

7. $(2000 \text{ pF})(10^{-6} \text{ μF/pF}) = 0.002$ μF

8. $\epsilon = \epsilon_r \epsilon_0 = (1200)(8.85 \times 10^{-12}$ F/m$)$
   $= 1.062 \times 10^{-8}$ F/m

9. $C = \dfrac{A\epsilon_r(8.85 \times 10^{-12})}{d}$
   $= \dfrac{(0.009 \text{ m}^2)(2)(8.85 \times 10^{-12} \text{ F/m})}{0.0015 \text{ m}}$
   $= 106.2$ pF

10. N750 indicates a *negative* temperature coefficient of 750 pF/μF°C, since there are 1,000,000 pF in one μF.
    $(-750 \text{ pF/μF°C})(0.1 \text{ μF})(25°C) = -1875$ pF

11. (a) False (b) True (c) True
    (d) True

12. Green: 5, 1st digit
    Blue: 6, 2nd digit
    Orange: 3, multiplier
    Gold: ±5%, tolerance
    $C = 56,000$ pF $\pm 5\%$

13. A: $1/C_T = (1/0.05 \text{ μF}) + (1/0.01 \text{ μF}) +$
       $(1/0.02 \text{ μF}) + (1/0.05 \text{ μF})$
       $= 1.9 \times 10^8$
       $C_T = 0.00526$ μF
    B: $C_T = (300 \text{ pF})(100 \text{ pF})/400 \text{ pF} = 75$ pF
    C: $C_T = 12$ pF$/3 = 4$ pF

14. $1/C_T = (1/0.025 \text{ μF}) + (1/0.04 \text{ μF}) + (1/0.1 \text{ μF})$
    $= 7.5 \times 10^7$
    $C_T = 0.0133$ μF
    $V_1 = (C_T/C_1)V_T = (0.0133 \text{ μF}/0.025 \text{ μF})250$ V
    $= 133$ V
    $V_2 = (C_T/C_2)V_T = (0.0133 \text{ μF}/0.04 \text{ μF})250$ V
    $= 83.13$ V
    $V_3 = (C_T/C_3)V_T = (0.0133 \text{ μF}/0.1 \text{ μF})250$ V
    $= 33.25$ V

15. A: $C_T = 100$ pF $+ 47$ pF $+ 33$ pF $+ 10$ pF
       $= 190$ pF
    B: $C_T = (5)(0.008 \text{ μF}) = 0.04$ μF

16. $X_C = \dfrac{1}{2\pi fC} = \dfrac{1}{2\pi(2 \times 10^6 \text{ Hz})(1 \times 10^{-6} \text{ F})}$
    $= 0.0796$ Ω

17. $X_C = V_{rms}/I_{rms} = 25$ V/50 mH $= 0.5$ kΩ
    $X_C = \dfrac{1}{2\pi fC}$
    $f = \dfrac{1}{2\pi X_C C} = \dfrac{1}{2\pi(500 \text{ Ω})(0.3 \times 10^{-6} \text{ F})}$
    $= 1.06$ kHz

18. $\mathscr{E} = \frac{1}{2}CV^2 = \frac{1}{2}(10 \times 10^{-6} \text{ F})(100 \text{ V})^2$
    $= 0.05$ J

Chapter 12

1. Sine wave

2. $Z = \sqrt{R^2 + X_C^2}$
   $X_C = \dfrac{1}{2\pi f C}$
   $= \dfrac{1}{2\pi(10 \times 10^3 \text{ Hz})(0.005 \times 10^{-6} \text{ F})}$
   $= 3.18 \text{ k}\Omega$
   $Z = \sqrt{(2.5 \text{ k}\Omega)^2 + (3.18 \text{ k}\Omega)^2} = 4.05 \text{ k}\Omega$

3. $X_C = \dfrac{1}{2\pi(20 \times 10^3 \text{ Hz})(0.005 \times 10^{-6} \text{ F})}$
   $= 1.59 \text{ k}\Omega$
   $Z = \sqrt{(2.5 \text{ k}\Omega)^2 + (1.59 \text{ k}\Omega)^2} = 2.96 \text{ k}\Omega$

4. $X_C = R$ at 45°
   $X_C = 10 \text{ k}\Omega$
   $1/(2\pi f C) = 10 \text{ k}\Omega$
   $f = \dfrac{1}{2\pi(10 \text{ k}\Omega)(100 \text{ pF})} = 159 \text{ kHz}$

5. $\tan\theta = X_C/R$
   $X_C = \dfrac{1}{2\pi f C} = \dfrac{1}{2\pi(5 \text{ kHz})(0.01 \text{ μF})}$
   $= 3.18 \text{ k}\Omega$
   $X_C/R = \tan\theta$
   $R = X_C/\tan\theta = 3.18 \text{ k}\Omega/\tan 60°$
   $= 3.18 \text{ k}\Omega/1.732$
   $= 1.84 \text{ k}\Omega$

6. $\theta = \arctan(X_C/R)$
   $X_C = \dfrac{1}{2\pi f C} = \dfrac{1}{2\pi(100 \text{ kHz})(100 \text{ pF})}$
   $= 15.9 \text{ k}\Omega$
   $\theta = \arctan(15.9 \text{ k}\Omega/10 \text{ k}\Omega) = 57.83°$

7. $\theta = \arctan(1000 \text{ }\Omega/500 \text{ }\Omega) = 63.4°$

8. $\theta = \arctan(X_C/R) = \arctan(X_C/2X_C)$
   $= \arctan 0.5 = 26.57°$

9. $V_R = \dfrac{R}{\sqrt{R^2 + X_C^2}} V_s$
   $= \dfrac{47 \text{ k}\Omega}{\sqrt{(47 \text{ k}\Omega)^2 + (30 \text{ k}\Omega)^2}} 10 \text{ V}$
   $= 8.43 \text{ V}$
   $V_C = \dfrac{X_C}{\sqrt{R^2 + X_C^2}} V_s$
   $= \dfrac{30 \text{ k}\Omega}{\sqrt{(47 \text{ k}\Omega)^2 + (30 \text{ k}\Omega)^2}} 10 \text{ V}$
   $= 5.38 \text{ V}$

10. $Z = \sqrt{R^2 + X_C^2} = \sqrt{(47 \text{ k}\Omega)^2 + (30 \text{ k}\Omega)^2}$
    $= 55.76 \text{ k}\Omega$
    $I = 10 \text{ V}/55.76 \text{ k}\Omega = 0.179 \text{ mA}$

11. $X_C = \dfrac{1}{2\pi(1 \text{ kHz})(0.005 \text{ μF})} = 31.83 \text{ k}\Omega$
    $\phi = 90° - \arctan(31.83 \text{ k}\Omega/33 \text{ k}\Omega)$
    $= 90° - 43.97° = 46.03°$

12. $V_{out} = \dfrac{X_C}{\sqrt{R^2 + X_C^2}} V_{in}$
    $= \dfrac{31.83 \text{ k}\Omega}{\sqrt{(33 \text{ k}\Omega)^2 + (31.83 \text{ k}\Omega)^2}} 5 \text{ V}$
    $= 3.47 \text{ V}$

13. $X_C = \dfrac{1}{2\pi(50 \text{ kHz})(0.1 \text{ μF})} = 31.83 \text{ }\Omega$
    $\theta = \arctan(X_C/R) = \arctan(31.83/75)$
    $= 22.996°$

14. $V_{out} = \dfrac{R}{\sqrt{R^2 + X_C^2}} V_{in}$
    $= \dfrac{75 \text{ }\Omega}{\sqrt{(75 \text{ }\Omega)^2 + (31.83 \text{ }\Omega)^2}} 12 \text{ V}$
    $= 11.05 \text{ V}$

15. No. This filter is a low-pass.

16. 5 kHz. This filter is a high-pass.

17. Figure 12-47:
    $I = 5 \text{ V}/Z = 5 \text{ V}/45.85 \text{ k}\Omega = 0.109 \text{ mA}$
    $P_{avg} = I^2R = (0.109 \text{ mA})^2(33 \text{ k}\Omega)$
    $= 0.392 \text{ mW}$
    $P_r = I^2 X_C = (0.109 \text{ mA})^2(31.83 \text{ k}\Omega)$
    $= 0.378 \text{ mW}$
    $P_a = I^2 Z = (0.109 \text{ mA})^2(45.85 \text{ k}\Omega)$
    $= 0.545 \text{ mW}$
    Figure 12-48:
    $I = 12 \text{ V}/Z = 12 \text{ V}/81.48 \text{ }\Omega = 0.147 \text{ A}$
    $P_{avg} = I^2 R = (0.147 \text{ A})^2(75 \text{ }\Omega) = 1.62 \text{ W}$
    $P_r = I^2 X_C = (0.147 \text{ A})^2(31.83 \text{ }\Omega)$
    $= 0.688 \text{ W}$
    $P_a = I^2 Z = (0.147 \text{ A})^2(81.48 \text{ }\Omega) = 1.76 \text{ W}$

18. Figure 12-47:
    $\theta = \arctan(X_C/R) = \arctan(31.83 \text{ k}\Omega/33 \text{ k}\Omega)$
    $= 43.97°$
    PF $= \cos\theta = \cos 43.97° = 0.72$
    Figure 12-48:
    $\theta = \arctan(X_C/R) = \arctan(31.83 \text{ }\Omega/75 \text{ }\Omega)$
    $= 23°$
    PF $= \cos\theta = \cos 23° = 0.92$

19. $I_R = V_s/R = 5 \text{ V}/25 \text{ k}\Omega = 0.2 \text{ mA}$
    $I_C = V_s/X_C = 5 \text{ V}/10 \text{ k}\Omega = 0.5 \text{ mA}$

CHAPTER 13

$I_T = \sqrt{I_R^2 + I_C^2} = \sqrt{(0.2 \text{ mA})^2 + (0.5 \text{ mA})^2}$
$= 0.539 \text{ mA}$

20. $\theta = \arctan(R/X_C) = \arctan(25 \text{ k}\Omega/10 \text{ k}\Omega)$
$= 68.2°$

Chapter 13

1. See Figure S–20.

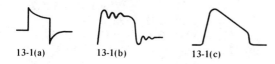

13-1(a)   13-1(b)   13-1(c)

FIGURE S–20

2. % tilt = $(\Delta V/V_A) \times 100$
   $= (2 \text{ V}/10 \text{ V}) \times 100 = 20\%$

3. $\tau = RC = (2 \text{ k}\Omega)(0.05 \text{ μF}) = 0.1 \times 10^{-3}$ s
   $= 0.1$ ms

4. $v_C = (0.63)(10 \text{ V}) = 6.3$ V

5. (a) $v_C = (0.86)(10 \text{ V}) = 8.6$ V
   (b) $v_C = (0.95)(10 \text{ V}) = 9.5$ V
   (c) $v_C = (0.98)(10 \text{ V}) = 9.8$ V
   (d) $v_C \cong (1)(10 \text{ V}) = 10$ V

6. $v_C = V_F(1 - e^{-t/\tau}) = 5 \text{ V}(1 - e^{-5/10})$
   $= 5 \text{ V}(1 - 0.607)$
   $= 1.97$ V

7. $v_c = (0.37)(15 \text{ V}) = 5.55$ V

8. See Figure S–21.

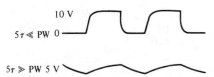

FIGURE S–21

9. See Figure S–22.

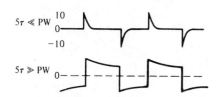

FIGURE S–22

10. $\tau = RC = (1 \text{ k}\Omega)(1 \text{ μF}) = 1$ ms. See Figure S–23. The time to reach steady-state is $5\tau = 5$ ms for repetitive pulses.

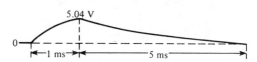

FIGURE S–23

11. See Figure S–24.

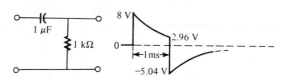

FIGURE S–24

12. $f = 0.35/t_r = 0.35/250 \text{ ns} = 1.4$ MHz

Chapter 14

1. $A = 1 \text{ mm}^2 = 1 \times 10^{-6} \text{ m}^2$
   $\phi = 1200$ μWb
   $B = \dfrac{\phi}{A} = \dfrac{1200 \times 10^{-6} \text{ Wb}}{1 \times 10^{-6} \text{ m}^2}$
   $= 1200$ tesla

2. $F_m = NI = (50)(100\text{mA})$
   $= 5000 \times 10^{-3}$ ampere-turns
   $= 5$ ampere-turns

3. $v_{ind} = N(d\phi/dt) = 50(25 \text{ Wb/s}) = 1250$ V

4. 1500 μH = 1.5 mH, 20 mH = 20,000 μH

5. $v_{ind} = L(di/dt) = (100 \text{ mH})(200 \text{ mA/s})$
   $= 0.02$ V

6. $L = (N^2 \mu A)/l$
   $N = \sqrt{Ll/\mu A}$
   $= \sqrt{(30 \text{ mH})(0.05 \text{ m})/(1.2 \times 10^{-6})(10 \times 10^{-5} \text{ m}^2)}$
   $= 3536$ turns

7. $I = V_{dc}/R_W = 12 \text{ V}/12 \text{ }\Omega = 1$ A

8. $L_T = 75 \text{ μH} + 20 \text{ μH} = 95$ μH

9. $1/L_T = (1/1 \text{ H}) + (1/1 \text{ H}) + (1/0.5 \text{ H})$
   $+ (1/0.5 \text{ H}) + (1/2 \text{ H}) = 6.5$
   $L_T = 0.154$ H

10. $f = 2$ MHz
    $X_L = 2\pi fL = 2\pi(2 \text{ MHz})(15 \text{ μH})$
    $= 188.5 \text{ }\Omega$

11. $X_L = V_{rms}/I_{rms} = 25 \text{ V}/50 \text{ mA} = 0.5 \text{ k}\Omega$
    $X_L = 2\pi f L$
    $f = \dfrac{X_L}{2\pi L} = \dfrac{500 \text{ }\Omega}{2\pi(8 \text{ }\mu\text{H})} = 9.95 \text{ MHz}$

12. $L = \dfrac{X_L}{2\pi f} = \dfrac{10 \text{ M}\Omega}{2\pi(50 \text{ MHz})} = 31.83 \text{ mH}$

Chapter 15

1. Sine wave

2. $X_L = 2\pi f L = 2\pi(10 \text{ kHz})(20 \text{ mH})$
    $= 1.26 \text{ k}\Omega$
    $Z = \sqrt{R^2 + X_L^2} = \sqrt{(3.3 \text{ k}\Omega)^2 + (1.26 \text{ k}\Omega)^2}$
    $= 3.53 \text{ k}\Omega$

3. $\theta = 45°$ when $X_L = R$
    $2\pi f L = 10 \text{ k}\Omega$
    $f = 10 \text{ k}\Omega/(2\pi \times 100 \text{ mH}) = 15.9 \text{ kHz}$

4. $\tan \theta = X_L/R$
    $X_L = 2\pi f L = 2\pi(5 \text{ kHz})(200 \text{ }\mu\text{H})$
    $= 6.28 \text{ }\Omega$
    $\tan 60° = 6.28 \text{ }\Omega/R$
    $R = 6.28 \text{ }\Omega/1.732 = 3.63 \text{ }\Omega$

5. $\theta = \arctan(X_L/R)$
    $\theta = \arctan(100 \text{ }\Omega/500 \text{ }\Omega) = 11.3°$

6. $\theta = \arctan(X_L/R), X_L = 2R$
    $\theta = \arctan(2R/R) = \arctan 2 = 63.4°$

7. $V_R = (R/\sqrt{R^2 + X_L^2})V_s$
    $= (47 \text{ k}\Omega/\sqrt{(47 \text{ k}\Omega)^2 + (20 \text{ k}\Omega)^2})120 \text{ V}$
    $= 110.42 \text{ V}$
    $V_L = (X_L/\sqrt{R^2 + X_L^2})V_s$
    $= (20 \text{ k}\Omega/\sqrt{(47 \text{ k}\Omega)^2 + (20 \text{ k}\Omega)^2})120 \text{ V}$
    $= 46.99 \text{ V}$

8. $Z = \sqrt{R^2 + X_L^2} = \sqrt{(47 \text{ k}\Omega)^2 + (20 \text{ k}\Omega)^2}$
    $= 51.08 \text{ k}\Omega$
    $I = 120 \text{ V}/51.08 \text{ k}\Omega = 2.35 \text{ mA}$

9. $X_L = 2\pi f L = 2\pi(1 \text{ kHz})(0.1 \text{ H}) = 628.3 \text{ }\Omega$
    $\theta = \arctan(X_L/R) = \arctan(628.3 \text{ }\Omega/1200 \text{ }\Omega)$
    $= 27.64°$

10. $V_{out} = V_R = (R/\sqrt{R^2 + X_L^2})V_{in}$
    $= (1.2 \text{ k}\Omega/\sqrt{(1.2 \text{ k}\Omega)^2 + (628.3 \text{ }\Omega)^2})6 \text{ V}$
    $= 5.32 \text{ V}$

11. $X_L = 2\pi f L = 2\pi(60 \text{ Hz})(0.2 \text{ H}) = 75.4 \text{ }\Omega$
    $\phi = 90° - \arctan(X_L/R)$
    $= 90° - \arctan(75.4 \text{ }\Omega/100 \text{ }\Omega)$
    $= 52.98°$

12. $X_L = 75.4 \text{ }\Omega$
    $V_{out} = V_L = (X_L/\sqrt{R^2 + X_L^2})V_{in}$
    $= (75.4 \text{ }\Omega/\sqrt{(100 \text{ }\Omega)^2 + (75.4 \text{ }\Omega)^2})5 \text{ V}$
    $= 3.01 \text{ V}$

13. At 1 kHz, because X_L is less.

14. (a) $P_{avg} = I^2R, I = V_{in}/Z_T$
    $Z_T = \sqrt{R^2 + X_L^2}$
    $= \sqrt{(100 \text{ }\Omega)^2 + (75.4 \text{ }\Omega)^2} = 125.2 \text{ }\Omega$
    $I = 5 \text{ V}/125.2 \text{ }\Omega = 0.0399 \text{ A}$
    $P_{avg} = (0.0399 \text{ A})^2(100 \text{ }\Omega) = 0.159 \text{ W}$
    (b) $P_r = I^2X_L = (0.0399 \text{ A})^2(75.4 \text{ }\Omega)$
    $= 0.120 \text{ W}$
    (c) $P_a = I^2Z = (0.0399 \text{ A})^2(125.2 \text{ }\Omega)$
    $= 0.199 \text{ W}$
    (d) $PF = \cos \theta$,
    $\theta = \arctan(X_L/R)$
    $= \arctan(75.4/100) = 37°$
    $PF = \cos 37° = 0.799$

15. $I_R = 5 \text{ V}/25 \text{ k}\Omega = 0.2 \text{ mA}$
    $I_L = 5 \text{ V}/40 \text{ k}\Omega = 0.125 \text{ mA}$
    $I_T = \sqrt{I_R^2 + I_L^2} = \sqrt{(0.2 \text{ mA})^2 + (0.125 \text{ mA})^2}$
    $= 0.236 \text{ mA}$
    $\theta = \arctan(R/X_L) = \arctan(25 \text{ k}\Omega/40 \text{ k}\Omega)$
    $= 32°$

Chapter 16

1. $\tau = L/R = 10 \text{ mH}/2 \text{ k}\Omega = 5 \text{ }\mu\text{s}$

2. $v_R = (0.63)(10 \text{ V}) = 6.3 \text{ V}$
    $v_{L(max)} = +10 \text{ V}, v_{L(min)} = -10 \text{ V}$

3. (a) $v_R = (0.86)(10 \text{ V}) = 8.6 \text{ V}$
    $v_{L(max)} = +10 \text{ V}, v_{L(min)} = -10 \text{ V}$
    (b) $v_R = (0.95)(10 \text{ V}) = 9.5 \text{ V}$
    $v_{L(max)} = +10 \text{ V}, v_{L(min)} = -10 \text{ V}$
    (c) $v_R = (0.98)(10 \text{ V}) = 9.8 \text{ V}$
    $v_{L(max)} = +10 \text{ V}, v_{L(min)} = -10 \text{ V}$
    (d) $v_R = 10 \text{ V}$
    $v_{L(max)} = +10 \text{ V}, v_{L(min)} = -10 \text{ V}$

4. See Figure S–25.

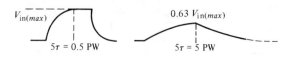

FIGURE S–25

CHAPTER 17

5. See Figure S-26.

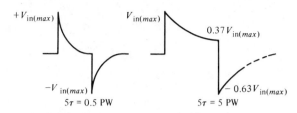

FIGURE S-26

6. $\tau = 10$ mH/10 Ω = 1 ms
$5\tau = 5$ ms
$v_{out(max)} = (0.63)8$ V = 5.04 V
See Figure S-27.

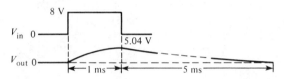

FIGURE S-27

7. See Figure S-28.

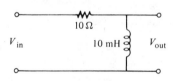

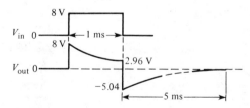

FIGURE S-28

8. (a) $V_{out} = (10\ \Omega/11\ \Omega)12$ V = 10.91 V
(b) $V_{out} = (1\ \Omega/11\ \Omega)12$ V = 1.09 V

Chapter 17

1. $k = 0.9$
2. $L_M = R\sqrt{L_1 L_2} = 0.9\sqrt{(10\ \mu\text{H})(5\ \mu\text{H})}$
    $= 6.36\ \mu\text{H}$

3. $N_s/N_p = 36/12 = 3$. It is a step-up transformer.

4. See Figure S-29.

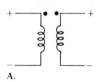   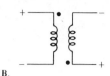

A.   B.

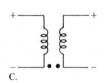   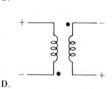

C.   D.

FIGURE S-29

5. $N_s/N_p = 15/10 = 1.5$
$V_s = 1.5V_p = 1.5(120$ V$) = 180$ V
6. $V_s = 0.5V_p = 0.5(50$ V$) = 25$ V
7. $N_s/N_p = V_s/V_p = 60$ V/240 V $= 0.25$
8. $I_s = (N_p/N_s)I_p = (1/0.25)(0.25$ A$) = 1$ A
9. $Z_p = (N_p/N_s)^2 Z_L = (1/5)^2(1$ k$\Omega) = 40\ \Omega$
10. $R_s = R_{\text{reflected}}$
    $= 10\ \Omega$ for maximum power to R_L
    $R_{\text{reflected}} = R_p = (N_p/N_s)^2(R_L)$
    $(N_p/N_s)^2(R_L) = 10\ \Omega$
    $R_L = 10\ \Omega/0.04 = 250\ \Omega$
11. 1: $V_1 = (N_s/N_p)V_s = (100/10)120$ V
        $= 1200$ V
    2: $V_2 = (10/10)120$ V $= 120$ V
    3: $V_3 = (5/10)120$ V $= 60$ V
12. 1200 V/2 = 600 V. The polarities are opposite.

Chapter 18

1. See Figure S-30 on page 738.
2. See Figure S-31 on page 738.
3. (a) $5 + j5 = \sqrt{5^2 + 5^2}\ \angle\arctan(5/5)$
        $= 7.07\ \angle 45°$
    (b) $12 + j9 = \sqrt{12^2 + 9^2}\ \angle\arctan(9/12)$
        $= 15\ \angle 36.87°$
    (c) $8 - j10 = \sqrt{8^2 + 10^2}\ \angle\arctan(-10/8)$
        $= 12.8\ \angle -51.34°$

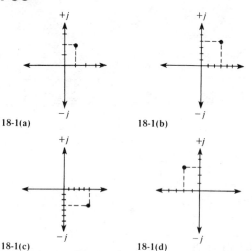

FIGURE S–30

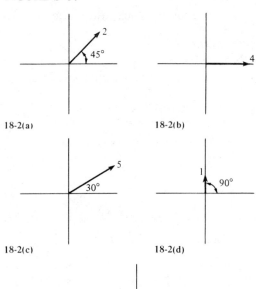

FIGURE S–31

(d) $100 - j50 =$
$\sqrt{100^2 + 50^2} \angle \arctan(-50/100)$
$= 111.8 \angle -26.57°$

4. (a) $1 \angle 45° = 1 \cos 45° + j1 \sin 45°$
$= 0.707 + j0.707$
(b) $12 \angle 60° = 12 \cos 60° + j12 \sin 60°$
$= 6 + j10.39$
(c) $100 \angle -80° = 100 \cos(-80°)$
$+ j100 \sin(-80°) = 17.36 - j98.48$
(d) $40 \angle 125° = 40 \cos 125° + j40 \sin 125°$
$= -22.94 + j32.77$

5. A: $\mathbf{Z} = R - jX_C = 50 \, \Omega - j30 \, \Omega$
B: $\mathbf{Z} = R + jX_L = 25 \, \Omega + j100 \, \Omega$

6. A: $\mathbf{Z} = 50 \, \Omega - j30 \, \Omega$
$Z \text{ magnitude} = \sqrt{(50 \, \Omega)^2 + (30 \, \Omega)^2}$
$= 58.31 \, \Omega$
B: $\mathbf{Z} = 25 \, \Omega + j100 \, \Omega$
$Z \text{ magnitude} = \sqrt{(25 \, \Omega)^2 + (100 \, \Omega)^2}$
$= 103.08 \, \Omega$

7. A: $\mathbf{I} = \dfrac{\mathbf{V}}{\mathbf{Z}} = \dfrac{10 \angle 0° \text{ V}}{58.31 \angle \theta° \, \Omega}$
$\theta = \arctan(-30/50) = -30.96°$
$\mathbf{I} = \dfrac{10 \angle 0° \text{ V}}{58.31 \angle -30.96° \, \Omega} = 0.172 \angle 30.96° \text{ A}$
B: $\theta = \arctan(100/25) = 75.96°$
$\mathbf{I} = \dfrac{\mathbf{V}}{\mathbf{Z}} = \dfrac{10 \angle 0° \text{ V}}{103.08 \angle 75.96° \, \Omega}$
$= 0.097 \angle -75.96° \text{ A}$

8. A: $\mathbf{Z} = \dfrac{RX_C}{\sqrt{R^2 + X_C^2}} \angle \left(-90° + \arctan\dfrac{X_C}{R}\right)$
$= \dfrac{(12 \, \Omega)(12 \, \Omega)}{\sqrt{(12 \, \Omega)^2 + (12 \, \Omega)^2}} \angle(-90° + 45°)$
$= 8.49 \angle -45° \, \Omega$
B: $\mathbf{Z} = \dfrac{RX_L}{\sqrt{R^2 + X_L^2}} \angle \left(-90° - \arctan\dfrac{X_L}{R}\right)$
$= \dfrac{(8 \text{ k}\Omega)(5 \text{ k}\Omega)}{\sqrt{(8 \text{ k}\Omega)^2 + (5 \text{ k}\Omega)^2}} \angle(90° - 32°)$
$= 4240 \angle 58° \, \Omega$

9. A: $\mathbf{I}_T = \dfrac{\mathbf{V}_s}{\mathbf{Z}} = \dfrac{5 \angle 0° \text{ V}}{8.49 \angle -45° \, \Omega} = 0.589 \angle 45° \text{ A}$
$\mathbf{I}_R = \dfrac{\mathbf{V}_s}{\mathbf{R}} = \dfrac{5 \angle 0° \text{ V}}{12 \angle 0° \, \Omega} = 0.417 \angle 0° \text{ A}$
$\mathbf{I}_C = \dfrac{\mathbf{V}_s}{\mathbf{X}_C} = \dfrac{5 \angle 0° \text{ V}}{12 \angle -90° \, \Omega} = 0.417 \angle 90° \text{ A}$
B: $\mathbf{I}_T = \dfrac{\mathbf{V}_s}{\mathbf{Z}} = \dfrac{5 \angle 0° \text{ V}}{4.24 \text{ k} \angle 58° \, \Omega}$
$= 1.179 \angle -58° \text{ mA}$

CHAPTER 19

$I_R = \dfrac{V_s}{R} = \dfrac{5 \angle 0° \text{ V}}{8 \text{ k} \angle 0° \text{ }\Omega} = 0.625 \angle 0°$ mA

$I_L = \dfrac{V_s}{X_L} = \dfrac{5 \angle 0° \text{ V}}{5 \text{ k} \angle 90° \text{ }\Omega} = 1 \angle -90°$ mA

10. It leads because the circuit is capacitive $(X_C > X_L)$.

11. $Z = \sqrt{R^2 - (X_C - X_L)^2} = \sqrt{(75 \text{ }\Omega)^2 + (40 \text{ }\Omega)^2}$
    $= 85 \text{ }\Omega$

12. $\dfrac{1}{Z} = \dfrac{1}{R \angle 0°} + \dfrac{1}{X_C \angle -90°} + \dfrac{1}{X_L \angle 90°}$
    $= \dfrac{1}{15 \text{ }\Omega} + \dfrac{1}{-j10 \text{ }\Omega} + \dfrac{1}{j20 \text{ }\Omega}$
    $= 0.067 + j0.1 - j0.05 = 0.067 + j0.05$
    $= 0.084 \angle 36.7°$ S

    $Z = \dfrac{1}{0.084 \angle 36.7° \text{ S}} = 11.9 \angle -36.7° \text{ }\Omega$

13. $X_C = \dfrac{1}{2\pi fC} = \dfrac{1}{2\pi(5 \text{ kHz})(0.2 \text{ }\mu\text{F})} = 159 \text{ }\Omega$

    $\dfrac{1}{Z} = \dfrac{1}{100 \angle 0° \text{ }\Omega} + \dfrac{1}{159 \angle -90° \text{ }\Omega}$
    $= \dfrac{1}{100 \text{ }\Omega} + \dfrac{1}{-j159 \text{ }\Omega}$
    $= \dfrac{1}{100 \text{ }\Omega} + j\dfrac{1}{159 \text{ }\Omega}$
    $= 0.01 + j0.0063 = 0.01182 \angle 32.2°$ S

    $Z = \dfrac{1}{0.01182 \angle 32.2° \text{ S}} = 84.6 \angle -32.2° \text{ }\Omega$
    $= 71.6 \text{ }\Omega - j45.1 \text{ }\Omega$

    $R_{eq} = 71.6 \text{ }\Omega$, $X_C = 45.1 \text{ }\Omega$,
    $C_{eq} = 1/(2\pi f X_C) = 0.71 \text{ }\mu\text{F}$

14. $\theta = \arctan(R/X_C) = \arctan(-100/159)$
    $= -32.2°$

15. $X_L = 2\pi fL = 2\pi(400 \text{ Hz})(15 \text{ mH}) = 37.7 \text{ }\Omega$
    $\dfrac{1}{Z_p} = \dfrac{1}{R_2 \angle 0°} + \dfrac{1}{X_L \angle 90°} = \dfrac{1}{30} + \dfrac{1}{j37.7}$
    $= 0.0333 - j0.0265$
    $= 0.0426 \angle -38.51°$ S

    $Z_p = \dfrac{1}{0.0426 \angle -38.51° \text{ S}} = 23.47 \angle 38.51° \text{ }\Omega$
    $= 18.37 \text{ }\Omega + j14.61 \text{ }\Omega$

    $Z_{eq} = R_1 + 18.37 \text{ }\Omega - j14.61 \text{ }\Omega$
    $= 68.37 \text{ }\Omega + j14.61 \text{ }\Omega$

    $R_{eq} = 68.37 \text{ }\Omega$, $X_{L(eq)} = 14.61 \text{ }\Omega$

    $L_{eq} = \dfrac{14.61 \text{ }\Omega}{2\pi(400 \text{ Hz})} = 5.8 \text{ mH}$

Chapter 19

1. $X_C = X_L = 500 \text{ }\Omega$

2. $f_r = \dfrac{1}{2\pi\sqrt{LC}} = \dfrac{1}{2\pi\sqrt{(20 \text{ mH})(12 \text{ pF})}}$
   $= 324.87 \text{ kHz}$
   $Z_T = R = 15 \text{ }\Omega$

3. $f_r + 10 \text{ kHz} = 334.87 \text{ kHz}$
   $X_L = 2\pi fL = 2\pi(334.87 \text{ kHz})(20 \text{ mH})$
   $= 42.08 \text{ k}\Omega$

   $X_C = \dfrac{1}{2\pi fC} = \dfrac{1}{2\pi(334.87 \text{ kHz})(12 \text{ pF})}$
   $= 39.61 \text{ k}\Omega$

   $Z = \sqrt{R^2 + (X_L - X_C)^2}$
   $= \sqrt{(15 \text{ }\Omega)^2 + (2.47 \text{ k}\Omega)^2} \cong 2.47 \text{ k}\Omega$

   $\theta = \arctan\left(\dfrac{X_L - X_C}{R}\right) = \arctan\dfrac{2.47 \text{ k}\Omega}{15 \text{ }\Omega}$
   $= 89.65°$

4. (a) $I_{max} = 1 \text{ V}/30 \text{ }\Omega = 33.33 \text{ mA}$
   (b) $f_r = \dfrac{1}{2\pi\sqrt{LC}} = \dfrac{1}{2\pi\sqrt{(8 \text{ mH})(330 \text{ pF})}}$
   $= 97.95 \text{ kHz}$
   (c) $V_R = (R/Z_T)V_s$
   $Z_T = R = 30 \text{ }\Omega$
   $V_R = V_s = 1 \text{ V}$
   (d) $X_L = 2\pi f_r L = 4.92 \text{ k}\Omega$
   $Q = X_L/R = 4.92 \text{ k}\Omega/30 \text{ }\Omega = 164$
   $V_L = QV_s = (164)(1 \text{ V}) = 164 \text{ V}$
   (e) $V_C = V_L = 164 \text{ V}$
   (f) $\theta = 0°$

5. $Z_{max} = \dfrac{L}{RC}$ at resonance
   $= \dfrac{2 \text{ H}}{(50 \text{ }\Omega)(0.02 \text{ }\mu\text{F})} = 2 \text{ M}\Omega$

6. $I = 5 \text{ V}/2 \text{ M}\Omega = 2.5 \text{ }\mu\text{A}$

7. $f_r = \dfrac{1}{2\pi\sqrt{LC}} = \dfrac{1}{2\pi\sqrt{(2.5 \text{ mH})(100 \text{ pF})}}$
   $= 318.3 \text{ kHz}$
   $X_L = 2\pi f_r L = 2\pi(318.3 \text{ kHz})(2.5 \text{ mH})$
   $= 5 \text{ k}\Omega$
   $Q = X_L/R = 5 \text{ k}\Omega/6 \text{ }\Omega = 833.33$
   $BW = f_r/Q = 318.3 \text{ kHz}/833.33$
   $= 381.96 \text{ Hz}$

8. $f_1 = f_r - (BW/2)$
   $= 318.30989 \text{ kHz} - 190.98 \text{ Hz}$
   $= 318.11891 \text{ kHz}$
   $f_2 = f_r + (BW/2)$
   $= 318.30989 \text{ kHz} + 190.98 \text{ Hz}$
   $= 318.5003 \text{ kHz}$

9. $f_r = \dfrac{1}{2\pi\sqrt{LC}} = \dfrac{1}{2\pi\sqrt{(500 \text{ mH})(10 \text{ }\mu\text{F})}}$

$= 71.2$ Hz
$X_L = 2\pi(71.2\ \text{Hz})(500\ \text{mH}) = 223.7\ \Omega$
$Q = X_L/R = 223.7\ \Omega/30\ \Omega = 7.56$
$\text{BW} = f_r/Q = 71.2\ \text{Hz}/7.56 = 9.42\ \text{Hz}$
$f_{C1} = 71.2\ \text{Hz} - (9.42\ \text{Hz}/2) = 66.49\ \text{Hz}$
$f_{C2} = 71.2\ \text{Hz} + (9.42\ \text{Hz}/2) = 75.91\ \text{Hz}$
At f_{C1}:
$X_L = 2\pi(66.49\ \text{Hz})(500\ \text{mH}) = 208.9\ \Omega$
$X_C = 1/[2\pi(66.49\ \text{Hz})(10\ \mu\text{F})] = 239.4\ \Omega$
$Z = \sqrt{R^2 + (X_C - X_L)^2}$
$= \sqrt{(30\ \Omega)^2 + (30.5\ \Omega)^2} = 42.78\ \Omega$
At f_{C2}:
$X_L = 2\pi(75.91\ \text{Hz})(500\ \text{mH}) = 238.5\ \Omega$
$X_C = 1/[2\pi(75.91\ \text{Hz})(10\ \mu\text{F})] = 209.7\ \Omega$
$Z = \sqrt{R^2 + (X_L - X_C)^2} = \sqrt{(30\ \Omega)^2 + (28.8\ \Omega)^2}$
$= 41.57\ \Omega$

10. $f_r = f_1 + (\text{BW}/2) = 8\ \text{kHz} + 1\ \text{kHz}$
$= 9\ \text{kHz}$

Chapter 20

1. $V_{out} = 0.707V_{max} = (0.707)(10\ \text{V}) = 7.07\ \text{V}$

2. $V_{OUT} = 10\ \text{V dc}$; the sine wave frequency is shorted to ground by the capacitor.

3. V_{out} is a 15-V peak-to-peak sine wave with a zero average value; the dc component is blocked by the capacitor.

4. $X_C = (1/10)2.5\ \text{k}\Omega = 250\ \Omega$
$C = \dfrac{1}{2\pi f X_C} = \dfrac{1}{2\pi(1\ \text{kHz})(250\ \Omega)}$
$= 0.6\ \mu\text{F}$
Use next highest standard value.

5. $V_{out(ac)} = \left(\dfrac{R}{\sqrt{R^2 + X_L^2}}\right)V_{in}$
$= \left(\dfrac{5\ \Omega}{\sqrt{(5\ \Omega)^2 + (100\ \Omega)^2}}\right)5\ \text{V}$
$= 0.25\ \text{V peak-to-peak}$
$V_{OUT(dc)} = V_{AVG} = 8\ \text{V}$
The output is a 0.25-V peak-to-peak sine wave riding on an 8-V dc level.

6. See Figure S–32.

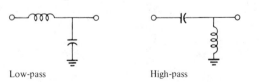

Low-pass   High-pass

FIGURE S–32

7. See Figure S–33.

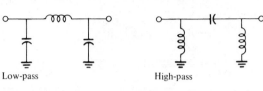

Low-pass   High-pass

FIGURE S–33

8. See Figure S–34.

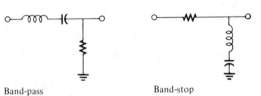

Band-pass   Band-stop

FIGURE S–34

9. See Figure S–35.

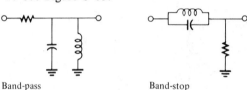

Band-pass   Band-stop

FIGURE S–35

10. $\text{BW} = f_r/Q = 5\ \text{kHz}/20 = 250\ \text{Hz}$
11. $20 \log(4/12) = -9.54\ \text{dB}$
12. $10 \log(5/10) = -3\ \text{dB}$
13. $f_c = \dfrac{1}{2\pi RC} = \dfrac{1}{2\pi(1\ \text{k}\Omega)(0.02\ \mu\text{F})}$
$= 7.96\ \text{kHz}$
14. $f_c = \dfrac{1}{2\pi(L/R)} = \dfrac{1}{2\pi(0.01\ \text{mH}/5\ \text{k}\Omega)}$
$= 79.58\ \text{MHz}$

Chapter 21

1. Half-scale (50 μA will produce full-scale deflection).

2. No shunt for 1-mA movement.
For 50-μA movement:
$R_{SH} = I_{MM} R_{MM}/I_{SH}$
$I_{SH} = 1\ \text{mA} - 50\ \mu\text{A} = 950\ \mu\text{A}$
$R_{SH} = \dfrac{(50\ \mu\text{A})(1\ \text{k}\Omega)}{950\ \mu\text{A}} = 52.63\ \Omega$

CHAPTER 21

3. $R_{INT} = (20{,}000\ \Omega/\text{V})(100\ \text{V}) = 2\ \text{M}\Omega$

4. $R_M = R_{INT} - R_{MM}$
   $= 2\ \text{M}\Omega - 1\ \text{k}\Omega = 1.999\ \text{M}\Omega$

5. $R_{INT} = (20\ \text{k}\Omega/\text{V})(1\ \text{V}) = 20\ \text{k}\Omega$
   $R_2 \parallel R_{INT} = \dfrac{(10\ \text{k}\Omega)(20\ \text{k}\Omega)}{30\ \text{k}\Omega} = 6.67\ \text{k}\Omega$
   $V_2 = (6.67\ \text{k}\Omega/16.67\ \text{k}\Omega)1.5\ \text{V}$
   $= 0.6\ \text{V}$ meter reading
   $V_2 = (10\ \text{k}\Omega/20\ \text{k}\Omega)1.5\ \text{V}$
   $= 0.75\ \text{V}$ unloaded
   The internal meter resistance in parallel with R_2 causes the voltage to be less than its actual value.

6. $R = 10 \times 1\ \text{k}\Omega = 10\ \text{k}\Omega$

7. The volts/div setting is 0.1 V/div. The wave form is 1.2 div vertically:
   $V = (1.2\ \text{div})(0.1\ \text{V/div})$
   $= 0.12\ \text{V}$
   The sec/div setting is 10 μs/div. The wave form has a period covering approximately two div:
   $T = (2\ \text{div})(10\ \mu\text{s/div}) = 20\ \mu\text{s}$
   $f = 1/T = 1/20\ \mu\text{s} = 0.05\ \text{MHz}$
   $= 50\ \text{kHz}$

8. Current; $6 \times 10\ \text{mA} = 60\ \text{mA}$ dc

9. $V_{pp} = (0.1\ \text{volts/div})(5\ \text{div}) = 0.5\ \text{V}$

10. $t = (1\ \mu\text{s/div})(10\ \text{div}) = 10\ \mu\text{s}$

11. $T = 1/f = 1/1\ \text{kHz} = 1\ \text{ms}$
    $t = (0.1\ \text{ms/div})(10\ \text{div}) = 1\ \text{ms}$
    One cycle is displayed.
    $t = (1\ \text{ms/div})(10\ \text{div}) = 10\ \text{ms}$
    Ten cycles are displayed.

INDEX

Admittance, 587–88
Air core, 471
Alternating current, 282
Alternation, 282
Amber, 2
Ammeter, 38, 158, 660–63
Ampere, 2, 24
Ampere-hour, 80
Ampere-turn, 460
Amplify, 4
Amplitude, 289–93, 305
Amplitude modulation (AM), 674
Angle, 570
Angular velocity, 304
Anode, 679
Apparent power, 392
Arctangent, 370
Atom, 22
Atomic number, 22
Atomic weight, 22
Attenuation, 633
Audio generator, 673
Automation, 8
Autotransformer, 559–60
Average power, 72–76, 355
Average value, 289, 291–92, 308–9, 433
AWG, 33
Axis, 570
  imaginary, 570
  real, 570

Back-off scale, 669
Balanced bridge, 201
Band-pass filter, 639–42
Band-stop filter, 642-46
Bandwidth, 614–17
BASIC language, 271, 694–97
Battery, 2, 691–93
  dry cell, 692–93
  lead-acid cell, 691–92
  types, 693
  wet cell, 691
Beam, electron, 679–80
Bleeder current, 196
Bode plot, 649
Branch, 130, 156, 262
Branch current method, 262–65
Breakdown strength, 33
Breakdown voltage, 331
Break frequency, 630
Bridge circuit, 201–3
Bypass capacitor, 632

Capacitance, 326–31
Capacitive reactance, 351–54
Capacitor, 326–58
  air, 340
  ceramic, 336
  charging, 327
  color code, 342
  coupling, 637
  discharging, 328
  electrolytic, 338
  energy storage, 354–55
  film, 339
  integrated circuit, 339
  mica, 336
  paper, 338
  polarized, 338
  stray, 470
  tantalum, 339
  testing, 357–58
  trimmer, 340
Carbon composition resistor, 27, 76
Cathode, 678
Cathode ray tube, 678–79
Center frequency, 616
Center tap, 557
Characteristic determinant, 265
Charge, 2, 24, 326–27
Chassis, 186
Chassis ground, 186
Choke, 478, 507, 632
Circuit, 36–37
  closed, 36, 116
  equivalent, 229–30, 594–97
  integrated, 6
  open, 36
  short, 37
Circuit breaker, 40
Circuit ground, 186
Circular mil, 34
Closed circuit, 36, 116
Coefficient, 265
Coefficient of coupling, 544
Coil, 463
Color code, 27, 29, 342
Communications, 7
Complex number, 570–73
  polar form, 573
  rectangular form, 572–73
Complex plane, 570
Computer, 4, 6
Computer program, 271, 274, 694–97
Conductance, 27, 135–36, 587

INDEX

Conductor, 33
Continuous wave, 674
Control grid, 678
Coordinate, 571
Copper, 22, 33
Core, magnetic, 468
Cosine, 296
Coulomb, 2, 24
Coupling, 390–91, 546, 637
CRT, 678–79
Current, 24, 50, 52–56, 62, 93, 178–81, 350–51, 379–87, 395, 443–44, 474–75, 497–98, 510, 609
Current divider, 150–54
Current source, 144–45, 216–21
Cutoff frequency, 615, 647–48
Cycle, 282

Damped sine wave, 409
d'Arsonval movement, 658
Decade of frequency, 649
Decibel (dB), 630, 646–47
Delta-wye, 247
Determinant, 265–67
Dielectric constant, 332–33
Dielectric strength, 333–34
Differentiator, 435–44, 446, 532–35
Diode, 3
DMM, 671
Dot convention, 547
DPDT switch, 43
Dry cell, 692–93
Duty cycle, 307
DVM, 671

Earth ground, 186
Effective value, 289, 291
Efficiency, 81–82
Electric circuit, 36
Electric field, 326
Electricity, 2
Electrodynamometer movement, 659
Electrolytic capacitor, 338
Electromagnetic induction, 461–63
Electromagnetism, 458–59
Electromotive force (emf), 25
Electron, 3, 23
Electron gun, 678
Energy, 25, 72, 77–78, 480
Equivalent circuit, 229–30, 594–97
Exponent, 12
Exponential, 416–22

Fall time, 306, 446
Farad, 3, 329–31
Faraday's law, 464–65
Film capacitor, 339
Filters, 387–91, 422, 435, 446, 504–6, 535–36, 630–50
  band-pass, 639–42
  band-stop, 642–46
  high-pass, 389–90, 435, 446, 505–6, 635–42
  low-pass, 388–89, 422, 445–46, 504–5, 535–36, 630–35
  π-type, 634, 639
  resonant, 640–42, 643–46
  T-type, 634, 638
Fixed capacitors, 336
Fixed resistors, 27
Flux, 457
Flux density, 457
Focus, 683
Free electron, 23
Frequency, 285–89, 307, 313, 615–17, 630, 647–48
  break, 630
  center, 616
  cutoff, 615, 647–48
  fundamental, 307, 313
  half-power, 617
Frequency modulation (FM), 674
Fundamental frequency, 307, 313
Fuse, 40

Galvanometer, 202
Generators, 673–77
  audio, 673–74
  function, 676
  programmable, 677
  pulse, 675–76
  rf, 674
  sweep, 674–75
Giga, 15
Grid, 678
Ground, 186–89
  chassis, 186
  earth, 186
Graphs, 416–22, 649–50
  exponential, 416–22
  logarithmic, 649–50

Half-power frequency, 617
Harmonics, 282, 313–14
Heater, 678
Henry (H), 3, 466, 468
Hertz (Hz), 3, 286

High-pass filter, 389–90, 435, 505–6, 635–42
Hypotenuse, 301, 369–70

Imaginary axis, 570
Impedance, 372–75, 395, 490–93, 510, 551–53, 581–88, 606–7, 612–14
Impedance matching, 553–54
Impedance triangle, 373–74
Induced current, 462
Induced voltage, 461–62, 465, 467, 526–28
Inductance, 466–68
Induction, electromagnetic, 3, 461–63
Inductive reactance, 476–78
Inductor, 463–64, 468–71, 479, 481
Instantaneous power, 355
Instantaneous value, 289–90
Insulator, 33
Integrated circuit, 6, 339
Integrator, 422–34, 443, 446, 528–31
Internal resistance, 214, 216–17
Iron-vane movement, 659
Isolation, 555–56
    dc, 555
    power line, 555–56

j operator, 570
Joule (J), 25, 72

Kilo (K), 15–16
Kilohm (k Ω), 53
Kilowatt (kW), 73
Kilowatthour (kWh), 78
Kilovolt (kV), 55
Kirchhoff's current law, 146–49, 262–65, 272–75
Kirchhoff's voltage law, 104–7, 262–65, 268–71

Ladder, 198–201
Lag, 296, 381–84, 498–501
Lead, 296, 501–3
Lead-acid cell, 691–92
Leading edge, 408
Leakage, 335
Left-hand rule, 459
Lenz' law, 466–67
Linear potentiometer, 31
Load, 80, 550, 665
Logarithm, 419
Loop, 262, 267
Low-pass filter, 388–89, 422 445–46, 504–5, 535–36, 630–35

Magnetic field, 456–57
Magnetomotive force (mmf), 460
Maximum power transfer theorem, 244–46
Medicine, 8
Mega (M), 15–16
Megawatt (MW), 73
Meghom (M Ω), 53
Mesh, 268
Mesh current method, 267–72
Meter movements, 658–60
    d'Arsonval, 658
    electrodynamometer, 659
    iron-vane, 659
Metric prefixes, 15
Mica capacitor, 336
Micro (μ), 15–16
Microampere (μ A), 54, 57
Microfarad (μ F), 330
Microprocessor, 6
Microvolt (μ V), 58
Microwatt (μ W), 73
Mil, 34
Milli (m), 15–16
Milliampere (mA), 54, 57
Millivolt (mV), 58
Milliwatt (mW), 73
Millman's theorem, 241–44
Multimeter, 671–73
Multiple-source circuit, 221
Multiple-winding transformer, 558–59
Multiplier resistance, 663
Mutual inductance, 544–45

Nano (n), 15–16
Negative, 2
Negative numbers, 570
Neutron, 22
Node, 262, 272
Node voltage method, 272–75
Norton's theorem, 237–41
Nucleus, 22

Ohm, 26
Ohmmeter, 38, 669–71
Ohm's law, 3, 50–64, 98–100, 142–44, 293, 353–54, 460, 477–78
Ohms-per-volt rating, 664
Open circuit, 36, 116, 156–57, 190
Open circuit voltage, 214
Oscillator, audio, 673
Oscilloscope, 678–84

INDEX

Output impedance, 673
Output resistance, 245
Output voltage, 194, 381, 383–84
Overshoot, 408

Parallel circuits, 130–60, 394–96, 509–11
   capacitance, 348–50
   inductance, 472–74
   resistance, 135–41
Parallel resonance, 611–14
Pass band, 630
Peak-to-peak value, 289–90
Peak value, 298–90
Period, 283–85
Permeability, 460
Permittivity, 332–33
Phase, 294, 382, 385–78, 610
Phase angle, 295, 375–78, 395, 493–96, 510
Phase shift, 295–99
Phasor, 300–305, 368–71, 572–73
Pi (π), 294
Pico (p), 15–16
Picofarad (pF), 330
Plate, 332
Polar form, 573
Polarity, 282
Polarized capacitor, 338
Pole, magnetic, 456
Positive, 2
Positive numbers, 570
Potential difference, 25
Potentiometer, 30–31, 112–13
   linear, 31
   tapered, 31
Power, 72–76, 114–16, 155–56, 244, 355–57, 391–94, 480, 507–9, 551
   apparent, 392
   average, 72–76, 355
   reactive, 355–56, 480–81
Power factor, 393, 508
Power supply, 80–82
Power triangle, 392, 508
Powers of ten, 11
Primary, 545
Printed circuit, 92
Proton, 22
Pulse, 306–9
Pulse generator, 675-76
Pulse repetition rate (PRR), 307
Pulse width, 306
Pythagorean theorem, 369

Quadrant, 571
Quality factor (Q), 618–21

Radian, 294
Ramp, 310–11
Rate of change, 287–88
Reactance, 351–54, 476–78, 579–80
   capacitive, 351–54
   inductive, 476–78
Reactive power, 355–56, 480–81
Real axis, 570
Rectangular form, 572–73
Rectification, 631
Rectifier, 668
Reflected impedance, 551–53
Relative permeability, 468
Reluctance, 460
Resistance, 26, 35, 51, 60–62, 579–80
Resistivity, 35
Resistor, 27, 29, 30, 76, 690
   carbon composition, 27, 76, 690
   fixed, 27
   variable, 30
Resonance, 606–14
   parallel, 611–14
   series, 606–10
Resonant circuit, 614–17, 622
Resonant frequency, 606–7, 612
Resonant Q, 618–21
Response, 368, 609, 646–50
rf choke, 507
Rheostat, 30
Right triangle, 301, 573
Ringing, 409
Rise time, 306, 446
rms value, 289, 291
Roll off, 649
Root mean square, 289, 291

Sawtooth wave form, 312
Schematic, 91, 131
Scientific notation, 11–15
Secondary, 545
Selectivity, 617
Self-inductance, 466–68
Semiconductor, 33
Sensitivity, meter, 660, 664
Series circuits, 94–118, 343–48, 471–72, 606–10
   capacitance, 343–48
   inductance, 471–72
   resistance, 94–118

Series-parallel circuits, 172–85
Series resonance, 606–10
Short circuit, 37, 190
Shunt, meter, 660
SI units, 10
Siemen (S), 27, 136
Signal, 282
Simultaneous equations, 263–67
Sine wave, 282, 490
Sinusoid, 282
Slope, 288, 310, 352
Source, 26, 101–3, 144–45, 214–21
Source resistance, 214
SPDT switch, 43
Spike, 438
Square wave, 308, 313–14, 368
Steady-state, 431, 441
Step, 305
Step-down transformer, 549
Step-up transformer, 548
Stray capacitance, 470
Stray inductance, 409
Superposition theorem, 221–28
Susceptance, 587–88
Sweep, 312
Sweep generator, 674–75
Switch, 41–43

Tangent, 370
Tank circuit, 611
Tap, 195, 557
Tapered potentiometer, 31
Temperature coefficient, 31, 335
Terminal, 26
Terminal equivalency, 218
Tesla (T), 457
Thermistor, 31
Theta (θ), 297
Thevenin's theorem, 229–37
Tilt, 409
Time base, 683
Time constant, 410–16, 522–26
Time response, 445–47, 535–36
Trailing edge, 408
Transformers, 545–49
   step-down, 549
   step-up, 548
   tapped, 557

Transformer construction, 560
Transformer ratings, 561
Transient time, 431–32
Transistor, 4, 216
Triangular wave, 311–12
Trigonometric functions, 370, 701–5
Trouble-shooting, 117–18, 158–60, 190–93
Tube, 4
Tuned circuit, 622
Turns ratio, 546

Undershoot, 408
Units, 10–11
Universal exponential curves, 421

Volt, 25
Voltage, 25, 51, 56–60, 62, 350–51, 379–81, 474–75, 497–98, 609
Voltage divider, 108–14, 194–97, 346–47
Voltage drop, 78–79, 133–34, 181–85
Voltage measurement, 186–89
Voltage source, 101–3, 214–16, 218–21
Volt-amperes (VA), 392
Volt-amperes reactive (VAR), 355–56
Voltmeter, 38, 663–67
VOM, 673

Watt (W), 72
Wave form, 282, 681
Wavelength, 11
Weber (Wb), 457
Wet cell, 691
Wheatstone bridge, 201–3
Winding capacitance, 470
Winding resistance, 469–70
Wiper, 112
Wire, 34–35
Wye-delta, 248

X_C (capacitive reactance), 351–54
X_L (inductive reactance), 476–78

Y (admittance), 587
Y (wye) network, 247

Z (impedance), 372
Zero-ohms adjustment, 669